PLEASE STAMP DATE DUE, BOTH BELOW AND ON CARD

DATE DUE	DATE DUE	DATE	DATE DUE

INSIDE THE SUN

ASTROPHYSICS AND SPACE SCIENCE LIBRARY

A SERIES OF BOOKS ON THE RECENT DEVELOPMENTS
OF SPACE SCIENCE AND OF GENERAL GEOPHYSICS AND ASTROPHYSICS
PUBLISHED IN CONNECTION WITH THE JOURNAL
SPACE SCIENCE REVIEWS

VOLUME 159
PROCEEDINGS

INSIDE THE SUN

PROCEEDINGS OF THE 121ST COLLOQUIUM OF THE
INTERNATIONAL ASTRONOMICAL UNION,
HELD AT VERSAILLES, FRANCE, MAY 22–26, 1989

edited by

GABRIELLE BERTHOMIEU
Observatoire de la Côte d'Azur, Nice, France

and

MICHEL CRIBIER
CEN Saclay, France

KLUWER ACADEMIC PUBLISHERS
DORDRECHT / BOSTON / LONDON

Library of Congress Cataloging in Publication Data

International Astronomical Union. Colloquium (121st : 1989 :
  Versailles, France)
    Inside the sun : proceedings of the 121st Colloquium of the
  International Astronomical Union, held at Versailles, France, May
  22-26, 1989 / edited by Gabrielle Berthomieu, Michel Cribier.
        p.   cm. -- (Astrophysics and space science library ; v. 159)
    ISBN 0-7923-0662-7 (alk. paper)
    1. Sun--Congresses.   I. Berthomieu, Gabrielle. II. Cribier,
  Michel. III. Title. IV. Series.
  QB520.I55 1989
  523.7--dc20   89-78506

ISBN 0–7923–0662–7

Published by Kluwer Academic Publishers,
P.O. Box 17, 3300 AA Dordrecht, The Netherlands.

Kluwer Academic Publishers incorporates
the publishing programmes of
D. Reidel, Martinus Nijhoff, Dr W. Junk and MTP Press.

Sold and distributed in the U.S.A. and Canada
by Kluwer Academic Publishers,
101 Philip Drive, Norwell, MA 02061, U.S.A.

In all other countries, sold and distributed
by Kluwer Academic Publishers Group,
P.O. Box 322, 3300 AH Dordrecht, The Netherlands.

Printed on acid-free paper

Printed in the Netherlands

TABLE OF CONTENTS

[*] Speaker during the Colloquium

PRÉFACE

Ce livre contient les exposés des orateurs invités au colloque IAU 121 « Inside the Sun » qui s'est tenu à Versailles du 22 au 26 mai 1989. Le but de « Inside the Sun » était d'accroître nos connaissances sur l'intérieur du Soleil et la physique qui lui est reliée, il a permis, espérons-le, un échange d'idées, de théories et d'informations entre les différents groupes. Cette conférence représente le premier pas vers une évaluation globale des idées importantes, tant théoriques qu'expérimentales, nécessaires à la compréhension de l'intérieur du Soleil. Les 175 participants de 24 pays sont la preuve qu'une communauté vivante, concernée par ce sujet existe bien. Quel endroit plus approprié que Versailles, la ville du Roi Soleil, Louis XIV, pour tenir, exactement deux siècles après les « Etats Généraux », cette sorte d' « Etats Généraux » du cœur du Soleil! A la fin de ce livre les deux petits textes historiques sont la réminiscence de cette conjonction spatio-temporelle. Enfin, bien que le temps soit resté magnifique durant toute la semaine, l'audience n'a jamais faibli.

Comme le faisait remarquer Jean Audouze dans son discours inaugural, le centre du Soleil est un sujet très interdisciplinaire. Dans les vingt dernières années, le problème des neutrinos solaires soulevé par l'expérience de Davis nous a conduit au choix suivant : soit repenser nos connaissances sur la structure du Soleil et l'évolution stellaire, soit modifier les propriétés fondamentales du neutrino avec toutes les implications que cela comporte pour les théories les plus achevées qui unifient toutes les forces fondamentales. Pendant le même laps de temps, l'héliosismologie s'est développée et a été capable de contraindre sévèrement la structure du Soleil presque jusqu'au centre. De nombreuses expériences, tant sur terre que dans l'espace, sont en cours ou en projet, et vont apporter des informations encore plus précises. Le champ magnétique, la rotation et les mouvements à grande échelle sont connus avec beaucoup plus de détails. Nous proposons d'appeler « heliophysiciens » tous ces spécialistes venus d'horizons divers qui se mettent en commun pour comprendre le problème de l'intérieur du Soleil. Une première illustration concrète : le comité d'organisation de la conférence réunissait des astrophysiciens et des physiciens des particules élémentaires.

Ces comptes rendus contiennent presque tous les textes des 30 orateurs invités et 7 des papiers soumis, sélectionnés et présentés pendant la conférence. Le plan est essentiellement celui suivi pendant la conférence. Nous commençons par la modélisation du Soleil, avec le modèle standard (qui a suscité des discussions très actives), et les exposés sur les opacités et l'équation d'état. Les résultats expérimentaux sur les neutrinos solaires précèdent les idées théoriques sur les masses des neutrinos. L'héliosismologie, présentée sous ses aspects expérimentaux et théoriques, permet depuis peu l'étude de la rotation interne du Soleil. La dynamo solaire et les phénomènes de transport sont inclus dans la même partie que les cycles solaires. Nous avions demandé aux auteurs de faire un effort de pédagogie dans leurs exposés écrits. Le rapport final de D. Gough fait le point sur tout ce qui reste à comprendre.

x

Trois sessions posters ont été organisées. La première session, présidée par James Rich (Saclay) était consacrée aux neutrinos solaires et aux modélisations du Soleil (26 contributions), la deuxième (29 contributions), animée par Eric Fossat (Nice) traitait l'héliosismologie et le problème de la diffusion, la troisième (36 contributions), co-organisée par Elisabeth Ribes (Meudon) et Françoise Bely-Dubau (Nice) couvrait la convection, la dynamo et le transport. Les posters restaient exposés en permanence et une très brève présentation orale au début de chaque session a permis, semble-t-il, des discussions très animées par la suite devant les panneaux. Solar Physics réunit dans un numéro spécial les papiers résultant des séances posters qui lui ont été soumis. A ce point il faut remercier les président(e)s de ces sessions qui se sont magnifiquement acquittés de cette tâche bien difficile.

Le comité d'organisation est heureux de remercier pour leurs appuis, tant financier que moral les organismes ou les sociétés suivantes : l'Institut de recherche fondamentale du Commissariat à l'énergie atomique (IRF-CEA), l'Union astronomique internationale (UAI), l' Agence spatiale européenne (ESA), le Centre national d'études spatiales (CNES), le Centre national de la recherche scientifique (CNRS), le Laboratoire d'astrophysique théorique du Collège de France, IBM, Matra Espace, Rank Xerox, la Mairie de Versailles, la Société européenne de physique (EPS), la Société française de physique, *the International Union of Pure and Applied Physics* (IUPAP)

Le comité scientifique international, présidé par E.Schatzman (Meudon) était composé de J.N.Bahcall (Princeton), R.M.Bonnet (ESA Paris), T.Brown (Boulder), R.Davis (Philadelphie), F.Deubner (Würzburg), W. Dziembowski (Varsovie), E.Fossat (Nice), D.O.Gough (Cambridge), H.Harari (Rehovot), T.Kirsten (Heidelberg), M.Koshiba (Tokyo), A.Maeder (Genève), T.Montmerle (Saclay), A.Renzini (Bologne), P.Roberts (UCLA), M.Spiro (Saclay) et G.T.Zatsepin (Moscou). Nous les remercions de toutes leurs suggestions et de leur participation active à l'élaboration du programme de la conférence.

Le comité d'organisation comprenait : F.Bely-Dubau (Nice), G.Berthomieu (Nice), J.Boratav (Saclay), M.Cassé (Saclay), M.Cribier (Saclay), W.Däppen (Saclay), B.Foing (Verrières-le-Buisson), E.Fossat (Nice), P.O.Lagage (Saclay), Y.Lebreton (Meudon), T.Montmerle (Saclay), E.Ribes (Meudon), J.Rich (Saclay), E.Schatzman (Meudon) et D.Vignaud (Saclay). Il souhaite exprimer ses remerciements très chaleureux à Jacqueline Boratav pour toute son aide, si précieuse, avant, pendant et après le colloque. Il remercie également Mmes Simone Roussiez, Dominique Brou et Mireille Kalifa ainsi que Jean-Pierre Soirat, Jacques Mazeau et Henri de Lignières pour leur appréciable soutien technique pendant les différentes phases de la conférence et de l'édition des comptes rendus.

Gabrielle BERTHOMIEU Michel CRIBIER

Editeurs

PREFACE

This volume contains the invited talks to the IAU colloquium 121 "Inside the Sun" held at Versailles, May 22-26, 1989. "Inside the Sun" aimed at increasing the knowledge on the solar interior and the underlying physics and hopefully contributed to the cross-fertilisation of ideas, theory and observations of all the groups. This conference was a first step to proceed to an assessment of all these current theoretical and experimental progress, in a global way. The 175 participants from 24 countries illustrate the existence of a lively community concerned by this subject. What site could be a better chance than Versailles, the city of Sun King Louis the XIVth, to hold exactly 200 years after the "Etats Generaux" a kind of "Etats Generaux" of what is happening in the heart of the Sun. At the end of this book, two texts are an illustration of this time-space conjunction! In spite of a wonderful weather during the whole week, the attendance was always very high.

As pointed out by Jean Audouze in his introductory talk, the interior of the Sun is a very interdisciplinary subject . These last decades, the solar neutrino problem raised by the Davis' experiment has led to the alternative of revising our view on the solar structure and stellar evolution theory or to explain it by fundamental properties of neutrinos, related themselves to the most achieved theories unifying all fundamental forces. At the same time, helioseismology developed greatly and was able to put very strong constraints on the structure of the Sun down to very near the centre. Many terrestrial and space experiments are under way or planned for the next years to obtain more accurate data. The magnetic field, the rotation and the large scale motions are also now observed in more details. We propose to call "heliophysicists" all these specialists from different fields which unite to understand fully the problem of the solar interior. To illustrate this, the organizing committee joined astrophysicists and particle physicists together.

The proceedings contain nearly all of the 30 invited papers, as well as 7 contributed papers presented during the Conference. We retain mostly for the proceedings, the plan adopted during the Conference. We thus begin by the solar modelling, including the Standard Solar Model - the related discussions were very active during the meeting - and the main inputs of physics (Equation of State, Opacities). The experimental results on solar neutrinos precede the recent theoretical ideas on neutrino masses. The subject of Helioseismology is presented both from a theoretical and experimental point of view, pointing out the recent results on the solar rotation and on the structure of the solar core. The problems of Solar Dynamo and of Transport processes are presented in connection with the Solar Cycles. The authors were asked to write their contribution from an educational stand-point. The final talk by D. Gough makes the inventory of questions that remain to be solved.

Three poster sessions were organized. The first was devoted to Solar Neutrinos and Solar Models (26 contributions), it was chaired by James Rich (Saclay), the second one (29 contributions), conducted by Eric Fossat (Nice) treated Helioseismology and Diffusion problems and the third one (36 contributions), co-chaired by Elisabeth Ribes (Meudon) and Françoise Bely-Dubau (Nice) covered the subjects of Convection, Dynamo and Transport. The posters were displayed during the whole Conference, and a brief presentation of the subject took place at the beginning of the session. The discussions were very active in front of the panels during all the breaks. Contributions presented during the poster sessions will be published in a special issue of Solar Physics. We want to thank the chairmen of these sessions who succeeded in the difficult task of allowing at the same time a short presentation of the subject, and of conducting an active discussion among the participants.

The Organizing Committee is glad to acknowledge the moral and financial support of : the *Institut de Recherche Fondamentale du Commissariat à l'Energie Atomique* (IRF-CEA), the International Astronomical Union (IAU), the European Space Agency (ESA), the *Centre National d'Etudes Spatiales* (CNES), the *Centre National de la Recherche Scientifique* (CNRS), the *Laboratoire d'Astrophysique Théorique du Collège de France*, IBM, *Matra Espace*, Rank Xerox, the *Mairie de Versailles*, the European Physical Society (EPS), the *Société Française de Physique*, the International Union of Pure and Applied Physics (IUPAP)

The members of the International Scientific Committee, chaired by E.Schatzman (Meudon) were J.N.Bahcall (Princeton), R.M.Bonnet (ESA Paris), T.Brown (Boulder), R.Davis (Philadelphia), F.Deubner (Würzburg), W. Dziembowski (Warsaw), E.Fossat (Nice), D.O.Gough (Cambridge), H.Harari (Rehovot), T.Kirsten (Heidelberg), M.Koshiba (Tokyo), A.Maeder (Geneva), T.Montmerle (Saclay), A.Renzini (Bologna), P.Roberts (UCLA), M.Spiro (Saclay) and G.T.Zatsepin (Moscow). We are very grateful to them for suggestions and active participation in the elaboration of the program.

The Organizing Committee was composed of F.Bely-Dubau (Nice), G.Berthomieu (Nice), J.Boratav (Saclay), M.Cassé (Saclay), M.Cribier (Saclay), W.Däppen (Saclay), B.Foing (Verrières-le-Buisson), E.Fossat (Nice), P.O.Lagage (Saclay), Y.Lebreton (Meudon), T.Montmerle (Saclay), E.Ribes (Meudon), J.Rich (Saclay), E.Schatzman (Meudon) and D.Vignaud (Saclay). This Committee wishes to express special thanks to Mrs Jacqueline Boratav for her dedicated work before, during and after the Colloquium. He wishes to thank Mrs Simone Roussiez, Dominique Brou and Mireille Kalifa as well as Jean-Pierre Soirat, Jacques Mazeau and Henri de Lignières for their valuable technical support in the different parts of the Conference and of the edition of these Proceedings.

Gabrielle BERTHOMIEU Michel CRIBIER

The Editors

INTRODUCTION

JEAN-CLAUDE PECKER
Collège de France
3 rue d'Ulm, 75231 Paris cedex 05
France.

ABSTRACT

1. Evolution of the modelling of the inside of the Sun, from "standard models" to the more empirical approach used now, and based on neutrino astronomy, solar sismology, and the interpretation of active phenomena. 2. New concepts introduced in the set of physical equations valid inside the Sun, under the pressure of new observations. 3. Similar trends seeming to affect in a comparable way the modelling of the solar cycle. 4. Other stars, and the improved knowledge of solar evolution may help to understand many observed features of rotation, magnetism, etc...Conclusion : the Sun is "one", - and it cannot safely be studied piecemeal.

1.   For a very long time, and until only quite recently, it was customary to consider that the only informations one had about the physical conditions inside the Sun were the global informations concerning our star, namely its mass, its radius, its luminosity. The spectrum was also, in a way, a set of information : but the chemical composition of the atmosphere was not necessarily representative of the evolved composition of the core, unless completed by the age of the Sun (as inferred from solar system studies) and a theory of its evolution, in particular an history of how mixing has been operating since the start of nuclear reactions.

From this limited amount of data, "models" were built of an evolved Sun. A given initial composition of the gaseous sphere, the game of thermonuclear reactions, the use of radiative equilibrium (where allowed), of hydrostatic equilibrium, of thermodynamic equilibrium, and wherever necessary, a proper account of convective instabilities (using then, in convective layers, the mixing-length theory, with a somewhat arbitrary ratio l /H) - this set of equations being completed by oversimplified boundary conditions at the "solar surface" - this was in essence enough to be happy with, once one has succeeded, the computation finished, in matching the values of mass, radius, luminosity, and in insuring that initial composition, age, and actual atmospheric composition were at least compatible.

One of the greatest challenges in solar modelling was therefore not getting the necessary observations, but performing with accuracy the solution of very difficult systems of equations. As the

1

G. Berthomieu and M. Cribier (eds.), Inside the Sun, 1–4.
© 1990 *Kluwer Academic Publishers. Printed in the Netherlands.*

author of an introduction is, I presume, free to express his own idiosyncrasies, I would define this period as dominated by numerical acrobatics, and regrettably far from physical insight or accurate data.

But it was clear, however, for almost any one in the field, that many phenomena, although well-known by the observers, were unduly ignored by the modelists. Rotation, and its latitudinal differential behaviour, convection, and the complex imbrication of convective cells and turbulent motions, magnetic active phenomena, and their obviously deep-rooted origin, cycles of activity and the general magnetic field of the Sun, - all these phenomena, and most of their interplays were known; but none was properly taken into account in modelling the inside of the Sun, and many were not even completely understood.

2. Fast progress came a few decades ago, from various new advances in techniques and in observations.

First of all, the quick development of high-speed performant computers allowed theoreticians to tackle more ambitious programs. But often, very gratuitous style exercises were performed, without enough attention being paid to the physical value of the game. At least, everything seemed to become possible...

Then the accuracy with which physical constants were known (cross sections of nuclear reactions, opacity tables, ionic and atomic processes), even improvements in such things as the equation of state, allowed the astrophysicists to work on safer grounds.

Finally, new phenomena were discovered. Not only was the explored electromagnetic spectral range considerably extended, allowing the discovery of such coronal features as coronal holes, bright X-ray points, etc..., but other messages from the Sun began to be deciphered ; the neutrino flux gives direct information concerning what happens in the Sun's core; and the oscillation frequencies spectrum can be inverted (as for sismic waves on Earth) and provides us with values of the sound velocity as a function of depth; even, in principle, it can also provide the solar physicist with the rotational velocity as a function of depth. But one can also state that abundances of ^3He in cosmic rays, or of ^7Li in the solar atmosphere may reveal much about the physical conditions at the bottom of the convective zone, and can be very useful clues.

Active phenomena, although still far from being understood in many of their more important aspects, give also clues during the various phases of the solar cycle, concerning the general behaviour of convective cells which act in dragging out magnetic tubes of force before they actually can emerge, and concerning the properties of the solar magnetic field embedded in the deep Sun. These clues request certainly much effort towards a real understanding of their meaning. But no theory can avoid to match these data, even if some of them are still ambiguous and perhaps compatible with quite different types of theories.

Thus, in a few decades, "modelling the Sun" became a completely different concept. From a very "abstract" approach, essentially a study a priori of gaseous spheres, based upon the solutions of the equations of internal structure, one has progressively been led to an almost "empirical" approach, where the ideal (standard) model serves only as a starting block, improved in its details on the basis of successive observational refinements.

3.   In this new approach, a close cooperation is necessary with the physicists, as contradictions appear between the refinements derived from different observational sources : as an example, we must note that the apparent depletion in the observed flux of neutrinos (compared with the expectations of standard models) has led to a stimulating interchange between physicists and solar physicists : has the solar structure to be modified ? or are the neutrinos behaving differently from what was previously thought ?

Improvements in the physics of the internal regions of the Sun are thus a by-product of this approach. Not only do they concern some basic physics (such as the neutrino physics), but also they imply the introduction of some concepts already known, but now thought to be relevant, in the equations of solar structure. For example, it is clear that <u>diffusive processes</u> have to play a part either by reintroducing in the core unprocessed matter, or by allowing some atoms, produced in the core, to reach the convective layers and to be observable in the spectra. But we have not yet reached a state of affairs where it is safe to assert quantitatively the effects of diffusion ... Clearly also, the phenomenological mixing-length theory has to be refined, to justify "the" good choice of the ratio l /H, and even to be replaced by some <u>exact hydrodynamical theory</u>. Above and below the convective layers, one has also to take into account <u>overshooting</u> : in itself, this process leads us to think that one cannot treat the transition layer between radiative inner or outer zones and convective zone as spherically symmetric. And departures from spherically-symmetric geometry yields to local instabilities which may well affect the transport of magnetic fields. Similarly, granulation affects the geometry of the photosphere, and abundance determinations may be sensitive to departures from spherically-symmetric geometry, - as they were already known to be sensitive to departures from local thermodynamic equilibrium. These sensitivities may, in their turn, affect the opacities.

4.   Modelling the cycle of activity is developing more slowly than modelling the averaged Sun. Still the progress follows similar lines. A few years ago, one computed with great accuracy dynamo models of the Sun, which were assumed to be linear, and, even under their non-magnetic aspects, grossly oversimplified. Nowadays, the highly non-linear character of the solar dynamo has been widely recognized. Assuming the observed characteristics of the emergence of active regions to be strongly coupled with the general features of the magnetic field, and with the hydrodynamical properties of the Sun, one can think of some partly theoretical, partly empirical description of the solar cyclic behaviour. There is a little doubt that much progress will come from that kind of approach in the years to come.

5.   For years, in the "old days" as well as more recently, much emphasis have been given, in an almost purely literary or philosophical perspective, to statements such as "the Sun is a star", or even "the active Sun is a star". True of course! ... But the present state of our knowledge of active

4

phenomena in stellar atmospheres, as well, on another side, as the theory of stellar evolution which affects all stellar layers, allow us to consider that the Sun is a body on its way from birth to the normal stellar life, and finally to its death as a main-sequence star. One has to understand not only how is the Sun behaving the way it does, but why. Conservation of energy, of momentum ... allow to reconstruct pre-stellar conditions, or to account for the distribution of magnetism and rotation in the Sun, or even in the solar system.

CONCLUSION

Therefore, one cannot escape the conclusion, obvious and trivial, but important, that the Sun is one object, in which the different layers are coupled with each other, in which the past conditions commands the future. One cannot treat a given solar region - say the convective layer - without taking into consideration the boundary conditions linked with what happens in the surrounding (radiative, in the chosen example) regions. One cannot understand one aspect of solar physics without some idea on the history of the involved matter and of its behaviour. In other terms, solar and stellar physicists cannot divide in overspecialized groups without risking to loose the very essence of their problem. We are well beyond the first approximation, well beyond the linearizations, in front of a highly coupled, highly non-linear set of physical equations. Solar physics cannot be reduced to a simple-minded modelling, according the "standard" line, whatever its merits as a good starting point.

AKNOWLEDGMENTS

This meeting has been co-organized, under the auspices of the I.A.U., by physicists and by astronomers, which, in itself, happily symbolizes the spirit of the new solar astrophysics. In particular, I would like to acknowledge the important work, behind or in front of the scene, of J. Boratav and of the Saclay physicists, M. Cribier, T. Montmerle and D. Vignaud. The chairman of the Scientific Organizing Committee has been Evry Schatzman. One should outline the fact that almost all aspects of the solar physics have been successfully tackled by Schatzman; he has been a novator in many of them. I want to thank him personally, for all what I have learnt from him, over more than four decades.

INSIDE THE SUN : UNSOLVED PROBLEMS

E.SCHATZMAN
Observatoire de Meudon
DASGAL
92195 Meudon-Cedex
FRANCE

ABSTRACT. As it is impossible to approach all the problems concerning the inside of the Sun, a number of questions will not be taken into consideration during the meeting. In this brief overview of the presently unsolved questions I shall insist on some special aspects of the solar properties : the variations of the solar radius, the generation of the solar wind, some interesting effects due to the presence of a strong gradient of ^{3}He, the history of the rotating Sun. The presence of the planetary system suggests that the Sun might have been a T Tauri star, with an accretion disc and may have started on the mains sequence as a fast rotating star. A sketch is given of the possible consequences.

1. Presentation of the symposium.

We understand the Sun and nevertheless we do not understand it . In other words, the gross properties of the Sun are clear : the energy generation, the radiative transfer, the general stability and instability properties, the convective zone and even the existence of a dynamo mechanism. But, when we want to have a quantitative agreement with the same precision than the observations, then we get into trouble. Do we have forgotten some major things in building up our models, or is it the physics which fails ? It is hardly necessary to recall, in presence of the best experts of the world, that the more information we have, the more difficult it is to understand what is really going on in the Sun : from the solar neutrinos to the shape of the circulation which drives the solar activity, with the funny names of *rolls* or *bananas*; from the distribution of the angular velocity inside the Sun, to the properties of the solar wind (so important for the loss of angular momentum), from the oscillation frequencies of the Sun to the surface abundances of the nuclearly processed elements.

In all these problems, we are facing difficulties which may come of all parts of physics. This is the reason for calling for this meeting, putting together physicists and astrophysicists. From the point of view of the astrophysicist, this is like asking the physicists to provide life-jackets before drowning.

I have discussed several times with particle physicists about the astrophysicist's statements concerning the inside of the stars. They are quite disturbed by the confidence which we show sometime when speaking of systems of 10^{57} particles. And, perhaps, they are right : part of this

5

G. Berthomieu and M. Cribier (eds.), Inside the Sun, 5–17.
© 1990 *Kluwer Academic Publishers. Printed in the Netherlands.*

meeting will be devoted to some of the collective behaviours which may turn out to be of the greatest importance for the understanding of the Sun. You will forgive me if I leave the problems of nuclear reactions, equation of state and opacities to the specialists ! The point is that I believe that there is still a lot to do about the motions inside the Sun and that we meet almost immediately non-linear problems, even when considering very slow motions like those induced by the circulation. I would like to take here my first example.

There will be a discussion about the turbulent diffusion mixing coefficient D_T in the radiative zone. Following here the suggestion of J.P.Zahn (1983), I shall accept the idea that differential rotation generates a 2-D turbulence on horizontal surfaces and that when the Rosby number Ro :

$$Ro = \frac{u}{l\,\Omega}$$

becomes larger than one it decays into a 3-D turbulence, with a turbulent diffusion coefficient :

$$D_T = \frac{4}{5} \frac{L\,r^3}{G\,M^2} \left| \left(1 - \frac{\Omega^2}{2\,\pi\,G\,\rho} \right) \right| \min\left[1, \frac{\Omega^2\,r}{g} (\nabla_{ad} - \nabla_{rad} + \nabla_{\mu})^{-1} \right],$$

The quantity under the sign min has the meaning of an efficiency coefficient and must be smaller than one. The boundary of the convective zone is reached when $(\nabla_{ad} - \nabla_{rad}) = 0$ (in a chemically homogeneous region). In the Sun, the quantity $(\nabla_{ad} - \nabla_{rad})$ becomes equal to $(\Omega^2 r/g)$ three kilometers below the boundary of the convective zone. The turbulent diffusion coefficient is very large, of the order of 9.10^5 cm^{2}s^{-1}. The condition on the Rosby number give the possibility of estimating the turbulent velocity, 1.8 cm s^{-1} and the scale of the turbulence, of the order of 5 kilometers. This has to be compared with the distance of penetration of convection from the convective zone. By the way, this is another controversial problem. Two arguments are in favour of a very small penetration : (i) the constrain which come from the presence of Lithium,so easily nuclearly processed , and suggests an exponential decrease of D_T with a vertical scale of just a few kilometers; (ii) a penetration with an exponential decrease of the velocity, with a vertical scale which, for high Rayleigh numbers Ra, varies like Ra^{-6} (Zahn et al, 1982,Massaguer et al,1984). However, this is valid only for an hexagonal planform with motion upwards along the axis of the cells. It is interesting to notice that these three vertical scales are of the same order of magnitude and are compatible with heliosismology. The heliosismology data are well interpreted by a discontinuity of the derivative of the gradient $(\partial \nabla^*/\partial\, r)$ at the boundary of the convective zone.

As far as mixing by diffusion is concerned, the quasi-singularity of D_T at the boundary of the convective zone is not very important. The WKB approximation does not suffer from the singularity, and the only important quantity is something like

$$\langle D_T \rangle = \left(\frac{1}{r_1 - r_2} \int_{r_1}^{r_2} \frac{d r}{\sqrt{D_T}} \right)^{-2}$$

which obviously does not show any singularity.

It is clear, that in order to obtain the exact efficiency of mixing, it is necessary to improve as much as possible the modelling of the turbulence in the radiative zone. This does not mean naturally that everything is said about turbulence in the convective zone!

2. One word on the convective zone.

The recent observations of Laclare (1987) of the variations of the radius of the Sun, and the analysis by Delache (1988) seem to confirm the XVII th century observations of Picard (1666-1682) recently discussed by E.Ribes (see the analysis of Ribes *et al* 1987). The variability of the apparent diameter can be due to a change of limb darkening (T.Brown, 1987, Brardsley *et al* 1989) or to a change of the surface temperature (Kuhn *et al* 1988). Atmospheric effects can also modify the apparent solar diameter. It should be mentioned that light scattering produces an apparent increase of the diameter : the measured diameter might depend on the variations of the amount of water vapour in the Earth atmosphere. However, a correlation between the solar sunspots, the diameter variability and the onset of stratospheric winds has been reported (Labitske, 1987; Ribes *et al* 1988). If confirmed, it would indicate a causal relationship between the solar cycle and the Earth atmosphere properties.

The anti-correlation between the solar radius and the solar activity (Delache et al 1986) can receive a simple explanation which has already been considered by Endal, *et al* (1985): the effect of the magnetic field. On a large scale, in the presence of a very entangled magnetic field, the compressibility γ is larger. A convective zone, starting from the same level, but with a larger compressibility, will have a greater thickness, given by :

$$\frac{\Delta H}{H} = \frac{\Delta \gamma}{\gamma (\gamma - 1)}$$

When the activity is weak, we can think that the magnetic field is located deep inside the convective zone, and as this corresponds to the higher temperature, we can expect the effect on the thickness of the convective zone,

$$H = \frac{\gamma - 1}{\gamma} \frac{\Re T}{g \mu}$$

to be the largest; on the contrary, during the phase of activity, when the magnetic field is located near the surface, the effect on the thickness of the convective zone is smaller. Such an explanation of this sort of <u>breathing</u> of the Sun is very attractive; it appears as being connected with the dynamo mechanism, but when looking at numbers, the situation does not seem so good. The change of γ is $(4/15)(B^2/8\pi P)$. In order to obtain a change of the thickness of the convective zone of about 400 km, it is necessary to have a magnetic pressure of about 0.005 P, which gives at the bottom of the convective zone about $3 \cdot 10^6$ Gauss, of the order of 100 times what is expected from the simple principle of equipartition with the density of cinematic energy. However,this of the order of magnitude which Durney(1988) expects from the effect of differential rotation on the magnetic flux tubes, and is compatible with an estimate of the magnetic field strength derived from an extrapolation of the value of the magnetic field in sunspots to its value at the bottom of the convective zone. It fits also with the estimates of Dziembowski *et al* (this meeting).Altogether, despite the fact that the physics seems reasonable, it is still necessary to build up a consistent theory.

3. Angular momentum.

The loss of angular momentum (Schatzman,1959,1962), which turns out to be of great importance for the history of the solar rotation and for the internal structure of the Sun, is governed by the rate of mass loss (the solar wind), and by the average surface magnetic field (the average being taken over spans of time much larger than the period of activity), its topology and its strength (Roxbugh 1983, Mestel and Spruit 1987). The solar dynamo just as any stellar dynamo, is dominated by non-linear effects. The growth of the magnetic field can reasonably be described by the linear effects, but the limit on the intensity of the magnetic field is essentially non-linear.

Recently, Kawaler (1988), Pinsonneault et al (1989) have worked out models of the solar spindown with a parametric expression of the mass loss rate. I noticed especially in the paper of Kawaler (1988) that there is no indication of a time dependence of the rate of mass loss. However, would it be only from the observational point of view, the rate of mass loss is not the same from a young main sequence star and for an old one, the activity depending on the Rossby number (Noyes, 1983; Mangeney and Praderie, 1984). Durney and Latour (1978) have avoided the problem by assuming that the velocity of escape is simply given, in order of magnitude, by $(GM/R)^{1/2}$.

In order to study stellar evolution with mass loss, Fusi-Pecci and Renzini (1976), have made some assumptions on the amount of mechanical energy which is carried out of the convective zone, based essentially on the analysis of Proudmann (1952) of the production of acoustic waves by the turbulence in the convective zone. Unno (1966) extended the work of Proudmann by considering also the monopolar and the dipolar contribution and not only the quadripolar. But it is well known that there is some difficulties in the theory of the stellar wind:
-it is difficult to fit the observations with the theoretical models;
-the production of the wind is due both to the injection of energy and to the injection of momentum;
-the process itself implies all kinds of dissipative mechanism, from shock waves to a variety of plasma dissipative effects in which the topology of the magnetic field is involved,
-the important quantity is the average mass loss, coupled with the strength of the magnetic field : the Alfvenic distance and the mass loss are not the same for the quiet Sun and for the active Sun.

What I claim here, is that the production of the solar and stellar winds is an internal structure problem; its source is in the convective zone with its turbulent motion; it involves these terms in the equations of motion which describe the compressible effects - which are usually neglected in the study of the convective zone - and finally it should take into account the presence of the magnetic field- and we know how difficult is the theory of the MHD turbulence.

We find here a fundamental problem. Magnetohydrynamics as well as hydrodynamics are based on deterministic equations. But can we say that the problem of mass loss and loss of angular momentum is also determined in the usual sense of the word determinism ? We see, in galactic clusters, that among stars which seem to obey to a common law, a few stars look anormal, just as if something different had append to them. Were the initial conditions different, with for example an anomalous initial angular momentum, or did they have a different history because they crossed a branching point which the other ones did not notice? did they have a different chaotic behaviour ? How does this apply to the Sun ? Billions of years ago, the Sun did not have the same period of rotation and if it had a periodic activity, the period was certainly not the same. Is it possible to see in billion years old sediments some trace of it, just as we see it in hundred millions years old ones ?

It is certainly a great temptation, in front of these difficulties, either to use an oversimplified model (and it is what I did ! : see section 5) or to use a completely phenomenological model, with adjustable parameters. This last method has the advantage that orders of magnitude can be derived from the comparison with the observations. But on the other hand, it hides entirely the possibility that some important physical effect has been forgotten.

4.Stability and instability.

It is well known that in the outer part of the Sun, ^{3}He is produced through the classical series of reactions (Schatzman, 1951 a, b):

$$^1H + {}^1H \Rightarrow {}^2D + e^+ + v$$

$$^2D + {}^1H \Rightarrow {}^3He + \gamma$$

and destroyed mainly by the reaction

$$^3He + {}^3He \rightarrow {}^4He + {}^1H + {}^1H$$

Building of ^{3}He only takes place in the outer half of the Sun, whereas destruction takes place in the inner half, the maximum concentration of ^{3}He being found at about one half of the solar mass or 0.3 solar radius. In the present Sun , the maximum concentration in mass is about $3 . 10^{-3}$.

I just want here to draw attention to a very unexpected effect of the ^{3}He gradient in the outer half of the Sun. Let us recall the usual definitions :

$$\nabla = \frac{d}{d \log P} \ ; \ \Delta\nabla = \nabla_{ad} - \nabla_{rad} \ ;$$

The 3-D turbulent flow induced by the differential rotation is inhibited by the μ-gradient (Zahn, 1983) if the non-thermal part of the Brunt-Väissälä frequency - which will persist even in the presence of strong thermal diffusion - is larger than the turnover frequency of the smallest eddies present in the turbulent spectrum, those of the Kolmogorof scale. This condition can be written :

$$\nabla\mu \ (\Delta\nabla + \nabla\mu) > \frac{4}{5} \frac{1}{\nu} \frac{L r^8 \Omega^4 H_P}{G^3 M^4} \tag{8}$$

With X_3 of the order of 10-3, the μ-gradient due to 3He turns out to be , in the present Sun, of the order of 1O-4. At that level, the stability condition (8) gives $\nabla\mu > 3$. 10-6 in the present Sun. With no mixing ,the μ-gradient due to 3He, would have stabilized the 3-D turbulence produced by the meridional circulation when the Sun was about two billion years old. On the other hand, it is possible to see that the rise of μ due to the production of 3He would not have stopped the meridional circulation. With a maximum concentration $Y_3 = 0.003$ reached at $t = 4.6$. 10^9 years, the μ-excess is $\Delta\mu = 0.0011$. In order to stop the meridional circulation, we should have :

$$\Delta\mu > \frac{L \Omega^2 R^3}{G^2 M M_r^2} r \ t_{Sun} \cong 0.015$$

and this condition is not fulfiled.

Beginning about two and a half billion years ago, diffusion of 3He towards the surface of the Sun concerned only the outer half of the hill of 3He concentration. It remains to check if this leads to a reconsideration of the constraint on diffusion coming from the surface abundance of 3He.

5. History of the rotating Sun.

5.1. BASIC ASSUMPTIONS

What we are concerned her are the initial conditions. A standard view is to consider the approach to the main sequence of a solar like star along the Hayashi track, with an initial rotational velocity close to the rotational breaking. The loss of angular momentum is such that the equatorial velocity hardly increases or even decreases, depending on the model of angular momentum losses and transport inside the star during the contraction (Pinsonneault et al, 1988), and the star reaches the main sequence with an equatorial velocity of the order of 20 km s^{-1}, as it appears to be the case for the slow rotators in the Pleiades (van Leuwen and Alphenaar, 1982; Soderblom, Jones and Walker,

1983; Stauffer et al, 1984; van Leeuwen, Alphenaar and Meys, 1987) and in α Per (Butler et al, 1987; Balachandran , Lambert and Stauffer 1988; Stauffer et al 1988) .

However, both the Pleiades and α Per contain fast rotators (up to 200 km s^{-1}), which are as well on the main sequence. It seems possible to compare the situation to the existence of two kinds of T Tauri stars (Cohen and Kuhi 1979), the wide lines T Tauri surrounded by a disk detectable by its infra-red emission (Mendoza 1968; Cohen and Kuhi 1979; Bertout, Basri and Bouvier 1988) and the narrow lines T Tauri which do not have a disk. When there is a disk it appears to be in strong interaction with the central star. The proportion of T Tauri with a disk seems to be of the order of 10% among all T Tauri stars. It seems possible to imagine that the late presence of a disk has the effect of feeding the central star with a large amount of angular momentum. This could explain the presence simultaneously in the Pleiades and in α Per of slow and fast rotators. Stars having lost their disk at an early phase of their evolution would have reached the main sequence at a moderate equatorial velocity.

The existence of the planetary system around the Sun strongly suggests that the Sun had a disk and that it can possibly have been a fast rotator.

It seems therefore necessary to consider two scenarios for the history of the rotating Sun, and to study the consequences both for the transfer of angular momentum and for the transport of passive contaminents.

5.2. STANDARD MODEL

The loss of angular momentum depends on several effects : the geometry of the magnetic field, which determines the fraction of the stellar surface which is occupied by open lines of force, the surface value of the magnetic field, which is related to the dynamo mechanism, and the rate of mass loss. The problem of the geometry of the magnetic field has been considered by Roxburgh (1984) and more recently by Mestel and Spruit (1987). The result depends on the choice of the structure of the magnetic field (dipole, quadrupole or a distribution of bipolar magnetic spots). The value of the magnetic field at the Alfvenic distance is then related to the surface value in a complicated way.

A parametric expression of the loss of angular momentum has been given by Kawaler (1988). We shall follow here, as Durney and Latour (1978) the simple assumption of flux conservation, $B \propto r^{-2}$. Similarly, the simplest assumption concerning the non-linear dynamo is to assume that the maximum rate of growth of the magnetic field is compensated by the losses due to buoyancy (Schatzman 1988). Finally, following Durney and Latour (1978) we shall assume, as shown by Parker (1975) that the velocity of escape of the stellar wind for solar like stars (Linsky 1985) is proportional to the velocity of escape from the surface of the star, $(G M / R)^{1/2}$.

We shall assume that the rate of loss of angular momentum is given by :

$$I\frac{d\Omega}{dt} = - K_F \, \Omega^{7/3}$$

The model of Schatzman (1988), does not give the exact value, but, without any parametric adjustment, provides nevertheless an excellent fit with the observed equatorial velocity of the Sun. We have the asymptotic expression :

$$\Omega = \left(\frac{4}{3} \frac{K_F t}{I} \right)^{-3/4}$$

with $(4/3)K_F = 1.16 \cdot 10^{44}$. It should be noticed that the $t^{-3/4}$ law, which differs from the Skumanich relation (1972) has been obtained by Bohugas et al (1976).

It is easy to obtain the order of magnitude of the angular velocity gradient which, at the boundary of the convective zone, is sufficient to carry away the flux of angular momentum. With a turbulent diffusion coefficient of the order of 1000 and the present rate of loss of angular momentum, it turns out that an angular velocity gradient

$$(d \, \Omega/d \, r) \cong 5.10^{-16}$$

is sufficient to carry away the angular momentum from the bottom of the convective zone.This corresponds however to a logarithmic gradient

$$(d \ln \Omega/d \ln r) = 8.6$$

The Richardson-Townsend condition of instability of the turbulent shear flow leads to a critical value :

$$(d \ln \Omega/d \ln r)_{crit} \cong 30$$

The vertical shear flow gradient does not allow the generation of turbulence, but it should be noticed that, compared to the results of heliosismology (Dziembowski et al 1989) this is very large. It would correspond to a a change of Ω by a factor 2 over 5% of the Solar radius ! We are meeting here one of the major problems concerning mixing inside the Sun : the turbulent diffusion coefficient for the transport of angular momentum is much larger than the turbulent diffusion coefficient for the transport of passive contaminents. When considering the quasi solid body rotation from 0.7 R_O to 0.3 R_O we have to introduce a turbulent diffusion coefficient at least of the order of :

$$D_T = (0.16R_O^2/t_O) \cong 6000$$

that is to say 6 to 10 times what is needed in order to explain Lithium burning. Tassoul and Tassoul (1989) introduce a turbulent diffusion coefficient $D_{T\Omega}$ which is 10 to 20 times the coefficient D_{TX}, and Pinsonneault et al (1989) meet the same problem. It is clear that the vertical shear flow, at the boundary of the convective zone, is at most

$$(d \ln \Omega/d \ln r) \cong 0.3$$

The physical question is then the following : what is the physical origin of this large turbulent diffusion coefficient ? Magnetic field is certainly the proper answer, as it is sufficient, in principle, to have a magnetic field such that the propagation of a perturbation at the Alfven velocity from the center to the surface takes less than $4.6 \cdot 10^9$ years $\cdot$ This corresponds to a magnetic field of the order of 10^{-5} gauss. However, the exact nature of the magnetic field in the radiative zone is not known, and the recent discussions of Spruit (1986) and Mestel and Weiss (1987) have just shown the difficulty of the problem.

With a larger angular velocity the turbulent diffusion coefficient due to differential rotation varies like Ω^2, but the rate of loss of angular momentum varies like $\Omega^{7/3}$. The gradient of angular momentum at the boundary of the convective zone varies like $\Omega^{1/3}$. A factor 100 on the angular velocity corresponds to an increase of the angular momentum gradient by a factor 4.6 only, and this is still below the critical value for the vertical shear flow instability.

In such conditions,the major effect seems to come from the differential rotation.

Numerical solutions have been given by Pinsonneault et al (1989), but it is worth giving an analytical approach, as it give a better insight to the physical problem.

5.3. THE LITHIUM PROBLEM.

After the introduction of turbulent diffusion mixing as contributing to the surface abundance of the elements (Schatzman, 1969), it became clear that this would affect the abundance of Lithium through Lithium burning (Schatzman, 1977). Scalo and Miller (1980) have shown that the abundance of Lithium in giants can be interpreted as the result of the dredge up taking place in stars which have destroyed their Lithium on the main sequence. Schatzman (1981) and Schatzman and Maeder (1981) have shown that the surface abundance of Lithium is a measure of the turbulent diffusion coefficient which explains the amount of Lithium burning . Schatzman (1983) noticed that, with a correct value of the masses of giants, it is possible to obtain an agreement between the results of Alschüler (19), taking into account the remarks of Scalo and Miller about the paper of Schatzman (1977), and the role of turbulent diffusion mixing in the surface destruction of Lithium. Baglin et al (1984) used the turbulent diffusion coefficient of J.P.Zahn in order to calculate the abundance of Lithium in the Hyades and in the Sun. The model turned out to lead to a number of contradictions and, as will be shown briefly in the following, it seems that the introduction of the proper dependence of the turbulent diffusion coefficient D_T on time can lead to a consistent explanation of the observations. This will be discussed again in the paper of A.Baglin (this meeting).

Coming back to the equation of evolution of the Lithium concentration,

$$\frac{\partial X_7}{\partial t} = \frac{1}{r^2} \frac{\partial}{\partial r} r^2 D_T \frac{\partial X_7}{\partial t} - K_7(r) X_7 \qquad (10)$$

it is possible to factorize the variables by replacing $K_7(r)$ by a step function, $K_7(r) = 0$ for $r > r_{Burning}$ and $K_7(r) = \infty$ for $r < r_{Burning}$, where $r_{Burning}$ is the burning level of Lithium. Writing

$$\Omega = \Omega_0 \left(1 + (t/t_O)\right)^{-3/4}$$

it is possible to introduce a new variable t',

$$dt' = (1 + (t/t_O))^{-3/2} \, dt$$

and then to use the eigenvalue equation (1). With the relation

$$\Omega = \Omega_0 \left(1 + \frac{4 K_F \Omega_0}{3 I}^{4/3}\right)^{-3/4}$$

and assuming that the distance h from the bottom of the convective zone to the Lithium burning level is small, we have (Schatzman 1988):

$$c = c_0 \exp\left[-0.3 \; \frac{2 L r^6 \Delta\nabla^{-1} \Omega_0^2}{G^2 M^3 h H_P \frac{K_F}{I}^{4/3} \Omega_0} \left(1 - \left(1 - \frac{4 K_F \Omega_0^{4/3}}{3 I} t\right)^{-1/2}\right)\right]$$

In the case of the Sun, this gives

$$c = c_0 \exp\left(-5.6\left(1 - \left(1 + \frac{t}{3.10^8 y}\right)\right)\right)$$

For $(t/10^{16}) = 12$, one obtains $c = 2.10^{-2} \, c_0$. A slight adjustment is necessary (which implies the model of the non-linear dynamo) and provides the proper value of the Lithium concentration.

This expression gives the decrease with time of the surface Lithium concentration. When h goes to zero, the Lithium concentration vanishes. For finite h, there is an asymptotic value of the concentration and this can explain the finite value of the Lithium concentration of the old population II stars (Spite and Spite,1982).

Solving the diffusion equation with the WKB approximation gives a better solution. The mass dependence of K_F and of the depth of the convective zone seem to provide a nice adjustment of the abundance of Lithium in the Hyades as a function of mass.

5.4. THE FAST ROTATING SUN

The presence, among fast rotators in α Per, of Lithium rich stars (Balachandran, 1988) (they are either not or slightly deficient) is quite remarkable.

Coming back to the expression of the turbulent diffusion coefficient, it can be seen that this coefficient do vanish at a certain level, given by

$$\Omega^2 = 2 \pi G \rho$$

The effect is a diffusion barrier. However, the effect of the diffusion barrier is important only when it is located below the bottom of the convective zone. This is possible only for the early spectral types. For a mass of the order of $0.9\ M_O$, $T_{eff}=5140\ °K$, this would imply $V_{equ}>230 kms^{-1}$. This means that for later types, say for $T_{eff}<5200\ °K$, it is necessary to find another explanation. The hypothesis of the presence of a disk provides simultaneously the explanation of the high equatorial velocity and of the presence of Lithium. The disk provides angular momentum <u>and</u> Lithium. The abundance of Lithium is then the result of a balance between the Lithium brought by accretion and the fast destruction inside the star. If p is the rate of Lithium destruction , X_7 the concentration in the disk, and X_7^* the concentration in the convective zone, we have

$$X_7^* = X_7 \frac{K_F \Omega^{4/3}}{4 \pi R^4 \left(\frac{r}{R}\right)^2 \rho\, H_P\, p + K_F \Omega^{4/3}}$$

For a $0.9\ M_O$ star, with $T_{eff} = 5140\ °K$, $H_P = 4.21 . 10^9$, $T = 2.33 . 10^6\ °K$,

$\rho = 0.41\ g\ cm^{-3}$, one finds $p = 4.63 . 10^{-14}\ s^{-1}$ for an equatorial velocity of 200 km s^{-1} or $\Omega = 3.71 . 10^{-4}$, and $(X_7^* / X_7) = 0.57$, which is quite a reasonable value.

From these quantities, it is possible to calculate the mass which has to be transferred from the disc to the star in order to satisfy the condition of conservation of angular momentum. It is about 0.03 solar masses and is quite compatible with the order of magnitude given by Bertout *et al* (1988)

After the disappearance of the disc the spin down process can start. There must be a very rapid spin down of the layers immediately below the convective zone, in such a way that the turbulent diffusion coefficient drops quickly, otherwise there would no Lithium left in these fast rotators. The propagation of the spin-down from the outside to the inside is certainly a characteristic of initially fast rotating stars. In the case of the Sun, this suggests, deep inside the Sun, a possible memory of a high initial angular velocity (Vigneron <u>et al</u> 1989). When looking at the neutrino problem, this should be kept in mind.

6. Comments.

As it has been said at the beginning, as soon as we try to obtain simultaneously a consistent theory and an agreement with the observational data we discover that new problems have to be solved . It is the aim of this conference, in making more precise the limits of our reliable knowledge, to help starting new directions of theoretical research, to suggest new observations, and perhaps to succeed not only in finding new results in fundamental physics but also in explaining a few more of the solar mysteries.

Acknowledgements.

I express my thanks for the discussions on these problems to A.Baglin, G.Michaud, E.Ribes, S. Vauclair, J.P. Zahn.

Bibliography.

Alschüler W.R., 1975, *Astrophys. J.*,**195**, 649.

Baglin A., Morel P., Schatzman E., 1985, *Astr. Astrophys.*, **149**, 309.

Balachandran S., Lambert D.L., Stauffer J.R., 1988, *Astrophys. J.* **333**, 267.

Beardsley B.J., Hill H.A., Cornuelle C.S. and Kroll R.J., *Bull. of the A.A.S.*, (in press).

Bertout C., Basri G., and Bouvier J., 1988 ,*Astrophys. J.* **330**,350

Bohugas J.,Carrasco L.,Torres C.A.O.,and Quast G.R., 1986, *Astr.Astrophys.*, **157**, 278.

Brown T.M., 1987, Ground-based observations related to the global state of the Sun, *NCAR proceedings*, P.Foukal ed., 176.

Butler R.P., Cohen R.D., Duncan D.K., and Marcy G.W., 1987, *Astrophys. J*, (*Letters*),**319**, L19

Cohen M., and Kuhi L.V., 1979, *Astrophys. J. Suppl.*,**41**,743

Delache Ph., 1988, *Adv. Space Research* , **8**, 119.

Delache Ph., Laclare F.,Sadasoud H.,1986, I.A.U. Colloquium 123,

Durney, 1988, private communiation

Durney B.R., and Latour J.,1978, *Geophys. Astrophys. Fluid Dynamics*,**9**,241

Dziembowski W.A., Goode Ph. R., Libbrecht K.G.,,1989, *Astrophys.J.*, **337**, L 53.

Endal A.S., Sofia S.and Twigg C.N.,1985, *Astrophys.J.* **290**,748

Fusi-Pecchi F., and Renzini A., 1976, *Astr. Astrophys.*,**46**, 447

Kawaler S.D.,1988, *Astrophys.J.*,**333**,236

Kuhn J.R.,Libbrecht K.G., and Dicke R.H.,1988, *Science,* **242**, 908.

Labitske F., 1987, *Geophysical Research Letters,* **14**, N° **5**, 535.

Laclare F.,1987, *C.R. Acad. Sci.*, **305**, série II, 451

Linsky J.F.,1985, *in* The Origin of Non-radiative Heating/Momentum in Hot Stars, *NASA Conference Publication 2358*

Massaguer J.M., Latour J.,Toomre J., and Zahn J.P.,*Astr. Astrophys.***140**,1

van Leeuwen F., and Alphenaar P., 1982, *E.S.O. Messenger*, No 28, p.15

van Leeuwen F., Alphenaar P., and Meys J.J.M.,1987, *Astr.Astrophys.* Suppl.**67**, 283

Mendoza V.E.E.,1968, *Astrophys.J.*,**151**,749.

Mangeney A. and Praderie F., 1984, *Astr. Astrophys.*, **130**, 143.

Mestel L. and Spruit H.,1987, *Mon. Not. R. Astr. Soc.*,**226**, 57.

Mestel L. and Weiss N., 1987, *Mon. Not. R. Astr. Soc.* **226**, 123

Noyes R.W., Hartmann L.W., Baliunas S.L., Duncan D.K., and Vaughan A.H.,1984, *Astrophys.J.* **279**, 763

Parker E.N.,1975,*Space Sci. Rev.* **4**, 666

Picard J., manuscripts D1, 14-16, Archives de l'Observatoire de Paris (1666-1682)

Pinsonneault M., Kawaler S., Sofia S., and Demarque P., 1989, *preprint*.

Proudmann I.1952,*Proc. Roy. Soc. London, A*,**214**, 119

Ribes E., Ribes J.C., and Barthalot R.1987, *Nature* , **332**, 689.

Ribes E., Ribes J.C., Vince I. and Merlin Ph., *C.R. Acad. Sci. (Paris),* **307**, série II, 1195.

Roxburgh I.W.,1983, *in Solar and Stellar Magnetic fields*, Reidel publisher, J.O.Stenflo ed., p.449

Scalo M.S., and Miller G.E., 1980, *Astrophys. J.,* **239**, 953.

Schatzman E., 1951 a, *C.R. Acad. Sci.* Paris.**232**, 1740

      1951 b, *Ann. d'Ap.* **14**, 294

Schatzman E., 1959, IAU Symposium N°10, The Hertzsprung-Russell diagram, J.L.Greenstein Ed., *Suppl. Ann. d'Ap.* N°**8**, p.129

Schatzman E., 1962, *Ann. d'Ap.* **25**, 18

Schatzman E. ,1969, *Astr. Astrophys.* **3**,331.

      1977, *Astr. Astrophys.*, **56**, 211.

      1981, CERN, 81-11.

Schatzman E., 1989, *in* Turbulence and non-linear dynamics in magneto- hydrodynamic flows, *International workshop, July 4-9, 1988, Cargèse (Corsica),France*, M.Meneguzzi, A.Pouquet,P.L. Sulem ed., Elsevier Science Publishers (North-Holland).

Schatzman E., and Maeder A., 1981, *Astr. Astrophys.*, **96**, 1.

Skumanich A.,1972, *Astrophys.J.*,**171**,565

Soderblom D.R.,Jones B.F.,and Walker B.F.,1983, *A pJ. (Letters)*,**274**,L37

Spite F. and Spite M.,1982, *Astr. Astrophys.*, **115**, 357

Spruit H., 1986, *in* The Hydromagnetics of the Sun, Proceedings of the Fourth European Meeting on Solar Physics, ESA SP-220, p. 21

Stauffer J.R.,Hartmann L.,Soderblom D.R..,and Burnham N., 1984, *ApJ.* **280**, 202

Stauffer J.R.,Hartmann L.,Burnham N. and Jones B.,1988, *Ap.J.* (preprint).

Tassoul J.L. and Tassoul M. 1989,*Astron. Astrophys.* , **213**, 397.

Unno W. 1966, *Trans. of the IAU, Vol XII B,* Academic Press, p.155.

Vigneron C., Schatzman E., Catala C., and Mangeney A. .Poster *this meeting*

Zahn J.P., 1983 in Astrophysical Processes in Upper Main Sequence Stars,*13th Advanced Course, Swiss Society of Astronomy and Astrophysics, Saas-Fee, 1983, B.Hauck and A.Maeder ed.*, p.253

Zahn J.P., Toomre J., Latour J., 1982, *Geophys. Astrophys. Fluid Dynam.* **22**, 159

PART 1

SOLAR MODELLING

THE STANDARD SOLAR MODEL

J.N.BAHCALL
Institute for Advanced Study
NJ 08540, Princeton
USA
M. CRIBIER
CEN SACLAY DPhPE/SEPh
F-91191 GIF-sur Yvette CEDEX
France

ABSTRACT. The main features of standard solar models, the logic of the calculations, and some of the important results concerning solar neutrinos experiments are given. The input parameters that cause the greatest uncertainties in the calculated neutrino fluxes are the nuclear rection rates, the chemical abundances, the radiative opacity, and the equation of state. This article is based, with permission of the publisher, on Chapters 1 and 4 of *Neutrino Astrophysics* by J. N. Bahcall, Cambridge University Press (1989).

1. Introduction

The Sun is an astronomical laboratory. Because of its proximity to the Earth, we are able to obtain information about the Sun that is not accessible for other stars. We can determine precise values for the solar mass, radius, geometric shape, photon spectrum, total luminosity, surface chemical composition, and age. In addition, astronomers have measured accurate frequencies for thousands of acoustic oscillation modes that are observed at the solar surface. These frequencies contain information about the solar interior. We are beginning to measure the spectrum of neutrinos produced by nuclear reactions in the solar interior. The geological records, the planets, comets, and meteorites, provide information about the past history of the Sun. Taken together, this treasure of experimental information provides a unique opportunity to test theories of stellar structure and evolution.

For two decades, the only operating solar neutrino experiment yielded results in conflict with the most accurate theoretical calculations. This conflict between theory and observation, which has recently been confirmed by a new experiment, is known as the solar neutrino problem. This problem can be stated simply. Both the theoretical and the observational results are expressed in terms of the solar neutrino unit, SNU, which is the product of a characteristic calculated solar neutrino flux (units: cm^{-2} s^{-1}) times a theoretical cross section for neutrino absorption (unit: cm^2). A SNU has, therefore, the units of events per target atom per second and is chosen for convenience equal to $10^{-36}s^{-1}$.

The predicted rate for capturing solar neutrinos in a ^{37}Cl target is

$$\text{Predicted rate} = (7.9 \pm 2.6)\,\text{SNU}, \qquad (1a)$$

where the indicated uncertainty represents the total theoretical range including three standard deviation (3σ) uncertainties for measured input parameters. The rate observed

G. Berthomieu and M. Cribier (eds.), Inside the Sun, 21–41.

by Davis and his associates in a chlorine radiochemical detector is

$$\text{Observed rate} = (2.3 \pm 0.75) \text{ SNU}, \tag{1b}$$

where the error is again a 3σ uncertainty [Davis (1989)].

There is no generally accepted solution to the discrepancy although a number of interesting possibilities have been proposed.

This discrepancy between calculation and observation has recently been confirmed by an independent technique using the Japanese detector of neutrino–electron scattering, Kamiokande II. The recent Kamiokande II result is [Hirata et al. (1989)]

$$\frac{\phi_{\text{observed}}}{\phi_{\text{predicted}}} = 0.39 \pm 0.09(stat.) \pm .06(syst.), \tag{2}$$

where the neutrino flux, ϕ, is from the rare ^{8}B solar neutrinos and the quoted error is the 1σ uncertainty.

The predictions used in Eqs. (1) and (2) are valid for the combined standard model, that is, the standard model of electroweak theory (of Glashow, Weinberg, and Salam) and the standard solar model.

The presentation given here is based upon the results of Bahcall and Ulrich (1988) and Bahcall (1989), as well as the references cited in this article. In a recent detailed study, Sienkiewicz, Bahcall, and Pacynski (1989) constructed a standard solar model using an independently developed numerical code. Adopting the standard model input parameters, the authors obtain 7.6 SNU for the calculated absorption rate in the ^{37}Cl experiment. The difference of 0.3 SNU between this result and the Bahcall-Ulrich value given earlier is due to the combined effect of several small differences in physical description that were to time-consuming to incorporate in the Sienkiewicz et al. code. Turck-Chièze et al. (1988) describe the result of a calculation in which they do not use modern standard input parameters for the ^{8}B nuclear production cross section and for the opacity. They choose to use a nuclear cross section that is different from that reported by the experimentalists in all the recent reviews [see for example Parker and Rolfs (1989) or Parker (1986)] and opacities that were computed more than a decade ago. Their results demonstrate that one can obtain different answers for the predictions of solar models (within the acknowledged uncertainties) by choosing different input parameters.

In section 2, the input parameters of the standard models are presented. The pp chain which is the dominant source of the energy for the sun is described in section 2.1. Neutrino interaction cross, section (2.2), permit the prediction of the rate in specific detectors. Chemical abundances (section 2.3), opacities (2.4), and the equation of states, (2.5), are described to indicate their contributions to the uncertainties of the calculations.

Section 3 describes the general method used in the computation of the solar standard model. The basic equations (3.1) with the main physical hypothesis are used in the calculation procedure (3.2). The main characteristics resulting from this computation are presented in section 4.

Prediced rates for the different solar neutrinos experiments are compared with the experimental results in section 5. The section 6 points out some features of the solar standard model related to helioseismology.

Table 1. **Some important solar quantities.** The measured
parameters are: photon luminosity, mass, radius, oblateness,
and age. All other quantities are calculated with the aid of
the standard solar model.

Parameter	Value
Photon luminosity $(L_\odot)$	3.86×10^{33} erg s^{-1}
Neutrino luminosity	$0.023 L_\odot$
Mass $(M_\odot)$	1.99×10^{33} g
Radius $(R_\odot)$	6.96×10^{10} cm
Oblateness	$\leq 2 \times 10^{-5}$
$[(R_{\mathrm{equatorial}}/R_{\mathrm{polar}}) - 1)]$	
Effective (surface) temperature	5.78×10^3 K
Moment of inertia	7.00×10^{53} g cm^2
Age	$\approx 4.55 \times 10^9$ yr
Initial helium abundance by mass	0.27
Initial heavy element abundance by mass	0.020
Depth of convective zone	$0.26 R_\odot (0.015 M_\odot)$
Central density	148 g cm^{-3}
Central temperature	15.6×10^6 K
Central hydrogen abundance by mass	0.34
Neutrino flux from pp reaction	6.0×10^{10} cm^{-2} s^{-1}
Neutrino flux from ^{8}B decay	6×10^6 cm^{-2} s^{-1}
Fraction of energy from pp chain	0.984
Fraction of energy from CNO cycle	0.016

2.   The input parameters

The major input parameters or functions that are used in a standard solar model are: nuclear parameters, solar luminosity, solar age, equation of state, elemental abundances, and radiative opacity.

Table 1 lists some of the main physical characteristics of the Sun. Of special importance for the solar neutrino problem are the accurately determined luminosity and mass, the initial heavy element to hydrogen ratio (Z), and an upper limit on the intrinsic solar oblateness [Dicke, Kuhn, and Libbrecht (1985)].

2.1   Nuclear energy generation and neutrino fluxes

The Sun shines by converting protons into α-particles. About 600 million tons of hydrogen are burned every second to supply the solar luminosity. Nuclear physicists have worked for half a century to determine the details of this transformation. The subject has been recently reviewed by Parker and Rolfs (1989). From the uncertain parameters in nuclear cross section, Bahcall and Ulrich (1988) derived uncertainties of 1.7 SNU for the ^{37}Cl experiment and 7 SNU for the ^{71}Ga experiments.

The main nuclear burning reactions in the Sun are shown in Table 2, which represents the energy-generating **pp chain**. This table also indicates the relative frequency with which each reaction occurs in the standard solar model.

The fundamental reaction in the solar energy-generating process is the proton–proton (pp) reaction. In the pp reaction, a proton β-decays in the vicinity of another proton

Table 2.   The pp chain in the Sun. The average number of pp neutrinos produced per termination in the Sun is 1.85. For all other neutrino sources, the average number of neutrinos produced per termination is equal to (the termination percentage/100).

Reaction	Number	Termination[†] (%)	ν energy (MeV)
$p + p \rightarrow {}^2H + e^+ + \nu_e$	1a	100	≤ 0.420
or			
$p + e^- + p \rightarrow {}^2H + \nu_e$	1b (pep)	0.4	1.442
${}^2H + p \rightarrow {}^3He + \gamma$	2	100	
${}^3He + {}^3He \rightarrow \alpha + 2p$	3	85	
or			
${}^3He + {}^4He \rightarrow {}^7Be + \gamma$	4	15	
${}^7Be + e^- \rightarrow {}^7Li + \nu_e$	5	15	(90%) 0.861
			(10%) 0.383
${}^7Li + p \rightarrow 2\alpha$	6	15	
or			
${}^7Be + p \rightarrow {}^8B + \gamma$	7	0.02	
${}^8B \rightarrow {}^8Be^* + e^+ + \nu_e$	8	0.02	< 15
${}^8Be^* \rightarrow 2\alpha$	9	0.02	
or			
${}^3He + p \rightarrow {}^4He + e^+ + \nu_e$	10 (hep)	0.00002	≤ 18.77

[†]The termination percentage is the fraction of terminations of the pp chain, $4p \rightarrow \alpha + 2e^+ + 2\nu_e$, in which each reaction occurs. The results are averaged over the model of the current Sun. Since in essentially all terminations at least one pp neutrino is produced and in a few terminations one pp and one pep neutrino are created, the total of pp and pep terminations exceeds 100%.

forming a bound system, deuterium (^{2}H). This reaction (number 1a in Table 2) produces the great majority of solar neutrinos; however, these pp neutrinos have energies below the detection thresholds for the ^{37}Cl and Kamiokande II experiments. Experiments with ^{71}Ga are sensitive primarily to neutrinos from the pp reaction. More rarely, a three-body reaction involving two protons and an electron initiates the reaction chain. While this reaction (number 1b in Table 2) occurs with a relative frequency of only one in 250, the resulting neutrino energy is larger by the equivalent of two electron masses, raising it above the threshold in the chlorine experiment. The deuteron produced by either of the initiating reactions is burned quickly by a (p,γ) reaction that forms ^{3}He (reaction 2 in Table 2). Reactions 1a and 2 occur in essentially all terminations of the pp chain in the Sun; reaction 1b occurs only rarely, in approximately 0.4% of all pp terminations. The richness and complications of the pp cycle begin at the next stage.

Most of the time, 85% in the standard solar model, the proton–proton chain is terminated by two ^{3}He nuclei fusing to form an α-particle plus two protons (reaction 3 of Table 2). No additional neutrinos are formed in this dominant mode.

About 15% of the time, a ^{3}He nucleus will capture an already existing α-particle to form ^{7}Be plus a gamma ray (reaction 4). It is the neutrinos formed after this process that are primarily detected in the ^{37}Cl experiment. Nearly always, the ^{7}Be nucleus will undergo electron capture, usually absorbing an electron from the continuum of ionized electrons (reaction 5). This branch produces neutrinos with energies of 0.9 MeV (90% of the time), which contribute small (but not negligible) fractions of the predicted standard model capture rate in the ^{37}Cl and ^{71}Ga experiments. There is no experiment in progress that isolates the contribution of the ^{7}Be neutrinos, although some suggestions for practical detectors have been made.

Most of the predicted capture rate in the ^{37}Cl experiment comes from the rare termination in which ^{7}Be captures a proton to form radioactive ^{8}B (reaction 7). The ^{8}B decays to unstable ^{8}Be, ultimately producing two α-particles, a positron, and a neutrino. The neutrinos from ^{8}B decay have a maximum energy of less than 15 MeV. Although the reactions involving ^{8}B occur only once in every 5000 terminations of the pp chain, the total calculated event rates for the ^{37}Cl and Kamiokande II experiments are dominated by this rare mode.

For ^{37}Cl, the ^{8}B contribution is most important because many of the neutrinos from this source are sufficiently energetic to excite a superallowed transition between the ground state of ^{37}Cl and the analogue excited state of ^{37}Ar (which closely resembles the ground state of ^{37}Cl). None of the more abundant neutrinos have enough energy to cause this strong analogue transition.

The last reaction in Table 2, number 10, is extremely rare, occurring about twice in every 10^{7} terminations of the pp chain. Nevertheless, the neutrinos from this reaction may be detectable in some direct counting electronic experiments (with, e.g., deuterium or ^{40}Ar) because they have the highest energies of any of the sources in Table 2.

The neutrinos from reaction 1b, which is initiated by three particles, p+e+p, are known as **pep** neutrinos. The neutrinos from the ^{3}He+p reaction are known as **hep** neutrinos.

The neutrino spectrum predicted by the standard model is shown in Figure 1, where contributions from both line and continuum sources are included. For Kamiokande II, only the ^{8}B and hep neutrinos (reaction 10) have enough energy to produce recoil electrons above the dominant backgrounds.

2.2 Neutrino interaction cross sections

The measured event rate in a solar neutrino experiment is the product of the neutrino flux times the interaction cross sections. The ingredients used in the calculation of neutrino absorption cross sections are discussed in detail in Bahcall (1978), Bahcall (1989). The results of this computation, including contribution from excited states and forbidden effects, as included in Bahcall and Ulrich (1988), is shown in Figure 2, for a ^{37}Cl target and ^{71}Ga target.

Uncertainties on the absorption cross section come from transitions to excited states and from forbidden corrections. The ^{37}Cl experiment benefits from the calibration of the transitions to excited states of ^{37}Ar using data from the decay of ^{37}Ca; the total uncertainties from absorption cross sections is 0.6 SNU. In the ^{71}Ga experiments, the uncertainties are dominated by transitions to excited states for all but pp (dominated by forbidden corrections) , ^{7}Be and ^{13}N sources (not enough phase space above threshold for excited states); the estimated uncertainty is asymmetric +16 SNU -11 SNU.

2.3 Chemical abundances

The chemical abundances of the elements affect the computed radiative opacity and hence the temperature–density profile of the solar interior. The joint efforts of many different

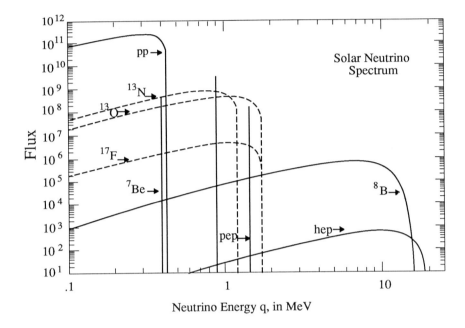

Figure 1. Solar neutrino spectrum. This figure shows the energy spectrum of neutrinos predicted by the standard solar model. The neutrino fluxes from continuum sources (like pp and ^{8}B) are given in the units of number per cm^2 per second per MeV at one astronomical unit. The line fluxes (pep and ^{7}Be) are given in number per cm^2 per second. The spectra from the pp chain are drawn with solid lines; the CNO spectra are drawn with dotted lines. [Reproduced with permission of the publisher from *Neutrino Astrophysics* by J. N. Bahcall, Cambridge University Press (1989).]

researchers has been ably summarized in two reviews one by Grevesse (1984), adopted by Bahcall and Ulrich (1988) and the other by Aller (1986).

The present composition of the solar surface is presumed, in standard solar models, to reflect the initial abundances of all of the elements that are at least as heavy as carbon. The fractional abundance by mass of elements heavier than helium is called the heavy element abundance and is traditionally denoted by Z. The corresponding abundances by mass of hydrogen and helium are denoted by X and Y.

The initial ratio by mass of elements heavier than helium relative to hydrogen, Z/X, is one of the crucial input parameters in the determination of a solar model. The fractional abundances of each of the elements are also important in determining the stellar opacity, which is closely linked to the predicted neutrino fluxes.

Table 3 lists the individual fractional abundances of the heavy elements that are recommended by Grevesse (1984) and Aller (1986). The two studies are in excellent agreement. The Grevesse (1984) value is $(Z/X)_{\text{Grevesse}} = 0.02765$ and for the Aller (1986) mixture $(Z/X)_{\text{Aller}} = 0.02739$. The difference between the value of Z/X used in old studies [Bahcall (1982)] and the current value of Grevesse (1984) and Aller (1986) is about 19%.

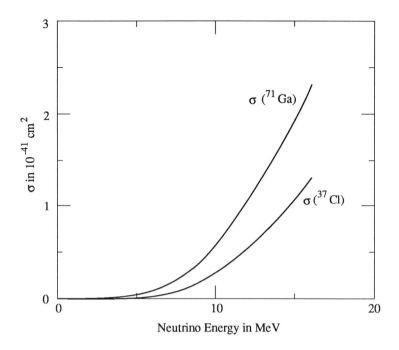

Figure 2. Neutrino absorption cross section. [Reproduced with permission of the publisher from *Neutrino Astrophysics* by J. N. Bahcall, Cambridge University Press (1989).]

Uncertainties due to chemical composition are 1.8 SNU and 10 SNU for ^{37}Cl and ^{71}Ga experiments respectively.

In order to employ these surface abundances in stellar interior calculations, two important but quantitatively plausible assumptions are made. First, the Sun is assumed to be chemically homogeneous when it arrives on the main sequence. Pre-main sequence models of solar type stars are convectively mixed [see Hayashi (1961, 1966)]. Second, the composition of the present solar surface is assumed to reflect the initial abundances of all elements at least as heavy as carbon. Nuclear burning, for material presently confined to the outer parts of the Sun, is negligible because the temperatures within the present convective zone are relatively low.

2.4 The radiative opacity

The transport of energy in the central regions of the Sun is primarily through photon radiation, although electron conduction contributes somewhat in the innermost regions and convection dominates near the surface. The calculated radiative opacity depends upon the chemical composition and upon the modeling of complex atomic processes. The calculations require, for the solar interior, the use of large computer codes in order to include all of the known statistical mechanics and atomic physics [see Huebner (1986)]. The primary source for accurate astrophysical opacities has been, for many years, the Los Alamos National Laboratory codes, presumably developed for related thermonuclear

Table 3.   Fractional abundances of heavy elements.

Element	Number fraction [Grevesse (1984)]	Number fraction [Aller (1986)]
C	0.29661	0.27983
N	0.05918	0.05846
O	0.49226	0.49761
Ne	0.06056	0.06869
Na	0.00129	0.00125
Mg	0.02302	0.02552
Al	0.00179	0.00198
Si	0.02149	0.02672
P	0.00017	0.00018
S	0.00982	0.01040
Cl	0.00019	0.00019
Ar	0.00230	0.00227
Ca	0.00139	0.00134
Ti	0.00006	0.00007
Cr	0.00028	0.00035
Mn	0.00017	0.00016
Fe	0.02833	0.02382
Ni	0.00108	0.00114
Total	1.000	1.000

applications [Cox (1989)]; another approach of this problem has been developped at the Lawrence Livermore Laboratory, as explained by Iglesias (1989).

Because the opacity determines in large part the temperature profile, the adopted opacity constitutes an important source of uncertainty for solar neutrino calculations : 0.5 SNU for ^{37}Cl and 3 SNU for gallium experiments. The typical uncertainty is less than 10% .

2.5   The equation of state

The equation of state, the relation between pressure and density, must include accurately the effects of radiation pressure and electron degeneracy [see, e.g., Rakavy and Shaviv (1967) or Schwarzschild (1958)], and screening interactions [according to the Debye–Hückel theory, see footnote 15 of Bahcall and Shaviv (1968)]. All of these effects can be included without unusual complications in a stellar interior code; the remaining recognized uncertainties do not significantly affect the calculated solar structure or the neutrino fluxes [see Bahcall *et al.* (1982) and Ulrich (1982)]. However, numerical experiments show that the computed neutrino fluxes are sensitive to hypothetical localized changes in the equation of state when the perturbations are introduced near 8×10^6K [Bahcall, Bahcall, and Ulrich (1969)].

3. General method

3.1 The ingredients

The standard solar model is calculated using the best physics and input parameters that are available at the time the model is constructed. Thus the set of numbers that correspond to the standard solar model vary with time, hopefully (nearly) always getting closer to the "true" standard model. In the quarter of a century that standard solar models have been used to compute neutrino fluxes, there have been many hundreds of improvements in the input parameters and in the description of the physics. A few seemingly esoteric upgrades of the codes made noticeable differences in the predictions of neutrino fluxes, but a number of the most difficult and careful investigations of new physics or input parameters resulted in little change in the calculated fluxes.

Some of the principal approximations used in constructing standard models deserve special attention since they have been investigated particularly thoroughly or often for possible sources of departure from the standard scenario.

(1) **Hydrostatic equilibrium.** The Sun is assumed to be in hydrostatic equilibrium; the radiative and particle pressures of the model exactly balance gravity.

$$\frac{\mathrm{d}P(r)}{\mathrm{d}r} = -\frac{GM(r)\rho(r)}{r^2}. \tag{3}$$

Observationally, this is known to be an excellent approximation since a gross departure from hydrostatic equilibrium would cause the Sun to collapse in a free-fall time, which is less than an hour. The pressure is the sum of the radiative and the (dominant) thermal pressure:

$$P(r) = \frac{a}{3}T^4 + \frac{1}{\mu}\frac{k\rho T}{m_{\mathrm{H}}}\left(1 + D\right). \tag{4}$$

$a = 4\sigma/c$ where σ is the Stefan-Boltzmann constant, k is the Boltzmann's constant and D represents easily calculated corrections for the degeneracy of electrons and for Debye–Hückel modifications to the equation of state. Pulsation, rotation, and pressure due to magnetic fields are all estimated to be unimportant for purposes of calculating solar neutrino fluxes.

(2) **Energy transport by photons or convective motions.** The equation governing energy transport is:

$$L_r = -4\pi r^2(ac/3)\frac{1}{\kappa\rho}\frac{\mathrm{d}T^4}{\mathrm{d}r}. \tag{5}$$

Here L_r is the energy per unit time that passes through a sphere of radius r and T is the temperature. The total opacity κ is the combination of a radiative and a conductive opacity: $\kappa^{-1} \equiv \kappa_{\mathrm{rad}}^{-1} + \kappa_{\mathrm{cond.}}^{-1}$. For solar interior conditions, the radiative opacity dominates the total opacity. For regions that are unstable against convective motions, the temperature gradient is taken to be the adiabatic gradient except near the surface (important for the helioseismological calculations) where mixing length theory is used. Additional transport due to acoustic or gravity waves is negligible in the standard solar model.

(3) **Energy generation by nuclear reactions.** The primary energy source for the radiated photons and neutrinos is nuclear fusion, although small effects of

contraction or expansion are included in the standard solar model. The standard codes include departures from nuclear equilibrium that are caused by the fusion processes themselves, for example, in the abundance of ^{3}He. The rate at which the luminosity is produced in spherical shells is the sum of nuclear ($\epsilon_{\text{nuclear}}$) and mechanical energy generation:

$$\frac{dL_r}{dr} = \rho(4\pi r^2)\left(\epsilon_{\text{nuclear}} - T\frac{dS}{dt}\right), \tag{6}$$

where S is the stellar entropy.

(4) **Abundance changes caused solely by nuclear reactions.** The primordial solar interior is chemically homogeneous in the standard model. Changes in the local abundances of individual isotopes occur only by nuclear reactions in those regions of the model that are convectively stable. Thermal and gravitational diffusion are not included at present, because they are estimated to be small over the lifetime of the Sun [see Cox, Guzik, and Kidman (1989)].

3.2 Calculational procedure

A standard solar model is the end product of a sequence of models. One begins with a main sequence star that has a homogeneous composition. Hydrogen burns in the deep interior of the model, supplying both the radiated luminosity and the local heat (thermal pressure) which supports the star against gravitational contraction. Successive models are calculated by allowing for composition changes caused by nuclear reactions, as well as the mild evolution of other parameters; the integration of the nuclear abundance equations involves some numerical complications that can be handled best by specialized techniques. The nuclear interaction rates are interpolated between the previous and new models and multiplied by a time step (usually of order 5×10^8 or 10^9 yr.), in order to determine the new chemical composition as a function of mass fraction included. The model at the advanced time is computed using the new composition. The models in an evolutionary sequence have inhomogeneous compositions; in the model for the present epoch, the innermost mass fraction of hydrogen is about one-half the surface (initial) value.

The stellar evolution models are constructed by integrating from the center outward and from the surface inward, requiring that the two solutions match at a convenient point that is typically at about $0.2M_\odot$. Only a relatively crude treatment of the solar atmosphere is required for computing accurate values for solar interior parameters. Even a 10% change in the outer radius of the model does not significantly affect the calculated neutrino fluxes [see Sears (1964) or Bahcall and Shaviv (1968), Eq. (3)]. From time to time, different works in the field have claimed a sensitivity. The difference between the most careful and the crudest treatment of the solar convection zone corresponds to at most a 2% change in the calculated solar neutrino fluxes [see Bahcall and Ulrich (1988), Section X.D]. (As different workers have adapted computer codes from other problems to the calculation of solar neutrino fluxes, they have sometimes reported a sensitivity of the fluxes to the atmospheric model. In all cases, the claimed sensitivity has disappeared as the computer bugs were removed, see Bahcall (1989).)

How does one proceed in practice? One begins by guessing initial values of X, the original homogeneous hydrogen abundance, and S, an entropy-like variable.[†] Typically, an evolutionary sequence requires of order five to seven solar models of progressively greater ages to match the luminosity and radius to the desired one part in 10^5.

The initial helium abundance of the model Y, is determined in the process of iteration. The other two composition parameters are fixed by the surface ratio of Z/X (heavy elements/hydrogen) that is taken from observations and by the fact that the sum of all the mass fractions is equal to unity, that is, $X + Y + Z = 1.0$.

A satisfactory solar model is a solution of the evolutionary equations that satisfies boundary conditions in both space and time. One seeks a model with a fixed mass $M_\odot$ and with a total luminosity (in photons) equal to $L_\odot$ and an outer radius $R_\odot$ at an elapsed time of 4.6×10^9 yr, the present age of the Sun. The initially assumed values of X (the hydrogen mass fraction) and S (the entropy-like variable) are iterated until an accurate description is obtained of the Sun at the present epoch.

The luminosity boundary condition has an especially strong effect on the calculated neutrino fluxes. The reason is that both the luminosity and the neutrino fluxes are produced by nuclear reactions in the deep solar interior.

4. Some characteristics of the standard model

There are a number of characteristics of the standard model that are of general interest. For example, the fraction of the photon luminosity that originates in the pp chain is 0.984; the corresponding fraction for the CNO cycle is 0.016. The net expansion at the present epoch corresponds to a luminosity fraction of -0.0003. The convection zone terminates at 1.92×10^6 K, corresponding to a radius of about $0.74 R_\odot$ and a density of $0.12\,\mathrm{g\,cm^{-3}}$; the convection zone comprises the outer 1.5% of the solar mass.[‡] One-half of the photon luminosity (or the flux of pp neutrinos) is produced within the inner $0.09 M_\odot$ ($R \le 0.11 R_\odot$); 95% of the photon luminosity is produced within the inner $0.36 M_\odot$ ($R \le 0.21 R_\odot$). The neutrino luminosity is 2.3% of the photon luminosity, which corresponds to an average of 0.572 MeV lost in neutrinos per termination of the pp chain. The pp chain is terminated 85.5% of the time by the ^{3}He–^{3}He reaction (number 3 of Table 2) and 14.5% of the time by the ^{3}He–^{4}He reaction (number 4 of Table 2).

Figure 3 illustrates some of the most interesting physical characteristics of the standard solar model. Figure 3a shows the fraction of the energy generation that is produced at different positions in the Sun. The energy generation peaks at a radius of $0.09 R_\odot$, which corresponds to about $0.06 M_\odot$. Figures 3b and 3c illustrate the distributions of temperature and density; the central values are, respectively, 15.6×10^6 K and $148\,\mathrm{g cm^{-3}}$. The peak of the energy generation occurs at a temperature of about 14×10^6 K and a density of about $95\,\mathrm{g cm^{-3}}$.

[†]Appendix A of Bahcall *et al.* (1982) defines S and discusses the initial steps in the construction of the model. S determines the adiabat of the convection zone. In earlier treatments of the problem, one adjusted the constant $K = P/T^{2.5}$, which gives the relation between pressure and temperature in the convective envelope [see Sears (1964) or Bahcall and Shaviv (1968)].

[‡]The precise parameters for the convective zone are unimportant for the solar neutrino problem although they are important for the calculation of the p-mode oscillation frequencies.

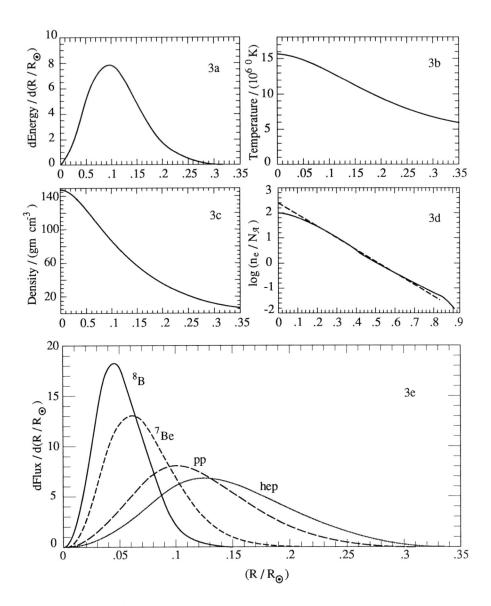

Figure 3. Radial profiles of physical parameters in the standard solar model. Fig 3a shows the fraction of the energy generation that is produced at each position. Fig 3b illustrates the temperature distribution. Fig 3c illustrates the density distribution. Fig 3d shows as a solid line the logarithm of the electron number density, N_e, divided by Avogadro's number, $N_{\mathscr{A}}$, as a function of solar radius. The dotted line is an exponential fit to the density distribution. Fig 3e represents the neutrino production as a function of radius for ^{8}B, ^{7}Be, pp and hep neutrinos. [Reproduced with permission of the publisher from *Neutrino Astrophysics* by J. N. Bahcall, Cambridge University Press (1989).

The dependence of the electron number density upon solar radius is shown in Figure 3d; these density is an important input for the MSW effect [Smirnov (1989)]. The equation of the dotted line (fit of a linear function) is :

$$n_e/N_A = 245 \exp(-10.54x) \text{ cm}^{-3}, \qquad (7)$$

where $x = R/R_\odot$. Note that the linear fit is not exact and the parameters depend upon where the fitting is done. In particular, the formula given in Eq. (7) gives a value of n_e that is too large by about a factor of 2.5 at the solar center.

Figure 3e shows where in the sun the different neutrinos originate. The comparison with the others figures shows clearly that the pp flux is produced in the same region than the energy. Because of its strong temperature dependence, the ^{8}B production is peaked at much smaller radii, $.05R_\odot$. ^{7}Be is intermediate between pp and ^{8}B. The hep production is the most extended, and reflect the ^{3}He abundance increase as one goes outward from the center.

Helium is increased in abundance with respect to hydrogen by nuclear burning in the solar interior. In the innermost region, the ^{3}He abundance is small because ^{3}He is burned rapidly by reactions 3 and 4 of Table 2. In the outermost region, no ^{3}He is produced by proton burning. Thus, there is a sharp peak in the ^{3}He abundance near $0.28R_\odot$. The helium mass fraction is highest in the interior as the result of hydrogen burning, while the heavy element abundance is constant everywhere, by assumption. In all of the modern calculations, the core of the Sun is convectively stable, although not by much.

The model of the present Sun has a luminosity that has increased by 41% from the nominal zero-age model (when the model Sun first reached quasistatic equilibrium on the main sequence) and the effective temperature has increased by 3%. The flux of ^{8}B neutrinos has increased dramatically; the contemporary flux is a factor of 41 times larger than the zero-age value.

The largest recognized contribution to the uncertainty in the inferred helium abundance is caused by the uncertainty in the initial value of Z/X and is of order a few percent. Standard solar models yield a well-defined value for the initial helium abundance:

$$Y = 0.27 \pm 0.01. \qquad (8)$$

This initial solar value of helium represents an upper limit to the primordial helium abundance at the beginning of the Big Bang. Three determinations of the helium abundance are in satisfactory agreement: the initial solar helium abundance, the present-day abundance of helium in the Galaxy's interstellar medium, and the preferred abundance based on cosmological considerations. All three quantities are equal to within the errors of their determinations, which are at least a few percent.

5. Solar neutrinos

Is there really a solar neutrino problem? The answer is yes if the difference between the predicted and the measured capture rates exceeds the range of the uncertainties. The answer is no if the uncertainties exceed the discrepancy between theory and observation.

The solar neutrino fluxes as well as the uncertainties, calculated from the standard solar model are shown in first row of Table 4.

The flux of the basic pp neutrinos can be calculated to an estimated accuracy of 2% using the standard solar model. Thus the pp flux, the dominant flux of solar neutrinos,

Table 4.   Calculated solar neutrino fluxes and predicted
capture rates for ^{37}Cl and ^{71}Ga detectors .

Source	Flux (10^{10} cm^{-2} s^{-1})	Capture rate ^{37}Cl	Capture rate ^{71}Ga
pp	6.0 (1 ± 0.02)	0.0	70.8
pep	0.014 (1 ± 0.05)	0.2	3.0
hep	8 × 10^{-7}	0.03	0.06
^{7}Be	0.47 (1 ± 0.15)	1.1	34.3
^{8}B	5.8 × 10^{-4} (1 ± 0.37)	6.1	14.0
^{13}N	0.06 (1 ± 0.50)	0.1	3.8
^{15}O	0.05 (1 ± 0.58)	0.3	6.1
^{17}F	5.2 × 10^{-4} (1 ± 0.46)	0.003	0.06
Total		7.9 SNU	132^{+20}_{-17} SNU

can be thought of as a reliable source, placed at an astronomical distance, which can be used for physical experiments on the propagation of neutrinos.

The production rate for the rare neutrinos from ^{8}B β-decay is sensitive to conditions in the solar interior, because of the relatively high Coulomb barrier for the ^{7}Be(p, γ)^{8}B reaction ($\sim$ 10 MeV compared to a mean thermal energy of 1 keV). The calculated flux of ^{7}Be electron capture neutrinos is intermediate in sensitivity between the pp and the ^{8}B neutrinos.

The calculated uncertainties are described in terms of a total theoretical range. The calculation of a *true* "three standard deviation level of confidence," cannot be done because the probability distribution is unknown for parameters that must be calculated, not measured (e.g., radiative opacity or higher-order corrections to neutrino cross sections). In practice, the meaning of the total theoretical range is that, if the true value lies outside this range, someone who has determined an input parameter (experimentally or theoretically) has made a mistake.

For measured quantities (e.g., nuclear reaction rates), we use standard 3σ limits to estimate the uncertainties. For theoretical quantities, we usually take the uncertainties in quantities that are calculated to be equal to the range in values in published state-of-the-art calculations, especially when this range exceeds (as it usually does) the published estimates of uncertainties. (Of course, the theory could be wrong in some fundamental way that would not be reflected in scatter in the values obtained by different treatments.) Quantities for which only one calculation is available require a more delicate judgment. For example, we have chosen to multiply the value of higher-order corrections to neutrino capture cross sections by three and call this the total uncertainty. It is possible that we assign relatively larger errors for experimentally determined parameters (for which the errors are more easily quantifiable) than we do for the calculated parameters such as the opacity. However, the adopted procedure is as objective as any we can think of and has the advantage of simplicity.

Figure 4 shows the predicted capture rates for the ^{37}Cl experiment as a function of the date of publication for each paper published by the author (1963 to 1988). The original

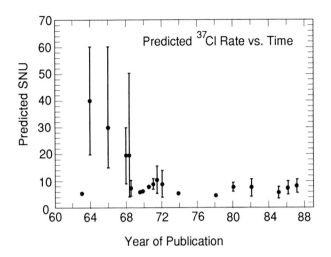

Figure 4.   Predicted capture rates as a function of time.   The published predictions of the author and his collaborators of neutrino capture rates in the ^{37}Cl experiment are shown as a function of the date of publication. [Reproduced with permission of the publisher from *Neutrino Astrophysics* by J. N. Bahcall, Cambridge University Press (1989).]

error bars are shown for every case in which they were published. All 14 values published since 1968 are consistent with the range given in Eq. (1.a).

The discussion up to this point has emphasized the theory of neutrino emission from the Sun. Fortunately, this theory can be tested observationally.

5.1   The ^{37}Cl experiment

The beautiful ^{37}Cl experiment of Davis and his collaborators was for two decades the only operating solar neutrino detector. The reaction that is used for the detection of the neutrinos is:

$$\nu_e + {}^{37}\text{Cl} \Rightarrow \text{e}^- + {}^{37}\text{Ar}, \tag{9}$$

which has a threshold energy of 0.8 MeV. The target is a tank containing 615 tons of C_2Cl_4), deep in the Homestake Gold Mine in Lead, South Dakota. The average rate at which ^{37}Ar is produced is

$$\text{Production rate} = 0.543 \pm 0.035 \text{ atoms day}^{-1}, \tag{10a}$$

of which a small part is background (from cosmic ray events),

$$\text{Background rate} = 0.08 \pm 0.03 \text{ atoms day}^{-1}. \tag{10b}$$

Subtracting the known background rate from the production rate yields the capture rate

$$\text{Capture rate} = (2.33 \pm 0.25) \text{ SNU}, \tag{10c}$$

which is due to solar neutrinos if all of the significant contributions to the background have been recognized. The errors quoted in these three observed rates are all 1σ uncertainties.

The ^{37}Cl experiment is discussed in more detail in this book [Davis (1989)].

5.2 The Kamiokande II experiment

The Kamiokande II experiment, which is located in the Japanese Alps, detects Cerenkov light emitted by electrons that are scattered in the forward direction by solar neutrinos. The reaction by which the neutrinos are observed is

$$\nu + e \rightarrow \nu' + e', \tag{11}$$

where the primes on the outgoing particle symbols indicate that the momentum and energy of each particle can be changed by the scattering interactions. For the higher-energy neutrinos ($> 5\,\mathrm{MeV}$, i.e., ^{8}B and hep neutrinos only) that can be observed by this process using available techniques, the scattering provides additional information not available with a radiochemical detector. Neutrino–electron scattering experiments furnish information about the incident neutrino energy spectrum (from measurements of the recoil energies of the scattered electrons), determine the direction from which the neutrinos arrive, and record the precise time of each event.

The preliminary results discussed at this Conference [Nakahata (1989), Hirata et al. (1989)], from the Kamiokande II detector, yield a ^{8}B neutrino flux that is approximately 0.39 of the standard model flux, about 3σ away from zero and from the standard model value. This result applies for recoil electrons with a minimum total energy of 7.5 MeV. A significant forward peaking of the recoil electrons is observed along the direction of the Earth–Sun axis. This result is of great importance since all of the previous observational results on solar neutrinos came from a single ^{37}Cl experiment.

5.3 Gallium detectors

Two radiochemical solar neutrino experiments using ^{71}Ga as a target are under way. The GALLEX collaboration uses 30 tons of gallium in an aqueous solution; the detector is located in the Gran Sasso Underground Laboratory in Italy [Kirsten (1989)]. The Soviet-American experiment, SAGE, uses about 60 tons of gallium metal; the solar neutrino laboratory is constructed underneath a high mountain in the Baksan Valley in the Caucasus Mountains of the Soviet Union [Gavrin (1989)].

The gallium experiments can furnish unique and fundamental information about nuclear processes in the solar interior and about neutrino propagation. The neutrino absorption reaction is:

$$\nu_e + {}^{71}\mathrm{Ga} \Rightarrow e^- + {}^{71}\mathrm{Ge}. \tag{12}$$

The germanium atoms are removed from the gallium and the radioactive decays of ^{71}Ge (half-life 11.4 days) are measured in small proportional counters. The threshold for absorption of neutrinos by ^{71}Ga is 0.233 MeV, which is well below the maximum energy of the pp neutrinos.

Table 4 shows the calculated contribution from individual neutrino sources to the predicted capture rate. Neutrinos from the basic pp reaction are expected, according to the standard model, to produce approximately half of the computed total capture rate. The other main contributors are ^{7}Be neutrinos, about one-quarter of the total rate, and ^{8}B neutrinos, about 10%.

6. Helioseismology

Helioseismology, like terrestrial seismology, provides information about the interior of the body under study by using observations of slight motions on the surface. The technique is analogous to striking a bell and using the frequencies of the emitted sound to make inferences about the bell's constitution. Leighton, Noyes, and Simon (1962) first discovered solar oscillations by studying the velocity shifts in absorption lines formed in the solar surface. They found that the surface of the Sun is filled with patches that oscillate intermittently with periods of the order of 5 minutes and velocity amplitudes of order $0.5 \, \mathrm{km \, s^{-1}}$. The oscillatory motion was subsequently detected in measurements of the solar intensity. The oscillations typically persist for several periods with a spatial coherence of order a few percent of the solar diameter.

We now know [Ulrich (1970) and Leibacher and Stein (1971)] that the Sun acts as a resonant cavity. Sound waves known as **p-modes** (or pressure modes, because the restoring forces is the compressibility of gases) are largely trapped between the solar surface and the lower boundary of the convection zone. The waves bounce back and forth between spherical-shell resonant cavities bounded on the outside by the reflections due to the density gradient near the solar surface and on the inside by refractions due to the increasing sound speed.

In order for a mode to resonate in the solar acoustic cavity, a half-integral number of waves must fit along the path leading from the solar surface to the base of the cavity. The depth of the cavity is fixed by the condition that horizontal wavenumber equals the total wavenumber (i.e., the vertical wavenumber becomes equal to zero), at which point the wave is refracted back towards the surface. The vertical wavenumber decreases with increasing depth in the Sun because the temperature rises in the inner regions. For many of the waves that have been most intensively studied, the base of the resonant cavity is close to the base of the convective zone.

For a given horizontal wavelength only certain periods will correspond to a resonance in the solar cavity. It was therefore predicted [Ulrich (1970)] and subsequently observed [Deubner (1975) and Rhodes, Ulrich, and Simon (1977)] that the strongest solar oscillations fall in a series of narrow bands when the results are displayed in a two-dimensional power spectrum that shows amplitude as a function of both period and horizontal wavelength. Just as in a musical instrument, the largest amplitudes correspond to standing waves that constructively interfere at the boundaries of the cavity. Solar rotation breaks the symmetry between otherwise degenerate modes and enables observers and theorists working together to make important inferences about the rate at which interior regions of the Sun are rotating.

Observations by Claverie *et al.* (1979) and by Grec, Fossat, and Pomerantz (1980), which utilized the integrated light from the entire solar disk, showed that the oscillations are globally coherent. The modes observed by these techniques provide the most important information currently available for the study of the deep solar interior since they penetrate most deeply toward the solar center.

The stability of the oscillations sets a limit on the precision of the helioseismological constraints that we can impose on the solar models. Present observations indicate that the frequencies of the oscillations can be measured to an accuracy of about two parts in ten thousand, which provides strong constraints. Indeed, some of the earliest detailed observational results led to the conclusion that the depth of the solar convection zone was somewhat greater than previously believed [Gough (1977) and Rhodes, Ulrich, and Simon (1977)].

The p-mode frequencies are insensitive to even a drastic change in nuclear energy generation. For example, the characteristic change in p-mode frequencies caused by switching off the ^{3}He + ^{4}He reaction (and all the higher-energy neutrino fluxes) is less than 0.01% [see Bahcall and Ulrich (1988)].

The frequencies of the **g-mode** [or gravity waves, with gravity as restoring force], which penetrate deeply into the stellar interior, exhibit a small sensitivity ($\sim 0.2\%$) to the hypothetical change in nuclear energy generation. There have been several reports suggesting that g-modes may have been detected in the Sun, but these claims are controversial.

The calculations using the standard solar model represent well the quantitative features of the solar p-mode frequency spectrum. However, there are small ($\sim$ a few tenths of a percent) discrepancies between observations and calculations of typical p-mode splittings, which are of order 10^2 μHz. The most significant discrepancy is the difference between the calculated and observed value of δ_{02} , the small (~ 10 μHz) frequency separation (for radial nodes n differing by 1) between the modes with spherical harmonic degrees $l = 0$ and $l = 2$. There are theoretical reasons for believing that this frequency separation, related to the gradient of the sound speed in the interior of the sun, is less susceptible to uncertain surface phenomena than are the much larger frequency splittings. Bahcall and Ulrich (1988) estimate the discrepancy to be at approximately the 3σ level of significance. The observed value for δ_{02} is between 8.9 μHz and 9.9 μHz [Pallé et al. (1987)], depending upon the pairs of radial nodes chosen. The value calculated with the standard solar model is 10.6 μHz.

A small gradient in the initial helium abundance can modify the calculated oscillation frequencies significantly and in the correct sense to improve the agreement with observations [see rows 14 through 17 of Table XX of Bahcall and Ulrich (1988)]. The calculated ^{8}B neutrino flux is thus only increased by about 15% by this specified ad hoc assumption regarding the composition gradient, an amount that is smaller than the currently estimated uncertainties in calculating the neutrino fluxes.

The histogram of the fractional contributions to the observed p-mode splitting is shown in Figure 5 for mass fractions from $0.05M_\odot$ to $1.0M_\odot$, corresponding to radial intervals from $0.08R_\odot$ to $1.0R_\odot$. The histograms for the production of neutrinos from ^{8}B decay and the generation of the solar luminosity (which is nearly the same as the histogram for the production of neutrinos from the pp reaction) are also displayed in Figure 5.

Figure 5 shows that the three observational quantities, p-mode oscillations, the solar luminosity (or pp neutrinos), and ^{8}B neutrinos, are primarily determined in different regions. Nearly all of the neutrinos from ^{8}B decay originate in the inner 5% of the solar mass. Almost 70% of the p-mode splitting comes from the outer 10% of the solar mass. The important regions for the generation of the solar luminosity, and the flux of neutrinos from the pp reaction, are intermediate in distribution between those for the p-mode splitting and those for the flux of ^{8}B neutrinos.

7. Conclusion

The studies of solar neutrinos and of p-mode oscillations are largely complementary. Both techniques are required in order to understand the solar interior and both kinds of studies have influenced work in the complementary field.

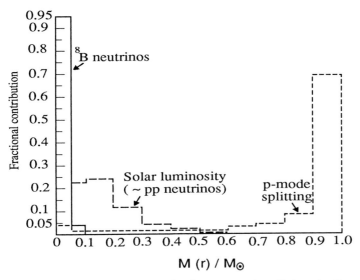

Figure 5. **Histogram of fractional contributions to p-mode splitting, the flux of neutrinos from ^{8}B decay, and the flux of neutrinos from the pp reaction.** Here $M(r)/M$ is the fraction of the solar mass interior to the point r. In order to resolve the ^{8}B neutrino emission, the width of the inner two histogram points is $0.05M(r)/M$, not $0.1M(r)/M$. [Reproduced with permission of the publisher from *Neutrino Astrophysics* by J. N. Bahcall, Cambridge University Press (1989).]

Solar models that involve WIMPs [see Spergel (1989)] may, according to some authors, improve agreement with p-mode oscillation measurements and reduce by a large factor the predicted ^{8}B neutrino fluxes. In the same time they could account for the dark matter needed to fill the universe.

A review of non-standard solar models, proposed in the context of the solar neutrino problems is given by Maeder (1989); the precision of most of the calculations in the literature, which is much less than for the standard solar model, makes difficult the comparison with observational results. Nevertheless comparison with the tight experimental constraints that are now being obtained in the study of helioseismology can greatly reduce the number of these "non-standard" solar models.

In the near future new data will be available. Solar neutrinos experiments will be sensitive to pp-neutrinos - Gallex or Sage - or able to measure in real time the energy spectrum [Spiro (1989)]. In helioseismology, with the advent of large networks of solar observatories around the earth [Hill (1989)] or in space [Bonnet (1989)], new constraints will be put on opacities [Cox (1989)].

If the solution of the solar neutrino problem is not in the field of astrophysics it could be in the properties of neutrinos. Indeed the problem takes another dimension with the MSW effect which is able to modify the energy spectrum of the neutrino emitted by the Sun [Smirnov (1989)]. This allows to test very small masses of neutrinos which are in the range of the predictions made by Grand Unified Theories [Harari (1989)].

Will solar neutrinos be the low-energy frontier of high-energy physics ? If MSW is the correct explanation, then information about the grand unification mass scale at 10^{15} GeV could be obtained from interaction of neutrinos driven by mass differences of 10^{-20} GeV !

References

Aller, L., H. (1986), in *Spectroscopy of Astrophysical Plasmas*, edited by A. Dalgarno and D. Layzer (Cambridge University Press, Cambridge, England), p.89.

Bahcall, J. N. (1989), *Neutrino Astrophysics*, (Cambridge University Press, Cambridge, England).

Bahcall, J.N. and G.Shaviv (1968) *Ap. J.*, **153**, 113

Bahcall, J.N., N.A. Bahcall, and R.K. Ulrich (1969) *Ap. J.*, **156**, 559.

Bahcall, J.N. (1978) *Rev. Mod. Phys.*, **50**, 881.

Bahcall, J.N., W.F. Huebner, S.H. Lubow, P.D. Parker, and R.K. Ulrich (1982) *Rev. Mod. Phys.*, **54**, 767.

Bahcall, J.N. and R.K. Ulrich (1988) *Rev. Mod. Phys.*, **60**, 297.

Bonnet, R.M. (1989) *These proceedings.*

Claverie, A., G.R. Isaak, C.P. McLeod, H.B. van der Raay, and T. Roca Cortes, (1979) *Nature*, **282**, 591

Cox, A.N., J.A. Gusik and R.B. Kidman (1989) *Ap. J.*, **999**, 999

R. Davis (1989) *These proceedings.*

Deubner, F.-L. (1975) *Solar Phys.*, **44**, 371.

Dicke, R.H., J.R. Kuhn and K.G. Libbrecht (1985) *Nature*, **316**, 687

Gavrin, V.N. (1989) *These proceedings.*

Gough, D.O. (1977), in *The Energy Balance and Hydrodynamics of the Solar Chromosphere and Corona,*edited by R.M. Bonnet and P. Delache (de Bussex, Clermont-Ferrand), p.3.

Grec, G., E. Fossat and M.A. Pomerantz (1980) *Nature*, **288**, 541

Grevesse, N. (1984) *Physica Scripta*, **T8**, 49.

Harari, H. (1989) *These proceedings.*

Hayashi, C. (1961) *Pub. Astro. Soc. Japan*, **13**, 450

Hayashi, C. (1966) *Ann. Rev. Astron. Astrophys.*, **4**, 171

Hirata, K. S., et al. (1989) KEK preprint 89-63.

Huebner, W.F. (1986) in *Physics of the Sun*, edited by P.A. Sturrock, T.E. Holzer, D.M. Mihala, and R.K. Ulrich (Dordrecht: Reidel)

Iglesias, C.A. (1989) *These proceedings.*

Kirsten, T. (1989) *These proceedings.*

Leibacher, J.W. and R.F. Stein (1971) *Astrophys. Lett.*, **7**, 191.

Leighton, R.B., R.W. Noyes, and G.W. Simon (1962) *Ap. J.*, **135**, 474.

Maeder, A. (1989) *These proceedings.*

Nakahata, M. (1989) *These proceedings.*

Pallé, P., J.C. Perez, C. Regulo, T. Roca Cortes, G.R. Isaak, C.P. McLeod, and H.B. van der Raay (1979) *Astron. Astrophys.*, **170**, 114

Parker, P. D. and Rolfs, C. (1989) *These proceedings.*

Rakavy, G., G. Shaviv and A. Zinamon (1967) *Ap. J.*, **150**, 131

Rhodes, E.J., Jr., R.K. Ulrich and G.W. Simon (1977) *Ap. J.*, **218**, 901

Sears, R.L. (1964) *Ap. J.*, **140**, 477.

Schwarzschild, M. (1958) *Structure and Evolution of the Stars* (Princeton University Press)

Sienkiewicz, R., Bahcall, J. N., and Paczynski, B. (1989), *Ap. J.* (December).

Smirnov, A. (1989) *These proceedings.*

Spergel, D.N. (1989) *These proceedings.*

Spiro, M. and D. Vignaud (1989) *These proceedings.*

Turck-Chièze, S., S. Cahen, M. Cassé and C. Doom (1988) *Ap. J.*, **335**, 415

Ulrich, R.K. (1970) *Ap. J.*, **162**, 993.

Ulrich, R.K. (1982) *Ap. J.*, **258**, 404.

EQUATION OF STATE AND IONIZATION OF DENSE PLASMAS

W. EBELING
Humboldt-Universität
Sektion Physik
1040 Berlin
DDR

ABSTRACT. In large regions of densities and temperatures the properties of plasmas are influenced by Coulombic forces. Here especially hydrogen and helium plasmas are studied. The region of nonideality is given and the main effects as screening, quantum corrections, level shifts and the cut off for the sum over states are discussed. Based on methods of classical and quantum statistics corrections to the ideal EOS are calculated. Interpolational expressions covering the whole density-temperature plane are given and phase transitions to highly ionized states are discussed. Level shifts due to nonideality are derived. Finally the ionization equilibrium and kinetics are considered theoretically focussing the attention on nonideality effects.

1. Introduction

From the point of view of plasma physics the sun is a big plasma ball which might be considered as a giant laboratory realizing plasmas in a broad range of state parameters. For example we meet dense plasmas with more than 10^{25} particles/cm^3 in the center where the temperature is above 10^7 K and on the other hand we find very diluted plasmas in the outer regions with temperatures below 10^4 K. For our crude physicists approach we model the sun as a spherical plasma consisting of about 90-93% hydrogen, 6-8% Helium and less

G. Berthomieu and M. Cribier (eds.), Inside the Sun, 43–58.

than 2% heavier particles, (all numbers refer to mol percents). We consider the plasma ball as being in mechanical equilibrium as well as in thermodynamical equilibrium with given distributions of temperature and densities along the radius. Hydrodynamic flows and other transport processes expect radiation will be neglected. However strong non-equilibrium with respect to nuclear and ionization reactions and with respect to radiating transitions is to be taken into account. From the physicists point of view the radiation from the sun which is absorbed and reemitted by the earth, constitutes the driving force of all selforganization processes on our planet [1]. A mechanism which we have called the photon mill [2] provides the earth with the export of entropy which is the necessary condition for the evolution of life, man, science and culture. We can easily estimate the flow of entropy from the system sun-earth to the sea of background radiation by assuming that in average 230 W/m^2 are absorbed on the surface in the earth and are reemitted. Assuming that the temperature of the incoming photons is about 5800 K and that the earth radiates as a black body of 260 K we get the entropy export flow

$$J_s = \frac{4}{3} \cdot \left(\frac{230}{5800} - \frac{230}{260}\right) \frac{W}{m^2 K} \approx -1 \frac{W}{m^2 K}$$

It is quite sure that any essential change of these numbers would have dramatic consequences for the conditions of life on our planet. the fundamental role of the sun for our life is a strong motivation for an increasing number of physicists, to provide the astrophysicists with better tools for an investigation of the basic physical processes in the sun. Here we restrict our consideration to a few physical properties and processes which are of relevance in this respect. We will start with a review of several methods of statistical thermodynamics and will develope then a new variant of the theory including an occupation number formalism which seems to be appropriate for the study of the sun. The idea of this approach is that each (internal) state of bound particles (atoms, molecules and ions) is treated as a separate species characterized by specific properties as e.g. an effective volume and by a relative occupation number. A similar approach was developed by

Rogers [3] and is being used now for extended calculations of the opacity of the sun [4] and of the EOS of the solar interior [5]. Another closely related approach was developed by Hummer, Mihalas, Däppen and others [6]. In difference to Rogers activity expansion method, our approach is based on a density expansion (canonical ensemble) [6,7] which has the advantage that generalizations to kinetic processes are more easy. Another advantage is the good convergence of density expansions in comparison to fugacity expansions [8,9].

2. Parameter regions and survey of theoretical methods

Let us consider a plasma consisting of n_e electrons and nuclei or ions with positive charges z_i.e (i = ... s) per unit volume. We define the total number of positive charges and charge averages by

$$n_+ = \sum_i n_i \ ; \ \langle z^p \rangle = n_+^{-1} \sum_i z_i^p n_i \tag{1}$$

By introducing the mean distance between electrons and positive charges

$$d_e = (\frac{3}{4}\pi n_e)^{\frac{1}{3}} \ ; \ d_+ = (\frac{3}{4}\pi n_+)^{\frac{1}{3}} \tag{2}$$

we define now two dimensionless correlation parameter

$$\Gamma_e = \frac{e^2}{\theta_e D_e} \ ; \ \Gamma_+ = \frac{e^2}{\theta_+ D_+} \langle z^{5/3} \rangle \langle z \rangle^{1/3} \tag{3}$$

Here θ_e and θ_+ are average kinetic energies which include degeneracy; in the classical case we have simply $\theta_e = \theta_+ = k_B T$. The curves $\Gamma_e = 1$, $\Gamma_+ = 1$ separate the regions of strong and weak correlations respectively. In addition we define a parameter which describes the strength of the electron-ion interactions by

$$\xi = 2\langle z \rangle (\frac{1}{\theta_e})^{1/2} \ ; \ \theta_e = \frac{p_e^{id}}{n_e} \tag{4}$$

where I is the ionization energy of hydrogen. Inside the line $\xi = 1$ we observe strong interactions e.g. in scattering and bound state effects. Typical parameters of the sun are

$$\Gamma_+ \leq .1 \; ; \; 0.01 \leq \xi \leq 100 \tag{5}$$

This shows that the matter in the sun is in conditions of weak nonideality but strong two-particle interactions. Another important parameter is the relative accupation of the available space by bound states

$$\eta = \frac{4\pi}{3} \sum_b n_b R_b^3 \tag{6}$$

where n_b and R_b are the density and the effective radius of the bound stable b. In the region $\eta \geq 1$ bound states break down due to the strong Pauli blocking effects which make bound states extremely unfavorable in the thermodynamic sense.

The existing theories [8-17] are mostly restricted to special regions in the density-temperatures plane. In the region of strong interactions $\xi \geq 10$ where atoms and molecules dominate, one of the most appropriate theories is the fluid variational theory [10]. In this approach one starts from the following decomposition of the free energy density

$$f = f_{ID} + f_{HC} + f_{VW} + f_{CO} \tag{7}$$

The coulombic part is zero in the region, where only neutrals exist. The hard-core part is approximated by standard expressions

$$f_{HC} = k_B T \frac{4\eta - 3\eta^2}{(1-\eta^2)^2} + corrections \tag{8}$$

The Van der Waals part is represented by perturbation theory

$$f_{vw} = 4\pi n^2 \int_R^\infty \Phi(r) g_{HS}(r,\eta) r^2 dr \tag{9}$$

Here $\Phi(r)$ is the effective interaction potentiel for the neutrals, a good choice seems to be exponential-six-potential [10,11]

$$\Phi(r) = \varepsilon \left\{ \frac{6}{\alpha\text{-}6} exp\left[\alpha\frac{r_m\text{-}r}{r_m}\right] - \frac{\alpha}{\alpha\text{-}6}(\frac{r_m}{r})^6 \right\} \qquad (10)$$

$$H_2, D_2 \ : \ \varepsilon/k_B = 36.4 \ K \ ; \ r_m = 3.49 \ A \ ; \ \alpha = 11.1$$
$$H_e \qquad : \ \varepsilon/k_B = 10.8 \ K \ ; \ r_m = 2.97 \ \text{Å} \ ; \ \alpha = 13.1$$

Further $g_{++s}(r)$ is the radial correlation function for hard spheres. The radius R is considered as a variational parameter of the theory. In real plasmas we have often several neutral species and therefore more elaborate theories for mixture of hard spheres with different diameters have to be used. In our calculations we have used the expressions developed by Mansouri [12] in combination with simple approximations for the attracting forces [18,19]

$$f_{HC} = f_{MANS} \ ; \ f_{VDW} = -\sum n_i n_j A_{ij}$$
$$A_{ij} = -4\pi \int_{R_i+R_j} \phi_{ij}(r)r^2 dr \qquad (11)$$

Let us still note, that in this approach electrons and nuclei have to be considered as spheres with zero radius [18,19].

Another limiting case of great importance is the classical Coulomb fluid consisting of heavy charged particles (protons, deutrons and higher charged nuclei) imbedded into a neutralizing uniform sea of electrons. For this special model many data from Monte Carlo studies are available [14,15]. The free energy density may approximated by the analytical formula for the free energy density

$$f_{CO} = -n_+ k_B T \left(\beta e^2\right)^3 1.447 \langle z^{5/3}\rangle\langle z\rangle)^{1/3} \qquad (12)$$

Instead of this simple expression more elaborate formulae may be used [13-15,18,19].

The methods described so far are restricted to special regions of the temperature-density plane. A more universal method is the Green functions approach developed by Martin, Schwinger,

Kadanoff, Baym, Abrikossov and others. In this method thermodynamic functions follow from the relation between density n, chemical potential and one particle Green function [7]

$$n(\mu) = V^{-1}\sum_{p} \int \frac{d\omega}{2\pi} f_1(\omega) Im G_1(p,\omega)$$

(13)

The Green function is expressed by Dysons equation

$$G_1 = G_1^0 + G_1^0 \Sigma_1 G_1$$

(14)

where G_1^0 is the Green function for free particles and Σ_1 is the self-energy which may be represented as an infinitive sum of diagrams. Within all these contributions we need a guidding idea to select out the relevant diagrams. For this we developed a special method, an occupation number formalism based on the "chemical picture" [7-9].

This may be described in short as follows : we approximate the selfenergy by ladder type diagrams

$$\Sigma_1^L G_1^0 = V G_2^L$$

(15)

All diagrams containing free particle propagators are completed by diagrams containing a propagator for the composites [16]. With these approximations for G_1 we arrive at expressions for the thermodynamic functions which contain also higher order correlations and n-particle composites

$$n(\mu) = \sum_{p} f_1(\varepsilon_1) + \sum_{kP} g_2(E_{kP}) + \sum_{p} \int d\omega\, g_2(\omega) \frac{d\sin \delta(\omega)}{d\omega}$$

(16)

where $\delta(\omega)$ is a generalized scattering phase. The energy levels contain first order perturbation shifts with free and bound state contributions

$$\varepsilon_1 = \frac{p^2}{2m} + \Delta^f(p) + \Delta^b(P)$$

(17)

$$E_{kP} = E_{kP}^0 + \Delta_k^f(p) + \Delta_k^b(P)$$

(18)

The essential step is now to identify the shifts for zero momentum with the chemical potentials of quasispecies (bound states)

$$\varepsilon_1 = \frac{p^2}{2m} + \Delta^f(p) + \Delta^b(P) \tag{19}$$

$$\mu_2(k) = \mu(1) + \mu(2) + E_k^0 + \Delta_k(0) \tag{20}$$

Each bound state k is considered as a separate chemical species [20]. After identifying the chemical potentials for the new species one goes to the free energy density by standard methods. Finally we have to require that the free energy density is minimized with respect to the division into the different chemical species

$$f(\beta, n_1, n_2^{(k)}, \ldots) = min$$
$$n = n_1 + \sum_k n_2^{(k)} + \ldots \tag{21}$$

The quasi-particle picture given here is consistent with the usual chemical picture [8]. However in different to earlier approaches now all the bound states are considered separately (compare [5,6] and [20]).

3. Thermodynamic properties of hydrogen helium systems

Since the matter constituting the sun is a mixture of hydrogen and helium we need an expression for the energy including at least double charges and three-particle bound states. We shall use eqs. (7-11) and (16-21) as the theoretical basis for constructing the free energy. Let us consider now several contributions separatly. The ideal free energy density has the form (in Rydberg units) :

$$
\begin{aligned}
f_{ID} = {} & f_{ID}^{(e)} + \tau \sum_i n_i \left\{ \left[ln\left(n_i \Lambda_i^3 / \tilde{g}_i \right) - 1 \right] + \sum_k n_k^{(ei)} \left[E_k^{(ei)} + ln\left(n_k^{(ei)} \Lambda_{ei}^3 / g_k^{\widetilde{(ei)}} \right) - 1 \right] \right. \\
& + \sum_k n_k^{(eei)} \left[E_k^{(eei)} + ln\left(n_k^{(eei)} \Lambda_{eei}^3 / g_k^{\widetilde{(eei)}} \right) - 1 \right] + \sum_{kj} n_k^{(eji)} \left[E_k^{(eji)} + ln\left(n_k^{(eji)} \Lambda_{eji}^3 / g_k^{\widetilde{(eji)}} \right) - 1 \right] + \ldots
\end{aligned}
\tag{22}
$$

Here the first contribution denotes the electronic part including degeneracy and

$$E_k^{(ei)}, E_k^{(eei)}, E_k^{(eji)}$$

the bound state energies of two or three particles respectively not including shifts. These shifts - which are strictly speaking already effects of nonideality - are defined by eqs. (17-20).

The quantum number k represents the energy states of hydrogen-like (ei), helium-like (eei) or molecular ion-like (eii) bound states. The reduced weight factor of the state k with the degeneracy g_k is according to a Brillouin-Planck-Larkin-renormalization [13,17] $\tilde{g}_k$:

$$\tilde{g}_k(T) = g_k\left[1 - exp(\beta E_k) + \beta E_k exp(\beta E_k)\right] \tag{23}$$

The sums over i and j in eq. (22) cover only the positive particles (nuclei); Λ is the De Broglie wave length.

All the densities refer to chemical species in the sense discussed above; they are not identical with the total densities of electrons and nuclei.

For the Coulombic part we have constructed in earlier work several Padé approximations which comprise the available analytical knowledge [18-22].

A special property of these Padé formulea is the smooth transition from a behaviour like $n^{1/2}$ at small densities to $n^{1/3}$ at high densities. Padé approximations are not unique; several variants may be found elsewhere [18-21]. In earlier work applications of these formulea to hydrogen plasmas [17,18], alkali plasmas [13], noble gas plasmas [19] and to electron-positron plasmas [21] were given. Here we concentrate our consideration on helium plasmas and hydrogen-helium plasmas.

In the calculations discussed below, we have used the following formula :

$$f_{CO} = f_{CO}^{(e)} - \tau n_+ \frac{q_0 \tilde{n}^{1/2} + 1000 \hat{n}^{3/2} e_1(\tilde{n})}{1 + q_1 (\tilde{n}\tau)^{1/2} + 1000 \hat{n}^{3/2}} - \tau n_+ \frac{Q_0 \tilde{n}^{1/2} + \bar{n}^{3/2} e_2(\hat{n}, r_s)}{1 + Q_1 (\tilde{n}\tau)^{1/2} + \bar{n}^{3/2} + Q_4 \tilde{n}^{1/2} ln(1 + Q_5/\tilde{n}^{1/2})}$$

$$q_o = 1.1816(z^2)^{3/2} \; ; \; \zeta_i = n_i/n_+ \; ; \; \gamma_i = m_e/m_i$$

$$q_0 q_1 = 0.696 \sum_{ij}^{+} \zeta_i \zeta_j z_i^2 z_j^2 (\gamma_i + \gamma_j)^{1/2}$$

$$e_1(x) = 1.447 \, x^{1/3} - 4.2944 \, x^{1/12} + 0.6712 x^{-1/12} + 0.2726 \, lnx + 2.983$$

$$\zeta_e = n_e/n_+ \; ; \; \bar{n} = n_e \Lambda_e^3 \; ; \; \tilde{n} = n_+ \beta^3 e^6 \; ; \; \hat{n} = \tilde{n}(z^{5/3})^3 (z)$$

$$Q_1 = [2.876 \zeta_e (z^2) - 1.666 \zeta_e^2 - q_0 q_1]/Q_0$$

$$Q_0 = 1.1816[((z^2) + \zeta)^{3/2} - (z^2)^{3/2} - \zeta^{3/2}] \; ; \; \zeta = n_e/n_+$$

$$Q_3 = .5236((z^3) - \zeta) - \{4(z^3 lnz) + ((z^3) - \zeta).ln[29.09((z^2) + \zeta)]\}/Q_0$$

$$Q_4 = .5236((z^3) - \zeta)^2/Q_0 \; ; \; Q_5 = exp(-Q_3/Q_4)$$

$$e_2(x,y) = \frac{.8511yx^{1/2}}{(1 + .3135x^{-1/2})(1 + 1.137y\tau^{1/2})} + \frac{yx^{1/3}(.0726 + .0161y)}{(1 + .0887y^2)} \; ; \; r_s = d_e/a_B$$

$$(24)$$

4. Ionization processes

The free energy and the composition are obtained by minimization of the free energy density after eqs. (7), (11), (21-23), (24) under the constraint that the total number of electrons is constant. Let us first consider helium plasmas. In order to describe the state of ionization we introduce several definitions :

α – part of electrons which are free,

α_2 - part of the nuclei which are free,

α_1 - part of the nuclei which have bound one electron,

α_0 - part of the nuclei which have bound two electrons.

We consider now α (degree of ionization of the electrons and α_2 (degree of double ionization of the nuclei as the independent quantities. The relations following from the balance for the electrons and nuclei

$$\alpha_0 = 1 + \alpha_2 - 2\alpha$$
$$\alpha_1 = 2\alpha - 2\alpha_2 \qquad (25)$$

may be used to determine the other quantities defined above. Figs.1-2 show the results obtained by minimization of the free energy at T = 30 000 k and T = 50 000 k as a function of the total density of nuclei.Looking at Figs. 1-2 we see that the degree of electron ionization α (number of free electrons related to the total number of electron) decreases with increasing density and after reaching a minimum increases again (n_η. total density of the nuclei). The latter effect is called pressure ionization. Beyond the minimum the density of free electrons is very high and the system will behave similar to a metal plasma. At intermediate densities of the nuclei (around 10^{22} - 10^{23} cm^{-3}) the free energy shows two minima what leads to two possible values of α and α_2 at fixed nuclei density n_η.

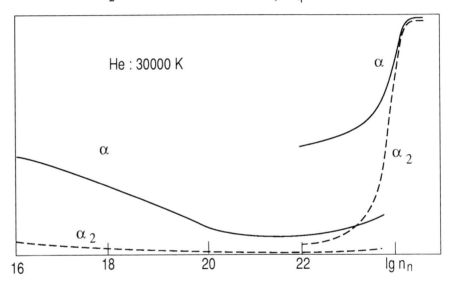

Fig. 1 *Degree of ionization of the electrons (α) and degree of ionization for the nuclei (α_2) for He-plasmas at T = 30 000 K.*

The existence of two minima of the free energy is always to be considered as a hint to phase transitions of first order. Based on the curves shown in Figs. 1-2 we would expect a plasma phase transition in helium plasmas at temperatures T ≤ Tc ≈ 55000 k. This is much higher than the estimate given by Hess [23] Tc ≈ 28500 which was obtained by consideration of single ionization only. We see here that the possibility of double ionization favours instabilities and phase transitions.

This tendency was observed already in an early work by Valuev, Medvedev and Norman [24]. Our consideration is more quantitative since it is based on more recent expressions for the free energy. Unfortunately even this is not yet enough for an exact calculation of the phase transition. Such a quantitative theory requires according to thermodynamics rules :

(i) Complete calculation of the free energy surface in dependence on all independent densities,

(ii) Construction of tangent planes connecting two minima which correspond to coexisting phases.

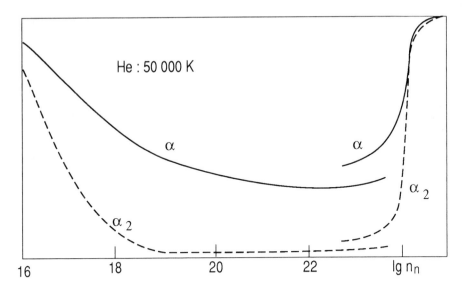

Fig. 2 *Degree of ionization for He at T = 50 000 K.*

In our calculations given above we have assumed that the two phases coincide in the mass densities and differ only in the degrees of ionization. This is only a first approximation to the real situation where the two phases may differ also in the mass densities.

Let us consider now mixtures of hydrogen and helium with a mole fraction of about 0.93 for the hydrogen component (Fig.3)

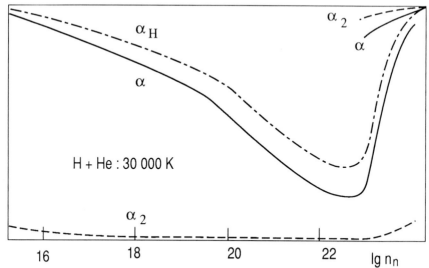

Fig.3 *Degrees of ionization for a hydrogen-helium mixtures (93% H and 7% He in mole percents).*

As Fig. 3 shows, we observe again a characteristic minimum in the degree of ionization for helium (α) and for hydrogen (α_H) at total nucleus densities 10^{22} - 10^{23} cm^{-3} and a strong increase of higher values. It is interesting to note that in a narrow density region (10^{23} cm^{-3}) at T = 30 000 k again two minima of the free energy, i.e. two values for the degree of ionization are observed. This points to the existence of a thermodynamical instability

$$\left[\frac{\partial lnp}{\partial \rho}\right]_T < 0 \tag{26}$$

leading to a plasma phase transition in H - He - plasmas with a critical point of about T $\approx$ 35 000 k. Let us mention that the estimate for pure H - plasmas is T_c = 16 500 k [18]. Again we see, that two-fold ionization favours instabilities and phase transitions. The relative high electron density leads to a strong decrease of the ionization energy for two-fold ionization and makes two-fold much more favourable than one predicts on the basis of the ideal theory. Having in mind that He has a rather high second ionization energy (54.1 eV) one may expect much stronger enhancement of two-fold ionization for the noble gas plasmas Xe (21.2 eV), Ne (41.1 eV) etc. and for the earth alkaline plasmas Ca (11.9 eV), Mg (15.0 eV), Hg (18.8 eV) etc. Let us still mention that for a dynamical theory of stars beside the traditional instabilities as eq. (26) also other instabilities as

$$\left[\frac{\partial lnp}{\partial \rho}\right]_s < \frac{4}{3} \tag{27}$$

are of fundamental interest. It remains open for further investigations to check our equation of state with respect of instabilities of the type (27).

Let us discuss now briefly the question of ionization rates following the results given in earlier papers [13, 21, 26-28].

We restrict the following consideration to two-particle bound states. From eqs. (17-18) follows a close connection between the interaction parts of the chemical potentials

$$\delta\mu_a = \frac{\partial(f-f_{ID})}{\partial n_a} \tag{28}$$

and the effective energy levels. By application to hydrogen-like bound states we get for the density-dependent energy from (18)

$$\widetilde{E}_K^{(ei)} = E_K^{(ei)} + \delta\mu_e + \delta\mu_i + \Delta_K^{(ei)} \tag{29}$$

From eq. (17) follows for the continuum edge

$$\varepsilon_1(0) = \widetilde{E}_\infty^{(ei)} = 0 + \delta\mu_e + \delta\mu_i \tag{30}$$

The difference between both expression yields the energy gap, between the continuum and the energy state k. The gap shift which increases with the density is given by solutions of the Bethe-Salpeter equation [27, 28]. A sufficiently simple approximation reads

$$\Delta_k^{(ei)} = \frac{z_i e^2}{r_D + r_k/2} \; ; \; r_k = \langle r \rangle_k \tag{31}$$

where r_D is the Debye radius and r_K the mean radius of the Bohr orbit in the state k [21, 28].

In a first approximation the ionization rates will follow an Eyring-type law, i.e. they will change exponentially with the gap. This assumption yields the following connection between the ionization rates, their ideal value and the gap shifts

$$\alpha = \alpha_{ID} exp \left[\frac{\Delta_k^{(ei)}}{\tau} \right] \tag{32}$$

A stronger derivation from kinetic equations was given in [27]. The relation (32) shows that the ionization rates increase exponentially with the nonideality.

Summarizing the results given in this paper we may state that the equation of state and the ionization processes are strongly influenced by nonideality effects in the region where the nonideality parameters given by eqs. (3-4) and (6) are sufficiently large.

References

[1] W. Ebeling, A. Engel, R. Feistel, Physik der Evolutions processe, Akademie Verlag, Berlin 1989

[2] R. Feistel, W. Ebeling, Evolution of Complex Systems, Kluwer Academic Publ., Dordrecht 1989

[3] F. A. Rogers, Phys. Rev. A38 (1988) 5007; Astrophys. J. 310 (1986) 723

[4] C.A. Iglesias, Invited Lecture at the IAU-Colloquium 121 and contribution to his volume

[5] W. Däppen, Y. Lebreton, F. Rogers, The Equation of State of the Solar Interior. Contr. to IAU-Colloquium 121,Versailles 1989

[6] D.G. Hummer, D. Mihalas, Astrophys. J. 331 (1988) 749 W. Däppen, L. Anderson, D. Mihalas, Astrophys. J. 319 (1987) 195;

D. Mihalas, W. Däppen, D.G. Hummer, Astrophys. J. 331 (1988) 815

[7] W. Ebeling, W.D. Kraeft, D. Kremp, G. Röpke, Physica 140A (1986) 160;

W.D. Kraeft, W. Ebeling, D. Kremp, G Röpke, Ann. Physik 46 (1988) 429

[8] W. Ebeling, Physica 43 (1969) 293; 73 (1974) 573

[9] W.D. Kraeft, W. Ebeling, D. Kremp, G. Röpke, Quantum Statistics of Charged Particle Systems, Plenum Press. New York 1986

[10] M. Ross, F.H. Ree, D.A. Young, J. Chem. Phys. 79 (1983) 1487

11] M. Ross, D.A. Young, Phys. Lett. 118A (1986) 463

[12] G.A. Mansouri et al., J. Chem. Phys. 54 (1971) 1523

[13] W. Ebeling, A. Förster, R. Redmer, T. Rother, A. Schlanges, Instabilities and Phase Transitions in Dense Hydrogen, Rare Gas and Alkali Plasmas, ICPIG-18, Invited Papers, Swansea 1987.

[14] S. Galam, J.P. Hansen, Phys. Rev. A14 (1976) 816

[15] H.E. Dewitt, Phys. Rev. A14 (1976) 129

[16] R. Redmer, G. Röpke, Physica 130A (1985) 523

[17] W. Ebeling, W.D. Kraeft, D. Kremp, Bound States and Ionization Equilibrium in Plasmas and Solids, Akademie-Verlag, Berlin 1976

[18] W. Ebeling, W. Richert, Phys. Lett. 108A (1985) 80; phys. stat. sol. (b) 128 (1985) 467

[19] W. Ebeling, A. Förster, W. Richert, H. Hess, Physica 150A (1988) 159

[20] W. Ebeling, Z. phys. Chem. (Leipzig) 271 (1990) N°2

[21] W. Ebeling, Contr. Plasmas Phys. 29 (1989) 165

[22] W. Ebeling, H. Lehmann, Ann. Physik 45 (1988) 529

[23] H. Hess, High Pressure Research 1 (1989) 203

[24] A.A. Valuev, I.T. Medvedev, G.E. Norman, Zh. eksp.teor. Fiz. (USSR) 59 (1970) 2228

[25] W.M. Tscharnuter, Instabilities in the Early Protosun Contr. to IAU-Colloquium 121, Versailles 1989; Formation of Viscous Protostellar Accretion Disks. Contr. to Conf. Accretion Disks, Garching 1989.

[26] M. Schlanges, Th. Bornath, D. Kremp, Phys. Rev. A38 (1988) 2174

[27] K. Kilimann, W.D. Kraeft, D. Kremp, Phys. Lett. A61 (1977) 393

[28] W. Ebeling, K. Kilimann, Z. Naturf. 44a (1989)

NUCLEAR REACTIONS IN THE SUN

CLAUS ROLFS
Institut für Kernphysik
Universität Münster, Münster
Germany

Due to unfortunate personal circumstances, the author could not provide a written version of his contribution presented at the Conference. He suggest the interested readers to refer to a recent review :

Nuclear energy generation in the solar interior

P.D. Parker and C. Rolfs

Chapter 2 in : *The Solar Interior and Atmosphere*

eds. A.N. Cox, W.C. Livingston and M.S. Matthews

Space Science Series Book (University of Arizona)

The Editors

G. Berthomieu and M. Cribier (eds.), Inside the Sun, 59.
© 1990 *Kluwer Academic Publishers. Printed in the Netherlands.*

SOLAR OPACITIES CONSTRAINED BY SOLAR NEUTRINOS AND SOLAR OSCILLATIONS

Arthur N. Cox
Theoretical Division
Los Alamos National Laboratory

ABSTRACT. This review discusses the current situation for opacities at the solar center, the solar surface, and for the few million kelvin temperatures that occur below the convection zone. The solar center conditions are important because they are crucial for the neutrino production, which continues to be predicted about 4 times that observed. The main extinction effects there are free-free photon absorption in the electric fields of the hydrogen, helium and the CNO atoms, free electron scattering of photons, and the bound-free and bound-bound absorption of photons by iron atoms with two electrons in the 1s bound level. An assumption that the iron is condensed-out below the convection zone, and the opacity in the central regions is thereby reduced, results in about a 25 percent reduction in the central opacity but only a 5 percent reduction at the base of the convection zone. Furthermore, the p-mode solar oscillations are changed with this assumption, and do not fit the observed ones as well as for standard models. A discussion of the large effective opacity reduction by weakly interacting massive particles (WIMPs or Cosmions) also results in poor agreement with observed p-mode oscillation frequencies. The much larger opacities for the solar surface layers from the Los Alamos Astrophysical Opacity Library instead of the widely used Cox and Tabor values show small improvements in oscillation frequency predictions, but the largest effect is in the discussion of p-mode stability. Solar oscillation frequencies can serve as an opacity experiment for the temperatures and densities, respectively, of a few million kelvin and between 0.1 and 10 g/cm^3. Current oscillation frequency calculations indicate that possibly the Opacity Library values need an increase of typically 15 percent just at the bottom of the convection zone at $3x10^6$K. Opacities have uncertainties at the photosphere and deeper than the convection zone ranging from 10 to 25 percent. The equation of state that supplies data for the opacity calculations fortunately has pressure uncertainties of only about 1 percent, but opacity uncertainties will always be much larger. A discussion is given about opacity experiments that the stars provide. Opacities in the envelopes of the Hyades G stars, the Cepheids, δ Scuti variables, and the β Cephei variables indicate that significantly larger opacities, possibly caused by iron lines, seem to be required.

1. INTRODUCTION

The stars have been an important opacity experiment. They have proved that the opacities that we use for calculating photon diffusion in stellar models are correct

61

G. Berthomieu and M. Cribier (eds.), Inside the Sun, 61–80.
© 1990 *Kluwer Academic Publishers. Printed in the Netherlands.*

enough to explain most of the features of stellar evolution. Nevertheless, many details remain to be explained, and the best available laboratory is the Sun. For 30 years precise solar models have been calculated. In the last 20 years, these models have mostly been constructed to study how to reduce the emergent neutrino flux to the level observed in the chlorine detector. In the past dozen years renewed interest has arisen because the solar oscillation frequencies have given new data for the internal solar structure. Today the solar structure can produce even more accurate constraints on stellar opacities.

This review discusses several interesting and current topics in solar opacity and structure that impact on the neutrino and oscillation problems. It also briefly touches on related opacity problems for several other classes of stars. For a review of how stellar opacities are calculated readers should consult the first comprehensive exposition by Cox (1965) or the modern update by Huebner (1986). Recent comprehensive tables that are of value in calculating stellar structure have been published by Cox and Tabor (1976). The Los Alamos Astrophysical Opacity Library by Huebner et al. (1977) allows Library users to calculate their own mixtures of hydrogen, helium, and the many other elements. Many tables have been calculated that way and passed around to other stellar astrophysicists all over the world. Hopefully, the Library is easy enough to use accurately, so that these tables are all consistent with each other.

Since the availability of the Library, there have been a few important specialized papers discussing improvements. Magee, Merts, and Huebner (1984) as well as Iglesias, Rogers, and Wilson (1987) have searched for opacity increases for the giant stars. It is hoped that the extensive equation of state (MHD) and opacity investigations by Mihalas, Hummer, Dappen, Seaton, and many others will produce a definitive set of tables for the future.

We discuss in this review the current issues for opacities at the solar center, the solar surface, and in the layers just below the convection zone. We also mention three other experiments that may help constrain opacities. These are the lithium problem of the Hyades G stars, the pulsation periods of the Cepheids and the δ Scuti variables, and the instability of the β Cephei variables.

Many researchers calculating stellar structure use the Los Alamos opacity tables directly using special interpolation procedures. This is what we do also in our pulsation studies at Los Alamos when the part of the star of interest is only the homogeneous composition stellar envelope. However, we have never interpolated between tables. For stellar models in advanced evolution stages, where the composition is changing throughout the model, we use the equation of state and opacity fits of Stellingwerf (1975ab) and Iben (1965 for the EOS and 1975 for the opacities). We find that this is adequate for most all studies, but maybe now solar structure studies have become so sophisticated that we need to interpolate between tables ourselves.

Opacities depend crucially of the details of the equation of state. Thus the Los Alamos equation of state results need to be just as comprehensive as those developed for the MHD equation of state. Not only do we need to know the internal partition function to calculate the degree of ionization of an element to know the mixture pressure and energy, but we also need that same data to enable us to calculate the bound-free and bound-bound photon absorptions that can occur. For many it may be a surprise that our equation of state data are available and agree closely with the MHD results. Just how closely is a topic now receiving interest for both equation of state and opacity studies.

There has been considerable activity lately on the question as to whether stellar opacities need to be increased because many weak lines, especially from iron, have been neglected. This is so even though iron is not a very abundant element in the solar composition. Evidence is growing for selected opacity increases, and in this review we will see that the exact accounting of the iron lines is important for the entire Sun, from the photosphere to the center.

2. THE SOLAR CENTER

Interest in the opacity at the solar center is due mostly to the solar neutrino problem. Two detectable sources, a line at 0.86 Mev and a spectrum extending to 17.98 Mev, respectively, for the ^7Be electron capture and the ^8B radioactive decay, are the main neutrinos that are effectively captured by ^{37}Cl. These fluxes, calculated from a recent solar model constructed by Cox, Guzik, and Kidman (CGK, 1989), produce 11 SNUs. This is more than other recent predictions near 8 SNUs, and considerably above the 2 SNUs (Davis, 1986) observed averaged over the last 20 years. Corrections for a known systematic pressure error in the Iben equation of state used for the CGK model reduce the central helium content, the central temperature, and the neutrino output to about the Bahcall and Ulrich (1988) 8 SNUs. But all recent predictions are larger than observations. Can solar opacities be blamed for this persistent prediction of too many neutrinos?

The answer for the last 20 years has been no. The opacity at the center of the Sun is approximately 1/3 free electron scattering, 1/3 free-free absorption of photons in the electric field of protons, alpha particles, and CNONe nuclei, and 1/3 due to a bound-free absorption edge and a bound-bound line of highly ionized iron. These absorption effects are simple enough, and the calculation of their attenuation reliable enough, that the opacity at the center of the Sun is probably known to an accuracy of 10 percent. Differences from model to model may be due more to composition changes rather than opacity uncertainties.

Figure 1 shows the monochromatic absorption coefficient versus photon energy at 15×10^6K and a density of 150 g/cm^3, conditions that are close to the solar center as calculated for current standard solar models. These data come from the EXOP opacity program, the immediate forerunner of the MOOP program that calculated the monochromatic absorption and scattering data for the Opacity Library. It has been used for this review because of its convenience in operation and in producing plots. The bound-free photoelectric absorption edge is caused by liberating the 1s electron from the iron atom with 2 1s electrons, 0.6 2s electrons and 1.7 2p electrons. This edge is at the mean position of individual edges for the several ionization stages present. The line is the sum of the 1s to 2p transitions in the iron ions that are 20, 21, 22, and 23 times ionized. The $1/\nu^3$ variation at photon energies below the absorption edge is the free-free absorption from all the ions in the mixture including those from iron. Above the edge the contribution of the free electron scattering becomes noticeable. The Rosseland mean weighting function peaks at $h\nu/kT = 7.0$, which is just over 9 kev. Significant weight is in the band of 3.5 to 18 kev, so one can roughly confirm that the mean opacity here is 1.18 cm^2/g.

The iron line transition 1s to 3p is not seen because the 3p level is destroyed by the large density of charges surrounding the iron ions. Shielding of the nucleus by the other bound electrons and the continuum depression puts the absorption edge at 7.3 kev instead of at 9.2 kev for the configuration with only one bound electron. The resonant scattering line at 3 kev and weak absorption lines near there are due to the 1s to 2p transitions in argon from ions with 1, 2, and 3 electrons attached.

Figure 2 gives the EXOP monochromatic absorption coefficients for a case at the same temperature and density where the iron is absent. This case has been considered by Dearborn, Marx, and Ruff (1987) as a possible way of reducing the central opacity in solar models, reducing the central temperature, and alleviating the solar neutrino problem. The mean opacity for this case is 0.92 cm^2/g. The iron line and edge are missing, and that results in the 25 percent opacity decrease.

The opacity for the case where all the Z elements are removed from the mixture is 0.80 cm^2/g. The contribution of completely ionized hydrogen and helium in producing free-free absorption and free electron scattering now produces the bulk of the opacity.

At Los Alamos we have been using the solar p-modes as an indicator of the internal solar opacity. That is, in addition to adjusting the mixing length in the standard mixing length convection theory to get the observed solar radius at the current age, and adjusting the original helium content to get the observed luminosity, we also adjust the opacity (at the bottom of the convection zone) to match the observed low degree p-modes. Figure 3 shows the observed minus calculated (O-C) frequencies for l=1 to 10 in the radial order band where oscillations are observed. The CGK case, with adjusted opacities to be discussed later, match the observations to within about 4 μHz out of the 3000 μHz for these modes. One important point is that opacity tables have not been directly used in the CGK work. Instead, a calibrated Iben (1975) fit is evaluated for the opacities.

The O-C values for the no-iron below the convection zones case are shown in figure 4. While the differences are actually smaller, there is a spread from one l value to the next. This gives a hint that things are worse than for the standard CGK case, but the difference between the two cases is slight.

Figures 5 and 6 show the small difference, which we call $\delta(n)$ for the two models with and without iron. This $\delta(n)$, the difference between the radial and quadrupole mode frequencies, has been discussed by many others (see Cox, Guzik, and Raby, CGR 1989), and it is a sensitive indicator of the central model conditions. Agreement of predictions with the observed frequency differences in the band of radial order 11 to 33, or, where the observations are more accurate, between 20 and 25, indicates that a model represents the actual solar center. We see that both models are in reasonable agreement with observations, with the standard CGK one (normal iron abundance) slightly better.

Collective effects have been frequently discussed as a way of reducing the electron scattering photon attenuation at the solar center. The most recent discussion is by Boercker (1987) who refines the earlier work of Diesendorf and Ninham (1969) and Diesendorf (1970). These latter authors have assumed that the angular dependence of the electron scattering integrated to zero, whereas for the non-vacuum solar conditions it does not. The result is a smaller opacity reduction. The correct reduction is about 25 percent, while the Los Alamos Astrophysical Opacity Library has a reduction at the solar center of about 20 percent, and the scattering in figures 1 and 2 above had no reduction at all. This matter seems to have been settled now, and its uncertainties are small, even though Bahcall and Ulrich (1988) calculate that the correct formulation gives opacities lower than the Opacity Library that is worth about a 9 percent reduction in the emergent calculated neutrino flux.

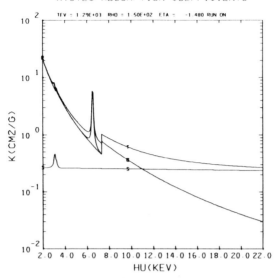

Figure 1. The monochromatic absorption coefficients for a solar mixture for the approximate solar center conditions of 15×10^6K and 150 g/cm^3. The iron bound-bound and bound-free transitions are seen together with the free-free absorptions from all the ions and the separately shown free electron scattering.

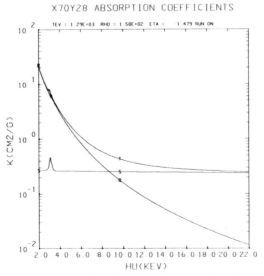

Figure 2. The monochromatic absorption coefficients for a solar mixture without iron for the approximate solar center conditions of 15×10^6K and 150 g/cm^3. The iron line and bound-free edge are now absent, and the opacity is reduced from 1.18 cm^2/g to 0.92 cm^2/g.

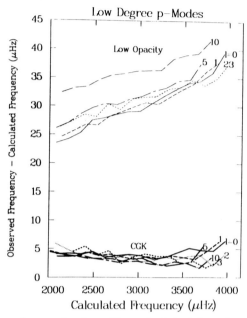

Figure 3. The observed minus the calculated low degree p-mode frequencies for the CGK and the low opacity models discussed in the CGR paper.

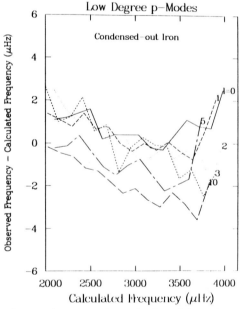

Figure 4. The observed minus the calculated low degree p-mode frequencies for the CGR condensed-out iron model.

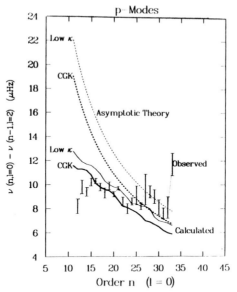

Figure 5. The $\delta(n)$ differences between the radial and quadrupole mode frequencies is plotted versus the radial order for the CGK and the low opacity models. The observations are given with error bars as published by Jimenez et al. (1988). Both the actually calculated eigenvalue frequency differences and the asymptotic theory variations are given.

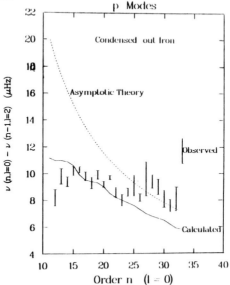

Figure 6. The $\delta(n)$ differences between the radial and quadrupole mode frequencies are plotted versus the radial order for the condensed-out iron model. Observations are given with error bars (Jimenez et al., 1988). Both the calculated eigenvalue frequency differences and the asymptotic theory variations are given.

There is one additional aspect of the collective effects. It seems that only the Los Alamos calculations have included electron degeneracy effects for the Debye radius that enters in the collective effects discussion. The above mentioned 20 percent Los Alamos Astrophysical Opacity Library opacity reduction discussed by Boercker includes this electron degeneracy effect.

Recently there has been this idea that there may be weakly interacting massive particles (WIMPs) that are numerous enough to supply the missing mass of the universe, or at least the missing mass around galaxies including ours. If these particles of about 5 times the proton mass really exist, then the Sun inevitably would collect many of them (one in 10^{11} protons) which would orbit in the inner 10 percent of the mass of the Sun. Even though they have an interaction cross section of only about $10^{-36} cm^2$ with ordinary matter, they are an important source for conduction of energy from the solar center to several times their orbit radius. Stuart Raby has taken the work of others such as Spergel and Press (1985) to give a refined expression for the WIMP or Cosmion opacity. This is:

$$\kappa_{Cosmion} = \kappa_0 (\frac{T}{T_0})^{3.5} (\frac{\rho}{\rho_0})^{-2} exp[\frac{r^2}{r_0^2}](r + r_0)/2r.$$

For the total opacity, including the Cosmion effects, the expression

$$1/\kappa_{total} = 1/\kappa_{Cosmion} + 1/\kappa_{photon}$$

is used. For the Cosmion opacity the values we used are $\kappa_0 = 10^{-3}$, $T_0 = 13 \times 10^6 K$, $\rho_0 = 200 \ g/cm^3$, and $r_0 = 0.0428$ solar radii.

Some workers have simulated the conduction of the Cosmions by considering them as an energy source some distance from the solar center instead of having them merely as another opacity (conduction) contribution. I understand that treating the Cosmion effects by the use of an effective opacity is more appropriate, and it does not lead to any instability problems that a few others (De Luca et al., 1989) have experienced.

With the κ_0 as given, corresponding to the above Cosmion abundance, the neutrino flux produces 2.2 (corrected to 1.5) SNU in the chlorine detector. A second Cosmion case has also been done by CGR (1989) with 9 times fewer Cosmions giving 15.9 times larger opacity, and that neutrino flux gives 4.4 (corrected to 3.5) SNU. This higher neutrino flux is just on the border of being compatible with observations. It is actually lower than the neutrino flux measured in the recent past.

Figure 7 gives again the O-C plot for the same low degree solar p-modes for the two Cosmion models. The opacity adjustment below the convection zone, to be discussed later, is still in the model. That means that this figure can be directly compared to figures 3 and 4 to see the effects of the Cosmions. The large spread of the differences between observation and prediction indicates that this model is not so good. Figure 8 plots the $\delta(n)$ between the radial and quadrupole modes for this standard Cosmion model and the one where the conduction by the Cosmions has been reduced by a factor of 15.9. These plots are in an Astrophysical Journal paper by CGR (1989). The reduced Cosmion number model with its marginally acceptable larger neutrino flux still does not fit the observed p-mode frequency differences very well.

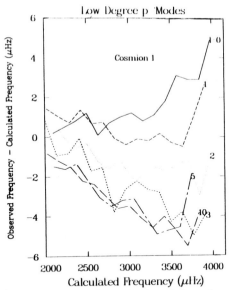

Figure 7. The observed minus the calculated low degree p-mode frequencies for the standard Cosmion model discussed in the CGR paper.

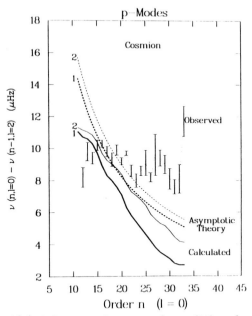

Figure 8. The $\delta(n)$ differences between the radial and quadrupole mode frequencies is plotted versus the radial order for the two Cosmion models. The observations are given with error bars as published by Jimenez et al. (1988). Both the actually calculated eigenvalue frequency differences and the asymptotic theory variations are given.

In a recent paper by Gilliland and Dappen (1988), it was suggested that WIMPs actually solve the solar neutrino problem. This is because the oscillation $\delta(n)$ values seem to match those observed for models that had considerable number of WIMPs and a suitably low neutrino flux. This result conflicts with the data given here. The explanation probably is that, if the O-C curve is rather positive, as shown in an extreme for our low opacity case in figure 3, then the $\delta(n)$ runs rather high as seen in figure 5. Gilliland and Däppen did not publish their O-C curve, but since they used the Eggleton, Faulkner, and Flannery (1973) equation of state without the coulomb corrections, and they used the old low Cox-Stewart (1970ab) opacities, it is likely that the O-C curve is mostly positive. Then they would have a high $\delta(n)$ curve, which they also did not publish. These authors would need WIMPs to lower the $\delta(n)$, and its mean value they do discuss, by about the amount they have found by using the older equation of state and opacity data.

A final matter concerning the opacity at the solar center is one discussed by Bahcall and Ulrich. During the solar evolution most of the nuclear energy is from the proton-proton reactions, but there is some cycling in the CNO process. This later process converts essentially all the carbon to nitrogen, which is slow to interact in an proton capture reaction. Even some of the oxygen is changed to nitrogen. Bahcall and Ulrich suggest that the opacity at the solar center is increased by 7.6 percent due to this processing. However, my recent detailed studies show that this is an incorrect result. Using a number of calculations, I find that the opacity is increased less than one percent by this composition change.

Table 1 gives some of the comparisons kindly calculated by Joyce Guzik for the Ross-Aller (1976) mixture. This is different from the mixture given by Cox and Tabor (1976) and used for figures 1 and 2. She has used the Opacity Library to show that when all the carbon is assumed to be nitrogen (but the oxygen is unchanged in its abundance), the opacity is increased by less than one percent. The track through the table following the central solar structure is 1.16 kev at a density of 100 and 1.35 kev at a density of 160. Only the Library temperatures have been used to avoid interpolation. Actually the CNO cycling effect gets smaller at higher temperatures and densities, merely reflecting the growing contribution of the free electron scattering. Such an effect would be expected by inspecting figures 1 and 2. All the bound-free edges of the CNO elements are at very low photon energies, and the only relic seen at the kilovolt photon energies and kilovolt temperatures is the free-free photon absorption in the electric fields of the CNO ions.

The extensive work over the last 20 years to reduce the neutrino outout has left us exactly where we were when we started. To solve the solar neutrino problem, I suggest that the neutrinos oscillate between their different types and interact with the electrons in the solar interior as Bethe (1986) and Rosen and Gelb (1986) have suggested. This so-called MSW effect requires neutrinos to have mass and to oscillate between the three different types, but that is not so unreasonable nowadays.

3. THE SOLAR SURFACE

Until the last few years, the opacity for the solar material at and just below the photosphere was obtained from the Cox and Tabor (1976) tables. In late 1979 Norman Magee kindly calculated a special low temperature table for the Ross-Aller (1976) composition using the same improved opacity program that was used to compile the Los Alamos Astrophysical Opacity Library (Huebner et al.,1977).

These opacities included molecules, but they are quite unimportant for solar models. Russell Kidman in 1983 then added Library opacities at the higher temperatures to form the Ross-Aller 1 table that is now widely used by solar modelers. The interesting thing is that even though molecules are included, the main reason for a doubling of the opacity over the Cox-Tabor values is that iron lines are now calculated in much more detail than in the Cox-Tabor tables.

Figure 9 plots the logarithm of the ratio of the opacities from EXOP, as used for the King IVa table, to those from the Ross-Aller 1 table. Both tables have essentially the same composition.

Figure 10 gives the monochromatic absorption coefficients versus photon energy for a temperature of 0.5 electron volt (5800K) and a density of 3×10^{-7} g/cm^3. The same plot using the Cox-Tabor data shows only a few scattered lines instead of the forest of iron lines shown in this figure and also actually seen in the solar photosphere spectrum. It is interesting that over ten years ago, Huebner (1978) felt it important to display this very same plot to show the strong influence of the iron lines on the solar photosphere spectrum. Apparently the contour diagram given by Cox (1983) and repeated by Magee, Merts, and Huebner (1984), which shows the effects of lines for astrophysical opacity mixtures, missed a small island with a large opacity increase at a temperature of 0.5 ev. Absorption lines are not important at temperatures lower than 4000K because of the large effects of water vapor molecules, and above 8000K (0.7 ev) hydrogen bound-free absorption becomes dominant.

While for solar models we have almost always used the Ross-Aller 1 opacities by calibrating the Stellingwerf (1975ab) fit, others have used the Cox-Tabor opacity tables. Thus we do not directly know the relative effects of these two opacity tables on p-mode frequencies. According to Christensen-Dalsgaard, however, p-mode frequencies fit somewhat better when the newer Ross-Aller 1 opacities are used in the solar model.

I have considered the relative nonadiabatic effects of the King IVa (Cox-Tabor using EXOP) and the Ross-Aller 1 tables for the p_9, $l=60$ mode observed at 3036 μHz. Figure 11 shows the work to drive pulsations per pulsation cycle versus zone number for the King IVa table in the model and in the oscillation eigensolution. These opacities were used (through the Stellingwerf fit) for the Kidman and Cox (1984) study of nonadiabatic effects on p-modes. The peak driving is at about 9000K. The mode decays approximately as inferred from the observed p-mode line widths in the spectrum. This can be seen from the figure because the integral over the driving (plotted per zone to make the integral easier to see) is slightly negative.

When the Ross-Aller 1 opacities are approximated by actually tripling the Stellingwerf opacity fit values, the pulsation driving is given by figure 12. Driving is now at a higher mass level where the temperature is again about 9000K, because the temperature gradient from the photosphere is steeper. In this case, however, the driving overwhelms the damping, and the mode is pulsationally unstable. Thus the overstability of the solar p-modes depends sensitively on the opacity and monochromatic absorption coefficients.

But care must be taken. Radiation transport effects are not included in these calculations either for the model structure or for the oscillation eigensolutions. Also convection luminosity, which carries perhaps 99 % of the emergent luminosity at the mass level where the p-modes are driven, is treated in our calculations as completely frozen-in. Therefore, a definitive statement about the pulsation driving is not possible. It seems that modern nonadiabatic calculations, for example the radiation transport work by Christensen-Dalsgaard and Frandsen (1983) or the

72

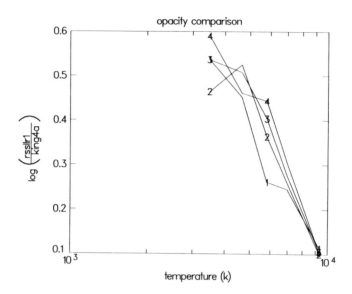

Figure 9. The logarithm of the ratio of the Ross-Aller 1 table opacities to the King IVa table opacities is plotted versus temperature for densities 3×10^{-8}, 1×10^{-7}, 3×10^{-7}, and 1×10^{-6} g/cm^3. At the solar photosphere, the new Opacity Library opacities are twice that from the Cox and Tabor tables.

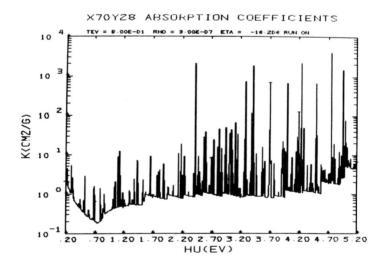

Figure 10. The monochromatic absorption coefficients for a solar mixture for the approximate solar photosphere conditions of 5800K and 3×10^{-7} g/cm^3. On top of the dominant negative hydrogen ion absorption one can see the myriad of iron lines coming from transitions from the M shell to higher levels.

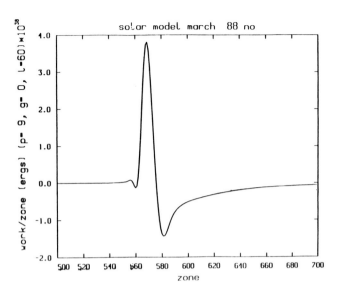

Figure 11. The work per cycle to drive oscillations is plotted versus zone number for the outer 700 zones of the 1700 mass zone CGK model using the Cox and Tabor King IVa opacity table. The peak driving at 9000K is not large enough to destabilize the mode, and this p_9 mode with l=60 decays roughly as actually observed.

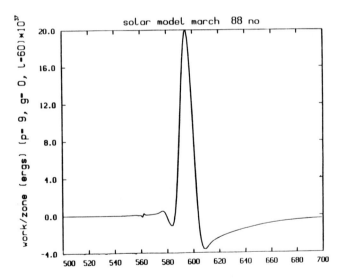

Figure 12. The work per cycle to drive oscillations is plotted versus zone number for the outer 700 zones of the 1700 mass zone CGK model using the Opacity Library Ross-Aller 1 opacity table to calibrate the Stellingwerf fit. The peak driving at 9000K is large enough to destabilize this p_9 mode with l=60.

radiation transport and convection studies by Balmford and Gough (1988), need to include the latest Ross-Aller 1 opacities in any case.

Modelers of stellar structure have needed complete opacity tables for many mixtures, but at the moment there are really only two modern ones. The Opacity Library allows opacity calculations only down to 11,600K temperature, because molecular contributions to the absorption become important at temperatures like 6000K. Mixtures named Ross-Aller 1 and Grossman 1 have been used to produce tables down to 2500K. This lack of tables is now being rectified by Weiss at Illinois and Munich who soon will have 20 additional tables, some with significantly different hydrogen and helium contents.

4. BELOW THE SOLAR CONVECTION ZONE

We have found that the solar p-modes are very sensitive to the opacity just below the convection zone. This sensitivity have been suggested by Christensen-Dalsgaard et al. (1985) and more recently by Korzennik and Ulrich (1989). CGK (1989) adjusted the opacity upward by 15 to 20 percent in the 2 to 7 million kelvin region to get almost perfect agreement with all the low degree p-modes. Recent work by Christensen-Dalsgaard, Lebreton, and Dappen (private communication) has shown that good agreement with the observed p-modes can be found without adjusting the opacity from the Los Alamos Astrophysical Opacity Library at all. Thus only a small or even zero opacity adjustment may eventually be agreed upon.

Figure 13 shows the monochromatic absorption coefficients versus photon energy for a temperature of 3×10^6K and a density of 0.4 cm^2/g, calculated by the EXOP opacity program. These conditions for the Cox-Tabor composition obtain just below the solar convection zone. The strong bound-free edge at 0.76 kev photon energy is from oxygen with 0.3 electron in the 1s level. The other strong bound-free edge is from neon at 1.18 kev with 0.9 electron in its K shell. The only other significant K edge is at 3.86 kev from argon with both of its 1s electrons attached, but the K edges of magnesium, aluminum, and silicon can just barely be seen. The carbon K edge is off scale to the left, and the iron K edge is of scale to the right. The edges at 1.4 kev are the L edges of iron with 4.7 electrons in the 2s and 2p levels.

What can cause the opacity uncertainty at these solar conditions? It does not seem that the CNONe ion hydrogen-like bound-free absorptions can be in error very much. Or can the small free-free absorption be uncertain? The electron scattering, with the iron resonance lines near 0.9 kev is small compared to other processes. It very well could be that the absorption lines that are all from transitions with the lower level being 2s or 2p in 15, 16, 17, 18, and 19 times ionized iron are much more numerous than we have indicated. The transitions to level 3s, 3p, and 3d are near 0.9 kev, while the 2 to 4 and 2 to 5 transitions are range from 1.0 to 1.4 kev, just up to the L edges. Note that the level 6 in iron is destroyed by the continuum depression from neighboring charged particles. There is also the possibility that with almost 0.7 electron in the 3 shell levels that there can be some 3 to 3 iron line transitions, which we are not considering. It is not at all unreasonable that the opacity might need to be increased 20 percent for this temperature and density. I predict that the missing opacity is due to the iron lines that are only partially displayed in figure 13.

We have calibrated the Iben (1975) opacity fit to calculate our solar models and to solve for our oscillation eigensolutions. For this case, we have found that a factor of 1.3 on a term called A_z and a factor of 2.0 on the term κ_e correct the Iben fit opacities to the Opacity Library values and then increase them by 15 to 20 percent in the temperature range of 2 to 7 million kelvin. Figure 14 shows the logarithm of the Ross-Aller 1 opacity to the Iben fit opacity ratio versus temperature for the three densities available in the Ross-Aller 1 table. The original Iben fit is 20 to 40 percent too low in the bottom layers of the convection zone and deeper to a temperature of $1.0 \times 10^7 K$. Figure 15 gives the fit calibrated as stated above relative to the Ross-Aller 1 table. Now the calibration factors make the opacity larger by 20 percent at $3 \times 10^6 K$ and less than 10 percent at $1.0 \times 10^7 K$.

Figure 3 shows the O-C plot for the two Iben fit cases discussed above. The original Iben fit gives very low opacities in the region below the convection zone, and the fit to the low degree p-mode frequencies is very poor. The dramatic improvement, providing that the equation of state is otherwise accurate, can be used to verify the accuracy of the solar opacities.

The $\delta(n)$ that corresponds to the low opacity case of figure 3 is given in figure 5. The larger figure 3 O-C values give larger $\delta(n)$ values. The sensitivity to the opacity decreases above $l=10$, because the higher degree modes have turning points nearer to the convection zone bottom or even in it. Then the amplitude in the deeper evanescent regions is not large enough to affect significantly the oscillation periods.

5. OPACITY EXPERIMENTS

For a long time I have been interested to see what the stars can tell about the opacity of their material. We have seen that the Sun can give information about opacities from its central regions, its surface structure, and its structure at and below the convection zone. At the center, we conclude that opacities are reasonably accurate, because we can accurately predict the actual observed low degree p-modes. If ever the g-mode periods of Hill and his colleagues can be confirmed, there will be an even stronger verification of the solar center opacities. These published g-mode periods are very near that expected for the standard solar model without any internal mixing during its life. It does seem that conduction by Cosmions is not supported by current g-mode period observations.

Solar surface Opacities are not strongly constrained by either p-mode periods or by their stability.

In the region below the convection zone, we have a rather interesting opacity experiment. CGK have suggested a needed increase, but Christensen-Dalsgaard and collaborators find that an increase is not so necessary. The issue of the importance of iron lines, seen at the convection zone bottom, arises again in the context of the G stars in the Hyades cluster and for classical variable stars.

There is a problem with the predictions for the lithium abundance in the Hyades cluster G stars. Stringfellow, Swenson, and Faulkner (1987) have brought attention to this problem that may be important for stellar opacities. It appears that the convection zone depth in current stellar models needs to be increased rather like that to predict correctly the solar p-mode frequencies and even the solar lithium depletion. A modest opacity increase would very much help this discrepancy with observations that show lithium depletion (presumably by nuclear processing) occurs

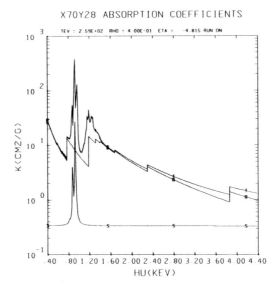

Figure 13. The monochromatic absorption coefficients for a solar mixture for the approximate solar conditions below the convection zone of $3x10^6$K and 0.4 g/cm^3. Absorption K edges of oxygen, neon, magnesium, aluminum, silicon, and argon can be seen. Also at 1.4 kilovolt photon energy weak closely spaced L edges of iron are visible.

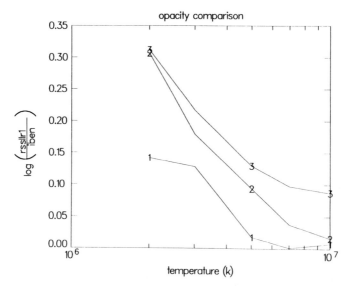

Figure 14. The logarithm of the ratio of the Ross-Aller 1 table opacities to the Iben fit procedure opacities is plotted versus temperature for densities $1x10^{-1}$, $1x10^0$, and $1x10^1$ g/cm^3. The new Opacity Library opacities are considerably larger than those from the Iben fit originally calibrated to the Cox-Stewart opacities.

for hotter G type stars than the experts predict with classical opacity values. Can the iron lines increase the opacity and solve this problem?

Andreasen and Petersen(1988) and Andreasen (1988) have pointed out that a large opacity increase suggested by Simon (1982) to solve a period ratio problem for the classical double-mode Cepheids can also help with period predictions in δ Scuti and RR Lyrae variables. This opacity increase has been discussed by Magee, Merts and Huebner (1984), and such a large increase over such a large temperature range was not found likely. Iglesias, Rogers, and Wilson (1987), however, have found that the forest of iron lines can increase the opacity by modest amounts at 20 ev (230,000K) temperature. Their latest unpublished results cite a factor of 2.2 opacity increase at a density of $10^{-5} g/cm^2$, but that temperature-density pair is found only in the yellow giant stars.

More recently Rozsnyai (1989) has proposed even more iron lines, and he finds very large opacity increases. A factor or 3 over that from the Simon model seems to exist at about 250,000K. This opacity increase starts at 150,000K and persists out to 800,000K. Maybe Rozsnyai has found the missing ingredient for Cepheid, δ Scuti, and RR Lyrae star models.

For years, there has been the problem of how the β Cephei variables pulsate. Investigations of many possible mechanisms has always shown, with careful study, that they cannot make these stars unstable, at least in the linear nonadiabatic theory. I now think that increased opacities may give a larger opacity derivative with respect to temperature and be the cause of B star pulsations!

I have increased B star opacities just as Rozsnyai has suggested. At 150,000K the opacities are those from the Opacity Library times a factor that linearly increases to three at a temperature of 170,000K. Then the opacities remain at three times the Library values up to 200,000K. A slow decrease is then followed out to a unity factor again at 800,000K. This profile matches reasonably well the results of both Rozsnyai and of Iglesias, Rogers, and Wilson, but it is at odds with the Los Alamos result of Magee, Merts, and Huebner (1984).

Let me recall an experiment that I did 25 years ago. In a mixture suggested by Aller (1961), I tripled the widths of all the lines in the opacity calculation. Changes of the order of only 30 percent were found, and iron was not an important component of the solar opacity. Nowadays the iron abundance is ten times larger in the solar mixture, and an opacity increase by tripling the line widths is much larger. The Rozsnyai increase is not unreasonable.

Figure 16 shows the work plot for a 400 zone model of a typical β Cephei variable at 12 $M_\odot$. With the rapid opacity increase with temperature, caused by the assumed sudden onset of the myriad of iron lines, the normal κ effect produces adequate driving in the pulsation driving region of 150,000K to 300,000K. Can these opacities be correct? It seems that the stars tell us so.

Not every B star is a β Cephei variable, even when they evolve into the strip defined by the many variables known. This fact is clearly seen in some galactic clusters where relative luminosities are not an unknown factor. I suggest that the overstability is indeed a function of the iron abundance. With a normal solar iron abundance, there may not be enough driving, but if there is only a slight iron enhancement (but not due to gravitational settling or radiation levitation that we calculate to be a small effect, however), then the iron lines can have an significant effect on the stellar opacity and stellar stability.

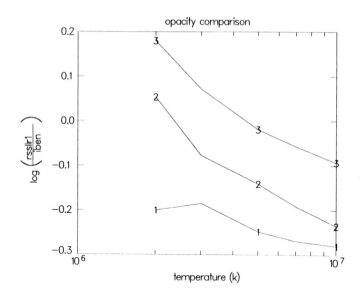

Figure 15. The logarithm of the ratio of the Ross-Aller 1 table opacities to the modified Iben fit procedure opacities is plotted versus temperature for densities 1×10^{-1}, 1×10^{0}, and 1×10^{1} g/cm^3. At these conditions, the modified Iben fit gives opacities 15 to 20 percent larger than the Library, and these opacities give a model that has oscillation frequencies very close to those observed.

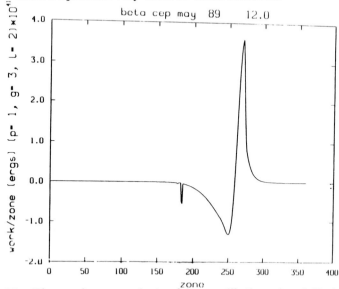

Figure 16. The work per cycle to drive oscillations is plotted versus zone number for the 400 zones of a β Cephei model using the opacities suggested by Rozsnyai. Peak driving at 170,000K in zone 270 is large enough to destabilize this $l=2$ mode,. The dip at 800,000K is due to the return to the normal Iben fit.

6. REFERENCES

Aller, L. H. 1961. *The Abundance of the Elements*, (New York: Interscience Publishers).

Andreasen G. K., 1988. Stellar consequences of enhanced metal opacity. I. An attractive solution of the Cepheid period ratio discrepancies. *Astron. Astrophys.*, **201**, 72.

Andreasen, G. K. and Petersen, J. O. 1988. Double mode pulsating stars and opacity changes. *Astron. Astrophys.*, **192**, L4.

Bahcall, J. N. and Ulrich, R. K. 1988. Solar models, neutrino experiments, and helioseismology. *Rev. Mod. Phys.*, **60**, 297.

Balmford, N. J. and Gough, D. O. 1988. Radiative and convective influences on stellar pulsational stability. *Seismology of the Sun and Sun-Like Stars*, ESA SP 286, ed. E. J. Rolfe, p. 47.

Bethe, H. A. 1986. Possible explanation of the solar neutrino puzzle. *Phys. Rev. Lett.*, **56**, 1305.

Boercker, D. B. 1987. Collective effects on Thomson Scattering in the solar interior. *Ap. J. Lett.*, **316**, L98.

Christensen-Dalsgaard, J., Duvall, T. L., Gough, D. O., Harvey, J. W., and Rhodes, E. J. 1985. Speed of sound in the solar interior. *Nature*, **315**, 378.

Christensen-Dalsgaard, J. and Frandsen, S. 1983. Radiative transfer and solar oscillations. *Solar Physics*, **82**, 165.

Cox, A. N. 1965. Stellar absorption coefficients and opacities. *Stars and Stellar Systems*, **8**, eds. L. H. Aller and D. B. McLaughlin (Chicago: University of Chicago Press) p. 195.

Cox, A. N. 1983. Stability problems with an application to early type stars. presented at Swiss Society of Astrophysics and Astronomy, Saas Fee, Switzerland, 1983, Mar 21-26.

Cox, A. N., Guzik, J. A. and Kidman, R. B. 1989. Oscillations of solar models with internal element diffusion. *Ap. J.*, **342**, 1187.

Cox, A. N., Guzik, J. A., and Raby, S. 1989. Oscillations of condensed-out iron and cosmion solar models. *Ap. J.*, submitted.

Cox, A. N. and Stewart, J. N. 1970a. Rosseland opacity tables for population I compositions. *Ap. J. Suppl.*, **19**, 243.

Cox, A. N. and Stewart, J. N. 1970b. Rosseland opacity tables for population II compositions. *Ap. J. Suppl.*, **19**, 261.

Cox, A. N. and Tabor, J. E. 1976. Rosseland opacity tables for 40 stellar mixtures. *Ap. J. Suppl.*, **31**, 271.

Davis, R. 1986. *Report to the Seventh Workshop on Grand Unification*, (ICOR-BAN '86, Toyoma, Japan), p. 237.

Dearborn, D. S. P., Marx, G., and Ruff, I. 1987. A classical solution for the solar neutrino puzzle. *Prog. Theo. Phys.*, **77**, 12.

DeLuca, E. E., Griest, K., Rosner, R., and Wang, J. 1989. On the effects of cosmions upon the structure and evolution of very low mass stars. *Ap. J. Lett.*, submitted.

Diesendorf, M. O. 1970. Electron correlations and solar neutrino counts. *Nature*, **227**, 266.

Diesendorf, M. O. and Ninham 1969. The effect of quantum correlations on electron-scattering opacities. *Ap. J.*, **156**, 1069.

Eggleton, P. P., Faulkner, J. and Flannery, B. P. 1973. An approximate equation of state for stellar material. *Astron. Astrophys.*, **23**, 325.

Gilliland, R. L. and Däppen, W. 1988. Oscillations in solar models with weakly interacting massive particles. Ap. J., 324, 1153.

Huebner, W. F. 1978. Proc. Informal Conf. on Status and Future of Solar Neutrino Research, BNL Rept. 50879, ed. G. Friedlander vol 1, p. 107.

Huebner, W. F. 1986. Atomic and radiative processes in the solar interior. Physics of the Sun, (Dordrecht: D. Reidel Publishing Company), 1, p. 33.

Huebner, W. F., Merts, A. L., Magee, N. H., and Argo, M. F. 1977. Astrophysical Opacity Library, Los Alamos Scientific Laboratory Report, LA-6760-M.

Iben, I. 1965. Stellar evolution I. The approach to the main sequence. Ap. J., 141, 993.

Iben, I. 1975. Thermal pulses; p-capture, α-capture s-process nucleosynthesis; and convective mixing in a star of intermediate mass. Ap. J., 196, 546.

Iglesias, C. A., Rogers, F. J., and Wilson, B. G. 1987. Reexamination of the metal contribution to astrophysical opacity. Ap. J. Lett., 322, L45.

Jiménez, A., Pallé P. L., Pérez, J. C., Régulo, C., Roca Cortés, T., Isaak, G. R., McLeod, C. P., and van der Raay, B. B. 1988. The solar oscillations spectrum and the solar cycle. Advances in Helio- and Asteroseismology, IAU Colloquium 123, eds. J. Christensen-Dalsgaard and S. Frandsen, p. 208.

Kidman, R. B. and Cox, A. N. 1984. The stability of the low degree five minute solar oscillations. in Solar Seismology from Space, eds. R. K. Ulrich, J. Harvey, E. J. Rhodes, and J. Toomre, (NASA Pub 8484), p. 335.

Korzennik, S. G. and Ulrich, R. K. 1989. Seismic analysis of the solar interior I. Can opacity changes improve the theoretical frequencies? Ap. J., 339, 1144.

Magee, N. H., Merts, A. L., and Huebner, W. F., 1984. Is the metal contribution to the astrophysical opacity incorrect? Ap. J., 283, 264.

Rosen, S. P. and Gelb, J. M. 1986. Mikheyev-Smirnov-Wolfenstein enhancement of oscillations as a possible solution to the solar neutrino problem. Phys. Rev., D34, 969.

Ross, J. E. and Aller, L. H. 1976. The chemical composition of the sun. Science, 191, 1223.

Rozsnyai, B. F. 1989. Bracketing the astrophysical opacities for the King IVa mixture. Ap. J., 341, 414.

Simon, N. R. 1982. A plea for reexamining heavy element opacities in stars. Ap. J. Lett., 260, L87.

Spergel, D. N. and Press, W. H. 1985. Effect of hypothetical, weakly interacting, massive particles on energy transport in the solar interior. Ap. J., 294, 663.

Stellingwerf, R. F. 1975a. Modal stability of RR Lyrae stars. Ap. J., 195, 441.

Stellingwerf, R. F. 1975b. Nonlinear effects in double-mode Cepheids. Ap. J., 199, 705.

Stringfellow, G. S., Swenson, F. J. , and Faulkner, J. 1987. Is there a classical Hyades lithium problem? BAAS, 19, 1020.

Table 1. Central Solar Opacities (cm^2/g), Ross-Aller Mixture

Density(g/cm^3)	1.00 kev Normal	1.00 kev No C	1.25 kev Normal	1.25 kev No C	1.50 kev Normal	1.50 kev No C
100	1.849	1.869	1.178	1.186	0.862	0.866
120	2.022	2.047	1.269	1.279	0.915	0.919
150	2.264	2.294	1.395	1.407	0.986	0.990
160	2.326	2.356	1.435	1.447	1.008	1.012

ASTROPHYSICAL OPACITIES AT LLNL

CARLOS A. IGLESIAS AND FORREST J. ROGERS
Lawrence Livermore National Laboratory
P.O. Box 808
Livermore, California 94550
USA

ABSTRACT. In an effort to reduce uncertainties in the theoretical radiative opacities a new code has been developed at LLNL which removes several of the approximation present in past calculations. Results from the new code with comparisons to other available opacity calculations are presented as well as experiments.

1. INTRODUCTION

The calculation of plasma radiative properties involves detailed knowledge of both atomic and plasma physics. As a result, the problem is complex and has been fully addressed by only a few groups. However, due to the complexity, approximations have been made in the past which may limit the accuracy of the results. Unfortunately, direct experimental measurements of the opacity are essentially nonexistent so that the success or failure of theoretical results must be determined indirectly. For example, the calculation of period ratios in Cepheids based on stellar models does not agree with experimental observations. It has been suggested by Simon[1] and later by Andreasen[2] that an arbitrary increase in the Rosseland mean opacity of the metals without increasing their abundance can explain the discrepancy. Comparison of observed and predicted p-mode oscillations in the sun can be improved by increasing the opacity[3]. The solar neutrino rate is sensitive to the opacity as well. There are other outstanding problems which could be solved by changes in the input opacity calculations. Probably the discrepancies will not all disappear with improved opacities, however, it will be easier to find the explanations if uncertainties in the opacities were reduced.

At present there are about ten theoretical efforts throughout the world developing radiative opacity codes. Although there is some overlap, the groups vary somewhat in approach and tend to emphasize different matter conditions. There are also some experimental efforts trying to measure directly the photon absorption[4]. These are difficult experiments and definite results are not presently available. Perhaps all this activity coupled to the astrophysical and astronomical community, which at present offer the best laboratories (stellar matter), will result in opacity calculations being better understood and uncertainties reduced.

Several reviews have appear in the literature[5] which discuss opacities and is not the purpose here to go over this ground, but rather to describe a new opacity effort at LLNL: the OPAL code. We shall consider temperatures sufficiently high (a few eV) that molecular absorption is negligible and photon energies low enough (less than about 10keV) that relativistic effects are

81

G. Berthomieu and M. Cribier (eds.), Inside the Sun, 81–90.
© 1990 *Kluwer Academic Publishers. Printed in the Netherlands.*

small. These conditions cover the relevant plasma temperature and photon energy ranges for computing opacities inside the sun. Under these circumstances the dominant absorption processes are electronic transitions in the field of ions; that is, line transitions (bound-bound), photoionization (bound-free), and inverse bremsstrahlung (free-free). At high temperatures nuclei are completely striped of any bound electrons and photon scattering from free electrons becomes important.

For conditions of interest here, collisions between the plasma constituents are sufficiently frequent that the plasma can be assumed to be in local thermodynamic equilibrium (LTE). Since the mean free path of an average photon is small compared to the scale of the matter temperature gradients inside a star, the photons are in equilibrium with the matter and have a black body spectrum at the material temperature. The transport of photons will then be well desscribed by the diffusion approximation with the diffusion constant given by the Rosseland mean opacity, K_R,

$$\frac{1}{K_R} = \int_0^\infty du \, \frac{1}{\tilde{K}(u)} \, \frac{\partial B(u,T)}{\partial T} \tag{1}$$

where the weighting function

$$\frac{\partial B(u,T)}{\partial T} = \frac{15}{4\pi^4} \, \frac{u^4 e^u}{(1-e^u)^2} \tag{2}$$

peaks at u (=photon energy/T) = 4 with T the matter temperature in units of energy. The extinction coefficient is defined by

$$\tilde{K}(u) = K_{abs}(u)(1-e^{-u}) + K_{sc}(u) \tag{3}$$

where K_{abs} is the absorption coefficient and K_{sc} is the scattering cross section,

$$K_{abs}(u) = \sum_{ijk} N_{ijk} \, \sigma_{ijk}(u)$$

$$\tag{4}$$

$$K_{sc}(u) = N_e \, \sigma_{sc}(u).$$

Here, N_{ijk} is the number of ions of charge j in electronic level k of element species i, $\sigma_{ijk}(u)$ is the absorption cross section for photons with energy u by those levels, and N_e the free electron number density with $\sigma_{sc}(u)$ the photon scattering cross section.

2. THEORY

2.1 Equation of State (EOS)

Every opacity calculation begins with the EOS to obtain the occupation numbers, N_{ijk}. Our EOS was developed by Rogers[6] and is based on an activity expansion for the grand canonical partition function. The approach relies on quantum statistical mechanics and does not require any *ad hoc* cutoffs so familiar in free energy minimization techniques. The well known problem in the latter is related to the internal partition function,

$$\sum_{j=0}^{N} g_j \exp\left(-\frac{E_j}{T}\right) \propto N^3 \qquad (5)$$

where the sum is over bound levels, g_j is the level degeneracy, and E_j the level energy. The divergence requires a phenomenological argument for truncating the series.

A proper solution to the problem recognizes that the partition function for the plasma involves a trace over <u>all</u> states. The many-body activity expansion developed by Rogers[6] and Ebeling et al.[7] not only avoids the *ad hoc* cutoffs but shows how the divergences are removed by including the scattering states. It starts from a description of the system in terms of electrons and nuclei interacting through the Coulomb potential and makes no assumptions about the internal states of composites (ions and atoms). This fundamental particle activity expansion is then renormalized to account for the composites. Ebeling will describe the EOS in another chapter of the present volume so that only brief remarks concerning the method will be given here.

For simplicity assume a plasma in LTE made up of protons and electrons interacting through the Coulomb potential. Following quantum statistical mechanics, one can write the pressure, P, in terms of an activity expansion and cluster coefficients,

$$\frac{P}{T} = \sum_{\alpha} z_{\alpha} + \sum_{\alpha}\sum_{\alpha'} z_{\alpha} z_{\alpha'}\, b_{2,\alpha\alpha'} + \cdots \qquad (6)$$

where z_{α} is the activity for species α, b_n's are the n-body cluster coefficients, and α is an electron or proton. As an example consider the electron-proton cluster coefficientt[8]

$$b_{2,ep} \propto Tr_{ep}\left\{\exp\left(-\frac{H_{ep}}{T}\right) - \exp\left(-\frac{H_{ep}^0}{T}\right)\right\} \qquad (7)$$

where Tr_{ep} is a trace over a complete set of electron-proton states, with H_{ep} and H^0_{ep} the electron-proton Hamiltonian with and without the Coulomb interaction, respectively. The cluster b_2 has a simple interpretation: It contains two-body interaction corrections to the pressure, the H_{ep} term, but one must be careful to subtract the non-interacting two body contributions already included in the ideal gas term, subtracted H^0_{ep} term. In this form it is easy to see that b_2 is not the two body partition function and to interpret it as such may lead to confusion.

Since the Coulomb interaction depends only on relative coordinates, the center of mass motion separates from the relative motion and we find after some manipulations[8]

$$
b_{2,ep} \propto \sum_{j(bound)} g_j \exp\left(-\frac{E_j}{T}\right) + \frac{1}{\pi}\sum_{l}(2l+1)\int_0^\infty dp \frac{d\delta_l(p)}{dp} \exp\left(-\frac{p^2}{2\mu T}\right) \quad (8)
$$

where E_j are the bound state eigenvalues of H_{ep}, $\delta_l(p)$ are the scattering state phase shifts with momentum p, and μ is the reduced mass. At this point it appears that $b_{2,ep}$ also diverges due to the bound state contributions. However, an integration by parts leads to

$$
b_{2,ep} \propto \sum_{j(bound)} g_j\left[\exp\left(-\frac{E_j}{T}\right)-1\right] + \frac{2}{\pi\mu T}\sum_{l}(2l+1)\int_0^\infty dp\, p\, \exp\left(-\frac{p^2}{2\mu T}\right)\delta_l(p) \quad (9)
$$

where use has been made of Levinson's theorem[8]

$$
\delta_l(p) = \pi(\text{number of bound states of quantum number } l) \quad (10)
$$

We can see in Eq. (9) that the contribution from the scattering states compensates for the leading divergence in the bound state contribution. Higher order Levinsons theorems have been proven for the Coulomb potential[9] and a second integration by parts yields

$$
b_{2,ep} \propto \sum_{j(bound)} g_j\left[\exp\left(-\frac{E_j}{T}\right)-1+\frac{E_j}{T}\right]
$$

$$
+ \frac{2}{\pi\mu^2 T^2}\sum_{l}(2l+1)\int_0^\infty dp\, \exp\left(-\frac{p^2}{2\mu T}\right)\left[\int dp\, p^2\, \delta_l(p)\right] \quad (11)
$$

and the bound state divergences are now fully compensated by the scattering states. This manipulations have redefined the continuum such that weakly bound states are treated with the scattering states in many-body perturbation. The sum over bound states in Eq. (11) is the so-called Plank-Larkin partition function which, of course, is not a partition function, but rather the bound state contribution from two-body bound states to the pressure.

There are other divergences in the activity expansion of the pressure which do not compensate. These are present even in classical Coulomb systems without bound states and are associated with the long-ranged Coulomb interaction. Their removal is well understood and involves summing certain classes of terms appearing in the expansion. The above procedures and their extensions to higher order cluster coefficients has been discussed extensively by Refs. 6 and 7

Even though the activity expansion provides a formalism for the EOS, it does not automatically yield occupation numbers for the plasma radiative properties. A final step described by Rogers[6] is necessary where a formal comparison of pressure expressions from the activity expansion and the free energy minimization was done. The results show that strongly bound state occupation numbers are described by Boltzman factors while weakly bound states require many-body corrections.

2.2. Absorption Cross Sections

The approach followed for computing the opacity is the so-called detailed configuration method in which every electronic configuration and corresponding term structure is considered explicitly. Such an approach requires vast amounts of atomic data. One possibility is to create a large data base with energy levels and cross sections. The data base, however, will not contain any density effects which may be important for high density plasma in stellar interiors. A second possibility, and the one chosen here, is to compute the atomic data "on-line". In the past, on-line calculations were too slow or not sufficiently accurate. Fortunately, we have developed a parametric potential method which is both fast and accurate[10]. The parametrization procedure required reproduction of experimental energy level data. Solving the Dirac equation with these parametric potentials provides wavefunctions and energies which in turn are used to compute the photon absorption cross sections. The parametric potentials have been developed for valence and inner shell electrons as well as multiply excited configurations. The accuracy of this method is comparable to single configuration, self-consistent-field calculations with relativistic corrections. The method affords the possibility of including density effects by suitable modifications of the Coulomb tail by plasma screening. It also allows for testing of various atomic physics issues by changing the atomic package in the code. Something not easily done with data bases. Finally, it is important to note that data bases are not "complete" and that on-line atomic calculations, much simplified over those in the data base,would still be necessary to supplement the data base approach. It is possible that a hybrid approach would be best where a very accurate, relatively small data base is supplemented with fast and reasonably accurate on-line method such as ours.

Previous opacity calculations have not included this level of accuracy in their atomic data, partly due to the limited computer facilities available at the time. Consequently, they have in some plasma conditions important for Cephied variables underestimated the opacity of the metals by large factors[11].

Below follows a brief description of how the three dominant absorption processes are computed in OPAL. It is important to realize that as in most calculations, it is assumed that many-body physics can be well described by an appropriate single particle representation.

2.2.1. *Bound-Bound Transitions.*

The accuracy obtained in calculating bound-bound transitions depends on the line location and associated oscillator strength. Using the parametric potential we can obtain 1% or better accuracy for the configuration averaged energies when compared to experiment for ions relevant to the solar interior. In order to match real experiments, it necessary to include the configuration term splitting. The angular momentum coupling is done using standard perturbation theory methods and includes either Russell-Saunders or intermediate coupling depending on the levels and nuclear charge. The term splitting results are not as accurate as the configuration averages but are better than 10% when compared to experiment. The strong oscillator strengths are also in the order of 10% accuracy, but very little experimental data are available. Of course, the comparisons are done for isolated ions and it must be emphasize that little is known of how ions behave under the extreme conditions of stellar interiors. Not all transitions are computed with term splitting. Transition from lower bound levels with quantum number greater than 5 are considered in the configuration average only. If in the future this "switch" is insufficient (perhaps in the much colder regions in the photosphere) it is easily change in the code to some higher value.

In nature spectral lines experience broadening, for example, Doppler effects, collisions, and natural lifetimes. In OPAL lines from single electron ions are treated with standard linear Stark theory[12]. These line shapes are in good agreement with experiment. In the near future we will include similar calculations for Helium-like and Lithium-like ions which have also compared well with experiments. For all other transitions we use Voigt profiles where the Gaussian width is due to doppler broadening and the lorentz width is due to estimated natural plus electron impact collisions.[12] The latter is computed using second order dipole approximation. These are not scaled hydrogenic results but use the same wavefunctions as in the atomic data calculations.

2.2.2. *Bound-Free Cross Sections.* These cross sections are computed explicitly for all levels with angular momentum quantum number less than 5. For levels with principal quantum number greater than 5, the configuration term structure is neglected. The resulting bound-free cross sections have compared well with experiments even for neutral atoms[6] except when configuration interaction effects are important. Such effects, however, are expected to be very small for solar interior calculations. For the remaining levels we use scaled hydrogenic cross sections.

2.2.3. *Free-Free Absorption.* Since the parametric potential model provides good results for bound-free cross sections where both bound and scattering states are required, we assume that the parametric potential method will also be valid for free-free calculations where only scattering states are necessary. We compute explicitly the dipole matrix elements except in some limiting regions where simpler approximations are valid. For example, for small photon energies elastic scattering cross sections are useful and easy to compute. Plasma screening effects are introduced into the electron-ion interaction. These effects are described by Rogers[6] and reduce to the Debye-Huckel[13] result for weakly coupled plasmas. This approach provides two improvements over calculations using Coulomb Gaunt factors. The first is corrections due to the plasma screening at small photon energies. The second is corrections at large photon energies where the scattering electron can penetrate the bound electron orbits and see a higher effective charge.

2.3 Scattering

The treatment of photon scattering from free electrons follows Boercker[14]. There, the transport cross section is computed including the many electron effects; that is,

$$\sigma_{sc}(k) = \sigma_t \int_{-1}^{1} d(\cos\theta) \, [1+\cos^2\theta][1-\cos\theta] \, S(k) \tag{12}$$

where

$$k = \frac{4\pi}{\lambda} \sin\left(\frac{\theta}{2}\right) \tag{13}$$

$$S(k) = 1 + h_{RPA}(k) + h_x(k) \tag{14}$$

where θ is the photon scattering angle, λ the photon wavelength, with h_{RPA} and h_x are the electron-electron correlation functions in the Random Phase Approximation and first order exchange, respectively.

3. Experiments

As mentioned earlier, absorption experiments are difficult. Firstly, they require LTE conditions which are difficult to obtain in laboratory plasmas. Independent measurements of the plasma density and temperature are necessary and care is require in order to avoid circular arguments. For comparison with theoretical calculations error bars would be quite useful in helping to discriminate between various models. Finally, the experiments will be restricted to low densities.

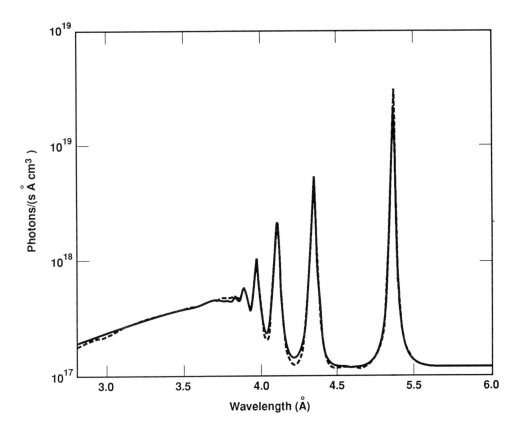

Fig. 1. Emissivities at $N_e=1.8 \times 10^{17}$ electrons/cm^3 and T=1.0×10^4K.
Dashed line: Wiese et al.[15]. Solid line: OPAL

In spite of all the difficulties there are some experiments with hydrogen. In Figures 1 and 2 we compare our results for the emissivity with experimental data by Wiese et al.[15]. In looking at the comparison one should note that the experiments were not in strict LTE and that the chosen plasma conditions were obtained by the experimentalist assuming LTE conditions. Similar experiments and comparisons have been done by other researchers[16,17].

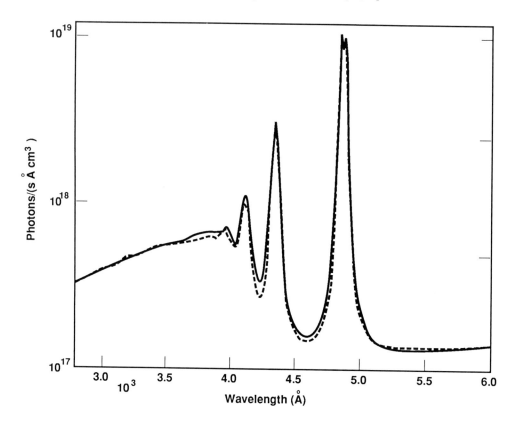

Fig. 2. Same as Fig. 1 with $N_e = 9.3 \times 10^{17}$ electrons/cm^3 and T= 1.33×10^4.

Unfortunately, the Wiese et al.[15] experiment is not a sensitive test for EOS or occupation numbers, but rather it is an excellent test for Stark broadening of Hydrogen spectral lines as originally intended by the authors. The reason is that Saha-type EOS formulations with principal quantum number cutoff above the Inglis-Teller[18] limit will reproduce reasonably well the experimental data[16]. The difficulty is in reproducing the line spectra near threshold where a careful theory, which is not presently available, would need to Stark mix many states with different quantum numbers plus the continuum. What has been done with some success in the past are phenomenological methods[16,17] that mimic the line broadening of overlapping lines. These methods are in effect "smoothing" procedures which conserve the oscillator strengths. Just as important, they restrict the line radiation to the region near threshold while standard line broadening theories have unphysical broadening to regions far from line center leaving a spectral window.

4. Opacity Results for the Sun

The Rosseland mean opacities are compared with the results from the Los Alamos Astrophysical Library published in Bahcall et al.[19]. The comparisons are presented in Table 1. The assumed solar mixture (labelled Ross-Aller'76) was obtained from Table IV of Ref. 19 and the Los Alamos Rosseland means from Table VI. We see no significant differences until the matter temperature drops below 7×10^6 K. As the temperature continues to drop, the opacity difference increases to approximately 18% at 1×10^6 K. Such increases are significant to Helioseismological data.

The opacity is also sensitive to uncertainties in the element mixture. For comparison, we did our calculations assuming the same temperatures and densities but assumed a more recent Aller[20] element abundance (labelled Aller'86). This mixture is richer in both Fe and Ne. In Table 1 one can see that near the solar center where Fe is important to the opacity there is a few percent increase in the Rosseland mean due to changes in the mixture when using the OPAL code. Similarly, near 3×10^6 K Ne is important and there is an 8% increase.

TABLE 1. Comparison of OPAL and Los Alamos[19] Rosselasnd mean opacities (cm^2/g) where X=0.35 and Z=0.0179.

T $(\times 10^6 K)$	ρ (g/cm^3)	Ross-Aller'76		Aller'86
		Ref.19	OPAL	OPAL
15.7	135.0	1.18	1.17	1.21
12.8	73.4	1.34	1.33	1.37
11.3	50.5	1.45	1.44	1.47
10.0	35.0	1.61	1.58	1.60
7.0	12.0	2.54	2.53	2.55
4.5	3.19	5.85	6.23	6.39
3.0	0.945	13.6	14.6	15.8
1.8	0.204	31.0	35.4	36.5
1.0	0.035	49.9	59.3	57.3

The differences between the results of the two codes are mostly due to the improved atomic physics package in OPAL. We have done some comparisons of the occupation numbers, N_{ijk}, but found small differences on the most relevant levels. There are also differences in the treatment of line broadening since the lorentz widths are not the same in the two codes. The subject of line broadening, in particular line wings, remains an open question in opacity calculations. The problem of line broadening of spectral lines in multi-electron ions has just begun to be explored.

5. Conclusions

The differences in the numerical results between the two codes may be interpreted as a measure of the uncertainties in opacity calculations for solar interior. However, a word of caution is necessary. Even though the codes were independently developed and several of the approximations in the previous calculations were removed, there are no experiments to guide the theory for such extreme temperature and density conditions. In the past, single particle representations of many-body problems have been successful, but usually experiments were necessary in order to pinpoint the important physics of the particular problem. Perhaps with very accurate astronomical experiments, such as the p-mode oscillation measurements, it will be possible to improve our understanding of EOS and absorption properties of very hot, dense matter.

6. References

1. N.R. Simon, Ap.J.(Letters)**260**, L87(1982).
2. G.K. Andreasen, Astron.Astrophys.**201**, 72(1988).
3. A. Cox, "Solar Opacities Constrained by Solar Neutrinos and Solar Oscillation," this volume.
4. C. Smith, J. Foster, and S. Davidson, in 6th APS Topical Conference on Atomic Processes in Hot , Dense Plasmas, September 28-October 2, 1987, Santa Fe, NM.
5. A. Cox, Stars and Stellar Structure, ed. L. Aller and D. McLaughlin(University of Chicago Press, Chicago) 1965; T.R. Carson and H.M. Hollingsworth, Mon. Not. R.Ast. Soc.**141**,77(1968); W.F. Huebner, Physics of the Sun, ed. P.A. Sturrock(Reidel Publishing)1986.
6. F.J. Rogers, Ap.J.**310**, 723(1986); and references therein.
7. W. Ebeling, D. Kraeft, and D. Kremp, Theory of Bound States and Ionizing Equilibrium in Plasmas and Solids,(Akademei-Verlag, Berlin)1976.
8. L. Landau and E. Lifshitz, Statistical Physics, (Pergamon Press, London)1958.
9. F.J. Rogers, Phys.Lett.A**61**,358,(1977); D. Bolle, Nucl.Phys.A**353**, 377c(1981).
10. F. Rogers, B. Wilson, and C. Iglesias, Phys.Rev.A**38**, 5007(1988).
11. C. Iglesias, F. Rogers, and B. Wilson, Ap.J.(Letters)**322**, L45(1987).
12. R.W. Lee, J.Quant.Spectros.Radiat.Transfer, 561(1988); H.R. Griem, Spectral Line Broadening by Plasmas(Academy Press,New York)1974.
13. P. Debye and E. Huckel, Z.Phys.**24**, 185(1923).
14. D.B. Boercker, Ap.J.(Letters)**316**, L95(1987).
15. W. Wiese, D. Kelleher, and D. Paquette, Phys.Rev.A**6**, 1132(1972).
16. W. Dappen, L. Anderson, and D. Mihalas, Ap.J.**319**, 1954(1987).
17. L. D'yachkov, G. Kobzev, and P. Pankratov, J .Phys.B**21**, 1939(1988), and references therein.
18. D. Inglis and E. Teller, Ap.J.**90**, 439(1939).
19. J. Bahcall, W. Huebner, S. Lubow, P. Parker, and R. Ulrich, Rev.Mod.Phys.**54**, 767(1982).
20. R. Aller, Spectroscopy of Astrophysical Plasmas, ed. A. Dalgarno and D. Layner (Cambridge University Press, Cambridge)1986.

CONSTRAINTS ON SOLAR MODELS FROM STELLAR EVOLUTION[1]

I. IBEN, JR.
The Pennsylvania State University
525 Davey Laboratory
University Park, PA 16802

ABSTRACT. This paper constitutes a brief and exceedingly elementary review of the consraints imposed on models of the solar interior by comparisons between theoretical models of evolving stars and the observed characteristics of stars, including the Sun.

1. Introduction

One of the dominant lessons to be learned from the science of stellar evolution is that, with a moderately simple set of approximations to the physics of matter and radiation under conditions expected in stellar interiors, one can account extremely well for the distribution in the Hertzsprung-Russell diagram of single stars in the field and in clusters which vary in heavy element composition by several orders of magnitude and range in age from 10^6yr to 10^{10}yr. The approximations include: spherical symmetry, no rotation, no acceleration in the pressure-gravity balance equation, initial homogeneity in composition, uniform mixing in convectively unstable regions, radiative and conductive flow in the diffusive approximation, no mixing in radiative regions, radiative opacity in the hydrogenic approximation, electron conductivity controlled by Coulomb scattering from randomly distributed nuclei, energy generation and particle transformation rates which are simple extrapolations of laboratory cross sections, the simplest of equations of state (radiation in a black body distribution, perfect gas for ions, perfect gas for free electrons with degeneracy allowed for, Saha equation with single term partition functions in regions of partial ionization, Coulomb forces taken crudely into account at high density if so desired), neutrino-loss rates from a well established theory of weak interactions, and an extremely simple Eddington-approximation surface boundary condition. The manner in which stars of

[1]Supported in part by the NSF grant AST 88-07773

G. Berthomieu and M. Cribier (eds.), Inside the Sun, 91–100.
© 1990 *Kluwer Academic Publishers. Printed in the Netherlands.*

near solar mass and initial composition evolve from formation to the final cooling phase following the exhaustion of all exploitable fuels is a very solidly founded consequence of comparisons between the observations and models of evolving stars which have been constructed by using the enumerated approximations. Such stars (1) contract onto the core hydrogen-burning main-sequence band, (2) live within the main-sequence band until having exhausted approximately 10 percent of their hydrogen fuel, (3) leave the main-sequence band to become giants with electron degenerate cores, (4) ignite and burn helium in the core as clump stars, (5) become asymptotic giant branch stars after exhausting helium in the core, (6) eject a shell of material which is caused to fluoresce by emanation of energetic photons from the compact but luminous remnant, and (7) finally evolve into white dwarfs.

In short, there is every reason to believe that models of the Sun which are constructed by using the same input physics that successfully accounts for many of the observable features of single stars in clusters and in the field are reasonably accurate representations of the interior structure of the Sun.

2.   The Sun as a Main-Sequence Star

From the fact that one sees emission and absorption lines in stellar spectra, one ascertains that matter at stellar surfaces is gaseous and that the distribution of radiated energy with respect to wavelength is very similar to that of a theoretical blackbody. From the location of the peak in the observed distribution, one may estimate a characteristic surface temperature. For the Sun, this characteristic temperature is of the order of 6000K. In the case of the Sun, one may also make use of the known angular diameter of the solar disk and of the mean temperature at the earth's surface to obtain another estimate of the temperature at the photosphere; this estimate is in agreement with the first one. Since energy flow requires a temperature gradient, one may infer that the temperature increases inward from the surface. Assuming that matter in the interior is gaseous, that the equation of state is that of a perfect gas, that the pressure gradient everywhere balances the gravitational acceleration, that the matter in the interior is of nearly homogeneous composition, and that the dominant element species are essentially completely ionized, one deduces that a star's central temperature is related to its mass M, radius R, mean molecular weight μ of constituent particles, and to fundamental constants (k = Boltzmann's constant, G = Newton's gravitational constant G, and M_H = mass of the hydrogen atom) by

$$kT \sim (GM/R)\ \mu M_H \sim 15 \times 10^6 K \times (M/R), \qquad (1)$$

where, on the far right, mass and radius are in solar units and μ has been chosen to accommodate the fact that by far the dominant species of element existing at stellar surfaces are hydrogen and helium. More sophisticated versions of this relationship may be constructed by taking into account the effect on the equation of state of radiation pressure, electron degeneracy, and Coulomb forces between charged particles, but, for stars of near solar mass, equation (1) turns out to be an excellent approximation.

From stars near enough to permit an estimate of distance by trigonometric parallax, one may translate the observed energy flux from the star into an estimate of the intrinsic bolometric luminosity L and use the expression

$$L = \sigma T_e^4 \times 4\pi R^2 \tag{2}$$

to estimate the stellar radius R.

The vast majority of nearby stars (when expressed in terms of space density = stars pc^{-3}) lie in a very tightly constrained main-sequence band in the Hertzsprung-Russell diagram (log L versus log T_e). The Sun lies comfortably within this band. From nearby stars in binary systems, one finds that there is a tight correlation between luminosity and mass for main-sequence stars. In particular, for stars of the Sun's mass or larger,

$$L \sim M^4, \tag{3}$$

where both L and M are in solar units. Along the main-sequence band mass and radius are related very roughly by

$$R \sim M^{0.7}. \tag{4}$$

Having an estimate of the mean interior density, $\rho \sim M/(4\pi R^3/3)$, and a first estimate of interior temperatures, one may verify that the assumption of complete ionization for the dominant element species, hydrogen and helium, over most of the interior is thoroughly justified for main-sequence stars, including the Sun.

Equations (1), (3), and (4) imply that the local rate of energy generation in the interiors of main-sequence stars more massive than the Sun increases very steeply with increasing temperature. In fact, these equations show that, within main-sequence stars of mass between 1 and 10 times the mass of the Sun, the mean rate of energy generation varies as about the 14th power of the central temperature. This dependence is in fair agreement with that expected for the CN cycle reactions, cross sections for which are measured in the laboratory; extrapolation to relevant stellar energies is quite straightforward.

For the Sun and less massive stars, the CN cycle reactions do not produce energy at a rate sufficient to account for the observations. However, there is strong theoretical and experimental basis for believing that the pp-reaction chains take up the slack. Unfortunately, the basic initiating reaction for these chains, the pp reaction itself, is not directly testable in the laboratory. Instead, one must rely on a calculation based on the theory of weak interactions normalized to fit with experimental results from β-decay experiments. Support for this reliance comes after constructing detailed main-sequence models which show that luminosity and mass are related in exactly the same way as in real main-sequence stars of the Sun's mass or less, viz.,

$$L \sim M^{2.2}. \tag{5}$$

Further insight into the structure of main-sequence stars of mass as large or larger than the Sun is obtained by supposing that energy is carried outward to the surface from the region of energy production predominantly by radiative diffusion. The effective cross section between photons and ionized atoms at any point is summarized in the opacity κ, which is a function of the temperature T, density ρ, and composition. The main sources of opacity are scattering by free electrons ($\kappa_e \sim$ 0.2-0.35 cm^2 g^{-1}), absorption by free electrons in the Coulomb potential of ionized nuclei, photoionization, and absorption between bound atomic levels (κ[free, bound-free]) $\propto \rho/T^{3.5}$). The photoionization and bound-bound contributions are strong functions of the abundances of heavy elements

The flux of energy carried by radiation is related to local conditions by

$$L(r)/4\pi r^2 = -(1/3) \ c \ \partial aT^4/\partial r \ 1/\kappa\rho, \tag{6}$$

where c is the velocity of light and a = 4σ/c is proportional to the Steffan-Boltzmann constant σ. Here, L(r) is the energy generated within a sphere of radius r and ρ and T are the local density and temperature, respectively. Combining equations (1) and (6) gives

$$L \sim A \ \mu^4 \ M^3 / \kappa, \tag{7}$$

where κ is a "mean" interior opacity and A is a combination of fundamental constants G, σ, c, and k that is of the order of unity when total stellar luminosity L and stellar mass M are in solar units. Comparison between equations (7) and (3) shows that the mean opacity is of the order of unity and that all three opacity sources contribute comparably to the mean opacity. The value of mean κ deduced in this way bolsters the estimate from an analysis of spectral lines that hydrogen and helium are the dominant element species in main-sequence star interiors (Eddington thought that iron was one of the main constituents of stars, as in the earth).

In the outer layers of low mass model stars, including solar models, the opacity in regions where hydrogen and helium are ionized is so large, that energy is carried through these regions predominantly by convection. The lower the mass of the star, the larger is the fraction of the model star's mass contained in a convective envelope. Models less massive than about one-third of the Sun's mass are completely convective. It is this change in mode of energy flow with decreasing mass, in conjunction with the temperature dependence of the pp chains, that is responsible for the difference in the mass-luminosity relationship for stars more massive than the Sun (equation [3]) relative to that for stars less massive than the Sun (equation [5]).

3. The Sun as an Evolved Star

Incorporating the full panoply of physics just sketched, one may construct models whose observable properties change in consequence of nuclear transformations in the deep interior. These transformations bring about a reduction in the number of free particles per gram, and, since pressure is proportional to the number of free particles, a balance between pressure forces outward and gravitational forces inward can be maintained only by contraction of the region in which the transformations are occurring most rapidly. With contraction comes heating, and with heating comes an increase in the rate of nuclear reactions. The increased flux of radiation through outer layers forces these layers to expand. Thus, as hydrogen is converted into helium in the deep interior, the radius and luminosity of the model star increase. The model star evolves through the observational main-sequence band defined by the bulk of nearby stars. Once hydrogen is exhausted over about 10 percent of the mass of the model, the model evolves ever more rapidly into the region in the Hertzsprung-Russell diagram occupied by observed red giants (such as, e.g., Capella).

There are two clusters in the disk of our Galaxy which contain stars whose positions in the Hertzsprung-Russell diagram beautifully confirm the theoretical picture painted by the evolutionary models of stars of mass near the Sun's mass and of surface composition near that of the Sun (it must be emphasized that the confirmation preceded the models). The two clusters are M67 and NGC188. Each cluster is far enough away from us that the distances between stars in each cluster are quite small compared with the distance of the cluster; therefore, the relative bolometric luminosity of one star relative to another in each cluster is known quite accurately. In each cluster, all of the stars less luminous than the Sun lie in the Hertzsprung-Russell diagram within the main-sequence band defined by nearby stars. In each cluster there is a well populated giant branch extending to luminosities several orders of magnitude brighter than the Sun at surface temperatures much less than the Sun's, and there is a sprinkling of stars between the main sequence and the giant branch. The luminosity at which the main sequence "turns off" towards the giant branch is larger in M67 than in NGC188 by a factor of about 2.5. The comparison with models proceeds as follows.

From the luminosity of the "turnoff point" one may estimate (using, e.g., equation [3]) that the mass of a star leaving the main sequence in M67 is about $1.25 M_\odot$ and that the mass of a star leaving the main sequence in NGC188 is about $1 M_\odot$. Knowing from model calculations that departure from the main sequence occurs when hydrogen has been exhausted over 10 percent of the mass of the model, noting that the time scale for burning a mass δM of hydrogen at luminosity L is proportional to δM/L, and making use of equation (3), we infer that stars near turnoff in NGC188 are older than stars near turnoff in M67 by a factor of about two. Knowing that approximately 24 MeV are liberated when four protons are converted into an alpha particle, we can then infer that stars near turnoff in M67 are approximately 5×10^9 yr old and that stars near turnoff in NGC188 are near 10^{10} yr old. One may construct arguments to show that the timescale for star formation in a cluster is far shorter than these estimated cluster ages. Then, by evolving models of different masses for 5×10^9 yr and plotting their positions in the Hertzsprung-Russell diagram one may construct a synthetic model cluster locus which may be compared with the locus defined by stars in M67. The agreement is impressive, including a distinct gap in the density of stars near cluster turnoff corresponding to a phase of rapid contraction in the models when hydrogen is exhausted over a finite region about the center which was convectively mixed during the main-sequence phase. The agreement of model cluster loci for an assumed age of 10^{10} yr with the locus defined by stars in NGC188 is also impressive, including the absence of a gap in stellar density near cluster turnoff and a near constant luminosity of stars between cluster turnoff and the base of the giant branch.

The lesson for the Sun is that, at an age of 4.6×10^9 yr (as determined by radioactive dating using ^{238}U and ^{206}Pb), it is only halfway through the main-sequence phase and it does not possess a convective core.

4. Solar Models, Solar Neutrinos, and Solar Oscillations

Having argued that there is every reason to expect that solar models constructed with standard input physics are not grossly in error, it remains to address the persistent discrepancy between predictions of "standard" solar models and results of solar neutrino experiments. The relation between a predicted flux, central temperature, assumed composition parameters, and choices of relevant nuclear cross section factors may be approximated by

$$F_\nu \sim 0.91 \text{ SNU y } T_c^{6.7} S_{34} (S_{11}S_{33})^{-0.5}$$

$$\times (1 + 8.1 \text{ y}^{0.6} T_c^{13.6} S_{17} S_{e7}^{-1}), \tag{8}$$

where F_ν is a predicted neutrino counting rate in the Davis Homestake mine experiment in units of 10^{-37} counts per ^{37}A atoms (= 1 SNU), T_c is the central temperature of the Sun in units of 15.23×10^6K , and S_{11}, S_{33}, S_{34}, S_{17}, and S_{e7} are center of mass cross section factors for the p(p,e$^+$)d, ^{3}He(^{3}He,2p)^{4}He, ^{3}He(^{4}He,γ)^{7}Li, ^{7}Be(p,γ)^{8}B, and ^{7}Be(e$^-$,ν)^{7}Li reactions in units of the most likely values for these cross sections. The parameter y is Y/0.25, where Y is the initial helium abundance by mass. The first term in expression (8) is due to neutrinos emitted in the reaction $^7Be(e^-,\nu)^7Li$ and the second term is due to neutrinos emitted in the $^8B(e^+,\nu)^8Be^*$ reaction. All other contributing reactions have been ignored. It is further the case that $T_c \propto \mu\ S_{11}^{-1/7} \propto y^{0.245}\ S_{11}^{-1/7}$ (see equation [1]), so that

$$F_\nu \sim 0.91\ SNU\ y^{2.65}\ S_{34}\ S_{33}^{-0.5}\ S_{11}^{-1.46}$$

$$x\ (1 + 8.1\ y^{3.97}\ S_{17}\ S_{e7}^{-1}\ S_{11}^{-1.95}). \qquad (9)$$

Note that the abundances of elements heavier than helium do not appear explicitly in either equation (8) or (9). The reason for this lies in equation (7), from which it is clear that, for a fixed luminosity, a decrease in the estimated opacity (which is in turn related to the assumed abundances of elements heavier than helium) requires a concomitant decrease in the molecular weight (i.e., a decrease in the abundance of helium). The particular normalization in equations (8) and (9) follows from a specific assumed relationship between opacity and the abundances of elements heavier than helium.

The appropriate value of Y is not directly known from the observations. In the solar wind, the ratio of ^4He to ^1H is on the average about 0.05, or only half of what is found at the surfaces of main-sequence stars where accurate analyses can be made. Presumably, the acceleration mechanism which drives the solar wind preferentially accelerates protons, whose charge to mass ratio is twice that of alpha particles. Thus, the value of Y relevant to the interior is actually found by constructing theoretical solar models which are constrained to be of age 4.6×10^9 yr and to have an interior heavy element composition equal to that estimated from photospheric spectral line strengths. With best estimates of center of mass nuclear cross sections and of abundances of heavy elements, the abundance by mass of ^4He in solar models is $Y \sim 0.25$-0.26, only ~ 0.03 larger than the Y predicted by Big Bang models. With this value of Y, predicted solar neutrino fluxes are typically 6-8 SNU.

From equation (1) it is evident that a decrease in the initial abundance of ^4He means a decrease in the calculated central temperature and, from equation (8) this means a decrease in the predicted neutrino flux, all other things being equal. However, even reducing Y to the Big Bang value, predicted values of F_ν are significantly larger than the observed long-term mean of $F_\nu \sim 2.3$ SNU. In some quarters, this has been taken as evidence that the mass of at least one type of neutrino (the electron neutrino, the muon neutrino, or the postulated tau neutrino) is not zero and that oscillations between the (two or three types of) neutrinos are induced by the interaction between the electron neutrino and the electrons in the Sun as the composite neutrino makes its way outward through the Sun (the Mikheyev-Smirnov-Wolfenstein process). This could account for the discrepancy between the predicted counting rate and the observed average counting rate.

However, the counting rate from the Davis experiment is not

constant, and has varied between essentially zero and about 4 SNU. Davis has suggested that the variation over the first two decades of the experiment is not necessarily stochastic and may be anticorrelated with the solar sunspot number, implying that the Sun's magnetic field may modify the characteristics of the electron neutrino. A neutrino magnetic moment is inconsistent with Majorana theory, except in the form of a "transition" magnetic moment; but theories are only theories until confirmed by experiment. It is significant that the first estimate of a solar neutrino flux from the Kamiokande experiment is consistent with the current counting rate from the Davis experiment, viz., ~ 4 SNU. If the two experiments continue to give consistent results and the variation with the solar cycle persists, there is a strong possibility that it is the physics of the neutrino and not a shortcoming of solar and stellar models that is at the bottom of the mystery.

Another component of standard stellar physics which explains the properties of a large class of stars and which is also relevant to the Sun is the theory of stellar pulsations. By allowing for dynamic motions in the linear approximation, it may be shown that, throughout most of the HR diagram, models are stable against acoustical pulsation in radial as well as non-radial modes. There are also two well defined boundaries to the red of which models are unstable to pulsation in either the fundamental radial mode or in the first overtone radial mode; the blue edges of these boundaries coincide precisely with the blue edge of a narrow strip in the HR diagram defined by large amplitude classical variables -- Cepheids, W Virginis stars, BL Her and RR Lyrae stars, delta Scuti stars, Am variables, and the white dwarf ZZ Ceti variables. The Sun does not lie within this strip and it is interesting to note that, as of this date, there do not appear to be any substantiated indications that the Sun is pulsating in radial modes (the fundamental radial mode should have a period between 55 and 65 min, and the first overtone should have a period of about 40 min). On the other hand, the Sun does oscillate in non radial modes at frequencies near 3 mHz; continuous excitation by coupling between these modes and turbulent convective motions (which show persistence time-scales of the order of 5-20 min) appears to be necessary to maintain the oscillations. An analysis of frequency splittings allows one to learn something about the variation of angular velocity with depth and current interpretations of the observations suggest that the Sun does not rotate significantly more rapidly in the interior than at the surface. Arguments about the existence of exotic particles such as WIMPs and about the accuracy of opacity estimates based on comparison between the observed mode structure and the mode structure given by detailed solar models remain unconvincing.

5.   What the Sun Teaches us about Other Stars

Thus far, an attempt has been made to show that a fairly simple theory of single star evolution gives a good account of the observed characteristics of real stars and that this implies that theoretical models of the Sun's structure are probably a very good first approximation.

The impression one might form is that rotation, including rotationally induced mixing of various sorts, magnetic fields, surface winds, and the surface activity which occupies the attention of most of the solar physics community are minor perturbations from the standpoint of the overall, global evolution of most stars, including the Sun.

However, when one looks more closely at stars other than the Sun, there are a number of situations which can most easily be understood only by invoking physical processes which have been left out in the simple picture, but which the solar experience tells us must be playing a role in the evolution of real stars, a goodly number of which are rotating far more rapidly than the Sun. In the Sun, rotation, magnetic fields, convection, acoustic and Alfen-wave energy transfer all play a role in establishing a corona and supporting a wind which, although it currently carries away mass at an insignificant rate, carries away angular momentum at a rate which, when used as a normalization in a somewhat heuristic theory of magnetic braking, predicts very high rates of angular momentum loss for rapidly rotating stars which also sport convective envelopes. Direct evidence for this type of magnetic braking comes from a comparison of typical $< v \sin i >$ values for stars of near solar mass in the Pleiades (of age about 10^8 yr) with $< v \sin i >$ values for solar mass stars in the Hyades (of age about 10^9 yr) and with the Sun's equatorial rotation velocity. To a reasonable approximation $< v \sin i > \propto (age)^{-1/2}$, as first pointed out by Skumanitch. Several quantitative indicators of magnetic activity follow this same relationship and one can therefore construct a heuristic algorithm for estimating the rate of angular momentum loss by a "magnetic stellar wind" (MSW) from solar type stars as a function of rotation rate. This algorithm has proved very useful in understanding the evolution of close binary systems in which at least one of the components is cool enough to have a deep convective envelope and shows evidence for surface magnetic activity. The MSW carries away angular momentum from the cool star, but tidal torques keep it spinning at a rate close to the orbital frequency; thus, orbital angular momentum decreases, forcing the stars to come closer together. This mechanism appears to be a major factor in forcing R CVn stars to evolve into Roche-lobe contact, following which the system is transformed into an Algol binary. It appears also to be a major driver of mass transfer in cataclysmic variables and in Algols and to be a mechanism whereby components in contact systems such as W UMa stars are forced to merge into single stars. This latter process offers an explanation for the presence of blue stragglers in old clusters which also contain contact binaries. In summary, a process not normally included in the simple theory of single star evolution plays a central role in the evolution of some binary stars.

Finally, it must be confessed that the simple theory has demonstrable shortcomings and that, in several cases, these are not minor shortcomings. There are mass-loss processes other than a MSW which are not included in the simple theory and which are known observationally to play an exceedingly important role in the evolution of single stars much brighter than the Sun. Mass loss from initially very massive main-sequence stars can be strong enough to convert such stars into Wolf-Rayet stars, which reveal at their surfaces material that has been highly processed by nuclear burning in an earlier phase of evolution. The mechanism for mass loss is possibly the transfer of photon momentum to atoms by resonance absorption from the continuum. Mass loss from bright and cool AGB stars terminates the AGB phase, converting the AGB star into a compact remnant which emits photons that cause the ejected material to fluoresce as a planetary nebula. The mechanism for mass loss in this case is possibly a hydrodynamic instability which occurs following a helium shell flash.

Mixing processes not included in the simple theory are known to occur. For example, the abundances of CNO elements and of light elements such as lithium and beryllium at the surfaces of low mass red giants are not quantitatively in accord with the expectations of the simple theory and suggest some form of extra mixing through formally radiative regions during both the main-sequence phase and the red giant stage. The lack of an extended giant branch in the Hyades cluster, suggests that stars only 1.6 times the mass of the Sun do not develop electron-degenerate cores before they ignite helium at their centers, in contradiction with the simple theory which suggests that stars of mass less than about $2M_0$ should develop such a core. The discrepancy can be resolved if it is supposed that matter within the formal convective core of stars of mass larger than $\sim 1.2M_0$ is mixed outward beyond the edge of the formal core by some form of convective overshoot, thus forcing the star to mimic a more massive star (as modelled by the simple theory which does not take convective overshoot into account).

In summary, although the simple theory of stellar evolution does an excellent job of describing qualitatively and often quantitatively many aspects of the global evolution of stars, there are situations in which the theory is inadequate. Despite the known shortcomings, it is the view of this reviewer that these shortcomings are not such as to invalidate the first-order picture of solar evolution which the simple theory provides.

CONVECTION IN THE SUN

JOSEP M. MASSAGUER
Departament de Física Aplicada
Univ. Politécnica de Catalunya
08034 Barcelona
Spain

ABSTRACT. The present knowledge of the dynamics and structure of the solar convection zone is reviewed with the aim of checking current assumptions and conjectures against laboratory experiments and numerical modeling of thermal convection. Buoyancy is the only forcing considered. Rotation and magnetic fields are explicitly avoided. Nor are departures from planar geometry considered , except as regards large scale structures. Local theories are reviewed in section §2, hydrodynamic models in §3, non-local theories in §4, the global structure of the convection zone is discussed in §5 and the flow patterns in §6.

1. Introduction

The solar convection zone is the site of very rich dynamics, the intricacies of which make it difficult to look through it into the deeper interior. Therefore, solar convection theory has often been approached with the avowed aim of obtaining a recipe for computing gradients and fluxes in places where a radiative gradient would be unstable. We shall take this view in the first part of the present review. Yet the solar convection zone is still the site of a highly structured magnetohydrodynamic flow, where buoyancy is thought to be one of the dominant external forcings. This view will be taken later on and buoyancy induced solar structures will also be reviewed.

The purpose of the present paper is to examine the solar convection zone as a whole, with laboratory and numerical experiments in mind. We shall attempt to stress similarities as much as possible and, as a consequence, we shall avoid discussing any process dominated by forcings other than buoyancy. Rotation and magnetic fields will not be considered at all, and the geometry will be assumed to be planar unless stated otherwise. However, as most of the laboratory experiments and numerical simulations in thermal convection are extremely dependent on details irrelevant to a stellar interior - the shape of the container, the boundary conditions, etc.- any extrapolation of numerical results or laboratory measurements is to be looked at carefully.

2. On local theories of stellar convection.

Local theories are widely used for modeling thermal convection in stars, as they are the simplest available algorithms, though some attempts at modeling stellar convection zones with non-local theories exist and will be reviewed in §3 and §4. Mixing-length scaling is widely accepted and consistency with it is required for any model, no matter whether it is local or non-local. However, local models can be derived without any reference to phenomenology. We shall take this view because

G. Berthomieu and M. Cribier (eds.), Inside the Sun, 101–115.
© 1990 *Kluwer Academic Publishers. Printed in the Netherlands.*

detailed physics often masks more fundamental similitudes.

In order to summarize the theory, let us begin by writing

$$\partial_r T = - \mathcal{F}(p, m, L, T) \tag{2.1}$$

where $\{p, m, L, T\}$ are the four dependent variables of stellar structure: pressure, mass, luminosity and temperature. In order to produce a convenient frame for a local theory we can define the Nusselt number N as

$$N = \frac{F_T - F_A}{F_R - F_A}$$

where $F_T = - k\, \partial_T$ is the total flux, $F_R = - k\, \partial_r T$ would be the flux if this convection were artificially inhibited, and $F_A = - k\, \partial_A$ is the flux transported radiatively by the adiabatic gradient. We shall also introduce for later use the definitions

$$\partial_T = - k^{-1} \frac{L}{4\pi r^2} \tag{2.2a}$$

$$\partial_A = (1 - \gamma^{-1}) \frac{T}{p} \partial_r p \tag{2.2b}$$

where $k = 4acT^3 / 3\kappa\rho$. Using the previous definition of N, $\partial_r T$ can be written as

$$\partial_r T = \frac{N-1}{N}\partial_A + \frac{1}{N}\partial_T \tag{2.3}$$

In (2.3) the radiative and convective equilibria are attained, respectively, for $N - 1 \ll 1$ and $N - 1 \gg 1$. Therefore, we do not expect significant differences from any theory giving $N - 1$ very large values inside the convection zone and very small ones, or zero, outside it. Any significant difference between two models concerns only the intermediate values of N.

2.1 DIMENSIONAL ANALYSIS

In a stratified medium the Nusselt number can be written as $N = N(Ra, \sigma, Z)$ - see, for instance, Massaguer & Zahn (1980)- where Ra and σ are, respectively, the Rayleigh and the Prandtl numbers

$$Ra = \frac{g\, d^4 (\delta / T)}{(k/\rho C_P)\, \nu}[-(\partial_T - \partial_A)]$$

$$\sigma = \frac{\nu}{k/\rho C_P}$$

In these expressions each variable retains its usual meaning with d being the thickness of the convection zone and δ / T the thermal expansion coefficient. The parameter Z introduced above is defined as a measurement of the density contrast across the convective layer, say $Z = d/H_P$ where $H_P = 1/\partial_r(lnp)$ is the pressure scale height. The choice of H_P as a local scale height instead of, say,

$H_T = 1/\partial_r(lnT)$ is indeed arbitrary, but is consistent with the theory, as H_P can be explicitly written in terms of the independent variables of the problem. This choice still deserves more substantial criticism as hydrodynamic models take Z as a global variable (Spiegel 1965). In the present derivation, Z as well as N have been arbitrarily turned into local variables.

The definitions of Ra and Z imply a knowledge of d, which is not a variable of the problem and cannot be included in a local model. Thus we must expect R and Z entering N as RaZ^{-4}. There is no direct test for the RaZ^{-4} law, but it displays the tendency expected from numerical computations. The only exception is the so-called up-solutions in the modal computations by Massaguer & Zahn (1980), but these solutions seem to be very peculiar - see Hurlburt, Toomre & Massaguer (1984) for a discussion.

Stellar plasmas are almost inviscid and largely conducting fluids, hence their Prandtl number σ is much smaller than one. It has been conjectured, and is widely accepted that under such circumstances the large scale motion is to be independent of the molecular viscosity. From the previous assumptions the Nusselt number is to be a function of the product $\Lambda = Ra\,Z^{-4}\,\sigma$, which can be explicitly written as

$$\Lambda = \frac{g\,H_P^4\,(\delta/\,T)}{(k/\rho\,C_P)^2}[-(\partial_T - \partial_A)] \qquad (2.4a)$$

So, we can finally write $N = N(\Lambda)$. A local theory is expected to provide such a relationship. It can be derived from phenomenology, from experiments - either numerical or laboratory - or from turbulence theory, as will be discussed below.

Mixing-length theories take a slightly different point of view. They assume $N = N(\Lambda_*)$, where Λ_* is the local value of Λ defined as

$$\Lambda_* = \frac{g\,H_P^4\,(\delta/\,T)}{(k/\rho\,C_P)^2}[-(\partial_r T - \partial_A)] \qquad (2.4b)$$

Then, from (2.3) we obtain the implicit equation $\partial_r T = -\mathcal{F}(p, m, L, T; \partial_r T)$ instead of (2.1). As an example of such an equation we can mention the well known Bohm-Vitense's cubic equation.

In the limit $\Lambda \ll 1$ most models agree with $N - 1 \sim \Lambda^2$ or $N - 1 \sim \Lambda_*^2$. As both limits will be shown to be equivalent, this seems to be a well established asymptotic law, but this is a limit displaying poor interest as it implies $\partial_r\,T \simeq \partial_T$. For the opposite limit, $\Lambda \gg 1$, mixing-length theories assume $N \sim \Lambda_*^{1/2}$. This law is the result of forcing the large scale dynamics - say the heat flux - to be independent of molecular conductivity. Yet experimental results would agree better with a $\Lambda^{1/3}$ power-law. For a discussion on power-laws, the reader is directed to Spiegel's (1971) review, but it can be anticipated that the $\Lambda_*^{1/2}$ and $\Lambda^{1/3}$ power-laws are asymptotically equivalent. By using the alternative definitions (2.4a,b) the Nusselt number can be written as $N = \Lambda\,/\,\Lambda_*$, and for a given law $N = N(\Lambda)$ we obtain $\Lambda_* = f(\Lambda)$.

Let us now assume, for simplicity, $N - 1 = Q\Lambda^n$ with $n > 0$. Then, for $\Lambda \ll 1$ and $\Lambda \gg 1$ the limits $\partial_r T = \partial_R$ and $\partial_r T = \partial_A$ are recovered from (2.3), no matter what the value of n is. And, in addition

$$\Lambda\,/\,\Lambda_* = 1 + Q\,\Lambda^n \qquad (n > 0) \qquad (2.5a)$$

An alternative derivation with $N - 1 = Q\Lambda_*^n$ would give

$$\Lambda \, / \, \Lambda_* \; = \; 1 \, + \, Q \, \Lambda_*^n \qquad (n \, > \, 0) \tag{2.5b}$$

Each one of the equations (2.5) can be solved for Λ_*, and $\partial_r T$ can then be obtained as a function of ∂_T and ∂_A. Therefore, given a power-law in terms of either Λ or Λ_*, the converse can be obtained. In particular, the mixing-length asymptotic power-law $N - 1 \propto \Lambda^{1/2}$ can be written as $N - 1 \propto \Lambda^{1/3}$, thus making it possible to establish contact between laboratory experiments and stellar convection theory.

As shown by Gough & Weiss (1976) a couple of power-laws, one for small and one for large Λ values are as good as any local model. To be precise, a simple local model can be written as

$$N - 1 \; = \; \begin{cases} Q_1 \Lambda_*^2 & \Lambda_* \leq \Lambda_c \\ Q_2 \Lambda_*^n & \Lambda_* \leq \Lambda_c \end{cases} \tag{2.6}$$

where the Q coefficients embody the so-called mixing-length parameter - i.e.: the ratio of the local scale height to H_P -, n takes the value $n = 1/2$ for mixing-length theories, but here it is kept free to allow for departures from them, and Λ_c is imposed so as to make $N - 1$ continuous. In practice, Q_1 can be taken to be zero and the sole concern of a local theory is to decide the values of Q_2 and n. These values can be obtained from phenomenology (Bohm-Vitense 1958), calibrated from stellar evolution (Gough & Weiss 1976), inferred from laboratory measurements or even computed from turbulence theory (Canuto, Goldman & Chasnov 1987).

At this point, it is worth mentioning that recent experiments of convection in low-Prandtl-number fluids (Heslot, Castaing and Libchaber 1987) suggest departures from the $\Lambda_*^{1/2}$ asymptotic power-law. Such a departures are still not well understood, so they can hardly be included in a phenomenological model. And they provide a good reason for restating local theories on a well established frame.

3. Hydrodynamic models.

Hydrodynamic models are usually derived from the dynamic equations (continuity, momentum and heat) as low order moment equations, often with *ad hoc* closures. Statistical relationships obtained from numerical experiments, such as those from Chan & Sofia (1989) for compressible convection, can be very useful for setting these closures, though few experiments have been devised with this purpose in mind, and most of the closures have been conjectured from phenomenological arguments.

Recently derived hydrodynamic models can be found in Xiong (1979, Unno, Kondo & Xiong 1985), van Ballegooijen (1982), Eggleton (1983), Kuhfuß (1986) and Tooth & Gough (1988). All these models are consistent with mixing-length scaling, which in turn can be viewed as a low order closure model. All these models attempt to give a detailed description of the dynamics of the convection zone, with their final goal being to describe more than just the mean stellar structure. Magnetic fields, stellar pulsation, etc. fall within their scope, so they are non-local and time-dependent. Unfortunately, only a few of these models have been checked against real stellar modeling or laboratory experiments and this is the very test for the underlying hypotheses.

More systematic closure techniques, such as those used in meteorology or engineering have not yet become popular among astrophysicists, possibly because of the difficulties in modeling turbulence in stratified fluids but, also, because of the temporal and spatial resolution they require. The work done by Cloutman & Whitaker (1980) and Marcus, Press & Teukolsky (1983) is pioneering in this respect, but more remains to be done.

A recent work that deserves more substantial comment is the attempt by Canuto, Goldman & Chasnov (1987) to extend the small scale turbulence theory of Kolmogorov and Heisenberg in order to model the largest scales of the flow. Their work follows an early attempt by Ledoux, Schwarzchild and Spiegel (1961) - see also Spiegel (1962). The model provides a detailed description of the - steady - kinetic energy spectrum, thus establishing a close connection with the more sophisticated modern turbulence theories, such as the so-called direct interaction approximation (Kraichnan 1964) and the renormalization group techniques (Yakhot & Orszag 1986).

The master equation for this model is the energy balance, written in terms of the turbulence spectrum as

$$\int_{k_0}^{k} F(k')n_s(k')dk' = \nu_t(k) \int_{k_0}^{k} F(k')k'^2 dk$$

where k_0 is lower cut-off wavenumber, $n_s(k)$ is the growth rate for the mode k, $F(k)$ is the average kinetic energy per wavenumber and ν_t is the eddy viscosity, which can be formally written as

$$\nu_t(k) = \int_{k}^{\infty} \frac{F(k')}{n_c(k')}dk$$

where $n_c(k)$ is a correlation frequency.

In order to solve the system, $n_s(k)$ is known from the linearized dynamic equations - i.e.: the dispersion relation -, therefore allowing us to include any body force - i.e.: magnetic fields, rotation, etc.- and $n_c(k)$ is to be given as a closure. By assuming the small scale closure of Heisenberg and Kolmogorov to be valid everywhere, the universal law for the inertial range $F(k) \approx \varepsilon^{2/3} k^{-5/3}$ is recovered for the small scales. But this closure has been shown to be inconsistent for large scales, where the energy balance equation itself imposes $\nu_t(k_0) = n_s(k_0)/k_0^2$ (Canuto, Goldman & Hubickyj 1984). If instead of that closure, the following closure is assumed, $\nu_t(k) = \gamma n_c(k)/k^2$, both the large and the small scales can be modeled self-consistently.

From this model, the asymptotic law $N \sim QRa^{1/3}$ can be recovered, and the value of Q has been computed with remarkable accuracy for convection in water, so extending local theory in the most natural way. Therefore, the model provides a promising frame for discussing hydrodynamics in the asymptotic ranges that conform the stellar regimes, possibly including any external forcing. Results concerning astrophysical plasmas are not that good, as the computed mixing-length coefficient Q_2 in (2.6) is off by a factor of two if compared with the calibration done by Gough & Weiss (1976). This failure seems to be of a very fundamental nature, as it is associated with the difficulty of properly modeling shear processes.

4. Non-local theories of convection

Local models are well suited to describe homogeneous or slowly varying media, as invariance properties constitute their conceptual framework. Boundaries or turning points -say the edge of the convection zone $\partial_T = \partial_A$ - are the genuine elements for breaking such an invariance. Local theories could be extended to include fixed boundaries by, say, turning (2.1) into $\partial_r T = -\mathcal{F}(p, m, L, T; z)$, with z being the distance to the boundary, much as in Prandtl's boundary layer theory. However nothing similar can be done if boundaries are moveable, or their position is not known *a priori*. Also, extending local models for dealing with convection beyond turning points requires changing $\mathcal{F}$ into a functional.

One of the simplest ideas for turning a local model into a non-local one was that of Shaviv & Salpeter (1973). Consistent with mixing-length theory they assumed that any parcel of fluid can travel a length $\ell = H_P$ before losing its identity. In their model the dynamics was reduced to a balance between buoyancy driving and kinetic energy production while the parcel was flying a distance ℓ.

Marcus, Press & Teukolsky (1983) suggested that a more realistic model should, in addition, include shear instabilities as a destabilizing mechanism. Convective motion from the unstable region stirs the neighbouring layers, thus eroding the stable layer. In a shearing motion the erosion will progress until it reaches a front edge where no fluid parcel can be overturned against buoyancy by shear stresses. The stability for this edge is given by the Richardson criterion $Ri > 0.25$, with the Richardson number Ri defined as

$$Ri = \frac{\mathcal{N}}{\partial_r U}$$

where U is the shear velocity, r is taken across the edge and $\mathcal{N}$ is the Brunt-Vaissala frequency defined by $\mathcal{N}^2 = gT^{-1}(\partial_r T - \partial_A)$. As discussed by Marcus, Press & Teukolsky, if shear stresses are to be stabilized by buoyancy, $\mathcal{N}^2$ at the edge is to be large and positive. Therefore, the temperature must change abruptly in order to give a large $\partial_r T$ value. This model introduces a new length-scale λ, defined by $\partial_r U \simeq U/\lambda$, with the result that the penetration depth may depend on the eddy size. The smaller the eddy size the larger the penetration depth.

A simple method of modeling convection with penetration by taking into account buoyancy and shear stresses is to assume for the edge of the convective zone a horizontal plane from which motion starts as plumes rising from heated point sources. Plume convection is a phenomenon well known in meteorology (see Turner 1973 for a review) and plume models have been widely tested from experiments in order to model entrainment of ambient material into the plume itself, a process which is a difficult one to model.

In Schmitt, Rossner & Bohn's (1984) work the plume model serves only to estimate the penetration depth below a locally modeled convective zone. However, the numerical simulations by Hurlburt, Toomre & Massaguer (1986) show that plumes are generated at the upper boundary of the convection zone, cross the unstable layer and finally die in the lower stable layer after some penetration. Thus it seems more appropriate to consistently model the whole convection zone, including penetration, with plumes.

In its simplest form a plume is an axisymmetrical mass flow diverging from a fixed point. Buoyancy work and entrainment of external material into the plume conform its dynamics through the balance of mass, momentum and buoyancy flux.

For a stratified fluid we can write

$$
\begin{aligned}
\partial_r \left(b^2 \rho v \right) &= \quad 2\alpha b \rho v \\
\partial_r \left(b^2 \rho v^2 \right) &= \quad b^2 g \, \rho' \\
\partial_r \left(b^2 \rho v S' \right) &= \quad -b^2 \rho v \partial_r S \; + \; 2b\alpha \rho v S'_{ext}
\end{aligned}
\tag{4.1}
$$

where $b = b(r)$ is the radius of the plume's cross section, α is an entrainment coefficient, S is the mean entropy profile, S' stands for entropy fluctuations in the plume and S'_{ext} is the entropy fluctuation in the ambient fluid. The convective flux can be written as $F_c = \rho v T S' + k \partial_r T + F_{ext}$, where F_{ext} is the external heat flux. A convenient modeling of S', S'_{ext} and F_{ext} closes the problem. If there is no entrainment of material, $\alpha = 0$, (4.1) can be reduced to Shaviv & Salpeter's model, but now with the fluid parcels being born at the boundaries, from where they fly across the whole layer.

5. Numerical simulations and laboratory experiments

In the previous sections we have reviewed the most relevant attempts at modeling stellar convection. None of these models have been derived from first principles and some knowledge of the phenomenology was required for their closure. Neither laboratory experiments nor numerical simulations can be easily extrapolated to the stellar case but they provide useful information about some particular aspects of the problem such as convection in a compressible or stratified fluid, penetrative convection or convection in low-Prandtl-number fluids. We shall review the most relevant of these results below.

5.1 CONVECTION IN COMPRESSIBLE OR STRATIFIED FLUIDS

Considerable efforts have recently been put into modeling convection in highly stratified fluids. Unfortunately neither laboratory experiments nor measurements in the Earth's atmosphere can provide the information required. In laboratory convection compressible fluids behave as if they were non-stratified -i.e.: Boussinesq fluids- with $Z \ll 1$, and the situation is quite similar for the Earth's atmosphere, where convection is confined to a region of height $Z \approx 1$. Therefore, phenomenology for highly stratified convection relies on numerical simulation.

In a layer spanning a large density contrast between top and bottom, $\rho_{top}/\rho_{down} \ll 1$, the local scale height at the top of the layer is much smaller than the depth of the layer itself, thus requiring large spatial resolution in numerical codes. Requirements of spatial resolution in compressible convection may come from two different sources: diffusion lengths, either conductive or viscous, and local scale heights. The former can be, and must be, conveniently parameterized to fit into today's computers. The latter simply cannot. Therefore, the bottleneck for hydrodynamic modeling of solar type stars is density stratification, with diffusion scales parameterized so as to be keep them smaller than local scale heights.

A recent attempt at modeling convection conveniently in a highly stratified layer is the three-dimensional numerical simulation of Chan & Sofia (1986, 1989). The latter assumed for the diffusion term Smagorinski's (1963) recipe, which is a well known parameterization procedure for incompressible fluids and one that can be used in the present context if the eddy diffusion lengths are taken to be much smaller than any local scale height. A very systematic discussion of subgrid scale turbulence with astrophysics in mind can be found in Marcus (1986). In which he also reports on the most dangerous flaws for unresolved numerical simulations. The main goal of Chan & Sofia's work was to assess or repudiate hypotheses underlying mixing-length theory or other closures used in building hydrodynamic models. Their results support most of the physics in mixing-length theory.

A major consequence of the stratification is the asymmetry between the upper and lower layers. As shown by Hurlburt, Toomre & Massaguer's (1984) two-dimensional numerical simulations, the asymmetry between both layers is not just a matter of scaling. Rescaling the layers by a local scale height, say H_P , would not be of any help. In the cellular flow which they obtained, the centre of the cell goes down while the density contrast is being increased, -i.e.: the separation between the upper streamlines becomes wider at the expense of the lower ones- thus becoming wider in places where H_P is smaller. Therefore, in a highly stratified atmosphere, mixing-length hypotheses are fulfilled only a number of local scale heights apart from the upper boundary.

A different description of these upper layers is proposed by Chan, Sofia & Wolf (1982) on the grounds of two-dimensional numerical simulations. They obtained for these layers some time-dependent eddies with sizes of the order of the local scale height, and recent three-dimensional numerical computations seem to support their results (Chan & Sofia 1989). However, their computations arouse some criticisms of a very general nature. First of all, the aspect ratio chosen for their box might be too small, and secondly, the choice of Smagorinski's recipe to parameterize the smaller scales results in an enlarging of the velocity boundary layers, from which the thickness of the upper boundary layer might become comparable to the pressure scale height, so producing a mismatch of both length-scales.

In Hurlburt, Toomre & Massaguer (1984) it was found that convection in a box of aspect ratio $A < 4$ with periodic lateral boundary conditions produces artificial time-dependence. This might also be the reason for the time-dependence seen in Ginet & Sudan (1987) as they took $A = 1$. In fact small aspect ratios might enhance pressure fluctuations, which is very dangerous when modeling convection in a compressible fluid.

5.1.1 *Anelastic approximation for compressible convection*

Numerical simulation of convection in a compressible fluid has to deal with propagation of sound waves. This is an unwanted effect as these waves impose very small time steps for the integration of the time-dependent equations whereas unless convection is nearly sonic, their contribution to the energy balance is negligible. Anelastic approximation is a second order asymptotic expansion in terms of the Mach number M (Gough 1969). Acoustic frequencies can be filtered out with the system still remaining energetically consistent at the chosen approximation order.

A widespread version of the anelastic approximation reduces the continuity equation to a divergence-free condition for the mass flux $\nabla.\rho v = 0$. Neglecting the eulerian time derivative of the density, $\partial_t \rho$, in the continuity equation is a crucial requirement for avoiding any pressure mode, but this might be inconsistent. If we split the density as $\rho = \bar{\rho} + \rho'$, where $\bar{\rho} = \bar{\rho}(z,t)$ is the horizontal average, the anelastic scaling implies $\rho'/\bar{\rho} = O(M^2)$ and the continuity equation splits as

$$\partial_t \bar{\rho} + \partial_z \overline{\rho v}_z = 0$$
$$\nabla.(\rho v)' = 0$$

with overbars and primes meaning, respectively, horizontal averages and fluctuations.

Ginet & Sudan (1987), by imposing the divergence-free condition for the mass flow realized some leakage of mass. Also Chan & Sofia (1989) measured in their three-dimensional fully compressible computations a non-zero vertical mass flux $\overline{\rho v_z}$, which is inconsistent with the assumption $\partial_t \bar{\rho} \simeq 0$. With stellar pulsation theory in mind this mean vertical mass flux has been called the radial mode by Gough (1969) and has been included by Latour, Spiegel, Toomre & Zahn (1976) in their derivation of the anelastic modal equations . Hence, the purely radial acoustic modes may couple thermal convection with radial pulsations.

5.2 PENETRATIVE CONVECTION AND OVERSHOOTING

Penetration and overshooting are names sometimes used as synonyms and sometimes not, but which attempt to describe different situations. Let us assume that the structure of a star is computed using a local model for convection. We may then ask how deep into the stable zone convection will penetrate if the structure of the star is kept unperturbed. In mixing-length terminology we should ask how far a parcel of fluid will *overshoot* the boundary of the convective zone. *Penetration* would be better used to describe those situations where convection has been computed self-consistently for the stable and unstable layers as a whole.

5.2.1 *Downward penetration*

As mentioned in §4 penetration into the stable layer can be controlled by two different mechanisms: buoyancy and shear. Shear depends critically on eddy sizes and buoyancy depends, through the Brunt-Vaissala frequency $\mathcal{N}^2$, on the relative stratification between the stable and the unstable layers

$$S = -\frac{[\partial_T - \partial_A]_{unstable}}{[\partial_T - \partial_A]_{stable}}$$

Penetration increases by increasing S and decreasing the shear length ℓ. In fact, as shown by Hurlburt, Toomre & Massaguer (1989), S and ℓ are somehow linked. By decreasing the stiffness of the stable layer (i.e.: increasing S), the penetration motion changes from being cellular (almost confined between flat boundaries) to being plume-like, with the velocity field being concentrated in narrow vertical vorticity sheets (i.e.: eddies with small ℓ values).

Cellular motion is possible only if the adjoining layers are very stable. In a Boussinesq fluid, and with the assumption of cellular convection, Zahn, Toomre & Latour (1982) estimated the distance Δ from the edge of the unstable zone to the first zero of the velocity to be proportional to the relative stability between the two layers, $\Delta \propto R/R_s$, where R and R_s are, respectively, the Rayleigh numbers of the unstable and stable layers. Massaguer, Latour, Toomre & Zahn (1984) generalized the $\Delta \propto R/R_s$ law for downwards penetration in a highly stratified fluid, but their results seem to overestimate penetration.

Hurlburt, Toomre & Massaguer (1989), from a two-dimensional numerical simulation in a compressible fluid, propose the more conservative law $\Delta \propto S^{1/2}$ for their plume-like penetration. Their plumes are laminar but they are still remarkably invariant with depth, which indicates that entrainment is contributing significantly to their dynamics, thus explaining why penetration is to be smaller in plumes than in cellular motion.

5.2.2 *Upward penetration*

The upward penetration in a highly stratified fluid is substantially different from the downward one. While the latter, roughly speaking, displays a Boussinesq physics, the former is dominated by non-Boussinesq effects. Going upwards the local scales shrink very fast and pressure fluctuations increase in magnitude until they overcome temperature in the buoyancy work. Buoyancy braking slows the motion down, so as to produce a very stable layer on top of the convection zone. This can explain why upward penetration is always cellular instead of being plume-like (Hurlburt, Toomre & Massaguer 1986). In spite of this enhanced stability, upward penetration can extend over a significant fraction of a local scale height, although measured in terms of the total layer thickness the unstable layer is very shallow.

5.2.3 *Downward penetration of small amplitude velocity fields*

So far, we have only considered large amplitude velocity fields, meaning by this a flow strong enough to flatten the entropy gradient. For the downward penetration this region includes, roughly speaking, the first countercell -i.e.: the region where the convective flux reverses its sign. Below this point the convective flux is so small that the temperature gradient cannot be distinguished from the radiative one. Below the first countercell the velocity field decreases its amplitude significantly, as do the transport coefficients. Time scales for turbulent diffusion and mixing become much larger than in the bulk of the convection zone, but diffusion and mixing can still be effective in these regions (Hurlburt, Toomre & Massaguer 1989).

In Hurlburt, Toomre & Massaguer (1986), much as in the ice-water experiments (Adrian 1975), the plume-like motion is time-dependent, thus forcing gravity waves in the stable layer. The Fourier spectra for these waves may be strongly dependent on the resonance properties of the whole cavity below the convection zone. Absorbing or reflecting walls can drastically change the amplitudes of the Fourier spectrum, but waves can also be absorbed or reflected selectively by the plumes themselves. As a result the feed-back between the gravity waves and the external forcing may be channeling energy towards some preferred frequencies. This might explain why the computed spectra for gravity waves is substantially different from that measured in the ocean (Phillips 1966).

6. Detailed structure of the solar convection zone

Current estimates give a depth for the solar convection zone of between twenty and thirty percent of the total radius of the star - i.e.: 150.000 km to 200.000 km. Although recent work by Christensen-Dalsgaard, Gough & Thomson (1989), based on the measurement of the sound speed in the solar interior by using helioseismological data, shows and adiabatic region delimitated by a sharp edge at a depth of twenty-three percent of the solar radius.

The Rayleigh number value based on the largest local pressure scale height, $H_P \approx 50.000$ km, and on a turbulent viscosity takes the value $R \approx 10^{12}$. And the Prandtl number based on the radiative diffusion coefficient is $\sigma \approx 10^{-9}$, thus giving $\Lambda = R\,\sigma > 10^3$ everywhere except, perhaps, near the photosphere - the reader must be aware that the Λ-value is independent of the assumed viscosity. Therefore, according to the critical Λ value estimated by Spiegel (1966), the solar convection zone is in a large Peclet number regime almost everywhere.

6.1 GRANULES

Granules are bright regions on the solar photosphere surrounded by dark lanes. Their mean horizontal size is $\ell \approx 1.400$ km and they last for some five to ten minutes. Bright regions are associated with upward motion, while dark lanes are associated with downward motion. The average vertical velocity, either up or down, is one kilometre per second and it is well correlated with the horizontal temperature difference, which shows a value of a hundred degrees.

The existence of a strong correlation between velocity and temperature has always been taken as the main argument in favour of the convective nature of the photospheric granules. However such a correlation only implies a very efficient transport of heat and does not tell us anything about the very nature of the driving mechanism. Forced convection is more efficient than natural convection but it is not buoyancy driven. The observation of a cellular structure has also been taken as an argument in favour of the convective origin for the granules, mostly because a turbulent motion, say shear driven, would display a continuous Fourier spectrum with the cut-off length scale being much shorter than the size presumed for the granules.

Two and three-dimensional numerical simulations for modeling granulation have been derived by Nelson & Musman (1977, 1978), Nordlund (1985) and Stein & Nordlund (1989). All these simulations give plausible results, but the geometrical similitude between the structures computed in the latter paper and the photographs of solar granules is so impressive that we can finally believe we understand granules. The pictures obtained from the numerical simulation strongly resemble a granulation photograph. However, in spite of the persistent optical impression of the granules showing a dominant size, their kinetic energy spectrum is smooth, without showing any appreciable peak. The conclusion, therefore, is that no preferred scale exists.

Something similar can be said of the measured spectrum once the resolution of solar observations has been sufficiently increased and low frequency pressure waves have been filtered. Muller & Roudier's (1985, Roudier & Muller 1986) observations show a continuum of sizes for the bright regions, with their transversal dimensions decreasing towards values much smaller than those presumed for granules. The power spectrum measured by the above mentioned authors clearly displays an inertial range, with a slope of $-5/3$ as corresponds to the famous Kolmogorov-Obukhov law - see Zahn's (1987) review. Therefore, observations and numerical simulations seem to agree with the granules being an optical effect associated to a privileged velocity-temperature correlation length.

Indirect evidence for the existence of a privileged correlation length can also be found in Roudier & Muller's (1986) work. By measuring the fractal dimension of the granular structures they revealed a change in dimension near the expected granular size. As discussed by Zahn (1987), at photospheric level and for length-scales close to the granules' size, the Peclet number is order one. Therefore at smaller scales the convective heat transport is expected to be very inefficient. The Nusselt number is $N \approx 1$ but, as discussed in Massaguer, Mercader & Net (1989),

velocity and temperature show a good correlation, possibly because they are linearly related. The granules with sizes larger than this are the more corrugated ones - i.e.: those of larger fractal dimension- possibly because of the enhanced turbulent transport. Their Peclet number value is larger than one, heat transport is more efficient, $N \gg 1$, but because of the non-linear relationship between them, temperature and velocity display a reduced correlation factor, and the contrast between hot and cold regions becomes smoother.

6.1.1 Vortices

Recent satellite observations made by Brandt et al. (1988), once conveniently filtered from low frequency acoustic modes, have shown the presence of a vortical flow. And vortices of size $\ell \approx 5.000$ km, corresponding to approximately three granules, have been identified. Photospheric rotational motions have been reported long before in sunspots. Yet what makes the subject new is the possibility of their being uniformly distributed in the photosphere, possibly in association with granules, and twisting magnetic flux tubes.

6.1.2 Mesogranules

November et al. (1981) have also found evidence of photospheric structures of sizes intermediate between granules and supergranules, what they have called mesogranules. It is a rather weak motion, as compared to granules and supergranules, with length-scales ranging between 5.000 km and 10.000 km and lifetimes of about two hours. They have been described as associations of granules by Kawaguchi (1980) but their physical origin is still unclear.

6.2 SUPERGRANULES

Supergranules are structures that extend well into the cromosphere. Their horizontal size is $\ell \approx 30.000$ km, approximately twenty times that of a granule, and they last for some twenty hours. The horizontal velocity ranges from three to five hundred metres per second and the vertical motion displays a slightly slower downward motion with the upward one being much slower. Horizontal temperature contrast is very small, with its upper limit being $\Delta T < 1 \ ^{0}K$. As an additional feature, the border of the supergranules is delimited by an intense magnetic field.

The flow described above for a supergranule is very close to what must be expected for cellular convection, and buoyancy driving has never been really questioned. A weak temperature difference across the cell is certainly not in contradiction with it. As, in addition, the presence of a magnetic field concentrated near the border of the cell increases thermal conductivity, such a small ΔT seems not to question a convective origin. More striking is the regularity of the hexagonal pattern observed. Low Prandtl number thermal convection is not the best context for such a regular flow as shear instabilities play a dominant role. Such a geometrical regularity is to be thought of as the result of some additional stabilization effects. As discussed in §5.1, a high-density contrast such as that found in the upper layers of the solar convection zone could be a candidate, but other contributions, like the stabilizing effect of a magnetic field, cannot be ruled out.

The most difficult question about supergranules is that concerning their size.

Several conjectures can be found in the literature but none of these conjectures seems to be soundly established. Neither laboratory nor numerical simulations support a cellular motion with a size much smaller than the layer depth. Therefore, the possibility of supergranules not being cellular structures has to be taken seriously. In fact, cellular convection is far from being the flow pattern most frequently seen in laboratory experiments. Even in the most classical Boussinesq problem transient bubbles or plumes do constitute the main flow. At this point it may be worth quoting the following description from the experimental work by Castaing *et al.* (1989). *In the central or interior region of the cell we envision motions on many scales up to that of the entire cell, consisting of convective eddies, and of thermals and plumes.*

In non-cellular Boussinesq convection we expect the largest eddies, of sizes comparable to the layer depth, to break into smaller eddies, thus cascading kinetic energy from small to large wavenumbers. In a highly stratified fluid, however, the situation might be different. From a patchy work with materials of very different origin and quality Zahn (1987) was able to build a power spectrum for the solar convection zone. If this spectrum is to be trusted, kinetic energy is being fed up into the motion at the supergranules size, not at the largest scale of motion. Again, much as for the granular flow, the real question about sizes does not concern scales of motion but coherence lengths. So as to produce the largest injection of buoyancy work, velocity and temperature have to be coherent. By writing the buoyancy work fed up to an eddy of size ℓ as

$$ E_B \approx \int_{r+\ell}^{r} \rho' v g \ dr $$

we can see that a large injection of buoyancy work requires coherence between ρ' and v, together with a large eddy size. Upper limits for ℓ are difficult to establish. There is no experience yet in modeling turbulence on highly stratified media, so we cannot establish any firm conclusion, but Chan & Sofia's (1989) results support the view that the pressure scale height is a convenient measurement of the coherence length. If so, buoyancy work has to show a maximum for eddy sizes $\ell \approx 50.000$ km, the largest pressure scale height, hence associating the supergranules with the deeper layers of the convection zone.

6.3 GIANT CELLS

The existence of giant cells was conjectured a long time ago by Simon & Weiss (1968). From laboratory experiments there is no doubt that a pattern with a horizontal scale of the order of the depth of the convection zone has to be present, but it is unclear whether it will show up as a cellular motion or not. The so-called torsional waves of Howard & LaBonte (1980) are the only pattern that fits such specifications, so they may be considered as a plausible candidate.

Until very recently there was widespread confidence in the large scale flow of convection in a rotating spherical shell being organized as elongated cells with their axis in the North-South direction, which have been called *banana cells*. Numerical as well as experimental work in fact support this conjecture (Hart *et al.* 1986, Gilman & Miller 1986), but they have never been observed. The observation of the so-called torsional oscillations has challenged this confidence (Snodgrass 1987, Snodgrass & Wilson 1987) much as the migrations of young sunspots (Ribes & Mein 1985, Ribes & Laclare 1988). The former authors describe the whole pattern as a set of toroidal vortices, one on top of the other, like *piled doughnuts*. Each one of these vortices would be a giant cell.

All these large scale flows show a slow equatorward drift along the solar-cycle, so making the magnetic field an obvious candidate for explaining the preference for a doughnut's structure instead of banana cells. In fact, a rotating spherical shell of thermally convecting fluid with an intense magnetic field concentrated in the lower portion of the shell can display such a toroidal pattern (Merryfield 1989).

7. Conclusions

In the present review we have examined the structure of the solar convection zone and the most relevant techniques for its modeling. As much as possible we have tried to avoid *ad hoc* hypotheses, so relying only on numerical results, laboratory experiments and observations. And we have shown how all these three pieces of material converge towards a better understanding of the physics of the solar convection zone.

I wish to thank Dr. J. P. Zahn for helpful discussions during the preparation of this paper. The present work was supported by the Dirección General de Investigación Científica y Técnica (DGICYT) under grant PS87-0107.

8. References

Adrian, R.J. 1975, *J. Fluid Mech.*, **69**, 753.
Böhm-Vitense E. 1958, *Z. Astrophys.*, **46**, 108.
Brandt, P.N., Scharmer, G.B. Ferguson, S., Shine, R.A., Tarbell, T.D., Title, A.M. 1988, *Nature*, **335**, 238.
Canuto, V.M., Goldman, I. and Chasnov, J. 1987 *Phys.Fluids*, **30**, 3391.
Canuto, V.M., Goldman, I. and Hubickyj, O. 1984, *Ap. J.*, **280**, L55.
Castaing, B., Gunaratne, G., Heslot, F.L., Kadanoff, L., Libchaber,A., Thomae, S., Wu, X.Z., Zaleski, S. and Zanetti, G. 1989, "Scaling of Hard Turbulence in Rayleigh-Benard Convection". *J. Fluid Mech*, (in press).
Chan, K.L., Sofia, S. and Wolff, C.L. 1982, *Ap.J.*, **263**, 935.
Chan, K.L., Sofia, S. 1986, *Ap.J.*, **307**, 222.
Chan, K.L., Sofia, S. 1989, *Ap.J.*, **336**, 1022.
Christensen-Dalsgaard, J., Gough, D.O. and Thomson, 1989, (poster, this meeting).
Cloutman, L.D. and Withaker, R.W. 1980, *Ap. J.*, **237**, 900.
Eggleton, P.P. 1983, *M.N.R.A.S.*, **204**, 449.
Gilman, P.A. and Miller, J. 1986, *Ap. J. Supplement Series*, **61**, 585.
Ginet, G.P. and Sudan, R.N. 1987, *Phys.Fluids*, **30**, 1667.
Gough, D.O. 1969 *J. Atmos. Science*, **26**, 448.
Gough, D.O. and Weiss N.O. 1976, *M.N.R.A.S.*, **176**, 589.
Hart, J.E., Toomre, J., Deane, A.E., Hurlburt, N.E., Glatzmaier, G.A., Fichtl, G.H., Leslie, F., Fowlis, W.W. and Gilman, P.A. 1986, *Science*, **234**, 61.
Heslot, F., Castaing, B. Libchaber, A., 1987, *Phys. Rev. A.*, **36**, 5870.
Howard, R. and LaBonte, B.J. 1980, *Ap.J.*, **239**, L33.
Hurlburt, N.E., Toomre, J., Massaguer, J.M. 1984, *Ap.J.*, **282**, 557.

Hurlburt, N.E., Toomre, J., Massaguer, J.M. 1986, *Ap.J.*, **311**, 563.

Hurlburt, N.E., Toomre, J., Massaguer, J.M. 1989, "Penetration and Mixing Below a Convection Zone" (preprint).

Kawaguchi, I. 1980, *Solar Phys.*, **65**, 207.

Kraichnan, R.H. 1964, *Phys. Fluids*, **7**, 1084.

Kuhfuß, R.M. 1986, *Astron. Ap.*, **160**, 116.

Latour, J., Spiegel, E.A., Toomre, J. and Zahn, J.P. 1976, *Ap.J.*, **207**, 233.

Ledoux, P., Schwarzchild, M., Spiegel, E.A., 1961 *Ap.J.*, **133**, 184.

Marcus, P.S. 1986, in *Astrophysical Radiation Hydrodynamics*. K.-H.A. Winkler and M.L. Norman (eds.), pp. 387-414. Reidel.

Marcus, P.S., Press, W. and Teukolsky, S.A. 1983, *Ap.J.*, **267**, 795.

Massaguer, J.M., Latour, J., Toomre, J. and Zahn, J.P. 1984, *Astron.Ap.*, **140**, 1.

Massaguer, J.M., Mercader, I., Net M. 1989, "Non-linear Dynamics of Vertical Vorticity in Low-Prandtl-Number Thermal Convection". *J. Fluid Mech.* (in press).

Massaguer, J.M. and Zahn, J.P. 1980, *Astron. Ap.*, **87**, 315.

Merryfield, W.J. 1989, (poster, this meeting).

Muller, R. and Roudier, T. 1985 in "High Resolution in Solar Physics". R. Muller (ed.) *Lecture Notes in Physics*, **233**, 242.

Nelson, G.D. and Musman, S. 1977, *Ap. J*, **214**, 912.

Nelson, G.D. and Musman, S. 1978, *Ap. J. Lett.*, **222**, 30.

Nordlund, A. 1985, *Solar Physics*, **100**, 209.

November, L.J., Toomre, J., Gebbie, K.B. and Simon, G.W. 1981, *Ap.J.*, **245**, L123.

Phillips, O.M. 1966, *The Dynamics of the Upper Ocean*. Cambrige.

Ribes, E. and Mein, P. 1985, in "High Resolution in Solar Physics". R. Muller (ed.), *Lecture Notes in Physics*, **233**, pp. 282-285. Springer.

Ribes, E. and Laclare, F. 1988, *Geophys. & Astrophys. Fluid Dynamics*, **41**, 171.

Roudier, T. and Muller, R. 1986, *Solar Phys.*, **107**, 11..

Schmitt, J.H.M.M., Rosner, R. and Bohn, H.U. 1984, *Ap.J.*, **282**, 316.

Shaviv, G. and Salpeter, E.E. 1973, *Ap.J.*, **184**, 191.

Simon, G.W. and Weiss, N.O. 1968, *Z. Astrophys.*, **69**, 435.

Smagorinski, J.S. 1963, *Mon. Weather Rev.*, **91**, 99.

Snodgrass, H.B. 1987, *Ap.J.*, **316**, L91.

Snodgrass, H.B. and Wilson, P.R. 1987, *Nature*, **328**, 696.

Spiegel, E.A. 1962, *J. Geophys. Research*, **67**, 3063.

Spiegel, E.A. 1965, *Ap. J.*, **141**, 1068.

Spiegel, E.A. 1966, in "Stellar Evolution" pp., 143-173. Plenum Press.

Spiegel, E.A., 1971, *Ann. Rev. Astron. Astrophys.*, **9**, 323.

Stein, R.F. And Nordlund, A. 1989, "Topology of convection beneath the solar surface". *Ap. J. Lett.*. (in press).

Tooth, P.D. and Gough, D.O. 1989, (poster, this meeting).

Turner, J.S. 1973, *Buoyancy Effects in Fluids*. Cambridge.

Unno, W., Kondo, M. and Xiong, D.R. 1985, *Publ. Astron. Soc. Japan*, **37**, 235.

van Ballegooijen, 1982, *Astron. Ap.*, **113**, 99.

Xiong, D.R. 1979, *Acta Astron. Sinica*, **20**; 238.

Yakhot, V., Orszag, S.A. 1986, *J. Sci. Comput.*, **1**, 3.

Zahn, J.P. 1987, in "Solar and Stellar Physics." *Lecture Notes in Physics*, **292**, 55.

Zahn, J.P., Toomre, J. and Latour, J. 1982, *Geophys. & Astrophys. Fluid Dynamics.*, **22**, 159.

COMMENTS ON SOLAR CONVECTION

Franz - Ludwig Deubner

Institut für Astronomie und Astrophysik
der Universität Würzburg
Am Hubland, D-8700 Würzburg

ABSTRACT

This contribution discusses observational aspects of the evolution of individual structures of solar convection.

It has been shown, that mesogranulation is a convective phenomenon that fits well into the gap between granulation and supergranulation. Apparently this observation justifies the view that the three members of the granulation family represent sections of a broad continuum of convective motions spanning the range of sizes from a yet unknown fraction of 1 Mm to about 50 Mm. Nevertheless, power spectra of velocity and brightness fluctuations exhibit three maxima, separated by intervals with significantly less power near 3 Mm and 7.5 Mm. Do these gaps give reasons for reconsidering the old idea, that each of the three characteristic scales has its own source layer at a certain depth in the convection zone?

Power spectra of the granular energy distribution near the observational limit of spatial resolution suggest a continuous transfer of kinetic energy to smaller eddies by turbulent decay of the larger scale elements. Morphological studies of granular evolution and a comparison of the observed spectral line bisectors with theoretical predictions seem to disprove this idea. These observations imply either that the turbulent cascade, if it exists, is buried in the spatially unresolved part of the power distribution, or that radiative losses ultimately limit the life time of individual granules on all scales.

1. Introduction

The condition for convection to occur in a star is well known: it requires that the local adiabatic temperature gradient $|dT/dr|_{ad}$ be smaller than the temperature gradient found in the ambient medium. In a solar type star with a convection zone in the outer part ($\sim 1/3$) of its envelope, the condition is fulfilled mainly through the decrease of γ, the ratio of the specific heats c_p and c_v caused by the ionization of H, He, and He^+ in successively deeper layers.

Observations of the sun in white light and in certain spectral lines formed in higher layers of the solar atmosphere have revealed the existence of at least three regimes of convective motions, with different typical sizes, called granulation (1" - 2"), mesogranulation (5" - 10"), and supergranulation (20" - 40"). Still larger systems of horizontal surface flows are being searched for, as they are believed to be of fundamental importance in theories of the solar activity cycle (solar dynamo).

3-D numerical simulations of convection based on realistic theoretical models of the solar envelope have made considerable progress in the last years (Nordlund, 1984; Stein et al., 1989; Steffen et al., 1989). It is now possible to compare directly the observed properties of

G. Berthomieu and M. Cribier (eds.), Inside the Sun, 117–124.

convective flows (e.g. spatial scale and life time, r.m.s. velocities and brightness contrast, changes of spectral line profiles) with the corresponding numbers and qualities derived from model simulations (Dravins et al., 1981; Wöhl and Nordlund, 1985; Steffen, 1989). However, due to practical limitations set by the available computing power, typical simulation studies have to contend with 3-D volumes which cover just a few granules in area and a few scale heights in depth.

In this situation, theoreticians presently tend to defend the hypothesis that in the (invisible) interior of the sun we should expect to find at each layer a situation similar to the one studied in the model simulation, which is characterized by extended horizontal flows and narrow downdraft funnels, and which is propagated upward from the bottom layers through essentially all of the convective envelope by selfsimilar expansion and ramification.

In the two chapters of this contribution we want to discuss first, whether the theoretical scenario of a broad continuum of convective scales is indeed supported by observation, with particular attention paid to the newest member of the granulation family (mesogranulation), which was only recently shown to be of genuine convective origin (Deubner, 1989). Later we shall turn to the effects caused by the transition from buoyancy driven structures to inertial eddies, and to the question whether the observations warrant recent speculation about a turbulent decay of convective motions in a Kolmogorovian fashion.

2. Mesogranulation - the missing link in a uniform spectrum of convective motions?

Although a mesoscale brightness structure was probably recognized in the photosphere as early as 1932 by Strebel and Thüring, it was only in 1981 through observations of radial velocities near the center of the disc that the existence of a system of vertical motions in the 5" to 10" range, presumably of convective origin, was made evident (November et al., 1981). The horizontal counterpart of this flow field became subsequently apparent in studies of proper motions of granules derived from white light high spatial resolution time lapse movies (November et al., 1987; Brandt et al., 1988).

The late discovery of the mesogranulation phenomenon can be understood if the comparatively low contrast of solar structures observed in this range of sizes is taken into consideration: At the photospheric level mesogranular r.m.s. brightness and velocity fluctuations are exceeded by those of granulation, whereas at higher levels (i.e. near the temperature minimum) the 5-min oscillations dominate the velocity signal. Nevertheless, several mesogranulation related effects studied in the past few years testify strongly for the presence of a powerful organized flow pattern.

Two-dimensional measurements of horizontal motions derived with the socalled local correlation technique (November et al., 1987) can be used to estimate the local divergence of the horizontal flow. A comparison of the divergence map with the distribution of magnetic flux reveals that flux concentrations are preferentially located at points or lines of lateral convergence, including alignments which are distinctly smaller than the wellknown supergranular network (Simon et al., 1988). Brandt et al.(1989) have shown, that properties of individual granules like life time, brightness, area, and rate of expansion are statistically correlated with the horizontal divergence in the sense that one would intuitively expect, namely of reinforcement of small scale convection by a similar but larger flow system. At the source points of the horizontal flow, presumably located in local upwellings, an enhanced birth rate of large granules is observed in granule movies.

These kinematic studies have recently been complemented by a comprehensive spatiotemporal analysis of the dynamical behaviour of non-oscillatory flows and brightness

fluctuations. A range of spatial scales between 1."7 and 200" was investigated spectroscopically at two levels in the lower photosphere simultaneously (Deubner, 1989). Using off center
observations on the disc it was confirmed, that the horizontal velocities of approximately 750
m/s derived from the changing granular brightness distribution by the local correlation technique do indeed represent a material motion rather than horizontally propagating waves, as was
initially suspected. Measurements of the phase lag between the observed brightness and velocity fluctuations, and of their coherence reveal the convective origin of the motions, and the
close similarity of mesogranulation to both granulation and supergranulation.

In fact, the phase lag between (upward) velocity and brightness is very close to zero for
small spatial scales (granules), and rises only slightly and very smoothly to about 20° in the
range of supergranules. The r.m.s. amplitude of the horizontal velocities decreases almost
monotonically from typical granular values (about 2 km/s) to approximately 500 m/s for supergranules. It is interesting to note that the ratio of vertical to horizontal r.m.s. velocities
decreases nearly linearly with increasing size of the convective elements; the ratio thus seems
to obey the scaling law implied by the continuity equation in cylindrical geometry.

The main question to be addressed in this brief discourse on properties of convective
motions concerns the existence of several discrete spatial scales of granulation. We do understand, why the different scales are best seen in different layers of the atmosphere, and by using
different techniques of observation. Velocity and brightness fluctuations of granulation decay
rapidly in the lowest 100 to 200 km of the photosphere and cannot be seen anymore at higher
levels. In contradistinction the Doppler effects due to meso- and supergranular motions are
more easily detected in medium strong spectral lines which provide the optimal S/N ratio, i.e.
higher in the atmosphere. Finally, the magnetic flux concentrations collected at the cell borders
by longer lived large scale motions are most easily seen in the bright cores of chromospheric
lines forming the chromospheric network. However, these distinctions would remain even if
the spectrum of sizes was perfectly uniform.

On the other hand, the observations of Deubner (1989) indicate gaps in the power and
coherence spectra near 4" and 10", separating the mesoscale from the larger and smaller size
convective elements. A confirmation of these observations by independent spatially two-
dimensional data is desirable. But the observational task is formidable, because of the wide
range (more than two orders of magnitude) of spatial scales and life times of the structures of
interest. Also the importance of long lasting high quality seeing to resolve the smallest elements, and of a representative coverage of the solar surface to achieve statistical significance
with the larger structures needs to be stressed.

A positive result would necessarily stimulate the question, whether the three different
scales can be traced back to certain source regions within the convection zone, as originally
suggested by Simon and Leighton (1964), or whether the repeated branching of downdraft funnels, as envisaged by Nordlund (1990), is capable of creating a stable non-uniform spatial
power spectrum by some kind of topological effect.

3. Has turbulent decay of granular convective motions been observed?

Although life and death of granular convection elements occur right before our eyes, the
physical processes which ultimately lead to the conversion of convective energy into heat
appear to be even less well determined than the ones discussed in the preceding chapter. The
central issue here is, whether convection decays through a cascade of smaller and smaller turbulent elements, and whether the spatial scales involved in this process are already being
observed in high resolution photographs and spectrograms.

This question has recently gained some new momentum through observations obtained by Roudier and Muller (1987) at the Pic du Midi Observatory, which exhibit a break in the area-perimeter relation of granulation in the continuum, and in the histogram of granule areas, at a size of 1."37. A discontinuity at 3" in the one-dimensional(!) power spectrum of granular velocities was presented by Muller (1989). At the short wavelength side of this discontinuity the slope of the observed power spectrum has a value of -5/3, corresponding closely to the Kolmogorov distribution of energy in a turbulent fluid.

To assess the significance of these observations for our question it is important to realize the preliminary character of the results published so far: The shape of the power spectra we are discussing here is easily altered by a number of effects, all of which need to be carefully considered.

a) Whatever the intrinsic shape of the solar energy distribution may be, the finite resolving power of the instrument and the effects of seeing in the atmosphere and in the spectrograph will inevitably degrade the results, depending on the circumstances. Independent knowledge of the actual point spread function of any given exposure is needed to correct the observed data. Not many observations are available which provide this kind of calibration (cf. Komm and Mattig, 1989; Deubner and Mattig, 1975).

b) Other motions like solar f- and p-modes, and internal gravity waves which are closely connected to the penetration of convective motions into the inertial regime (Deubner and Fleck, 1989), are superposed on, or interfere with the convective velocity and brightness fields. Not all of them can be removed by filtering the data in the wavenumber and frequency domain (Title et al., 1989; Deubner, 1988), since - at least in the most interesting region, the low photosphere - the spectral contribution functions extend across the border line between convectively stable and unstable layers. In this region, the quasistationary primary motions cannot be readily separated from secondary velocity fields, like gravity waves.

c) It is necessary to use two-dimensional spatial information to obtain accurate power spectra. In principle one may convert one-dimensional data under the assumption of azimuthal isotropy to yield two-dimensional spectra. But their statistical stability is often insufficient, and spurious peaks as well as slopes with the wrong gradient are frequently found.

d) But even using a nearly ideal data set which is supposedly free of, or has been corrected for the effects just mentioned, one could hardly expect to retrieve the intrinsic power distribution from the observed spectra, unless the transfer effects of the solar atmosphere are properly taken into account. Since the measured variables (brightness as a proxi of temperature, or velocity) vary not only laterally but also as a function of height in the atmosphere, the inevitable summing of the variables along the line of sight may influence the resulting power law depending on the size and therefore also on the horizontal scale of the structure.

The development of granular structure has first been described in detail by Mehltretter (1978), and more recently by Title et al.(1989), who based their study on a time series of white light pictures obtained from the SOUP data after subsonic filtering. The two studies agree that splitting of granules into two or more fragments is not the only process marking the end of life of selected individual granules. Others disappear by fading, they merge, or they are "squeezed out of existence" by their stronger neighbors. The observed fragmentation does not in general continue on smaller scales; rather, many fragments subsequently become the centers of new full scale granules of the next generation.

The spatial coherence of the brightness and velocity distribution of granules has recently been studied by Deubner (1988) in a high resolution time series of (spatially one-dimensional) spectra of the CI 5380 and the FeI 5383 line. He found that velocity and brightness fluctuations (after subsonic filtering!) are well correlated down to element sizes as small as ~0."8 in the lowest photosphere. At this point the transfer function of the combined atmospheric and spectrographic seeing is expected to decline rapidly due to the comparatively long exposure time of 3s. But there is no obvious break in the coherence spectra at longer wavelengths, e.g. near the critical value of 3" reported by Muller (1989), and no substantial change can be detected at 1."37, where Roudier and Muller (1987) have seen a break in the area-perimeter relation.

A very informative pictorial representation of spectral line profiles is the "bisector", i.e. a line which connects the mid-points in between the red and blue wings of the actual profile. This representation contains in a single curve the basic information on the depth dependence of those variables that contribute to an asymmetric shape of the spectral line. It facilitates greatly a direct comparison of observed and theoretical profiles.

Synthetic bisectors have been calculated for this purpose along with the 3-D studies of convection simulation (Dravins et al., 1981; Steffen, 1989) that we have mentioned in the introduction. Of these, the bisectors published by Steffen (1989), which were derived from a stationary model of convection, are the ones that agree best with observed bisectors from high resolution spectra obtained under excellent seeing conditions (Mattig et al., 1989). Synthetic and observed bisectors are displayed in Figures 1 and 2.

The preeminent property of the synthetic profiles is the dependence of the curvature and a monotonic variation of the inclination of the bisectors as a function of continuum brightness. In other words: in Steffen's model, the shape of spatially resolved profiles from the bright center of a granule is systematically and distinctly different from the shape of intergranular profiles. It is not surprising that a stationary model would yield this kind of results; but then, what are the consequences of the impressive similarity of the simulation results with the observed data?

In the range of spatial scales larger than 0."8 which are well resolved by modern high resolution observations it appears that brightness and velocity are distributed in a well organized fashion, not only with regard to their horizontal distribution but also at any given position as a function of depth in the atmosphere. It is difficult to imagine an atmosphere composed of turbulent eddies with sizes in the vicinity of 1" or less, which is capable of producing such a smooth and well defined behaviour of the bisectors. It seems therefore unlikely, that turbulent decay of solar convective power occurs within this range of spatial scales.

4. Conclusions

The results of extensive theoretical modelling and the data from improved high resolution observations of solar photospheric convection are rapidly converging. There is no doubt that within the limits of present days computer capacity 3-D computer simulations have helped immensely to understand convective processes and to predict their signature on the sun and other stars.

Nevertheless, several questions touching on fundamental subjects like the internal structure of the convection zone, and the detailed energy balance near the border zone to stable stratification, remain to be answered, and further high resolution observations of supreme quality will be necessary along with further refinements of the diagnostic tools to make further progress. It apprears that a sophisticated study of the distribution of convective element sizes may provide answers valuable as a diagnostic of interior structure within the convection zone. Certainly the large body of information available in the form of power-, phase-, and coherence spectra of quasi stationary motions should be utilized to gauge present theories of convection.

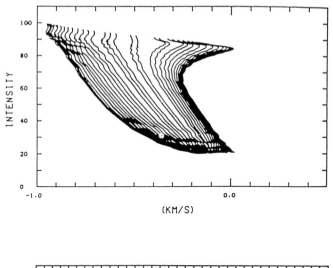

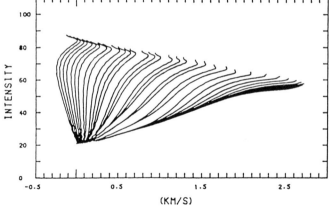

Fig. 1 **Top:** Series of computed bisectors of a strong FeII line. From left to right, the solar surface area over which the spectrum is integrated is expanded to progressively include contributions of darker granulation elements. **Bottom:** Similar as in the top figure, but starting with the darkest surface elements (at the extreme right), and advancing by including increasingly brighter elements. Velocities are scaled relative to the laboratory position. *Courtesy M. Steffen*

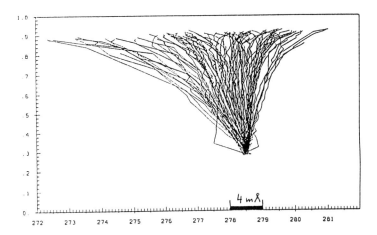

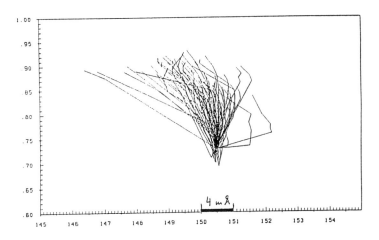

Fig. 2 Top: Series of observed bisectors from profiles of the strong FeI 6494.9 line, covering a range of 16 ". **Bottom:** Series of observed bisectors from profiles of the weak line FeI 6494.5, as in the top figure. All bisectors have been shifted to the same line core position. *Courtesy W. Mattig, A. Hanslmeier, and A. Nesis*

References

Brandt,P.N., Scharmer,G.B., Ferguson,S.H., Shine,R.A., Tarbell,T.D., Title, A.M.: 1988, *Nature* **335**, 238

Brandt,P.N., Scharmer,G.B., Ferguson,S.H., Shine,R.A., Tarbell,T.D., Title, A.M.: 1989, in *High Resolution Solar Observation* (O.von der Lühe, ed.) Sacramento Peak,Sunspot N.M. (in print)

Deubner,F.-L.: 1988, *Astron.Astrophys.* **204**, 301

Deubner,F.-L.: 1989, *Astron.Astrophys.* **216**, 259

Deubner,F.-L., Fleck,B.: 1989, *Astron.Astrophys.* **213**, 423

Deubner,F.-L., Mattig,W.: 1975, *Astron.Astrophys.* **45**, 167

Dravins,D., Lindegren,L., Nordlund,Å.: 1981, *Astron.Astrophys.* **96** , 345

Komm,R., Mattig,W.: 1989, in *High Resolution Solar Observation* (O.von der Lühe, ed.) Sacramento Peak,Sunspot N.M. (in print)

Mattig.W., Hanslmeier,A., Nesis,A.: 1989, in *Solar and Stellar Granulation* (R.J.Rutten, G.Severino, eds.) Kluwer Acad. Publ.,Dordrecht. p.187

Mehltretter,J.P.: 1978, *Astron.Astrophys.* **62**, 311

Muller,R.: 1989, in *Solar and Stellar Granulation* (R.J.Rutten, G.Severino, eds.) Kluwer Acad. Publ., Dordrecht. p.101

Nordlund,Å.: 1984, in *Small Scale Dynamical Processes* (S.Keil, ed.) Sunspot N.M. p.181

Nordlund,Å.: 1990, in *Solar Photosphere: Structure, Convection, and Magnetic Fields.* IAU Symp. **138** (in print)

November,L.J., Toomre,J., Gebbie,K.B., Simon,G.W.: 1981, *Astrophys.J.* **245**, L 123

November,L.J., Simon,G.W., Tarbell,T.D., Title,A.M., Ferguson,S.: 1987, in *Theoretical Problems in High Resolution Solar Physics II* (G.Athay, D.S.Spicer, eds.) NASA Conf.Publ. **2483 R**, p.121

Roudier,Th., Muller,R.: 1987, *Solar Phys.* **107**, 11

Simon,G.W., Leighton,R.B.: 1964, *Astrophys.J.* **140**, 1120

Simon,G.W., Title,A.M., Topka,K.P., Tarbell,T.D., Shine,R.A., Ferguson,S.H., Zirin,H.: 1988, *Astrophys.J.* **327**, 964

Steffen,M.: 1989, in *Solar and Stellar Granulation* (R.J.Rutten, G.Severino, eds.) Kluwer Acad. Publ., Dordrecht. p.425

Steffen,M., Ludwig,H.G., Krüß,A.: 1989, *Astron.Astrophys.* **213**, 371

Stein,R.F., Nordlund,Å., Kuhn, J.R.: 1989 in *Solar and Stellar Granulation* (R.J.Rutten, G.Severino, eds.) Kluwer Acad. Publ., Dordrecht. p.381

Strebel,H., Thüring,B.: 1932, *Zs.Ap.* **5**, 348

Title,A.M., Tarbell,T.D., Topka,K.P., Ferguson,S.H., Shine,R.A.: 1989, *Astrophys.J.* **336**, 475

Wöhl,H., Nordlund,Å.: 1985, *Solar Phys.* **97**, 213

ON THE ACCURACY OF SOLAR MODELLING

S. TURCK- CHIÈZE
Service d'Astrophysique
C.E.N. Saclay
91191 Gif sur Yvette
FRANCE

ABSTRACT. The confrontation between theoretical predictions and observations requires an estimate of the uncertainties of these predictions. Recent results on nuclear reaction rates and photospheric abundances are analysed, in a classical model framework. The role of opacities in the determination of the helium content, the neutrino fluxes prediction and the adiabatic sound speed is discussed. Comments on extra phenomena as mass loss, turbulent mixing and WIMPS are also presented, in the light of very recent seismological results.

1. Introduction

The estimate of the accuracy of classical solar modelling is fundamental if we consider the solar reference model as a basis for discussing the role of extra phenomena, neglected in the classical theory of stellar evolution, as turbulent mixing, rotation, magnetic field...In this conference, the complexity of the parametrisation of such dynamical effects in stars has been evoked. But, very recently, the impressive interpretation (Christensen-Dalsgaard et al 1985, 1988b, Gough and Kosovishev 1988) of the 5 mn acoustic mode observations (Grec et al 1983, Duvall et al 1988) in terms of radial distribution of the sound speed or density, added to the multiplication of the neutrino detections (Davis et al 1968, Koshiba 1988, Kirsten 1986, Barabanov et al 1985) push us to reemphasize the idea that the Sun is a privileged place to quantify the relative importance of these phenomena. In paper 1 (Turck-Chièze et al 1988), we have discussed the characteristics of an updated reference model and compared different published models. Most of the noticed differences arise from different choices of the input parameters (nuclear reaction rates, equation of state, initial abundances and opacities). So I shall comment on the role of these ingredients and reevaluate the resulting uncertainties of solar modelling. The second part will be devoted to the constraints on the extra phenomena set by the radial distribution of the sound speed deduced from observations. As examples, I shall discuss the effect of turbulent mixing, mass loss and WIMPS.

2. Present status

2.1 THE ROLE OF THE INGREDIENTS

Figure 1 shows the radial behaviour of two important quantities influential on neutrinos production and seismology: temperature and mean molecular weight, and the respective role of the ingredients versus depth. In the outer 3% of the solar radius, the mean molecular weight increases rapidly, all the elements are partially ionised, or even neutral, molecules

125

appear, so a refined equation of state (Däppen and Lebreton 1989, Christensen-Dalsgaard et al 1988a) is a necessity together with detailed opacity calculations including molecules and non adiabatic estimate of the sound speed. Convection dominates the outer 27% of the Sun , so opacity do not play a dominant role there (the temperature gradient is taken as adiabatic), and physical understanding of the dynamical effect of convection is crucial to go beyond the limitations of the reference model regarding the surface abundances of lithium and boron (Cayrel et al 1984) and the extention of the convective zone (Vorontsov 1988, Shibahashi and Sekii 1988). This region is certainly the most difficult to simulate today but is very crucial for evolved stars. In the case of the Sun, the age and the radius are precisely known, so the mixing length parameter α is determined along with the approximate size of the convective zone and we can partly disconnect this external region from the internal one.

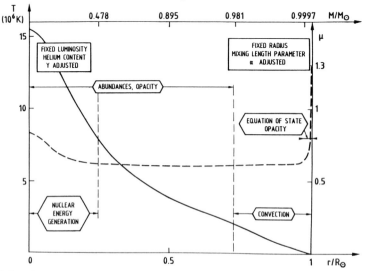

Figure 1: Radial distribution of the temperature (T—) and the mean molecular weight (μ - -) of our reference model (paper1) and sites of privileged importance of the ingredients.

Thus I shall concentrate, in this paper, on the inner 70-75% radiative part of the Sun, which has been thoroughly studied this last 20 years. Here, the equation of state plays a minor role: the gas is nearly perfect, hydrogen and helium are completely ionised and the effect of the other elements is smaller than 1% . The behaviour of the "intermediate" region between nuclear core and convective zone is dominated by the opacity coefficients. The central part is influenced by the nuclear reactions rates and the abundances through the opacity coefficients. The initial helium content, unobservable in the photosphere, is adjusted to fit the present luminosity. So, abundances, opacities and nuclear reaction rates are tightly correlated in this inner 25 % of the solar radius.

2.2 THE REFERENCE MODEL

An updated reference model offers the opportunity to choose the direction on which effort must be stressed in the classical framework. The model of Turck-Chièze et al (1988) based upon a Paczynsky code includes new nuclear data: neutron lifetime determined by Bopp et al (1986), the recent value of Krauss (1987) for the rate of the reaction $^3He(^3He,p)^4He$ and the value recommended by Filippone (1986) for ^{3}He$(\alpha,\gamma)^7$Be. The opacities are derived from the Los Alamos library (Huebner et al 1977) using the relative abundances of Ross and Aller (1976), the equation of state includes the Coulomb interaction in the Debye-Huckel approximation. The initial abundances are taken from Aller (1986) and the adopted ratio of the metal abundance to hydrogen (by mass) Z/X is 0.028±0.03 according to Aller (1986) and Meyer (1987). The main results are the following:

initial helium content $Y = 0.273 \pm 0.012$ $\quad$ $r_{BCZ} = 0.73 \, R_\odot$

$T_C = 15.5 \, 10^6 \, K$ $\qquad$ $\rho_C = 150 \, g \, / \, cm^3$ $\qquad$ $T_{BCZ} = 2.02 \, 10^6 \, K$

neutrino capture rates (SNU) $\quad$ ^{71}Ga: 124 ± 5 $\qquad$ ^{37}Cl: 5.7 ± 1.3

The chlorine prediction is about three times greater than the experimental average (1968-1986) of the Davis experiment but nearly consistent with the result of 1987 (4.2± 0.7 SNU) (Davis 1989).

The acoustic frequencies of this model have been computed (Turck-Chièze, Däppen and Cassé 1988). The absolute low degree p-modes are systematically underestimated by about 10 µHz compared to the observations (Grec, Fossat and Pomerantz 1983, Duvall et al 1988). The calculated $\Delta\nu_{20}$ (see previous references for definition) is about 1µHz greater than the observed one. Pressure p-modes are sensitive to the whole structure of the Sun and particularly to the outer which is estimated with less care than the inner part. Since the accuracy of the data and of the inversion technique improves so rapidly, I prefer to rely on the adiabatic sound speed which allows to explore different depths of the Sun. Figure 2 illustrates the difference between our reference model and the sound speed derived from the observations by Christensen Dalsgaard, Gough and Thompson (1988). The relative difference in the square of the adiabatic sound speed is lower than 2% in the radiative region. This implies a variation of the ratio T/µ not greater than 2%, with possible greater effect on T and µ individually. Assuming a polytropic behaviour (P=ρ^γ), one can deduce the sign of the variation of ρ and its amplitude assuming a nearly perfect gas ($c^2 = 5/3 \, P/\rho$). We notice a polytropic exponent of 1.2 in our models, suggesting density variation < 10%.

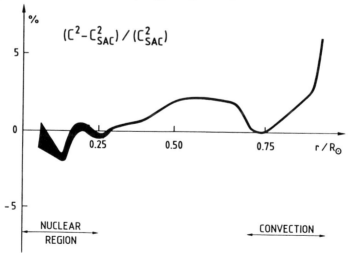

Figure 2: Comparison of the square of the adiabatic sound speed deduced from the observations by Christensen Dalsgaard et al (1988b) and that of paper 1.

2.3 PRESENT ACCURACY OF THE REFERENCE MODEL

In paper 1, the different contributions to the uncertainty on the predicted neutrino capture rates and helium content are discussed. The corresponding logarithmic derivatives of the sound speed will be published elsewhere (Turck-Chièze and Däppen 1989). Except for the gallium prediction, where the accuracy is quite good, we have pointed out two main sources of unaccuracy: the ratio Z/X and the central opacities. We stressed also the important role of the $^7Be(p,\gamma)$ reaction rate for the chlorine prediction which is, in fact, the key reaction for chlorine (70% of the neutrinos produced) and water (100%) detectors (Turck-Chièze 1988).

2.3.1 The nuclear reaction rates and the case of $^7Be(p,\gamma)$ ^8B.

All the reactions of the p-p chain and the CNO cycle have been remeasured recently, most of the cross sections are determined at a level of 5%. (Rolf 1989). Each individual experiment has more limited accuracy but one usually uses the average value of all the different experiments and the corresponding error bar. Moreover they are most generally performed at higher energies than the astrophysical range (typically 20 keV for central solar conditions), so a theoretical extrapolation is necessary. As an illustration I discuss the case of $^7Be(p,\gamma)$. The experimental situation is summarised in Parker (1986) and recalled in table 1. The mean value is recommended by Filippone (1983). But as usual, the compilation is presented in terms of the astrophysical factor $S_{17}(0)$ derived with, in each case, the crude theoretical extrapolation of Tombrello (1965). This theoretical calculation has been revised by Barker (1980,1983) and more recently by Kajino (1988) (figure 3). These two calculations, including not only s-state (as in the case of Tombrello) but also d-state, are in agreement together and lead to a reduction of $S_{17}(0)$ of 15%. **We adopt the mean experimental value recommended by Filippone (1986) but the most recent theoretical results. $S_{17}(0)$ is reduced to 0.021 keV-b. The chlorine neutrino capture rate is consequently reduced by 13%. This choice constitutes one of the differences with Bahcall and Ulrich work (1988).** We note that the experimental results are very scattered and remeasurement on the whole range 100 keV- 4 MeV would be useful. It is the reason why we adopt an error bar of 15%.

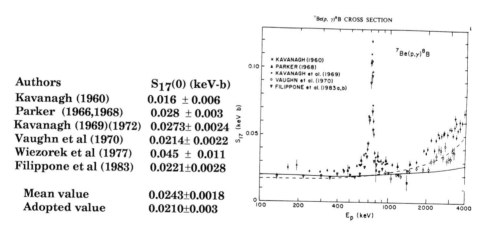

Authors	$S_{17}(0)$ (keV-b)
Kavanagh (1960)	0.016 ± 0.006
Parker (1966,1968)	0.028 ± 0.003
Kavanagh (1969)(1972)	0.0273 ± 0.0024
Vaughn et al (1970)	0.0214 ± 0.0022
Wiezorek et al (1977)	0.045 ± 0.011
Filippone et al (1983)	0.0221 ± 0.0028
Mean value	0.0243 ± 0.0018
Adopted value	0.0210 ± 0.003

Table 1

Figure 3. *Experimental measurements and theoretical extrapolation of the astrophysical factor with Tombrello (1965) (-) and Barker (1980) (- -) calculations.*

Very recently, Assenbaum et al (1987) have noticed that electron screening affects the measured cross sections at low energy, and increases exponentially, the experimental value. Since, the nuclei are mainly bare in the central Sun, this effect leads to an overestimate of S(0). It could concern $^3\text{He}(^3\text{He},2\text{p})^4\text{He}$ (6%) and $^3\text{He}(\alpha,\gamma)^7\text{Be}$ and has never been taken into account in the estimation of the uncertainties on the neutrino capture rates.

2.3.2. Role of abundances and effect of opacity.

The initial composition of the Sun is assumed to be similar to the present observable photospheric one. During hydrogen burning, the p-p chain dominates the production of energy and the CNO cycle contribution is of only 1.75%. So only elements up to oxygen participate to the nuclear reactions. On the other hand, the opacity, which regulates the energy flux in the radiative region of the star and determines its structure, is largely influenced by the heavy elements (up to iron), they contribute to 40% of the total opacity in the center and to 90% in what we call the "intermediate" radiative region. In this context, the determination of the solar composition is influential on the central structure of the Sun. *In our computation, we fix Z/X according to the observations as Bahcall and Ulrich (1988).* As a consequence, the initial helium content and the metallicity cannot be chosen independently: if the helium content increases, the metallicity must decrease, so that Z/X remains equal to the observed value. This constraint requires to interpolate between two opacity tables $\kappa = \kappa\,(\,\text{T},\rho,\text{X},\text{Z})$. Consequently, the transformation of carbon and oxygen into nitrogen during hydrogen burning is taken into account, at the first order, as an enhancement of the metallicity by 3%. This method is justified by the similarity of the opacity behaviour of C,N,O for central solar conditions. *This CNO evolution is treated differently in Bahcall and Ulrich paper(1988).* In the poster entitled " **The influence of metallicity on the opacity coefficients in the solar modelling"** (Courtaud et al **1989**), we discuss two sources of uncertainties on the opacity coefficients : the relative composition of the Sun at birth and the accuracy of the calculations. In the nuclear central part, the accuracy of the calculations is good (5%), the possible variations of the opacity could arise from the determination of the abundances. We observe variations of the central structure using Ross and Aller (1976) composition in the opacity calculation with Z/X = 0.028 or Anders and Grevesse new compilation (1989) with Z/X = 0.0273. We notice also that the unexplained discrepancy between photospheric (Fe/X=$4.68\pm0.33\ 10^{-5}$) and meteoritic (Fe/X = $3.24\pm0.075\ 10^{-5}$) iron values could lead to opacity variations up to 13% and a related reduction of the chlorine neutrino capture rate of 1 SNU. In the intermediate region,the accuracy of the opacity calculations is at a level of 20% and could increase the temperature of the bottom of the convective zone by 10%, with few effect on the central structure.

Electron collective effects must be taken into account in the calculation of the central solar opacities. It is not clear if this process is included in the library of Los Alamos 1977. We have always considered that it was not the case, following the discussion of Bahcall et al (1982) and Huebner (1986). In our computation, this correction is added according to the expression : $\kappa = \kappa - 0.07\,(1+\text{X})\ \text{cm}^2\text{g}^{-1}$ (Bahcall et al 1982). *This point must be clarified in the near future to avoid counting this effect twice or not at all.* A suppression of this additive correction increases the chlorine neutrino capture rate by 0.6 SNU and the helium content by 0.006. Boercker (1987) independently, has reconsidered this correction to the Thomson scattering and concludes that the different estimates lead to an error on the total opacity of $\pm2.5\%$ at the center of the Sun.

In summary, the principal ingredients are determined at a level of 5%-10%accuracy. A list of reasons has been evoked to justify small variations of the central temperature and perhaps 30% on the chlorine neutrino flux. But it seems difficult to find classical effects which reduce this flux by a factor 3 or 4. So room is left for extra phenomena.

3. Effect of other phenomena

As a first approach, we have considered two additional physical effects (Turck-Chièze and Däppen (1989):-an hypothetical mass loss in the early life of the Sun
                    - the influence of WIMPS in the inner part of the star (see **poster untitled "WIMPS and solar evolution codes" by Giraud-Héraud et al)**.

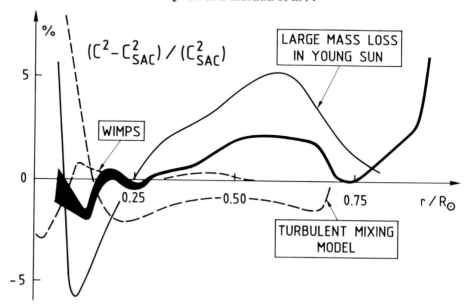

Figure 4 : Radial distribution of the squared sound speed for models including extra phenomena.

Figure 4 shows that, if one believes the extraction of the sound speed from the helioseismological data and the corresponding error bar, all the models presented here are strongly constrained or excluded:
- The present Sun (and especially its global oscillations) is unconsistent with a total initial mass loss greater than 0.2 $M_\odot$ (Turck-Chièze, Däppen and Cassé 1988), the neutrino predictions are not reduced. In the case of initial 2 $M_\odot$ (fig. 4), an initial convective core could persist at the present day (if overshooting is included), which produces a modification of the structure, at low radius, of too large amplitude but suggestive of the one deduced from the observation.
- Even if for some special conditions of mass and interaction cross section, the introduction of WIMPS appreciably reduces the neutrino capture rate as expected, the sound speed profile of the model leads to a marked disagreement with observation, at low radii. This was not predictable by looking only to the integrated value of Δv_{20}.
- Concerning turbulent mixing, the difficulty comes from the determination of the temporal and spatial evolution of this phenomenon. From the examination of the work of Schatzman et al (1981) it appears now clearly that the proposed effect is certainly surestimated, due to an overestimate of the chlorine capture rate prediction. Nevertheless, if the consequent effect on the sound speed is excessive, the inner profile is interestingly not far from that deduced from the observations.

4. Comments

In the inner radiative region of the Sun, most of the input parameters are now estimated within 10% accuracy or even less, possible new sources of unaccuracy are critically examined: electron screening effect in cross section measurements at low energy, electron collective effect in opacity calculations, role of partially ionised heavy elements in opacity calculations. In the classical framework, the internal structure of the Sun seems more and more established and only small variations of temperature and density are possible. The 27-30% external part is more questionable due to dynamical effects as convection; detailed equation of state and nonadiabatic estimate of sound speed are a necessity for the 3% outer.

We consider that the predicted ^{8}B neutrino flux accuracy is not yet well established and propose different possible sources of uncertainties: the measurement of ^{7}Be (p,γ) ^{8}B reaction rate (15%) and the determination of the central opacities (20%) due to evolution of detailed abundances. We emphasize the importance, for this quantity, to solve the discrepancy between photospheric and meteoritic iron abundance.

The confrontation of the classical theory of the Sun evolution with the observations shows that the thermodynamical quantities of an updated reference model are representative of the real Sun at a level of 10% accuracy in the radiative region: neutrino fluxes detection mainly constrains the temperature of the central region (r < 0.15 RO), helioseismology constrains the inner density in the same region. Following Kosovishev (1989), and our recent results on neutrino fluxes, a certain compatibility seems to exist between seismological interpretation and present chlorine detection. Is temporal evolution of these two kinds of information an help to interpret a larger reduction of neutrino flux as was observed before?. The study of dynamical effects is now largely encouraged by the seismological constraints in the central part. The hypothesis of WIMPS seems more questionable but central limited mixing and very weak mass loss cannot be excluded.

5. Acknowledgements

This work was initiated four years ago by M. Cassé and is the fruit of a large collaboration. The stellar evolution code was provided by J. P. De Greve, B. de Loore and C. Doom, the Los Alamos opacities by A. Noels and R. Papy, the nucleosynthesis was performed by N. Prantzos, the seismological interpretation was done with W. Däppen. More recently, a collaboration was engaged with D. Courtaud, G. Damamme, E. Genot and M. Vuillemin to critically discuss the opacity determination, and with A. Bouquet, Y. Giraud-Héraud, J. Kaplan, F. Martin, C. Tao for the study of the effect of WIMPS. Basic results were obtained with S. Cahen and C. Doom. I am very grateful to all of them.

6. References

Aller,L. H.,1986, in *Spectroscopy of Astrophysical Plasmas,* ed A. Dalgarno and D. Layzer (Cambridge: Cambridge University Press) p 89.
Anders, E. and Grevesse, N., 1989, *Geochim. Cosmochim. Acta*, **53**, 197.
Assenbaum, H.J., Langanke, K., and Rolfs, C. 1987, *Z. Phys. A-Atomic Nuclei,* **327**, 461.
Bahcall,J.N., Huebner,W.F., Lubow,S.H., Parker,P.D., and Ulrich,R.K. 1982, *Rev. Mod. Phys.*, **54**, 767.
Bahcall, J. N., and Ulrich, R. K. 1988, *Rev. Mod. Phys.*, **60**, 297.
Barabanov, I. R., et al 1985, in *Solar Neutrinos and Neutrino Astronomy*, ed. M. L. Cherry, K. Lande, and W. A. Fowler (New York:AIP), p. 175.
Barker, F. C. 1980, *Aust. J. Phys.*, **33**, 177.

132

Barker, F. C. 1983, *Phys. Rev. C*, **28**, 1407.

Boercker, D. B. 1987, *Ap. J. (Letters)*, **316**, L95.

Bopp, P., Dubbers, D., Hornig, L., Klemst, E., Last, A., and Schultze, H. 1986, *Phys. Rev. Lett.*, **56**, 919.

Cayrel, R., Cayrel De Strobel, G., Campbell, B., Däppen, W. 1984, *Ap. J.*, **283**, 205.

Christensen-Dalsgaard, J., Duvall, T. L., Jr, Gough, D. O., Harvey, J.W., Rhodes, E. J., Jr, 1985, *Nature*, , **315**, 378.

Christensen-Dalsgaard, J., Däppen, W., Lebreton, Y. 1988a, *Nature*, **336**, 634.

Christensen-Dalsgaard, J., Gough, D. O., Thompson, M. J. 1988b, in *Seismology of the Sun and Sun-like stars*, **ESA SP-286**, 493.

Courtaud,D., Damamme,G., Genot,E., Vuillemin,M., Turck-Chièze,S. 1989, This conference.

Däppen, W. and Lebreton, Y. 1989, Astr. Ap., in preparation. and references therein.

Davis, R., Jr., Harmer, D. S., and Hoffman, K. C. 1968, *Phys. Rev. Letters*, **20**, 1205.

Davis, R. 1989, These proceedings.

Duvall, T.L. , Harvey, J. W., Libbrecht, K.G., Popp, B.D., and Pomerantz, M. A. 1988, *Ap. J.*, **324**,1171.

Filippone, B. W. , Elwyn, A. J., Davids, C. N., Koetke, D. D. 1983, *Phys. Rev. Lett.*, **50**, 412. and *Phys. Rev. C*, **28**, 2222.

Filippone, B. W. 1986, Ann. Rev. Nucl. Part. Sci.,36, 717.

Giraud-Héraud,Y., Kaplan, J., Martin, F., Tao, C., Turck-Chièze, S., 1989, Solar Phys.,This conference.

Grec, G., Fossat, E. and Pomerantz, M. A., 1983, *Solar Phys.*, **82**, 55.

Gough,D.O., and Kosovishev,A.G. 1988, in *Seismology of the Sun and Sun-like stars*, **ESA SP-286**, 195.

Huebner, W. F., 1986, in *Physics of the Sun*, Vol 1, ed. P. A. Sturrock (Dordrecht: Reidel), p. 33.

Huebner, W. F., Merts, A.L., Magee, N. H., Jr., and Argo, M.F. 1977, Los Alamos Sci. Lab. Rept., LA-6760-M.

Kajino, T., Bertsch, G. F., and Barker, F. C. 1988, in *Clustering Aspects in Nuclear and Subnuclear Systems,* to appear in Suppl. J. Phys. Soc. Japan

Kirsten, T. 1986, in *86 Massive Neutrinos in Astrophysics and in Particle Physics*, ed. O. Fackler and J. Tran Thanh Van (Ed. Frontières, Gif sur Yvette) p119.

Koshiba, M. T. 1988, in *5th Force and Neutrinos Physics* , ed. O. Fackler and J. Tran Thanh Van (Ed. Frontières, Gif sur Yvette) p 215.

Kosovishev, A.G. 1989, These proceedings.

Krauss, A., Becker, H.W., Trautvetter, H.P., and Rolfs, A. 1987, *Nucl. Phys. A*, **467**, 273.

Meyer, J. P. 1987, in *Origin and Distribution of the Elements,* ed. G. J. Matthews (Singapore: World Scientific), p 337.

Parker, P. D. McD. 1986, in *Physics of the Sun*, Vol 1, ed. P. A. Sturrock (Dordrecht: Reidel), p. 17.

Rolf, K. 1989, These proceedings.

Ross, J.E., and Aller, L. H. 1976, *Science*, **191**, 1223.

Schatzman, E., Maeder, A. 1981, *Astr. Ap.*, **96**, 1.

Shibahashi, H., and Sekii, T. 1988, in *Seismology of the Sun and Sun-like stars*, **ESA SP-286**, 471.

Tombrello, T. A. 1965, *Nucl. Phys.*, **71**, 459.

Turck-Chièze, S. 1988, in *5th Force and Neutrinos Physics* , ed. O. Fackler and J. Tran Thanh Van Ed. Frontières, Gif sur Yvette) p 165.

Turck-Chièze, S., Cahen, S., Cassé, M., Doom, C. 1988, *Ap. J.*, **335**, 415.

Turck-Chièze, S., Dappën, W. , Cassé, M. 1988, *in Seismology of the Sun and Sun-like stars*, **ESA SP-286**, 629.

Turck-Chièze, S., Däppen, W., 1989, in preparation.

Vorontsov S. V. 1988 *in Seismology of the Sun and Sun-like stars*, **ESA SP 286**, 475.

A LOOK ON NON–STANDARD SOLAR MODELS

André Maeder
Geneva Observatory
CH-1290 Sauverny, Switzerland

ABSTRACT. A short review of some of the non–standard models proposed in these last two decades is presented. Their main physical assumptions are shown, as well as the way they meet or do not meet the various observational constraints.

1 PHILOSOPHY OF NON–STANDARD MODELS

The French physicist J.M. Levy–Leblond (1981) in a paper entitled "Eloge des theories fausses" has distinguished four kinds of false theories: 1. The adherent ones, which are correct over a sizeable but still limited range of conditions. 2. The different ones, which are internally coherent but finally in disagreement with nature. 3. The aberrant ones, which contain big mistakes and errors. 4. The "siderant" ones, which are just scientific phrasing without scientific meaning. Of course, the non–standard models considered here do not belong to the last two categories! Apart from joke and as emphasized by Levy–Leblond, it is very didactical to understand why a non–standard theory may fit or not; a better comprehension of the physics is gained in this way.

Numerous non–standard solar models have been proposed, mainly stimulated by the solar neutrinos problems. They cannot all be right, but anyhow the study of their properties and why they fit or do not fit the observational requirements gives us a deeper insight and understanding of solar and stellar physics. Also, it is quite likely that some features of today's non–standard models will be retained in the standard models of tomorrow.

On the whole we must say that the study of many non–standard models has also increased our confidence in the standard model (cf. Bahcall, 1989). Some deviation from it will probably be found and confirmed in the future, but it is unlikely to be large.

Among other recent reviews on non–standard models we may quote that by Newman (1986) and the one by Bahcall and Ulrich (1988).

2 MODELS WITH ADDITIONAL EFFECTS IN CLASSICAL PHYSICS

This is the broadest and, in my opinion, most interesting class of non–standard models. Some of their properties will certainly be incorporated as basic ingredients in future, particularly regarding transport processes and rotation.

2.1 Models with transport in absence of rotation

Thermal diffusion and gravitational settling is just standard physics. These processes have been included in the solar model by Gabriel et al. (1983), and more recently by Cox et

G. Berthomieu and M. Cribier (eds.), Inside the Sun, 133–144.

al. (1989), using a procedure devised by Iben and McDonald (1985). Cox et al. find that the helium and metal abundances changed from $Y = 0.30$ to 0.263, and from $Z = .02$ to .179 at the solar surface during its past evolution. At the center the effects are of the order of 10^{-3}. Clearly, no major change occurs neither for the solar neutrino flux, nor for the p–mode frequencies which are still too small by $10 - 20 \, \mu Hz$. Anyhow, the effects on surface abundances and opacities are significant.

The possibility that the positive He–3 gradient produces instabilities leading to mixing was examined by Dilke and Gough (1972); cf. also Ulrich and Rood (1973). According to Christensen–Dalsgaard et al. (1974), the He–3 gradient may produce unstable low order g–modes, but the issue regarding mixing remains uncertain.

The case of nuclear instabilities in the Sun was studied with negative conclusions by Schwarzschild and Härm (1973) and Gilliland (1985) for the radial case and by Rosenbluth and Bahcall (1973) in the non radial case.

2.2 Models with rotational mixing

A high interior ratio of centrifugal force to gravity may lower the central T and thus the neutrino flux. However, the data on solar oblateness place some constraints (cf. Rood and Ulrich, 1974) on the many suggestions made along that line in the past. Rotationally induced mixing is certainly a more promising possibility in which rotation could influence solar evolution.

2.2.1 Models with rotationally induced turbulent diffusion

The concept of turbulent diffusion was introduced by Schatzman (1977) and applied in a series of papers by himself and coworkers. The idea is the following one. According to Schatzman (1962) the slow rotation of late type stars results from the loss of angular momentum by stellar winds. This leads to differential rotation, the interior spinning faster than the surface. The increase of the angular velocity with depth is limited by the hydrodynamical instabilities, which cause an outward diffusion of angular momentum. The external convective zone of the Sun is highly viscous and rotates nearly at the surface rate. Schatzman (1977) suggested that the shear between the rapidly rotating interior and the external convective zone is unstable and leads to turbulence. The angular momentum is transported from core to surface, therefore reducing the differential rotation rate until stability is restored. Further loss of angular momentum again generates a shear flow with a large angular velocity gradient, thus re–establishing the turbulent regime. Schatzman suggests that the Sun adjusts itself in such a way that it is always locally at the edge of shear flow instability. As it is difficult to predict the level of the turbulence and its transport efficiency, he considers turbulent mixing as an isotropic diffusion process, the effective diffusivity D being larger than the microscopic viscosity by a constant factor called Re*. The diffusion equation is:

$$\rho \frac{\partial X_i}{\partial t} = \operatorname{div}(\rho D \operatorname{grad} X_i)$$

with

$$D = \operatorname{Re}^*(\nu_{\mathrm{rad}} + \nu_{\mathrm{mol}})$$

where ν_{rad} and ν_{mol} are respectively the radiative and molecular viscosities.

Since these first tentative investigations much progress has been made in the understanding of the instabilities and their occurrence in rotating stars (Zahn, 1983, 1985; Spruit, 1985;

cf. also Tassoul and Tassoul, 1989). Zahn proposes the following very interesting mixing mechanism: when a star does not rotate cylindrically, the surfaces of constant pressure and the isentropic ones no longer coincide and the equilibrium state is called baroclinic; in the very small wedge between a surface of constant entropy and the corresponding horizontal surface horizontal motions may become unstable, generating turbulence. The turbulence is two–dimensional, at least for scales where the Coriolis force is greater than the inertial one (i.e. for scales which satisfy the Rossby criterion: $\Omega > 2.5\,u/l$). The usual cascade of turbulence towards small scales leads to a three–dimensional turbulence as soon as the Rossby condition is violated. This is this small scale 3–D tail of the 2–D turbulence which is responsible, according to Zahn, for the vertical diffusion of the angular momentum and chemical elements. Assuming that the kinetic energy of differential rotation, fed itself by advection of the angular momentum, is converted into turbulent energy, Zahn obtains the following value of the turbulent diffusion coefficient D, due to the 3–D turbulence:

$$D = \mathrm{Re}^{*}\,\nu$$

with

$$\mathrm{Re}^{*} \simeq \frac{K}{\nu}\,\frac{\Omega^{2}r}{g}\,\left(\nabla_{\mathrm{ad}} - \nabla_{\mathrm{rad}}\right)^{-1}$$

where r is the distance to the center, g the gravity at distance r, Ω the angular velocity, $\nabla_{\mathrm{ad}} - \nabla_{\mathrm{rad}}$ the local subadiabatic gradient in the radiative zone and K is the thermal diffusivity. This holds in the local instability approximation (i.e. it is not valid near the center). The simulation of models with this Re^{*}–expression is difficult, because we do not know the spatial distribution of the differential rotation Ω in the Sun and other main sequence stars. The value of Ω should depend on the radius (r) and on the latitude (Λ) considered. However, Zahn (1973) has shown that an Ω–dependence on Λ would imply a rapid re–distribution of angular momentum along equipotential surfaces by finite amplitude shear instabilities, so as to make Ω constant on such surfaces. This prediction was beautifully confirmed by the analysis of the frequency splitting of solar p–models (cf. Brown, 1985), which shows that, for growing depths, the solar latitudinal differential rotation is much smaller than at the surface, the rotation being roughly constant on spheres with a rotation rate close to the surface equatorial value. Zahn (1983) also showed that the vertical μ–gradient can prevent the 2–D turbulence from becoming 3–D, and this effect was generally accounted for in the models.

Various applications of the turbulent diffusion model were made to account for the abundances of Li, Be and B at the surface of the Sun (Vauclair et al., 1978), the solar neutrino flux (Schatzman and Maeder, 1981), the ratio $({}^{12}\mathrm{C}/{}^{13}\mathrm{C})$ at the surface of red giant stars (Genova and Schatzman, 1979; Bienaymé et al., 1984), the solar and stellar Li-abundances (Baglin et al., 1985), the He-3 abundance and the frequencies of solar oscillations (cf. Lebreton and Maeder, 1987).

Presently, as strong constraints on the possibility of mixing in the Sun, we may mention the following ones:

^{3}He: Bochsler et al. (this meeting; see also Geiss, 1972) show that the abundance of ^{3}He at the solar surface cannot have increased by more than 15% over the solar lifetime. This beautiful result places a very strong constraint on mixing processes in the deep interior. Indeed, as shown by Bochsler et al. the Reynolds number Re* cannot be larger than about 30 in the interior. Equivalently, if

$$\mathrm{Re}^* = \frac{8r}{\alpha^2 H_p}$$

as it was represented by Lebreton and Maeder (1987), $\alpha \simeq 1.3$, leading to diffusion coefficients 146, 225, 321, 538, 853 cm^2/sec at $Mr/M =$ .77, .85, .90, .95, .98. Also, in this model of the Sun, the μ–gradient would have stopped mixing in regions interior to $M_r/M = 0.76$.

^{7}Li: Baglin et al. (1985) show that it is necessary to call for a diffusion process in the outer layers to account for the smooth decrease of the Li–abundance for stars of decreasing masses in the Hyades. Thus, contrarily to a frequently made assumption, the overshooting below the external convective zone is not an appropriate mechanism for explaining the low Li abundance in the Sun, because one would then have to call for very different overshooting parameters for stars of different masses. On the contrary, turbulent diffusion with a unique parameter is shown to be able to account for the Li abundances in stars of various masses.

Solar p-modes: In order to explain by turbulent diffusion the low solar neutrino flux, it would be necessary to assume a rather high diffusion coefficient of the order of Re* = 100 (cf. Schatzman and Maeder, 1981). However, a number of authors have shown that such a high diffusion would be in contradiction with the results on the low degree solar acoustic modes (cf. Ulrich and Rhodes, 1983; Provost, 1984; Berthomieu et al., 1984; Cox et al., 1985; Christensen–Dalsgaard, 1986; Lebreton and Maeder, 1987; Lebreton et al., 1988). In particular, the splitting of low degree modes is much too large in models with a diffusion. However, as shown by Lebreton and Maeder, diffusion with a moderate coefficient Re* and with inclusion of the μ-gradient inhibition does not change the splitting values.

Figure 1 shows the difference $\delta c^2/c^2$ betwen a strongly mixed (Re* = 100) and a standard model (cf. Christensen–Dalsgaard, 1986) where c is the sound velocity. As $c \sim (T/\mu)^{1/2}$, the high central peak is due to the lower μ of the diffusion model, which is compensated by a higher μ and a lower c outwards from $r/R = 0.15$. It is interesting to notice that the shape in Fig. 1 is very reminiscent, but with a lower amplitude, of the difference between the observed and the standard model (cf. Gough and Kosovichev, 1988).

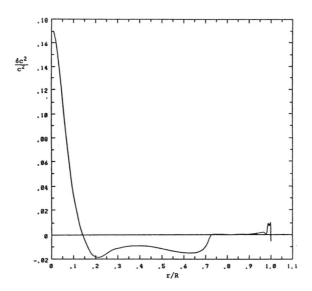

Figure 1: Difference $\delta c^2/c^2$ between a strongly mixed $(Re^* = 100)$ and a standard model, c being the sound velocity (from Christensen–Dalsgaard, 1986).

Solar g-modes: Mixing leads to flatter μ and temperature gradients in the Sun. Thus, the restoring force in radiative regions is smaller and so is the Brunt–Väisälä frequency. Thus, the period spacing P_o of the gravity modes in partially mixed models is increased with respect to the standard case (cf. Berthomieu et al., 1984; Cox et al., 1985; Lebreton and Maeder, 1987). The observed values of Po lie in the range of 36 to 41 minutes, while standard models give values of Po between 34 and 37 minutes. This discrepancy is compatible with a value Re* of about 20 in deep interior regions $(r/R \simeq 0.15)$.

Thus, it is very interesting to notice that the He–3 abundance and solar oscillations place the same rather severe limit on diffusive mixing in the deep interior. However, it is important to realize that if the mixing would effectively be at the limit of Re* = 20 or 30, very significant consequences would result for the evolution of the Sun and low mass stars. In particular, the lifetimes of the low mass stars would be increased, with non–negligible consequences for the age determinations.

2.2.2 Models with rotational mixing and transport of angular momentum

The rotational velocity and the internal distribution of the angular velocity $\Omega(r)$ in the Sun starts becoming basic observational constraints to be added to the traditional M, L, R, T_{eff} abundances. Since long valuable attempts to properly modelize the internal $\Omega(r)$ have been made (cf. Endal, 1987; Sofia, 1987; and reference therein). Recently Pinsonneault et al. (1989) have developed a new rotating stellar evolution code and applied it to the Sun. As specific basic ingredients they incorporate the loss of angular momentum J via magnetic winds by adopting some parametrisation of dJ/dt in terms of Ω, M, R and mass loss rate $\dot{M}$. The redistribution of angular momentum by various processes, such as Eddington circulation, Goldreich–Schubert–Fricke instability and shear instability, is accounted for; the mixing of chemical species resulting from these instabilities and circulation motions are also considered. This modellisation of complex processes of course contains free parameters that the authors adjust on the observations.

These models of the rotating Sun contain several very interesting results among which we may mention:

— In order to reproduce both the observations or abundances and rotation, the diffusion coefficient of chemical elements has to be only a small fraction (i.e. about 5%) of the diffusion coefficient of the angular momentum (cf. Spruit, 1987; Schatzman, 1987; see also Tassoul and Tassoul, 1989).

— The angular velocity Ω is nearly constant in the outer layers ($r/R > 0.5$). A central fast spinning core (with rotation period $< 12h$) is present and these features are found for a wide range of global parameters.

— The present Ω at the surface of low mass stars is weakly dependent on the initial angular momentum. The reason is that for fast rotation the loss of angular momentum at the surface is also higher, as well as the redistribution process.

— However, the value of the initial angular momentum strongly influences the present values of the abundances of some elements, in particular Li and to a lesser extent Be and He-3 in the Sun. The reason is that for a higher initial angular momentum, the instabilities and transport processes are also more active. Rotation also influences CNO abundances in red giants after the first dredge–up. The necessity of an additional main sequence destruction of lithium in order to explain the observed Li abundances in red giants has also been shown recently by Charbonneau et al. (1989).

Recently, Tassoul and Tassoul (1989) have calculated the secular changes of the solar inner rotation, taking into account the solar wind torque as well as the gradual reduction of the eddy diffusivities in the chemically inhomogeneous inner regions. In particular, they show that a moderate amount of eddy–like or wave–like motions is sufficient to keep the inner and outer parts of the Sun's radiative interior rotating nearly uniformly. Thus, there is no reason, according to Tassoul and Tassoul, to assume the existence of a large magnetic field that could enforce the uniform rotation by itself. They also suggest a ratio of eddy diffusivity to eddy viscosity of the order of 5 to 10%, as quoted above.

2.3 Models with magnetic fields

The inclusion of magnetic pressure $B^2/8\pi$ could in principle lower the central temperature. But a proper corresponding balance of the helium and metal contents has to be made and then, as shown by Iben (1969) and Bahcall and Ulrich (1971), there is no longer a reduction of central T and neutrino flux to be expected.

The magnetic tension term $B \cdot (\nabla B)$ has to be included and this requires a modellisation of the field geometry. Models with a magnetic field of the Cowling type were made by Ulrich and Rhodes (1983). They showed that the tension term leads to a reduction of central pressure and neutrino flux. However, the field has to be very high, i.e. $3 \cdot 10^8$

Gauss for a decrease of the neutrino flux of 30%. An attractive feature is that the p–mode frequencies, and spacing as well, are increased in the sense required by observations.

2.4 Models with accretion or high mass loss

Accretion was topical for a time with the suggestion by McCrea (1975) that ice ages on the earth may be due to the entering of the solar system into a dense interstellar cloud. Both the effects of extinction and of the release of gravitational energy should be accounted for. Current estimates of the mass accreted by the Sun in its evolution give (cf. Talbot and Newman, 1976) values of the order of $10^{-4} M_\odot$. Thus, this effect is unlikely to have had great consequences.

The idea that the Sun might have experienced a relatively high mass loss was put forward by Willson et al. (1987) and Guzik et al. (1987). They suggest that main–sequence stars located in the pulsation instability strip may experience mass loss rates in excess of $10^{-9} M_\odot y^{-1}$. Thus typically, starting on the zero age sequence with an initial mass of about $2 M_\odot$, the Sun would reach its present mass in about $10^9 y$. There is neither detailed physical process demonstrated to be able to do that, nor any direct observational confirmation of the advocated mass loss rates. The authors propose this high mass loss to be responsible for the existence of blue stragglers; the consequences for the age determination and history of the Galaxy are also emphasized.

A fortunate and unvoluntary consequence of this model is to demonstrate how powerful and constraining some of the available observational data are.

— ^3He: The new and reliable determination of the ^3He abundance at the solar surface by Bochsler et al. (1989) shows that the ^3He enrichment cannot be larger than 15% during the solar life, while the models by Guzik et al. predict an enrichment of nearly an order of magnitude. There is no way to circumvent this difficulty.

— Li, Be, B: These elements would have been very early completely destroyed at the solar surface in the model of the mass–losing Sun. Guzik et al. (1987) advocate for spallation reactions to synthesize these elements at their present level in the Sun. Indeed, if this were true, the Li abundance of low mass stars in clusters would increase with cluster ages. The observations by Hobbs and Pilachowski (1988; see also Twarog and Twarog, 1989) show just the opposite: Li regularly declines for old age clusters. Moreover, the model by Guzik et al. predict complete destruction of Li in the solar–like stars of the Pleiades, while the observations on the contrary show an absence of depletion with respect to the interstellar medium.

— Solar oscillations: The course of the sound speed in the interior of a solar model having experienced a high mass loss is incompatible as shown by Turck–Chieze et al. (1988) with the helio–seismological results. The limit placed by the inversion technique on the total amount of mass lost by the Sun is about $0.2 M_\odot$. This is much less constraining than the Li–data in clusters, which tell us that the Sun may not have lost more than about 3% of its mass during the first hundreds of million years. Otherwise Li would already be entirely depleted in clusters in the age range from Pleiades to Hyades, which is not the case.

In conclusion, three different valuable observations let no room for a highly mass–losing solar model.

2.5 Models with effects resulting from pre–main sequence evolution

Some effects in pre–main sequence evolution, such as a separation of the elements in the solar nebula, could still influence the present structure of the Sun. We may range the inhomogeneous initial solar models in this category, although their authors frequently just aimed at providing some trial and test calculations, which often are very instructive.

A low interior metallicity Z (with a proper account of the helium abundance in order to match the solar luminosity) leads to a lower solar neutrino flux (e.g. Iben, 1969; Ulrich and Rhodes, 1983; Bahcall and Ulrich, 1988). However, as shown for example by these last authors (cf. also Christensen–Dalsgaard and Gough, 1980; Ulrich and Rhodes, 1983), the low Z model increases the disagreement between the observed and calculated frequency of low degree p–modes.

The consequences of a high interior helium content Y were also explored (e.g. Iben, 1969; e.g. Bahcall and Ulrich, 1988). The following interesting result was shown by Bahcall and Ulrich. Some changes of the Sun's composition, for example an increase in Y by 0.025 between $Mr/M_\odot = 0.2$ to 0.4, can considerably improve the agreement for solar oscillations with little change for the neutrino flux, which illustrates the great specificity of the various constraints.

Another consequence of pre–main sequence evolution concerns the possibility of survival of convective cores in the Sun. It was occasionally suggested that a central convective core may be sustained for some time by convective overshooting (e.g. Roxburgh, 1987). In this connection, it would be quite interesting to examine how the presence and temporary survival of a convective core at the beginning of main sequence evolution depend on the proper or improper equilibration of He–3 and CN elements as a result of pre–main sequence evolution.

3 EXOTIC MODELS

Such models either call for "new" physics, not (or maybe not yet) generally accepted or not confirmed by experience or, as for the model of the Sun with a central black hole, they bring a well accepted concept in a context where it is highly unusual. However, it is very hazardous to attempt to further characterize these models. Perhaps the only common feature to all these exotic models is that they generally have been proposed with a very specific aim in mind (e.g. solving the neutrino problem) and that the authors have not submitted their models to the widest range of possible observational tests.

3.1 Models with modified gravity

Many solar models in the line of *Dirac's Large Number Hyphthesis* were made (see for example Pochoda and Schwarzschild, 1964; Maeder, 1977). In the case proposed by Dirac, where the gravity constant G varies like $1/t$, where t is the cosmic time, the Sun would have been more tightly packed together in the past and would look more evolved. The explanation of the present solar luminosity would require a much too low initial He–content. Other cases were considered in the framework of the scale covariant cosmology (cf. Canuto et al., 1977; see also Maeder, 1979), but the solar tests were generally less constraining than astrometric measurements in the solar system, which generally do not favour such theories.

Central gravity in stars could also be modified by *the hypothetic presence of central black holes* remaining from the Big Bang (cf. Hawking, 1971). Clayton et al. (1975b) considered black holes of $10^{-5} M_\odot$, assumed to radiate at their Eddington luminosity. The auxiliary energy source provides about half of the solar luminosity and the central temperature is thus reduced, which leads to a low solar neutrino flux. The various implications of this model in other evolutionary phases (e.g. star formation) have not been worked out.

The effects of *the so-called fifth force*, which would reduce the gravity constant G over short ranges ($\lesssim$ 200 m), have been examined by Gilliland and Däppen (1987) for their consequences on solar and stellar evolution. They find that the changes to solar structure, neutrino fluxes and oscillation frequencies are within current observational and theoretical uncertainties. Stellar lifetimes could, however, undergo significant changes.

3.2 Non standard effects affecting nuclear physics

There are so many explicit or implicit hypotheses made in stellar microphysics, which are generally no longer questioned and are adopted once for ever, that any critical discussion in this context is most welcome. Such is the case for the study by Clayton et al. (1975a) of possible *depletion of the high energy tail of the Maxwell distribution*. Even a very weak depletion could highly reduce the solar neutrino flux, and Clayton's et al. work illustrated the need for many body physics in the study of long–range Coulomb interactions.

More exotic was the proposal by Libby and Thomas (1969) of *fusion catalysis by free quarks* which, even if present in stellar interiors, would not appreciably affect the energy production in the Sun (cf. Salpeter, 1970).

The question of hypothetic *small violations of Pauli Principle* was raised by Plaga, (1989).

Very valuable exploratory calculations were performed by Bahcall and Ulrich (1988) who examined the consequences of *non–standard nuclear physics* on solar structure, solar oscillations and neutrino flux, by arbitrarily modifying the rates of some nuclear reactions. Such simulations allowed them to enlighten the most critical factors for the various reactions considered.

3.3 Modified energy transport

The consequences of changes in the process of energy transport have been studied by Newmann and Fowler (1976): more transport means a lower T–gradient and thus a lower neutrino flux. More recently Faulkner and Swenson (1988) examined the consequences of the existence of isothermal core in stars, by arbitrarily multiplying the opacity coefficient by a factor 10^{-3} in the central 5% of the solar mass. They found that stellar evolution on the main sequence is then determined by the attempt of violation of the Schönberg–Chandrasekhar limiting mass fraction. These studies were made in relation with the WIMP models, a presently interesting and topical class of non–standard models into which we shall not enter here since it is largely covered by other presentations at this meeting. These models assume exotic particles, the WIMPS, affecting the energy transport in stars. Massive non–baryonic particles were also considered for their effects as internal energy sink by Finzi and Harpaz (1989).

4   CONCLUSIONS

To the major question "Do we need to change the standard model?" one would tend to answer "yes probably, but not by very much". In the case of extra–mixing, for example, one has learned that whatever mixing may be present in the solar interior, it is not likely to be larger than as given by $Re^* = 20 - 30$ in the deep interior which places rather severe limits on the motion of chemical species in the profound solar interior. On the whole, it is impressive how robust the standard model has been found over the last twenty years, despite its very simple recipes about convection and energy transport.

Nowadays, solar and stellar evolution is fortunately subject to a larger number of observational tests, which at the same time are more accurate than those done some 30 years ago. Let us mention:

1. Values of $L_\odot$, $R_\odot$ at given mass and age.

2. Oscillation data on acoustic modes.

3. Oscillation data on low gravity modes.

4. Account for surface angular velocity Ω, in relation with oscillation data on the angular velocity $\Omega(r)$ in the solar interior (cf. Brown, 1985).

5. Solar oblateness (cf. Hill, Stebbins, 1975).

6. Solar depletion in Li by a factor of 200 (cf. Muller et al., 1975).

7. Solar depletion in ^{8}Be by a factor 2 to 5 (cf. Molaro and Beckman, 1984).

8. Maximum enrichment in ^{3}He of 15% during the solar life (cf. Bochsler et al., 1989).

9. Maximum increase of ^{13}C/^{12}C of 7% during the solar life (cf. Harris et al., 1987).

10. Solar neutrino flux (cf. Bahcall, 1989).

11. Solar magnetic field.

Related stellar data:

12. Sequences of old clusters in the HR diagram; both the shape of the sequence and the star frequency along the sequences have to be examined.

13. Distribution of rotational velocities of main sequence stars and red giants in old clusters.

14. Abundance ratios of C/N, ^{13}C/^{12}C of red giants in old clusters (cf. Brown, 1987; Lambert and Ries, 1981).

15. Li in red giants of old clusters (cf. Charbonneau and Michaud, 1989).

To my knowledge, none of the non standard models have been submitted to all these tests. We would even say that, in the case of the standard models also, these tests have neither been all considered simultaneously. Ideally, they should all be performed before a new solar model can be accepted. These observational data play an essential role with respect to futur progress in stellar evolution, either as censors of theory or as guides for future research.

5 REFERENCES

Baglin A., Morel P., Schatzman E.: 1985, *Astron. Astrophys.* **149**, 309
Bahcall J.N.: 1989, *these Proceedings*
Bahcall J.N., Ulrich R.K.: 1971, *Astrophys. J.* **170**, 593
Bahcall J.N., Ulrich R.K.: 1988, in *Rev. Modern Physics* **60**, 297
Berthomieu G., Provost J., Schatzman E.: 1984, *Nature* **308**, 254
Bienaymé O., Maeder A., Schatzman E.: 1984, *Astron. Astrophys.* **131**, 316
Bochsler P., Geiss J., Maeder A.: 1989, *these Proceedings*
Brown J.A.: 1987, *Astrophys. J.* **317**, 701
Brown T.M.: 1985, *Nature* **317**, 591
Canuto V., Adams P.J., Hsieh S.H., Tsiang E.: 1977, *Phys. Rev.* **D16**, 1643
Charbonneau P., Michaud G., Proffitt C.R.: 1989, *Astrophys. J.*, in press
Christensen–Dalsgaard J.: 1986, in *Seismology of the Sun and the Distant Stars,* Ed. D.O. Gough, Reidel Publ. Co., p. 23
Christensen–Dalsgaard J.: 1988, Proc. Symp. *Seismology of the Sun and Sun–like Stars,* ESA SP–286, p. 431
Christensen–Dalsgaard J., Dilke F.W.W., Gough D.O.: 1974, *Monthly Notices Roy. Astron. Soc.* **169**, 429
Christensen–Dalsgaard J., Gough D.O.: 1980, *Nature* **288**, 545
Clayton D.D., Dwek E., Newman M.J., Talbot R.J.: 1975a, *Astrophys. J.* **199**, 494
Clayton D.D., Newman M.J., Talbot R.J.: 1975b, *Astrophys. J.* **201**, 489
Cox A.N., Guzik J.A., Kidman R.B.: 1989, *Astrophys. J.*, in press
Cox A.N., Kidman R.B., Newman M.J.: 1985, in *Solar neutrinos and neutrino Astronomy,* Ed. M.L. Cherry et al., American Institute of Physics, p. 93
Dilke F.W.W., Gough D.: 1972, *Nature* **240**, 262, 293
Endal A.S.: 1987, in *The internal solar angular velocity,* Ed. B.R. Durney, S. Sofia, Reidel Publ. Co., p. 131
Faulkner J., Swenson F.J.: 1988, *Astrophys. J.* **329**, L47
Finzi A., Harpaz A.: 1989, *these Proceedings*
Gabriel M., Noels A., Scuflaire R.: 1983, in *Europhysics Study Conference on Oscillations as a probe of the Sun's interior,* Catania
Geiss J.: 1972, *Solar Wind,* Proc. ASILOMAR Conf., NASA SP–308, 559
Genova F., Schatzman E.: 1979, *Astron. Astrophys.* **78**, 323
Gilliland R.L.: 1985, *Astrophys. J.* **290**, 344
Gilliland R.L., Däppen W.: 1987, *Astrophys. J.* **313**, 429
Gough D.O., Kosovichev A.G.: 1988, Proc. Symp. *Seismology of the Sun and Sun–like Stars,* ESA SP–286, p. 195
Guzik J.A., Willson L.A., Brunish W.M.: 1987, *Astrophys. J.* **319**, 957
Harris M.J., Lambert D.L., Goldman A.: 1987, *Monthly Not. Roy. Astron. Soc.* **224**, 237
Hawking S.W.: 1971, *Monthly Not. Roy. Astron. Soc.* **152**, 75
Hill H.A., Stebbins R.T.: 1975, *Astrophys. J.* **200**, 471
Hobbs L.M., Pilachowski C.: 1988, *Astrophys. J.* **334**, 734
Iben I.: 1969, *Ann. Phys.* **54**, 164
Iben I., MacDonald J.: 1985, *Astrophys. J.* **296**, 540
Lambert D.L., Ries L.M.: 1981, *Astrophys. J.* **248**, 228
Lebreton Y., Berthomieu G., Provost J.: 1988, in IAU Symp. **123**, 95

Lebreton Y., Maeder A.: 1987, *Astron. Astrophys.* **175**, 99

Levy–Leblond J.M.: 1981, *L'esprit de sel,* Ed. Fayard

Libby L.M., Thomas F.J.: 1969, *Nature* **222**, 1238

Maeder A.: 1977, *Astron. Astrophys.* **56**, 359

Maeder A.: 1979, in *Physical Cosmology,* Ed. R. Balian et al., North Holland Publ. Co., 1980, p. 534

McCrea W.H.: 1975, *Nature* **255**, 607

Molaro P., Beckmann J.: 1984, *Astron. Astrophys.* **139**, 394

Müller E.A., Peytremann E., de la Reza R.: 1975, *Solar Physics* **41**, 53

Newman M.J.: 1986, in *Physics of the Sun,* Ed. P.A. Sturrock, Reidel Publ. Co., p. 33

Newman M.J., Fowler W.A.: 1976, *Astrophys. J.* **207**, 601

Pinsonneault M.H., Kawaler S.D., Sofia S., Demarque P.: 1989, *Astrophys. J.* **338**, 424

Plaga R.: 1989, *these Proceedings*

Pochoda P., Schwarzschild M.: 1964, *Astrophys. J.* **139**, 587

Provost J.: 1984, in IAU Symp. **105**, p. 47

Rood R.T., Ulrich R.K.: 1974, *Nature* **252**, 366

Rosenbluth M.N., Bahcall J.: 1973, *Astrophys. J.* **184**, 9

Roxburgh I.W.: 1987, in *The internal solar angular velocity,* Ed. B.R. Durney, S. Sofia, Reidel Publ. Co., p. 1

Salpeter E.E.: 1970, *Nature* **225**, 165

Schatzman E.: 1962, *Ann. Astrophys.* **25**, 18

Schatzman E.: 1969, *Astrophys. Lett.* **3**, 331

Schatzman E.: 1977, *Astron. Astrophys.* **56**, 211

Schatzman E.: 1987, in *The internal solar angular velocity,* Ed. B.R. Durney, S. Sofia, Reidel Publ. Co., p. 159

Schatzman E., Maeder A.: 1981, *Astron. Astrophys.* **96**, 1

Schwarzschild M., Härm R.: 1973, *Astrophys. J.* **184**, 5

Sofia S.: 1987, in *The internal solar angular velocity,* Ed. B.R. Durney, S. Sofia, Reidel Publ. Co., p. 173

Spruit H.C.: 1985, in *The internal solar angular velocity,* Ed. B.R. Durney, S. Sofia, Reidel Publ. Co., p. 185

Talbot R.J., Newmann M.J.: 1976, *Astrophys. J. Suppl.* **34**, 295

Tassoul, J.–L., Tassoul, M.: 1989, *Astron. Astrophys.* **213**, 397

Turck–Chieze S., Däppen W., Cassé M.: 1988, Proc. Symp. *Seismology of the Sun and Sun-like Stars,* ESA SP–286, p. 629

Twarog B.A., Anthony–Twarog B.J.: 1989, *Astron. J.* **97**, 759

Ulrich R.K., Rhodes J.: 1983, *Astrophys. J.* **265**, 551

Ulrich R.K., Rood R.T.: 1973, *Nature Phys. Sci.* **241**, 111

Vauclair S., Vauclair G., Schatzman E., Michaud G.: 1978, *Astrophys. J.* **223**, 567

Willson L.A., Bowen G.H., Struck–Marcell C.: 1987, *Comments Astrophys.* **12**, 17

Zahn J.P.: 1973, in *Stellar Stability and Evolution,* Eds. P. Ledoux, A. Noels, A.W. Rogers, Reidel Publ. Co., 1974, p. 185

Zahn J.P.: 1983, in *Astrophysical Processes in Upper Main Sequence Stars,* 13th Advanced Course Saas–Fee, Publ. Geneva Observatory

Zahn J.P.: 1985, in *The internal solar angular velocity,* Ed. B.R. Durney, S. Sofia, Reidel Publ. Co., p. 201

SOLAR COSMIONS

DAVID N. SPERGEL
Princeton University Observatory
Princeton, NJ 08540 USA

ABSTRACT. Two of the outstanding problems in astrophysics are the solar neutrino problem and the missing mass problem. The "solar cosmion", a weakly interacting massive particle, could solve both problems. Several particle physics models have been suggested for the solar cosmion.

The solar cosmion may have other interesting astrophysical effects. It will alter the predicted helioseismology spectrum, effect horizontal branch evolution and may alter the mass-radius relationship in low mass stars. These considerations constrain solar cosmion properties.

Several laboratories have begun an active experimental search for the solar cosmion. The UCSB-UCB-Saclay silicon experiment in the Oroville mine has already placed stringent limits on solar cosmions that couple to matter through spin-independent interactions. A planned Saclay experiment may either detect or rule out the existence of "solar cosmions".

The Sun is an powerful laboratory for exploring particle physics beyond the standard model. Even if "solar cosmions" do not exist, the Sun can help "illuminate" the search for other weakly interacting particles posited as solutions to the missing mass problem.

1. Solar Cosmions and the Solar Neutrino Problem

For over 20 years, Ray Davis' solar neutrino experiment has been hinting that something is missing in our basic understanding of either solar or neutrino physics (for review, see Bahcall 1989).

In recent years, it has been popular to blame the neutrino as the cause of the solar neutrino problem (Wolfenstein 1979, Mikheyev and Smirnov 1985, Okun *et al.* 1986). However, the fault may lie not with the neutrino but inside the Sun. The solar models are based on standard physics: the atomic and ionic cross-sections are either measured in the laboratory or calculated from well-established theory; the nuclear reaction rates are derived from experimental values. However, a new physical assumption could alter the Sun's thermal profile and change the predicted SNU flux.

G. Berthomieu and M. Cribier (eds.), Inside the Sun, 145–152.

My collaborators and I suggested that a mechanism for altering energy transport in the stellar interior. We posited the existence of a new particle that carried much of the energy in the core of the Sun. Bill Press and I suggested that a weakly interacting particles could be extremely efficient at energy transport (Spergel and Press 1985). After publication, we learned that John Faulkner and Ron Gilliland had considered this possibility several years earlier, but were dissuaded from publishing their results. [Most of their conclusions were summarized in Steigman et al. (1978). Their full paper finally appeared seven years later (Faulkner and Gilliland 1985)].

Particles with cross-sections of order 10^{-36} cm^2 are ideal for transporting energy in the Sun. In the conductive (large cross-section) regime, energy transport scales as the mean free path. As the cross-section decreases, the cosmion travels through a larger temperature gradient between collisions and is thus more effective at transporting energy. In the small cross-section regime, collisions are so rare that the energy transport scales as the collision rate. The cross-over between these two regimes occurs at the optimal cross-section for energy transport: when $\sigma \approx 6 \times 10^{-36}$ cm^2 and the cosmion's mean free path is its orbital radius. The cosmion can deposit a large fraction of its kinetic energy at aphelion and can increase its kinetic energy at perihelion.

The cosmions are extremely efficient at energy transport. The timescale for the cosmion to transfer energy from the center of the Sun to a scale height, the free fall time (≈ 100 s), is much shorter than the timescale for photons to diffuse the same distance, the Kelvin-Helmholtz time ($\approx 10^6$ years). In fact, only 10^{-11} cosmions per baryon is sufficient to significantly alter energy transport in the solar core and lower the predicted SNU flux to the observed value.

The net effect of the cosmion on the temperature distribution in the Sun is to cool the central core of the Sun while heating the region near the aphelion of the typical orbit (Spergel and Press 1985, Nauenberg 1986 and Gould 1987a, Gould 1989). The scale height of the cosmion distribution can be estimated by equating the cosmion's thermal energy with its potential energy,

$$r_x = 0.13 \left(\frac{m_p}{m_x} \right)^{1/2} R_\odot$$

Most of the B^8 neutrinos are produced in the inner 0.05 $R_\odot$, while most of the Sun's luminosity is produced in the inner 0.2 $R_\odot$. Thus a cosmion with mass between 2 and 10 GeV will reduce the B^8 neutrino production rate without reducing the solar luminosity or affecting the production rate of pp neutrinos. Hence the predicted count rate from a solar model cum cosmions for the pp-neutrino sensitive ^{71}Ga experiment does not differ significantly from a standard model. Cosmions more massive than 10 GeV will be too centrally concentrated to affect the thermal structure in most of the ^{8}B neutrino producing region (Gilliland, Faulkner, Press and Spergel 1986).

The Sun will capture weakly interacting particles from the galactic halo. The escape velocity from the Sun's surface is 617 km/s, while the escape velocity from the core is over 1000 km/s. A halo cosmion with typical velocity 30 km/s will fall into the Sun where it can be captured through a single collision as long as its mass is less than ~ 50 proton masses. Thus the solar capture rate is approximately the cross-sectional area of the Sun, $\pi R_\odot^2$, divided by the typical cosmion halo velocity, times the escape velocity squared, since gravitational focusing is always important in any elastic capture processes. Press and Spergel (1985) discuss these effects and find that the capture rate is sufficient for the Sun to accumulate a significant number of cosmions in the solar lifetime. If we multiply the capture rate by the lifetime, we find that we can achieve a significant concentration of cosmions relative to baryons,

$$\frac{n_x}{n_b} \simeq 3 \times 10^{-10} \left(\frac{\rho_x}{1 M_\odot/pc^3} \right) \left(\frac{v_{esc}}{\bar{v}} \right) \left(\frac{m_p}{m_x} \right) \min \left[\left(\frac{\sigma}{\sigma_{crit}} \right), 1 \right]$$

Recall that a concentration of 10^{-11} of cosmions with cross-section of 4×10^{-36} cm^2 will resolve the solar neutrino problem. If the cosmions compose the halo $\left(\rho_{HALO} \approx 10^{-2} M_\odot/pc^3 , \ v_{HALO} \approx 300 km/s \right)$, then their cross-section must be within a factor of 2 of σ_{crit}. If cosmions compose the disc $\left(\rho_{DISC} \approx 10^{-1} M_\odot/pc^3 \right.$ and $\left. v_{DISC} \approx 50 km/s \right)$, then they can resolve the solar neutrino problem, if their baryon scattering cross-section is between 10^{-37} and 10^{-34} cm^2.

Since cosmions alter the solar thermal structure, they affect the seismology of the Sun. Solar seismology, which measures the sound speed as a function of radius, might detect the variations in density and temperature induced by cosmion energy transport. Däppen et al. (1986) and Faulkner et al. (1986) suggest that cosmions can eliminate the discrepancy between the observed p-wave spectrum and the standard solar model. Bahcall and Ulrich (1988) argue that this discrepancy may not be not significant. Cosmions would have more dramatic effects on the still unobserved g-wave spectrum, which is more sensitive to the core conditions. (For more details, see J. Faulkner's paper in the same proceedings).

Most of the cosmions in the Sun are tightly bound: their typical velocities, $\sqrt{3kT/2m_x} \approx 300$ km/s, are much less than the escape velocity from the core $v_{esc}^2 = 1400$ km/s, so scatterings that produce $v \geq v_{esc}$ are rare. In the conclusion of Spergel and Press (1985), the evaporation rate is estimated as the fraction of cosmion distribution with energy sufficient to escape divided by the time to repopulate the tail. More detailed calculations show that evaporation is negligible for cosmions with $m_x > 4m_p$ (Griest and Seckel 1987, Gould 1987a). Since the core is optically thick for cosmions with larger cross-sections ($\sim 10^{-34}$ cm^{-2}) , they do not evaporate unless their mass is less than 2 m_p (Gilliland *et al.* 1986).

Annihilation can also reduce the number of cosmions in the core of the Sun. If the cosmion is a Majorana particle, it is its own anti-particle and will self-annihilate. If the cosmion is a Dirac particle and the Sun contains both it and its anti-particle in equal numbers, annihilation will also reduce its solar abundance. The cosmion annihilation timescale in the Sun can be estimated,

$$ t_{ann} = (n_x \sigma_{ann} v)^{-1} = \left(\frac{n_p}{n_x} \right) \left(\frac{\sigma_{ann}}{\sigma_{bx}} \right) t_{coll} \quad , $$

where σ_{ann} is the cosmion annihilation cross-section and σ_{bx} is the cosmion-baryon scattering cross-section. If the cosmion is to resolve the Solar neutrino problem,. $t_{coll} \approx t_{dynamical} \approx 100$ seconds and $n_p/n_x \approx 10^{11}$. Most of the attractive particle physics cosmion candidates, photinos, scalar neutrinos, massive and Dirac neutrinos, have scattering cross-sections less than or on the order of their annihilation cross-sections; this implies $t_{ann} < 10^{13}$ seconds, much shorter than the age of the Sun (Krauss, Freese, Spergel and Press 1986). Cosmions are more centrally concentrated than baryons; this enhances their annihilation rate and exacerbates the problem.

There are several possible ways of avoiding this annihilation problem. If there is a net asymmetry between cosmions and anti-cosmions (perhaps, equal to the asymmetry between baryons and anti-baryons), or if $\sigma_{ann} << \sigma_{b,x}$, then cosmion annihilation is suppressed.

Several particle physics models have been proposed for the solar cosmion (Gelmini *et al.* 1986; Raby and West 1988, Ross and Segré 1988). In these models, the cosmion couples to baryons through either a higgs, Z or some new exchange particle. The effects of the cosmion on stellar evolution depends upon the nature of this interaction. If the cosmion couples to nucleons through a spin-dependent interaction, then it will be difficult to detect in the laboratory and have minimal effects on the later stages of stellar evolution. If the cosmion has a spin-independent interactions, then it has a much larger scattering cross-section with helium and heavier nuclei than with hydrogen. As we will see in the next two sections, this leads to observable effects in the later stages of stellar evolution.

2. Solar cosmions in other stars

Solar cosmions will be captured not only by our Sun, but also by other stars moving through the galactic halo (see e.g. Bouquet and Salati 1987). Massive stars accumulate very few cosmions during their brief lives, however, in low mass stars, cosmion energy transport produces subtle, but potentially observable effects.

DeLuca *et al.* (1989) found that solar cosmions slightly alter the mass-radius relationship in low mass stars. Careful observations of a low mass binary system could potentially confirm or contradict the solar cosmion solution to the solar neutrino problem.

Faulkner and Swenson (1988) followed the evolution of low mass stars with modified energy transport. The solar cosmions isothermalized the core of these stars during the end of their main sequence evolution. These isothermal cores could no longer support themselves and began to collapse, driving the star off the main sequence. Faulkner and Swenson (1988) concluded that solar cosmions could reduce estimates of globular cluster ages.

Renzini (1986) suggested that cosmions would significantly alter energy transport in the cores of horizontal branch(HB) stars. Horizontal branch stars burn helium to carbon in their convective cores. Renzini claims that cosmions will isothermalize the cores of HB stars, thus suppressing core convection. Without core convection and semi-convection, the core would rapidly exhaust its supply of Helium and evolve off of the horizontal branch. Shortening the horizontal branch lifetime relative to the asymptotic giant branch lifetime would produce a discrepancy with star counts.

Spergel and Faulkner (1988) made analytical estimates of solar cosmion capture, survival and energy transport in an HB star. They noted that solar cosmions with spin-dependent couplings do not interact with the helium and are ineffectual at energy transport. Cosmions with spin-independent interactions have their numbers depleted through evaporation during the helium flash and the subsequent HB phase. For most parameters, these cosmions have too large a cross-section to efficiently transport energy in HB stars.

Dearborn *et al.* (1989) investigated the effects of solar cosmions on HB star evolution. They assumed that the solar cosmion parameters were optimal for altering HB evolution and used a numerical code to simulate cosmion energy transport. They found that an HB star with solar cosmions can not be in thermal equilibrium: the core oscillates between two phases. During the active phase, it rapidly burns helium and expands on a dynamical timescale. The expansion overshoots and turns off core helium burning. Without helium burning, the core contracts until it achieves the high density needed to re-enter the active phase. These oscillations do not alter HB lifetimes, however, in their simulations, HB stars with cosmions burn $\sim 1/2$ magnitude brighter than standard HB stars.

3. Experimental WIMP searches

3.1. DETECTING WIMPs IN THE LABORATORY

Solar cosmions share the epithet "WIMP" (Weakly Interacting Massive Particles) with several other particles including heavy neutrinos and supersymmetric relics.

Heavy neutrinos have the oldest pedigree of the WIMP candidates. They arise in four generation models: if the heaviest neutrino is stable and its mass is ~ 2 GeV, then it can comprise the missing mass. LEP and SLC will soon test the viability of these models.

Supersymmetry offers a particularly attractive dark matter candidate. In supersymmetry, there is a new conserved quantum number: R- parity. This implies that there will be a lowest mass particle with charge R=1 and that charge conservation will imply that it is stable. In most theories, this lightest supersymmetric particle (LSP) is a linear combination of photino, higgsino and zino interaction eigenstates. The photino is the fermionoic supersymmetric partner of the photon. The higgsino is the partner of the Higgs and the zino is the partner of the Z-boson. In minimal supersymmetry models, there is a large range of parameter space within which the LSP can close the universe (Ellis et al. 1984).

WIMPs are potentially detectable through their direct interactions in the laboratory. (Goodman and Witten 1985, Wasserman 1986, Drukier et al. 1986). If WIMPs comprise the halo, then their number density is ~ 0.1 cm^{-3} and their flux is $\sim 10^7$ cm^{-2} s^{-1}. A tiny fraction of this incident flux will collide with a nucleus in an experiment and deposit ~ 1 keV of kinetic energy. The experimental challenge is detecting this rare event.

Germanium spectrometers, originally designed for double β decay experiments, provided the first limits on weakly interacting halo dark matter (Ahlen et al. 1987, Caldwell et al. 1987). These experiments, with their low energy threshold and ultra-low backgrounds, are well suited for WIMP detection. Silicon detectors can push these limits to lower masses and cross-section (Sadoulet et al. 1988). The Saclay-UCB-UCSB silicon experiment may soon either rule out or detect solar cosmions with spin-independent interactions.

Detecting cosmions with spin-dependent coupling will require new technologies. These particles have their largest interaction rate per gram with hydrogen and couple extremely weakly to silicon and germanium. Rich and Spiro have suggested that a hydrogen-filled time-projection chamber (TPC) could be used search for these elusive solar cosmion candidates.

3.2. DETECTING WIMPs IN THE SUN

Even if WIMPs do not solve the solar neutrino problem, the Sun's interior may still be an important particle physics laboratory. Any WIMP in the halo, not just the solar cosmion, can be captured by the Sun. The WIMP number density will accumulate until balanced by WIMP annihilation.

When WIMPs annihilate, their annihilation products are likely to include high energy neutrinos. These GeV neutrinos will stream towards the earth, where they can be detected in underground experiments (Silk, Olive and Srednicki 1986). The Frejus, IMB and Kamiokande detectors have all searched for these high energy neutrinos. So far, they have failed to detect an excess number of events coming from the direction of the Sun.

The current limits on WIMPs from non-detection of high energy (GeV) neutrinos from the Sun already constrain several particle physics candidates (Roulet and Gelmini 1988). Detectors, coming on line in the next few months, may rule out several solar cosmion candidates. In the coming years, sensitivities are likely to improve and WIMPs may be either detected or ruled out.

ACKNOWLEDGEMENTS

This research is supported in part by NSF PHY88-05895, NSF AST88-58145 (PYI) and an A.P. Sloan Fellowship.

REFERENCES

Ahlen, S., Avignone, F.T., III, Brodzinsky, R., Drukier, A.K., Gelmini, G. and Spergel, D.N., *Phys. Lett.*, **B195**, 603 (1987).

Bahcall, J.N., *Neutrino Astrophysics*, Cambridge Univ. Press, Cambridge, UK (1989).

Bahcall, J.N. and Ulrich, R.K. *Rev. Mod. Phys.*, **60**, 297 (1988).

Bouquet, A. and Salati, P., "Life and death of cosmions in stars", LAPP-TH-192/87 (1987).

Caldwell, D.O., Eisberg, R.M., Goulding, F.S., Grumm, D.M., Sadoulet, B. Smith, A.R. and Witherell, M.S., *Phys. Rev. Lett.*, **59**, 419 (1987).

Däppen, W. Gilliland,R.L. and Christensen-Dalsgaard, J. *Nature*, **321**, 229 (1986).

Dearborn, D., Raffelt, G., Salati, P., Silk, J. and Bouquet, A., CfPA-TH-89-008 (1989).

DeLuca, E.E., Griest, K., Rosner, R. and Wang, J., FERMILAB-PUB-89-49-A (1989).

Ellis, J., *et al.*, *Nucl. Phys.*, B238, 453 (1984).

152

Faulkner, J. and Gilliland, R.L. *Ap. J.*, **299**, 994 (1985).

Faulkner, J. Gough, D.O. and Vahia, M.N. *Nature*, **321**, 226 (1986).

Faulkner, J. and Swenson, F.J. *Astrophys. J.*, **329**, 81 (1988).

Gelmini, G.B. Hall, L.J. and Lin, M.J., *Nucl.Phys.B* **281** , 726 (1986).

Gilliland, R., Faulkner, J., Press, W.H. and Spergel, D.N., *Ap.J.*, **306**, 703 (1986).

Goodman, M.W. and Witten, E., *Phys. Rev. D*, **31** , 3059 (1985).

Gould, Andrew *Ap. J.*, **321**, 560 (1987a).

Gould, Andrew *Ap. J.*, **321**, 571 (1987b).

Gould, Andrew, IAS preprint AST-89/20, submitted to *Ap. J.* (1989).

Griest, K. and Seckel, D., *Nucl.Phys.B* **283**, 681 (1987).

Krauss, L.M., Freese, K., Spergel, D.N. and Press, W.H., *Ap. J.*, **299**, 1001 (1985).

Mikheyev, S.P. and Smirnov, A.Yu., *Nuovo Cimento*, **9C**, 17 (1986).

Nauenberg, Michael, *Phys. Rev.*, **D36**, 1080 (1987).

Okun, L.B., Voloshin, M.B., and Vysotskii, M.I., *Sov. J. Nucl. Phys.*, **44**, 546 (1986).

Press, W.H. and Spergel, D.N., *Ap. J.*, **296**, 663 (1985).

Raby, S. and West, B.G., *Phys. Lett.*, **202B**, 47 (1988).

Renzini, Alvio, *Astronomy and Astrophysics*, **171**, 121 (1986).

Ross, G. and Segré, E., *Phys. Lett.*, **197B**, 45 (1987).

Roulet, E. and Gelmini, G., "Cosmions, Cosmic Asymmetry and Underground Detection", SISSA-106/EP/88 (1988).

Sadoulet, B., Rich, J., Spiro, M. and Caldwell, D.O., *Ap. J. (Lett.)*, **324**, L75 (1988).

Spergel, D.N. and Faulkner,J., *Ap. J. Lett.*, **331**, L21 (1988).

Spergel, D.N. and Press, W.H., *Ap.J.*, **294**, 679 (1985).

Srednicki, M., Olive, K.A. and Silk, J., *Nucl.Phys.B* **279**, 804 (1987).

Steigman, G., Sarazin, C.L., Quintana, H. and Faulkner, J. *A.J.*, **83**, 1050 (1978).

Wasserman, I., *Phys. Rev. D*, **33**, 2071 (1986).

Wolfenstein,L., *Phys. Rev. D*, **20**, 2634 (1979).

L'ISOTOPE ³HE DANS LES ÉTOILES. APPLICATION À LA THÉORIE DES NOVÆ ET DES NAINES BLANCHES

Evry SCHATZMAN
Observatoire de Meudon
DASGAL
92195 Meudon Cedex France

During the active discussions on solar models, intervenants were refering to a point raised by Evry Schatzman in the early fifties and published only in the *Comptes rendus de l'Académie des Sciences, séance du 7 mai 1951*. John Bahcall raised the idea to reproduce it in these Proceedings, and the author found this idea interesting.

The Editors

G. Berthomieu and M. Cribier (eds.), Inside the Sun, 153–154.

1740   ACADÉMIE DES SCIENCES.

SÉANCE DU 7 MAI 1951.

ASTROPHYSIQUE. — *L'isotope* ³He *dans les étoiles. Application à la théorie des Novæ et des naines blanches.* Note de M. **Evry Schatzman**, présentée par M. André Danjon.

1. La probabilité de la réaction

(1) $$^3He + ^3He \rightarrow \, ^4He + p + p$$

est très grande. La largeur du niveau Γ, telle qu'il résulte des données de Fermi et Turkevitch, rassemblées par Alpher et Hermann ([1]), est de $5,8.10^7$ eV. Si n_3 est le nombre de particules ³He par unité de volume, le nombre de réactions par seconde est

(2) $$p_2 = n_3^2 \, 1,4 . 10^{-11} \, T_s^{-\frac{1}{2}} \, 10^{-11,69 T_s^{-\frac{1}{3}}}$$

avec $T_s = 10^{-8} \, T$.

On peut ainsi montrer que l'évolution normale se fait suivant les réactions suivantes :

(3)
$$\begin{cases} ^1H + ^1H \rightarrow \, ^2D + \beta^+, \\ ^2D + ^1H \rightarrow \, ^3He, \\ ^3He + ^3He \rightarrow \, ^4He + p + p. \end{cases}$$

On peut alors montrer qu'une faible accumulation de ³He est possible dans les étoiles à condition que la température soit assez basse et la densité assez élevée.

2. *Novæ.* — *a.* La rapidité de la réaction (1) entraîne la possibilité de la formation d'une onde de détonation.

À titre d'exemple on peut montrer que 0,01 g de ³He par gramme de matière à la densité 1000 peut donner lieu à une onde de détonation d'épaisseur 7 km se propageant à la vitesse de 700 km/s⁻¹, à la température de cent millions de degrés.

b. L'accumulation de ³He provoque une augmentation progressive de l'exposant ν_{eff} de la loi de température du débit d'énergie. Lorsque ν_{eff} atteint une certaine valeur $\nu_{critique}$ l'étoile devient vibrationnellement instable, et une explosion peut se produire.

On trouve ainsi la relation

(4) $$t \sim \frac{E}{L}$$

entre le débit d'énergie par seconde L et le débit total d'énergie E au cours d'une explosion et la période des récurrences. Cette relation est vérifiée de façon satisfaisante depuis les étoiles SS Cygni jusqu'aux Novæ. Elle découle directement de l'hypothèse faite sur l'origine des explosions.

3. *Stabilité vibrationnelle des naines blanches.* — Dans les naines blanches, la durée de vie des noyaux ³He est très courte. Au cours d'une pulsation, la réaction (1) va se trouver varier en quadrature de la pulsation et ne pas contribuer à l'instabilité vibrationnelle. L'exposant effectif ν_{eff} devient alors de l'ordre de 2, inférieur par conséquent à la valeur critique 2,6 trouvée par Ledoux et Sauvenin Goffin ([2]).

Les conclusions de ces auteurs doivent donc être modifiées pour tenir compte du rôle nouvellement découvert de la réaction (1). Les naines blanches sont donc vibrationnellement stables. La théorie de l'auteur ([3]) en tire donc une nouvelle confirmation.

([1]) *Rev. Mod. Phys.*, **22**, 1950, p. 153.

([2]) *A. J.*, III, 1950, p. 611.

([3]) Schatzmann, *Le spectre des naines blanches et leur débit d'énergie*, Copenhague, 1950.

PART 2

NEUTRINOS

SOLAR NEUTRINO PROJECTS

M. SPIRO , D.VIGNAUD
DPhPE/SEPh
CEN Saclay
F 91191 Gif-sur-Yvette

ABSTRACT. An overview of the solar neutrino projects is given, with an emphasis on the complementarity of the different experiments (gallium, indium, heavy water,...) to solve the solar neutrino problem that was raised by the chlorine and the Kamiokande results. The separation of the different sources of neutrinos in the Sun would contribute significantly to the astrophysical understanding of the Sun. Some of the planned experiments could be able to pinpoint neutrino oscillations (within a wide range of parameters) almost independently of solar models. Projects which are particularly sensitive to a variation of the neutrino flux with time are also discussed.

1. Introduction

Solar neutrino detection is a challenge for astrophysics (test of the standard model of the Sun and of the stars) and for particle physics (the observed deficiency may be due to neutrino oscillations). In this section we briefly summarize neutrino properties, neutrino production in the Sun and the different types of solar neutrino detectors.

There are three flavours of neutrinos, ν_e , ν_μ and ν_τ , with a generic name ν_x . They interact with matter either by producing their charged lepton partner (e^-, μ^- and τ^- respectively) via W^+ exchange, which is called charged current interaction, or via Z^o exchange, which is called neutral current interaction. These processes are shown in Fig. 1. If neutrinos have a mass ($m_1 \neq m_2 \neq m_3$) the mass eigenstates ν_1, ν_2 and ν_3 may be different from the weak interaction eigenstates ν_e , ν_μ and ν_τ as it is observed for quarks. In this case the flavour eigenstates are related to the mass eigenstates by mixing angles and there are oscillations between the different flavours. The parameters of the oscillation between two flavours are the squared mass difference Δm^2 and the mixing angle $\sin^2 2\theta$.

The Sun produces pure ν_e , via the four main reactions :

$$
\begin{aligned}
&\text{p p} \rightarrow \text{d e}^+ \, \nu_e & &\nu_{pp} \\
&\text{p e p} \rightarrow \text{d} \, \nu_e & &\nu_{pep} \\
&{}^7\text{Be e}^- \rightarrow {}^7\text{Li} \, \nu_e & &\nu_{Be} \\
&{}^8\text{B} \rightarrow {}^8\text{Be}^* \text{e}^+ \, \nu_e & &\nu_{B}
\end{aligned}
$$

G. Berthomieu and M. Cribier (eds.), Inside the Sun, 157–169.
© 1990 *Kluwer Academic Publishers. Printed in the Netherlands.*

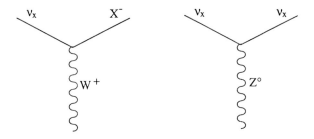

Figure 1 : Diagrams for neutrino interaction. a) charged current. b) neutral current. ν_x means ν_e , ν_μ or ν_τ and X^- means e^-, μ^- or τ^-.

These reactions are well known from nuclear and particle physics. The corresponding ν_e energy spectrum is displayed in Fig. 2. It extends to 0.420 MeV for ν_{pp} and 14 MeV for ν_B . On the contrary, the relative amount of these contributions may depend on details of the Sun model : central temperature, opacities, cross sections,... [1]. Only the ν_{pp} contribution is almost model independent, since it is fixed by the solar luminosity.

Solar neutrino experiments detect neutrinos via :
- charged current interactions : $\nu_x + (A,Z) \rightarrow X^- + (A,Z+1)$. Such experiments are only sensitive to ν_e . In the case where ν_e oscillate and become ν_μ or ν_τ , they cannot be detected by this process since the threshold for producing a muon (m = 106 MeV) or a tau (m = 1780 MeV) is well above the maximum solar neutrino energy. The produced electron is almost isotropic and does not give information on the neutrino direction. However the electron energy spectrum reflects directly the neutrino energy spectrum : $E_e = E_\nu - E_{threshold}$.
- neutral current interactions : $\nu_x + A \rightarrow \nu_x + A^*$. The detection is insensitive to the neutrino flavour. It integrates all types of neutrinos.
- elastic scattering on electron : $\nu_x + e^- \rightarrow \nu_x + e^-$. This reaction can occur via both charged and neutral current for ν_e , (see Fig. 3), and only via neutral current for ν_μ and ν_τ . Moreover the cross section for the charged current process is about 6 times larger than for the neutral current process. This means that, contrary to intuition, the elastic scattering of ν_e on electrons proceeds mainly via charged current process. An advantage of this reaction is that, for kinematical reasons, the scattered electron keeps the direction of the neutrino. This property is a great help for background reduction. The counterpart is that the electron energy spectrum does not reflect the neutrino energy spectrum.

Solar neutrino projects focus on three main physics goals.
1. Observation and separation of the neutrinos coming from the different sources in the Sun : ν_{pp} , ν_{pep} , ν_{Be} , ν_B . This can be achieved by combining the results of various radiochemical or real time experiments (chlorine, gallium, Kamiokande, indium, heavy water,...).
2. Oscillations of neutrinos from one flavour to another, between their production

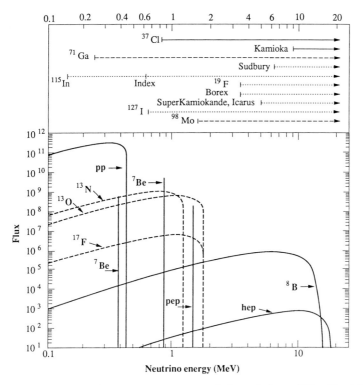

Figure 2 : Solar neutrino energy spectrum (adapted from [1]). Neutrino fluxes from continuum sources are in $cm^{-2} s^{-1} MeV^{-1}$. Line fluxes are in $cm^{-2} s^{-1}$. The insert above gives the sensitivity interval of the different detectors above the threshold. Full lines : existing detectors. Dashed lines : detectors in installation. Dotted lines : projects.

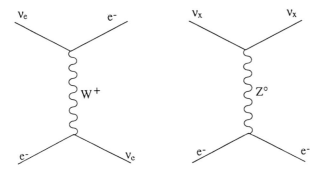

Figure 3 : Diagrams for neutrino electron interaction. a) charged current. b) neutral current. ν_x means ν_e , ν_μ or ν_τ .

place to the detector location. This problem can be addressed in some of the solar neutrino projects **independently** of solar models by looking at i) the measurement of the total ν_{pp} flux, which is solar model independent (gallium), ii) a possible distortion in the ν_B energy spectrum (Sudbury, Borex, Icarus), iii) the ratio between charged current and neutral current interactions which is well predicted from particle physics only (Sudbury, Borex, Icarus), iv) day/night effects induced by the MSW mechanism.

3. Look at possible variations of the neutrino fluxes with time. This can be achieved by high statistics experiments (Sudbury, SuperKamiokande, iodine) for time scale variations of a few years (day/night, annual variations, solar cycle correlation). Larger time scales can rely only on geochemical experiments (molybdenum).

Some projects which did not progress significantly in the last years are not quoted. The reader is referred to the review by Kirsten [2] for a more exhaustive list of solar neutrino projects.

2. Measurement of the ν_{pp}, ν_{Be} and ν_B contributions

There are now two existing solar neutrino experiments (chlorine and Kamiokande), two funded (Gallex and Sage), and several projects at a different stage of design. Table 1 shows most of the possible targets for which the neutrino capture cross section is well known and which can then provide constraints on the different neutrino fluxes coming from the Sun. It will be completed by table 2. The different reaction thresholds are also presented in Fig. 2.

reaction	reaction threshold	experimental technique	ν contribution
$\nu_e + {}^{37}Cl \rightarrow {}^{37}Ar + e^-$	0.814 MeV	radiochemical	ν_{Be} and ν_B
$\nu_e + e^- \rightarrow \nu_e + e^-$	none	Cerenkov, H_2O	ν_B
$\nu_e + {}^{71}Ga \rightarrow {}^{71}Ge + e^-$	0.233 MeV	radiochemical	ν_{pp} , ν_{Be} , ν_B
$\nu_e + {}^{115}In \rightarrow {}^{115}Sn + e^-$	0.128 MeV	scintillator	ν_{Be} , ν_{pep}
$\nu_e + {}^{19}F \rightarrow {}^{19}Ne + e^-$	3.5 MeV	scintillator	ν_B
$\nu_e + D \rightarrow e^- + p + p$	1.44 MeV	Cerenkov, D_2O	ν_B

Table 1 : Main solar neutrino detectors.

The radiochemical Davis **chlorine** detector in Homestake (600 tons) [3] counts ${}^{37}Ar$ atoms every two months. The real time **Kamiokande** experiment (fiducial volume of 680 tons in 2140 tons of water) [4] detects Cerenkov light emitted by electrons with a detection threshold of 9 MeV. These two experiments, which detect mainly ν_B , have provided results which constitute the solar neutrino problem. An upscale version of the chlorine experiment is planned in USSR (3000 tons of C_2Cl_4 in the Baksan Underground Laboratory). In Japan, a significant extension of Kamiokande is proposed (SuperKamiokande) : 50000 tons of pure water with thousands of photomultipliers to detect Cerenkov light. The threshold for electrons could be lowered to 5 MeV, giving about 20 solar neutrinos per day in a 22000 tons fiducial volume.

The two radiochemical **gallium** experiments (30 tons in the form of $GaCl_3$ for Gallex in the Gran Sasso [5] and 60 tons of metallic Ga for Sage in Baksan [6]) are underway and should provide their first results within one year. Their main objective is the detection of ν_{pp} from the primordial pp fusion reaction. Their results are expected impatiently.

The idea of an **indium** target is from Raghavan [7]. The threshold is so low (128 keV) that it is very sensitive to ν_{pp} . However the natural radioactivity of ^{115}In (E_{max}=494 keV) is a formidable background in the low energy region and none of the many projects could fight against it. The detection of ν_{Be} and ν_{pep} should be much easier. Taking this in mind, the indium target has been revived recently [8]. This collaboration (Index) plans to use a scintillator detector (10 tons of In) to measure in real time the ν_e coming from the two neutrino line sources in the Sun (ν_{Be} and ν_{pep}). The detector would consist in plastic scintillating fibers surrounded by 3.5 μm of indium, and would be placed in a large tank of ultrapure water. Another solution would use directly liquid scintillator doped with indium. They expect about 50 events a year. The measurement of the ratio between these two lines gives a strong constraint on solar models and is almost free of systematics and uncertainty on capture cross section. Low temperature indium detectors are currently being investigated by various groups, with different approaches (superconducting junctions, superconducting granules) [9]. These are unfortunately still far from a real detector with several tons of indium.

A **fluorine** experiment, using a scintillator technique, sensitive only to ν_B , is now under study in Moscow [10]. The major difficulty is the separation of the ^{19}Ne signal which decays with a 20 s lifetime from the natural radioactivity background.

Three real time experiments (Sudbury, Borex and Icarus) aim to measure the ν_B contribution both in charged current and in neutral current. They can also detect the elastic interaction on electrons. Their main characteristics are displayed in table 2.

experiment	reaction	detection threshold	signature	events/yr
Sudbury 1000 tons D_2O Cerenkov	ν_e D $\rightarrow$e$^-$ p p	6.5 MeV	e > 5 MeV $(D_2O) - (H_2O)$	9750
	ν_x D $\rightarrow \nu_x$ p n	2.2 MeV	n capture on ^{35}Cl $(D_2O + NaCl) - (H_2O)$	2800
	ν_e e$^- \rightarrow \nu_e$ e$^-$	5 MeV	e > 5 MeV Sun direction	1100
Borex 2000 tons (200 t ^{11}B) scintillator	ν_e ^{11}B $\rightarrow ^{11}$C* e$^-$ ^{11}C$^* \rightarrow ^{11}$C γ	6 MeV	e > 3.5 MeV no γ or γ(2,4.3,4.8 MeV)	2300
	ν_x ^{11}B $\rightarrow ^{11}$B* ν_x ^{11}B$^* \rightarrow ^{11}$B γ	4.5 MeV	γ (4.4 or 5 MeV)	130
	ν_e e$^- \rightarrow \nu_e$ e$^-$	3.5 MeV	e > 3.5 MeV no Sun direction	1550
Icarus I 200 tons liquid argon drift chamber	ν_e ^{40}Ar $\rightarrow ^{40}$K* e$^-$ ^{40}K$^* \rightarrow ^{40}$K γ	11 MeV	e > 5 MeV + γ 2.1 MeV	100
	ν_x ^{40}Ar $\rightarrow ^{40}$Ar* ν_x	6 MeV	γ 6.1,7.8,9.6 MeV	20
	ν_e e$^- \rightarrow \nu_e$ e$^-$	5 MeV	e > 5 MeV Sun direction	80

Table 2 : Real time charged and neutral current sensitive detectors.

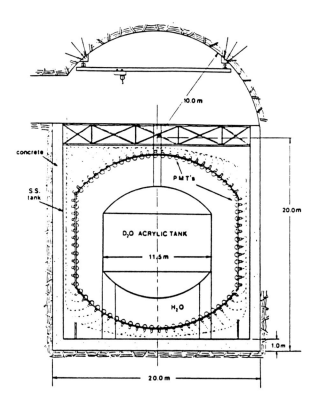

Figure 4 : Conceptual design of the Sudbury neutrino detector. Neutrinos interacting in the heavy water produce relativistic electrons which emit Cerenkov light. This light is detected by phototubes which cover 40 % of the surface.

The **Sudbury** project (Canada-USA-UK) [11] consists in 1000 tons of heavy water D_2O surrounded by 4 m of purified light water H_2O (see Fig. 4). The Cerenkov light emitted by the electrons is detected by photomultipliers as in the Kamiokande experiment. The detector which is almost funded will be installed in a deep mine near Sudbury in Canada (2070 m underground). The main difficulty of this experiment is to reduce the backgrounds at a very low level. This needs in particular the use of low activity materials : less than 10^{-15} g/g of U and Th; the high energy gamma rays from the U and Th chains can photodissociate the deuterium, emitting a neutron which can fake a neutral current process. This purity problem is essential for all similar experiments. The result is obtained by subtracting the H_2O signal to the D_2O signal, the internal target being filled alternatively with the two liquids. The addition of NaCl for some run should allow the measurement of the neutral current process with deuterium dissociation, by looking at the neutron capture by ^{35}Cl.

The **Borex** project [12] is a large tank containing 2000 tons of borated liquid scintillator (200 tons of ^{11}B) and immersed in pure water. It could be installed by an USA-Italy collaboration in the Gran Sasso Underground Laboratory (about 3300 m of water equivalent).

A major difficulty is also to obtain a very pure liquid scintillator. The charged current reaction is signed by an electron in coincidence with a photon. When there is no photon there is an ambiguity with the elastic scattering on e^- .

The **Icarus** project (Italy-USA) [13] is really ambitious with 3000 tons of liquid argon which would also be installed in the Gran Sasso. A major problem consists in the ability to drift ionization electrons over large distances, which needs among other things to have very pure argon. A smaller project, Icarus I, using 200 tons of liquid argon is being developed as a first step. This necessary step will allow the detection of solar neutrinos, but at smaller rate (about 200 / year) than Sudbury or Borex.

The use of ^{13}C as a target for solar neutrino detection in scintillation counters has been recently proposed by Arafune et al. [14]. The threshold is around 3 MeV and there are similarities with Borex about NC and CC detection. More work is however needed before writing a proposal.

The radiochemical experiments (Cl,Ga) measure only the integrated number of inter-actions over the whole ν_e energy spectrum above threshold. A combination of experiments is then necessary to separate the different contributions. In the standard solar model [1] the gallium detectors are sensitive to a linear contribution of ν_{pp} (56%), ν_{Be} (26%) and ν_B (11%) and the chlorine detects a linear combination of ν_{Be} (14%) and ν_B (77%) (the remaining small contributions are mainly due to the CNO cycle).

On the other hand Kamiokande or Sudbury, Borex, Icarus, or the fluorine experiment directly provide, with somewhat different thresholds, only the ν_B contribution.

In principle one should be able, by combining the gallium, the chlorine and the "only ν_B " sensitive experiments to disentangle all three contributions. However, by doing so, one expects rather large errors, especially on ν_{Be} . This last contribution would be best measured by an indium experiment which would provide a good ν_{pep} /ν_{Be} ratio (about 7% in the standard model but weakly model dependent).

3. Do neutrino oscillate ?

The idea that the solar neutrino problem is due to vacuum neutrino oscillations has been raised just after the first chlorine results in 1968. The main difficulty with that explanation is that a factor 3 reduction in the observed ν_e flux needs a maximum mixing between the three neutrino flavours. Although still possible, this is not the favoured scenario. The discovery of the MSW effect in 1985 was a real breakthrough, allowing a much more elegant and much less constrained explanation. Using the Wolfenstein formalism for neutrino propagation in matter, Mikheyev and Smirnov showed that an adiabatic transformation of solar ν_e into a mass eigenstate only weakly coupled to electrons could take place in the Sun [15]. This effect can lead to strong ν_e flux suppressions in a large range of the $(\Delta m^2 , \sin^2 2\theta)$ plane .

It is known from quantum mechanics that for infinite density the propagation eigenstates are the flavour eigenstates (ν_e and ν_μ in the two-flavour case), while at zero density (in the vacuum) these are the mass eigenstates (ν_1 and ν_2). The presence of a charged current diagram for ν_e and not for ν_μ breaks the symmetry between them and the eigenvalue of the total hamiltonian is larger for the ν_e than for the ν_μ at infinite density. What happens when density varies from infinity to a small or null value ? The adiabatic theorem states that the instantaneous eigenstate of propagation goes smoothly from ν_e to ν_2 provided the density decreases sufficiently smoothly. The almost level crossing is for a given value ρ_R of the density (called the resonant density, because the rate of change between ν_e and ν_2 has a resonant shape for this value). This is illustrated in Fig. 5a, following Bethe [16]. If $\nu_2 = \nu_e \sin\theta + \nu_\mu \cos\theta$ the probability that ν_2 appears as a ν_e is then $\sin^2\theta$ which means that the smaller the mixing angle, the larger the reduction flux.

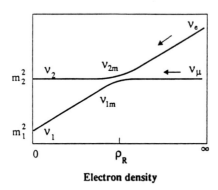

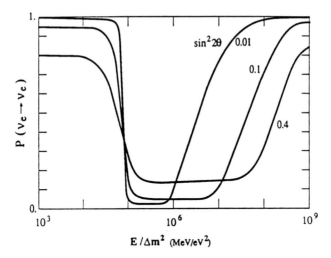

Figure 5 : a) Eigenvalues of the propagation eigenstates in the matter, ν_{1m} and ν_{2m}, as a function of the electron density. The level crossing is for a given value ρ_R (the resonant density). b) Probability for a neutrino ν_e created in the center of the Sun to escape from the Sun as a ν_e . The calculation is done as a function of $E/\Delta m^2$ for different values of the mixing angle.

Two physical conditions determine the ν_e energy region in which the flux is reduced by this factor $\sin^2\theta$ [17]. The minimum energy condition E_{min} (MeV) $= 10^5 \cos 2\theta$

$|\Delta m^2|\,(eV^2)$ is given by the electron density of the central Sun which must be higher than the resonant density. The maximum energy condition $E_{max}\,(MeV) = 2\,10^8\sin2\theta\tan2\theta$ $|\Delta m^2|\,(eV^2)$ is the adiabatic condition for the transformation to happen smoothly. Fig. 5b shows the solar ν_e flux suppression as a function of $E/\Delta m^2$ for different mixing angles. The smaller θ, the smaller the energy interval with a suppression. If θ becomes too small, E_{max} becomes smaller than E_{min} and there is no more effect.

A day/night effect would be surprising as far as the solar neutrino flux is concerned. For detectors which are sensitive only to ν_e, this is no longer true. Let assume that the neutrino oscillation parameters are in the range where solar neutrinos are affected between their production place in the Sun and their detection on earth. There may be an amusing effect when neutrinos arrive during the night : if they have to cross the earth there may be regeneration of ν_e. This effect has been studied in particular in ref. [18]. There is a region in the $(\Delta m^2, \sin^2 2\theta)$ plane around $\Delta m^2 = 10^{-5} - 10^{-6}eV^2$ (which depends on the solar neutrino energy) where the day/night difference may be as large as a factor 3, which should be relatively easy to observe.

	$\sin^2 2\theta$		
	0.002	0.02	0.2
$\Delta m^2 = 10^{-4}$	$E_{min} = 10$ MeV $E_{max} = 40$ MeV ν_B spectrum NC/CC ratio	$E_{min} = 9.9$ MeV $E_{max} = 405$ MeV ν_B spectrum NC/CC ratio	$E_{min} = 8.9$ MeV $E_{max} = 4.5$ GeV ν_B spectrum NC/CC ratio
$\Delta m^2 = 10^{-5}$	$E_{min} = 1$ MeV $E_{max} = 4$ MeV none	$E_{min} = 1$ MeV $E_{max} = 40$ MeV NC/CC ratio day/night effect	$E_{min} = 0.9$ MeV $E_{max} = 450$ MeV NC/CC ratio day/night effect
$\Delta m^2 = 10^{-6}$	$E_{min} = 0.1$ MeV $E_{max} = 0.4$ MeV ν_{pp} suppression	$E_{min} = 0.1$ MeV $E_{max} = 4$ MeV ν_{pp} suppression	$E_{min} = 0.09$ MeV $E_{max} = 45$ MeV ν_{pp} suppression NC/CC ratio
$\Delta m^2 = 10^{-7}$	$E_{min} = 0$ $E_{max} = 0.04$ MeV none	$E_{min} = 0$ $E_{max} = 0.4$ MeV ν_{pp} suppression	$E_{min} = 0$ $E_{max} = 4.5$ MeV ν_{pp} suppression
$\Delta m^2 = 10^{-8}$	$E_{max} = 0$ none	$E_{max} = 0.04$ MeV none	$E_{min} = 0$ $E_{max} = 0.45$ MeV ν_{pp} suppression

Table 3 : $(\Delta m^2, \sin^2 2\theta)$ plane with sensitivity regions to different aspects of the MSW effect. E_{min} and E_{max} correspond to the neutrino energy interval in which the MSW suppression is maximum. (Δm^2 values are in eV^2).

Table 3 illustrates the regions of $(\Delta m^2, \sin^2 2\theta)$ plane where the MSW effect can affect one or another solar neutrino experiment. It corresponds to a large range of the neutrino oscillation parameters which is practically not accessible by any other non solar neutrino experiment. Grand unified theories are really in favour of this region for the neutrino mass and mixing angle parameters (see [19]) which still enhances the interest for solar neutrino detection.

The main feature of the MSW effect is a reduction of the ν_e flux, ν_e being transformed in ν_μ or ν_τ . But the reduction is not uniform at all. The ν_e energy spectrum is modified in a way which depends on the neutrino oscillation parameters. There are then several possibilities to evidence an effect, almost independently of solar models :

- suppression of the integrated ν_e flux.
- modification of the shape of the ν_e energy spectrum.
- modification of the NC/CC ratio. Indeed ν_μ and ν_τ coming from ν_e oscillation can induce neutral current interactions, modifying the NC/CC ratio.
- day-night differences.

These different approaches are listed in table 3 in the regions where they give a significant effect. We can now give more details, trying to isolate the MSW modifications which can lead to an interpretation almost independent of solar models.

The neutrino flux suppression may be observed in the case of ν_B sensible detectors. It is possible to interpret in this way the chlorine and Kamiokande experiments, which leads to the famous triangular region in the $(\Delta m^2 , \sin^2 2\theta)$ plane (see for example [15,18]). However the ν_B flux is strongly model dependent and a model independent proof of the MSW effect would be the observation of a modified ν_e spectrum. This can be achieved by Sudbury, Borex and Icarus.

The neutrino flux suppression can also be observed in the forthcoming gallium experiments [5,6]. A global reduction flux would be model dependent. But these experiments are sensitive to the primordial ν_{pp} whose flux is directly connected to the solar luminosity and practically model independent. A value below the predicted ν_{pp} flux (i.e. below 70 SNU) would be a probable evidence for the MSW effect, which is now the only serious explanation for such a deficit.

The indium experiment which detects the ν_{pep} / ν_{Be} ratio would be sensitive to a variation of this ratio which is not strongly model dependent. A large statistics and a small error would be needed to do so.

When ν_e are transformed into ν_μ or ν_τ they become sterile for detectors which are sensitive only to ν_e (all radiochemical detectors for example). The neutral current of ν_μ and ν_τ has the same cross section as that of ν_e . An increase of the number of the observed NC or of the ratio NC/CC would clearly favour the MSW effect. The Sudbury, Borex and Icarus experiments have still the possibility of doing this.

Finally regeneration of ν_e into the earth could induce a day/night flux variation which would also be a unique and convincing evidence for MSW neutrino oscillations. Radiochemical experiments may have difficulties to do this. However the integration over days still induce some seasonal effects which could be observed by large statistics experiments as the chlorine in Baksan. It is very unlikely that the gallium experiments could observe something. The real time experiments are better adapted for such a purpose. Though Kamiokande is real time, the solar neutrino signal, after background subtraction, is not. Sudbury or SuperKamiokande should be better placed. In this case one year of running should be sufficient to observe a significant effect.

Moreover, one can see from table 3 that in the regions of the $(\Delta m^2 , \sin^2 2\theta)$ plane where no distortion of the energy spectrum can be seen and where only the neutral/charged current ratio can be used as model independent evidence for neutrino oscillations, we get sizeable day/night effects (there may be a factor 3 difference between day and night). Such effects are likely to be much easier to detect than the neutral currents.

4.   Variation of the neutrino flux with time

The interesting effects issued from the MSW mechanism and showing day/night differences were discussed in the previous section. The idea that the solar neutrino flux could vary slowly with time is far from evident because solar neutrinos are produced at the very center of the Sun (less than 0.3 solar radius). In this region the 11-yr solar cycle should not be seen since it is due to magnetic phenomena which affect almost exclusively the convection zone. Moreover one does not expect important variations of this flux since the last 100 million year. All these things have however to be investigated.

The first (and alone) possible experimental evidence for a variation of the solar neutrino flux comes from Davis in the chlorine experiment [3]. By comparing since 1970 the measured ^{37}Ar signal with the number of sunspots, which determines the 11-yr solar cycle, Davis seems to observe an anticorrelation : when the ^{37}Ar signal is low, the external activity of the Sun is maximum, and vice-versa. It is however too early to draw a definite conclusion, since the statistical significance of these observations is still marginal. As stated before there is no simple explanation to this phenomenon. This is why further investigation is needed, which needs mainly a larger statistics. Three forthcoming experiments are better placed to do this : Sudbury, the chlorine experiment in Baksan and SuperKamiokande. The answer to this question will not be immediate : about 10 years of data taking will be necessary to confirm or infirm the Davis suggestion. A new idea for an experiment using **iodine** has been proposed one year ago by Haxton [20] and could contribute to understand this point. The reaction (ν_e ^{127}I $\rightarrow$ ^{127}Xe e$^-$) has a threshold of 664 keV and ^{127}Xe decays with a lifetime of 36.4 d. This radiochemical experiment, which could use any suitable iodine-bearing liquid is very similar to the chlorine one and is sensitive to ν_{Be} and ν_B . 380000 l of methylene iodide would give about 20 times more ^{127}Xe atoms than the ^{37}Ar atoms in the Davis chlorine experiment. No proposal has been yet written, but the idea remains attractive since it duplicates relatively closely the chlorine experiment, but with a different target.

There is a last class, the geochemical experiments, which may give information on the solar neutrino flux integrated over several million years. Solar ν_e are absorbed by ^{98}Mo (threshold = 1.7 MeV) which gives ^{98}Te which has a period of 4.2 million years. The present abundance of ^{98}Te in 1000 tons of a **molybdenum** ore is being measured by a Los Alamos group [21] which counts ^{98}Te atoms (about 10^7) using a dedicated mass-spectrometer. The analysis is in progress and should give results within few months. It has been shown [22] that the expected result should yield the same value for the ν_B flux as is determined by contemporary observations using the chlorine and Kamiokande detectors. Uncertainties on the ν_e capture cross section will however put limits on the interpretation. A similar experiment, **Lorex**, is planned to detect ^{205}Pb obtained from neutrino absorption in ^{205}Tl [23]. The reaction threshold is very small (54 keV) and the experiment is mostly sensitive to ν_{pp} . Unfortunately there are large uncertainties on the absorption cross section. The ore (lorandite or $TlAsS_2$) comes from the Allchar mine near the border between Yugoslavia and Greece, and is about 10^7 years old.

5.   Conclusion

The ultimate goals of solar neutrino astronomy are to infer, from the rates of all neutrino producing reactions, whether the Sun behaves like it is supposed to do and to determine whether neutrino parameters like mass, mixing angle, lifetime and magnetic moment influence neutrino propagation to the Earth. To do this we need measurements in real time of the flux, flavour and energy spectrum of all individual sources of solar neutrinos. This unfortunately cannot be achieved in a single experiment. However, by performing various radiochemical, real time and geochemical experiments, one may hope to extract the basic information in a foreseeable future.

The main problems encountered in the present various projects are :
- size and cost of the experiments
- background
- uncertainties in some theoretical interaction rate of neutrinos.

However the growing interest in this field makes such a program more realistic than it was a few years ago.

Acknowledgements : It is a pleasure to thank J.N.Bahcall, R.Barloutaud, M.Cribier, T.Kirsten, J.Rich and C.Tao for many discussions.

References

[1] J.N.Bahcall, these Proceedings
J.N.Bahcall and R.K.Ulrich, Rev. of Mod. Phys. 60 (1988) 297

[2] T.Kirsten, Proc. of the 13th Int. Conf. on Neutrino Physics and Astrophysics, Boston, June 1988, p.742

[3] R.Davis, these Proceedings

[4] M.Nakahata, these Proceedings
K.S.Hirata et al., Phys. Rev. Lett. 63 (1989)16

[5] T.Kirsten, these Proceedings

[6] V.N.Gavrin, these Proceedings

[7] R.S.Raghavan, Phys. Rev. Lett. 37 (1976) 259

[8] Bell Labs-IN2P3-Oxford-Pennsylvania-London-Saclay-Munich, Report, July 1989

[9] N.E.Booth, in Superconducting and Low-Temperature Particle detectors, G.Waysand and G.Chardin ed., Elsevier Science Pub. (1989) p.69
L.Gonzalez-Mestres and D.Perret-Gallix, Moriond meeting on neutrinos and exotic phenomena, Les Arcs (France) (1988)

[10] I.R.Barabanov, G.V.Domogatsky, G.T.Zatsepin, Proc. of the 13th Int. Conf. on Neutrino Physics and Astrophysics, Boston, June 1988, p.331

[11] G.T.Ewan et al., Sudbury Neutrino Observatory Proposal, SNO 87-12, October, 1987
G.Aardsma et al., Phys. Lett. B194 (1987) 321
H.B.Mak, Poster presented at this Conference

[12] R.S.Raghavan and S.Pakvasa, Phys. Rev. D37 (1988) 849
R.S.Raghavan et al., Design concept for Borex, AT&T Bell Labs report 88-01, March 31, 1988
S.Bonetti et al., Poster presented at this Conference
T.Kovacs et al., to appear in Solar Physics

[13] J.N.Bahcall, M.Baldo-Ceolin, D.Cline and C.Rubbia, Phys. Lett. B178 (1986) 324
L.Bassi et al., Icarus I : an optimized, real time detector of solar neutrinos, Proposal, March 21, 1988

[14] J.Arafune et al., Phys. Lett. B217 (1989) 186

[15] A.Yu.Smirnov, these Proceedings
S.P.Mikheyev and A.Yu.Smirnov, Nuovo Cimento 9C (1986) 17
L.Wolfenstein, Phys. Rev. D17 (1978) 2369

[16] H.A.Bethe, Phys. Rev. Lett. 56 (1986) 1305

[17] J.Bouchez et al., Z. Phys. C32 (1986) 499

[18] M.Cribier et al., Phys. Lett. B182 (1986) 89
A.J.Baltz and J.Weneser, Phys. Rev. D37 (1988) 3364
A.Dar et al., Phys. Rev. D35 (1987) 3607

[19] H.Harari, these Proceedings

[20] W.C.Haxton, Phys. Rev. Lett. 60 (1988) 768

[21] G.A.Cowan and W.C.Haxton, Science 216 (1982) 51

[22] J.N.Bahcall, Phys. Rev. D38 (1988) 2006

[23] See Proc. of the Int. Conf. on Solar Neutrino Detection with ^{205}Tl , Nucl. Instr. and Meth. in Phys. Research A271 (1988)

REPORT ON THE HOMESTAKE CHLORINE SOLAR NEUTRINO EXPERIMENT

R. Davis, K. Lande, C.K. Lee
University of Pennsylvania, Phila., PA 19104

B.T. Cleveland
Los Alamos National Lab., Los Alamos, NM 87545

J. Ullman, Herbert Lehman Col., Bronx, NY 10468

ABSTRACT. A report on the results obtained from the chlorine radiochemical solar neutrino experiment in the Homestake mine, Lead, SD. Over the period 1970-1988 a neutrino capture rate of 2.3 $\pm$ 0.3 SNU was observed. This rate is discussed in relation to the theoretical standard solar model, the results from the Kamiokande II experiment, and variations in the solar neutrino flux.

1. INTRODUCTION

This report will be concerned with the latest experimental results from the Homestake chlorine experiment, a radiochemical neutrino detector based upon the neutrino capture reaction, ν_e + $^{37}Cl \rightleftarrows e^-$ + ^{37}Ar (35 day half life). The detector contains 615 m tons of C_2Cl_4 , and the ^{37}Ar is removed by circulating helium through the liquid by a pump-eductor system (1). A measured volume of ^{36}Ar or ^{38}Ar carrier is used to measure the efficiency of argon recovery and finally the entire recovered argon sample is placed in a proportional counter to measure the radioactive ^{37}Ar produced. Argon-37 is identified by measuring the pulse amplitude (2.82 kev Auger electrons) and pulse rise-time (2). The experiment is located at a depth of 4100 $\pm$ 200 hectograms cm^2 standard rock, in the Homestake Gold Mine in Lead, South Dakota. The detector has operated continuously since 1970.3 except for a 1.4 year period from 1985.4 to 1986.8 when both liquid circulating pumps were out of commission. Except for this period, the observations were continuous and extractions of the ^{37}Ar activity were carried out at an average rate of 5 per year. Observations were extended over this long period for a number of

171

G. Berthomieu and M. Cribier (eds.), Inside the Sun, 171–177.
© 1990 Kluwer Academic Publishers. Printed in the Netherlands.

reasons: to obtain an improved result with reduced errors, search for short and long period variations in the [37]Ar production rate, measure the neutrino flux in the event of a supernova within 10 Kpc, and to insure that observations from the chlorine experiment overlapped those of new solar neutrino experiments. In this brief report we will compare our results with the standard solar model, and the recent observations from the Kamiokande II detector, and discuss the question of a variation in the observed [37]Ar production rate.

2. AVERAGE [37]Ar PRODUCTION RATE

The events observed in the proportional counter that had the correct energy (fwhm 2.82, 25% resolution) and pulse rise time were analysed by a maximum liklihood method into a decaying component with a 35 day half-life and a constant counter background (3). A growth factor for a 35 day radioactive product and the counting efficiency was applied to obtain the [37]Ar production rate. A plot of the individual measurements is shown in figure 1, and the average [37]Ar production rate is listed in table 1 for the entire observing period 1970.3 - 1988.3, and also for the period of overlap with the Kamiokande II observations, 1986.8 - 1988.3. Additional data up to the present time are being analysed and will be presented at a later date.

Table 1. Summary of [37]Ar Production Rates
(Atoms/day in 615 tons C_2Cl_4)

Period of Observation	1970.3-1988.3	1986.8-1988.3
Average [37]Ar production rate	0.518 $\pm$ 0.036	0.87 $\pm$ 0.13
Cosmic ray background	0.08 $\pm$ 0.03	0.08 $\pm$ 0.03
[37]Ar above known backgrounds	0.438 $\pm$ 0.047	0.79 $\pm$ 0.13
[37]Ar Production rate in SNU*	2.33 $\pm$ 0.25	4.2 $\pm$ 0.7

*Solar Neutrino Unit = 10^{-36} interactions sec^{-1} target $atom^{-1}$ or 5.31 [37]Ar atoms/day in the 615 tonne Homestake detector.

There is a cosmic ray muon background of 0.08 $\pm$ 0.03 [37]Ar atoms per day that must be subtracted to obtain the rate to be attributed to neutrinos. This background rate was derived from exposing smaller tanks of C_2Cl_4 at higher levels in the mine and deriving the rate for the full volume of C_2Cl_4 at the full depth of the experiment (4). An experiment is in progress that uses the photonuclear process ^{39}K ($\mu^{\pm}$, $\mu^{\pm}$ np) ^{37}Ar (5). The new results with the potassium experiment are similar to the ones used here, though somewhat lower, approximately 0.05 ^{37}Ar per day.

3. COMPARISON WITH THE STANDARD SOLAR MODEL AND THE KAMIOKANDE II EXPERIMENT

The net rate from the chlorine experiment for the entire period 1970.3 to 1988.3 is 2.33 $\pm$ 0.25 SNU (1σ error). It is this value that can be compared to the standard solar model. The most recent calculations are those of Bahcall and Ulrich (6) who obtained 7.9 $\pm$ 2.6 (3σ error) SNU and Turck-Chieze et al (7) who obtained 5.8 $\pm$ 1.3 (1σ error) SNU. These calculations differ chiefly because the separate groups chose different parameters for the solar opacities and the cross-section of the Be (P,γ) ^{8}B reaction. It is well known that the chlorine experiment is primarily sensitive to the flux of the energetic neutrinos from ^{8}B decay, 0-15 Mev. Assuming the usual solar sources and fluxes, approximately 77 percent of the rate in ^{37}Cl should be produced by ^{8}B decay neutrinos.

Recently the Kamiokande II experiment has measured the flux of solar neutrinos above 9.3 Mev (8). This detector is an imaging water Cherenkov detector system that observes neutrinos by ν_e - e^- elastic scattering, a process that has a favorable angular distribution of the recoil electrons with respect to the neutrino direction of energetic neutrinos. They compare the rate of events from the direction of the Sun with those from all other angles to obtain the ^{8}B solar neutrino flux. The rate observed is 0.45+0.15 (1σ error) times the ^{8}B flux predicted by Bahcall and Ulrich using a Monte Carlo derived shape for angular distribution (6). The Kamiokande II results were obtained from January 1987 through May 1988 and correspond to the data period 1986.8 to 1988.3 when the chlorine experiment observed a rate of 4.2 $\pm$ 0.7 SNU, see table 1 and figure 1. These two very different solar neutrino detectors are considered to be in essential agreement assuming that both experiments are observing primarily the same solar neutrino source. The high rate observed recently by the chlorine experiment is probably the result of a variation in the measured solar neutrino flux. Both experiments at the present time observe rates close to those calculated by the standard solar model of Turck-Chieze et al (7).

A low signal rate in these experiments can be attributed to resonance mixing of neutrino flavors suggested by Mikheyev and Smirnov based upon matter

174

Figure 1. ^{37}Ar Production Rates in the Homestake C_2Cl_4 Solar
            Neutrino Detector.

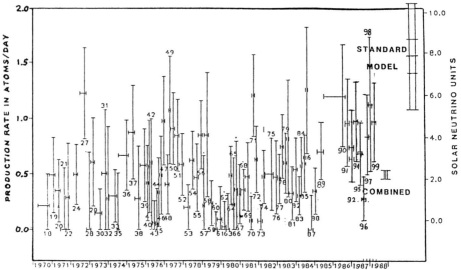

Figure 2.   Comparison of Time Variation of 5 Extraction Running
            Averages of Measured Solar Neutrino Flux with
            Number of Sunspots and Solar Diameter.

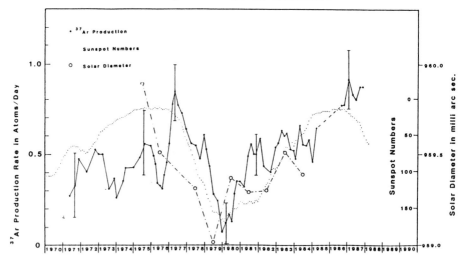

oscillations of Wolfenstein (9). This process, called the
MSW effect, arises from a difference in the scattering of
electron neutrinos (ν_e) and muon or tauon neutrinos (ν_μ, ν_τ)
with electrons. The process depends upon the difference
in the masses between the neutrino types (or flavors), the
mixing angle θ between the phases of neutrino eigenstates,
and the electron density in the Sun. If leptons, $e^\pm$, $\mu^\pm$, $\tau^\pm$
and their neutrinos, are not conserved then electron
neutrinos (ν_e) produced in the core of the Sun could be
converted to ν_μ or ν_τ in passing through the Sun. The
neutrino masses and mixing angles are unknown, but over a
wide range of these parameters the flux of ν_e could be
greatly reduced. These processes have not been observed in
accelerator experiments, or cosmic ray studies. Perhaps the
best opportunity for observing neutrino mixing is by solar
neutrino studies.

4. IS THE SOLAR NEUTRINO FLUX CONSTANT?

The observed ^{37}Ar production rate exhibits an anti-
correlation with the solar activity cycle. This matter has
been pointed out in several earlier reports (10). Figure 2
shows a 5 point running average of 78 individual ^{37}Ar
production rates measured over the last 18 years. The
running average shows an anti-correlation with the solar
activity cycle as measured by sunspot occurrences and
correlates with the solar diameter measurements of Laclare
(16). The rate was highest at solar minimum: in 1977 the
average was 4.1 + 0.9 SNU and in 1986.8 - 1988.3 the
average was 4.2 + 0.7 SNU. During solar maximum in 1979.5
to 1980.7 the rate was 0.4 + 0.1 SNU. It is of course of
great interest to see if the ^{37}Ar production rate is again
low at solar maximum of cycle 22. Furthermore, it will be
interesting to see if the Kamiokande II experiment observes
a low rate in 1990-1991.

It is unlikely that the neutrinos from the core of the
Sun would exhibit a large change in flux correlated with
the solar activity cycle. There have been two suggestions.
Voloshin, Vysotsky and Okun (11) suggested the spin of a
left handed electron neutrino ν_e (L) could be rotated into a
right handed sterile ν_e(R) in passing 10^{10}cm through
transverse solar magnetic fields of a few thousand gauss,
if the neutrino had a magnetic moment of 10^{-10} - 10^{-11} μ_B
(Bohr magnetrons). This mechanism was considered unlikely
by theorists when first suggested because a neutrino with a
small mass would have a correspondingly small magnetic
moment, $\sim 10^{-17}$ μ_B. However, the magnitudes of the internal
magnetic fields in the Sun are unknown. They could be very
much larger at the base of the convective zone. Also a

neutrino could have an acceptably larger transition magnetic moment (11,12). It was pointed out by Lim and Marciano (13) and Akhmedov (14) that the combined effect of matter and magnetic fields would produce a spin flavor transition, ν_e (L) → $\overline{\nu}_\mu$(R), into non-detectable muonic anti-neutrinos. Calculations of these effects (15) show a large reduction in detectable neutrinos occurs in the convective zone of the Sun. It was pointed out by Voloshin, Vysotskii, and Okun (11) that an experimental test of these mechanisms could occur as a result of the fact that the Sun's axis of rotation is inclined 7 1/4 degrees with respect to the plane of the ecliptic. Twice a year the neutrinos reaching the earth come uneffected through the field-free solar equator (5 June, 5 December) and twice a year the neutrinos pass through the magnetic fields at higher latitudes (N or S, 5 March, 5 September) causing a greater loss of ν_e flux. The experimental data from 1979 – 1982 appears to show this effect (17).

ACKNOWLEDGEMENT. This work was supported by the National Science Foundation and grants from the University of Pennsylvania Research Foundation and from the CUNY Research Foundation of the City College of New York.

5. REFERENCES

1. Davis, R., Harmer, D.S., and Hoffman, K.C. (1968), Phys. Rev. Lett. 14, 20.

2. Davis, R. (1978), 'The Status and Future of Solar Neutrino Research' 1, 1, in G. Friedlander (ed.), Brookhaven National Laboratory Report BNL 50879; Rowley, J.K., Cleveland, B.T. and Davis, R. (1985), in M.L. Cherry, W.A. Fowler, and K. Lande (eds.), American Inst. Phys. Conf. Proc. No. 126, p.1; Bahcall, J.N. and Davis, R. (1976), Science 191, 264.

3. Cleveland, B.T. (1983), Nucl. Inst. and Methods 214, 451.

4. Wolfendale, A.W., Young, E.C.M. and Davis, R. (1972), Nature, Phys. Sciences 238, 1301; Cassiday, G.L. (1973), Proc. 13th Inst. Cosmic Ray Conf. 13, 1958; see calc. Zatselim, G.T. et al (1981), Soviet J. Nucl. Phys. 33, 200.

5. Fireman, E., Cleveland, B.T., Davis, R. and Rowley, J.K. (1985), in M.L. Cherry, W.A. Fowler and K. Lande (eds.) American Inst. Phys. Conf. Proc. No 126, p.22.

6. Bahcall, J.N. and Ulrich, R.K. (1988), Rev. Modern Phys. 60, 297; report Inside the Sun Conference.

7. Turck-Chieze, S., Cahen, S., Casse, M. and Doom, C. (1988), Astrophys. J. 335, 415; report Inside the Sun Conference.

8. Hirata, K.S. et al (Kamiokande group) (1988), Phys. Rev. Lett. 63, 16 (see also Phys. Rev. D38, 448); report Inside the Sun Conference.

9. Mikheyev, S.P. and Smirnov, A. Yu. (1985), Soviet J. Nucl. Phys. 42, 913 (see the excellent review in Soviet Phys. Uspkhi 30, 759 (1987) summarizing calculations by many authors); Wolfenstein, L. (1978), Phys. Rev. D17, 2369.

10. Rowley, J.K. et al ref (2); Davis, R., 7th Workshop on Grand Unification/ICOBAN-86, Toyama, Japan, p.237, J. Arafune (ed.); Davis, R., Cleveland, B.T. and Rowley, J.K., 20th Int. Cosmic Ray Conference, Aug. 2-15, 1987 Moscow, 4, 328; Neutrino 88 Conference, Boston, MA, 5-11 June 1988, J. Schneps (ed.).

11. Voloshin, M.B., Vysotskii, M.I. and Okun, L.B. (1986), Soviet J. Nucl. Phys. 44, 440; (1986) Soviet Phys. 64, 446.

12. Fukugita, M. and Yanagida, T. (1987), Phys. Rev. Lett. 58, 1807; Babu, K.S. and Mohapatra, R.N. (1989), Phys. Rev. Lett. 63, 228.

13. Lim, C.-S. and Marciano, W. (1988), Phys. Rev. D37, 1368.

14. Akhmedov, E. Kh. (1988), Soviet Nucl. Phys. 48, 382; (1988) Phys. Lett B213, 64.

15. Minakata, H. and Nunokawa, H. (1989), Phys. Rev. Lett. 63, 121 and Akhmedov, E. Kh., (1989), Zh. Eksp. Teor. Fiz. 95, 1195; and 95, 442.

16. Delache, P., Laclare, F. and Sadsaoud, H. (1985), Nature 317, 426; report Inside the Sun Conference.

17. Veselov, A.I., Vysotskii, M.I. and Yurov, V.P. (1987), Soviet J. Nucl. Phys. 45, 865.

THE KAMIOKANDE SOLAR NEUTRINO EXPERIMENT

THE KAMIOKANDE-II COLLABORATION

K. S. HIRATA, T. KAJITA, T. KIFUNE, K. KIHARA, M. NAKAHATA,
K. NAKAMURA, S. OHARA, N. SATO, Y. SUZUKI, Y. TOTSUKA AND
Y. YAGINUMA
Institute for Cosmic Ray Research, University of Tokyo, Tanashi, Tokyo 188, Japan

M. MORI, Y. OYAMA, A.SUZUKI, K.TAKAHASHI, H.TAKEI[1] AND T.TANIMORI
National Laboratory for High Energy Physics (KEK), Tsukuba, Ibaraki 305, Japan

M. KOSHIBA
Tokai University, Shibuya, Tokyo 151, Japan

T. SUDA AND T. TAJIMA
Department of Physics, Kobe University, Kobe, Hyogo 657, Japan

K. MIYANO, H.MIYATA AND M. YAMADA
Niigata University, Niigata, Niigata 950-21, Japan

Y. FUKUDA, K. KANEYUKI, Y.NAGASHIMA AND M.TAKITA
Department of Physics, Osaka University, Toyonaka, Osaka 560, Japan

E. W. BEIER, L. R. FELDSCHER, E. D. FRANK, W. FRATI, S.B.KIM,
A. K. MANN, F.M.NEWCOMER, R VAN BERG AND W.ZHANG
Department of Physics, University of Pennsylvania, Philadelphia PA 19104, U. S. A.

Presented by M. NAKAHATA

ABSTRACT. The observation of ^{8}B solar Neutrinos in the Kamiokande-II detector is presented. Based on 450 days of data in the time period of January 1987 through May 1988, the measured flux obtained with $E_e \geq 9.3$ MeV was 0.46 ± 0.13 (stat) ± 0.08 (sys) of the value predicted by the standard solar model. The detector and analysis methods were improved since June 1988 and the background level has been decreased by a factor of about three since then.

1. Introduction

The discrepancy in the solar neutrino flux between the ^{37}Cl experiment [1] (2.1 ± 0.3 SNU; 1 σ error) and the prediction by the standard solar model(SSM)[2] (7.9 ± 2.6 SNU;

---

[1] On leave from Niigata University

G. Berthomieu and M. Cribier (eds.), Inside the Sun, 179–186.

3 σ error) is known as the missing solar neutrino problem. There have been many possible explanations to this discrepancy.[3, 4, 5] In order to solve this problem we need more observational data on solar neutrinos. Moreover, the recent result from the ^{37}Cl experiment gives a high neutrino flux[6] (4.2 ± 0.7 SNU) and an anti-correlation of the solar neutrino flux with solar activity is suggested.[6] In order to check these results several independent experiments should be performed.

The Kamiokande-II is an imaging water Cherenkov detector, which detects ^{8}B solar neutrinos by neutrino-electron scattering, $v_e \, e^- \rightarrow v_e \, e^-$, and yields information on the neutrino arrival time, the direction, and the energy spectrum.

2. Detector

The schematic view of the Kamiokande-II detector is shown in Fig.1. It is described in detail in ref.[7]. The detector volume contains 2140 tons of water, which is viewed by an array of 20-in.-diameter photomultipliers tubes (PMT) on a 1×1-m^2 lattice on the surface. The photocathode coverage amounts to 20 % of the total inner surface. The attenuation length of Cherenkov light in the water is usually in excess of 45 m. The inner detector is completely surrounded by a water layer (anti-counter) of thickness ≥ 1.4 m. The anti-counter is an absorber of γ-rays from surrounding rocks and a monitor of cosmic ray muons. The fiducial volume of the detector is 680 tons, which is 2 ~ 3 m inside from the wall of the detector.

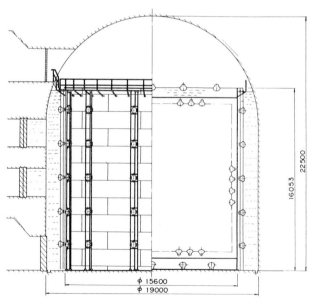

Figure 1. Schematic view of the Kamiokande-II detector. The inner detector is viewed by 948 20-in.-diameter PMTs. The anti-counter surrounds the inner detector and is viewed by 123 PMTs. Dimensions in the figure are in millimeters.

The detector was triggered by at least 20 hit PMT's within 100 nsec. Charge and time information for each PMT above threshold was recorded for each trigger. The trigger accepted 7.6 (6.7) MeV electrons with 50 % efficiency and 10 (8.8) MeV electrons with 90 % efficiency over the fiducial volume of the detector. These numbers were the values before October 1987 (October 1987 - May 1988).[1] The raw trigger rate was about 0.6 (1.2) Hz of which 0.37 Hz was cosmic-ray muons. The remaining rate was due to radioactive contamination in the detector and external γ-rays. The Cherenkov light of an electron of 10 MeV total energy fired ~ 26 PMTs. The energy calibration was performed with γ-rays of energy up to 9 MeV from the reaction Ni(n, γ)Ni, with electrons from μ decays, and with spallation products by cosmic ray muons. From these calibrations, the absolute energy normalization was known to be better than 3 %. The energy resolution was expressed by 22 % $/(E_e/10 \text{ MeV})^{0.5}$ and angular resolution for an electron of 10 MeV was 28°, respectively.

3. Data Selection

The first search for ^{8}B solar neutrinos was carried out on the 450 days of data taken from January 1987 through May 1988.

First of all, events that satisfy following three criteria were selected:

(1) The total number of photoelectrons (p.e.) per event in the inner detector should be less than 100, corresponding to a 30 MeV electron.

(2) The total number of photoelectrons in the outer detector is less than 30 for ensuring containment of an event.

(3) The time interval from the preceding event should be longer than 100 μsec to exclude electrons from μ decays.

The event rate after the selection is shown in (a) in Fig.2.

The vertex positions and the directions of the selected events were reconstructed with an algorithm based on the time and position of hit PMTs. The rms vertex position resolution is 1.7 m for 10 MeV electrons. By limiting the events in the fiducial volume of 680 tons, the backgrounds of external γ-rays are reduced. As shown in (b) in Fig. 2, the fiducial volume cut reduces the event rate by an order of magnitude.

Most of the remaining events were found to be due to unstable spallation products of through-going muons in the detector. These beta-decay events have time and spatial correlation with the preceding cosmic ray muons. Furthermore, these preceding muons often accompany energetic cascade showers. These spallation events were reduced by criteria which take into account these features. The details of the criteria was described elsewhere.[8] They reduced the event rate with $E_e \geq 10$ MeV by 70 % (Fig.2 (c)) whereas the introduced dead time of 10.4 %.

One of the remaining background after the reduction of spallation backgrounds was external γ-rays which are not completely eliminated by the fiducial volume cuts. To reduce them, events were rejected which had vertex positions near the edge of the fiducial volume (outer 1 m layer) and directions inward (cosine of the angle relative to the normal to the nearest wall > 0.67). This cut further reduced the event rate ~ 40 %, and introduced an additional dead-time of 13 %.

---

[1] The trigger thresholds are 6.1(5.2) MeV with 50 % efficiency from June 1988 through April 1989 (since May 1989), respectively.

The whole processes of the data analysis were performed also to solar neutrino signals generated by a Monte Carlo simulation and an expected spectrum from SSM was calculated, which is shown in Fig.2 (e).

Figure 2. Differential energy distribution of low energy events (a) in the total mass of 2140 tons; (b) in fiducial mass of 680 tons; (c) after the spallation cut; (d) after remaining γ-ray cut; (e) Monte Carlo prediction of SSM after all cuts.

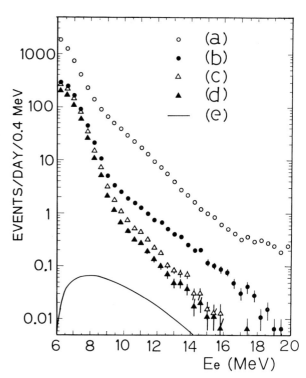

4. Intensity of ^{8}B Solar Neutrino Signal

The selected event sample was then tested for a directional correlation with the sun. Figure 3 shows the $\cos\theta_{sun}$ distribution of the events with $E_e \geq 9.3$ MeV and $E_e \geq 10.1$ MeV, in which $\cos\theta_{sun} = 1$ corresponds to the expected direction to the earth from the sun. One sees a clear enhancement near $\cos\theta_{sun} = 1$. The solid histograms in the figure gives the signal expected from a Monte Carlo simulation based on the SSM. The figure indicates that the observed signal is less than the expectation from SSM. The energy spectrum of the ^{8}B solar neutrino signal is obtained by fitting the $\cos\theta_{sun}$ distribution with a flat background plus an expected angular distribution of the signal for the data in each energy bin. The obtained energy spectrum is shown in Fig.4 with the expectation from the SSM

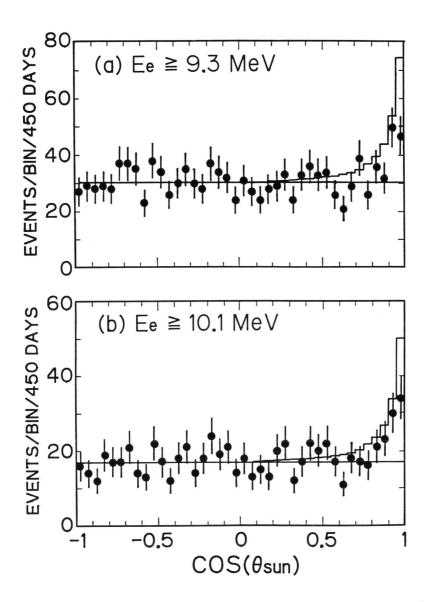

Figure 3.   Distributions in $\cos\theta_{sun}$, cosine of the angle between the reconstructed direction of an electron and the direction from the sun to the earth, for the events with (a) $E_e \geq$ 9.3 MeV and (b) $E_e \geq$ 10.1 MeV.

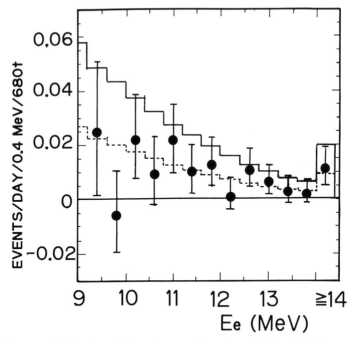

Figure 4. Energy distribution of the solar neutrino signal. The solid histogram shows the prediction from SSM. The dotted histogram shows the best fit to the data.

by a solid histogram. The flux of the ^{8}B solar neutrinos is obtained by fitting the electron energy spectrum by scaling the expected spectrum. The obtained flux was :

$$\frac{\text{Kam-II data}}{\text{SSM}} = 0.46 \pm 0.13(\text{stat}) \pm 0.08(\text{sys}) \,.$$

The best fit of the spectrum is shown by the dotted histogram in Fig. 4. The systematic error was determined mainly by the uncertainties in energy calibration and angular resolution.

This result is statistically consistent with the value obtained by the ^{37}Cl experiment in essentially the same time period.[6] The result of the Kamiokande-II confirmed the missing solar neutrino problem.

5. Present Status

The sources of the remaining backgrounds after all cuts were : (a) radioactivities in the detector materials; (b) spallation products and external γ-rays which were not completely eliminated by the above cuts, where backgrounds of (a) and (b) dominated at relatively lower energy ($E_e \leq 9$ MeV) and at higher energy ($E_e \geq 9$ MeV), respectively. These backgrounds can be reduced by improving the energy resolution, vertex position resolutions and methods of cuts. Therefore, we have performed following improvements since June 1988.

(1) Gain of the PMTs were increased by a factor of two.

The effects of the PMT gain increase were : (i) ~ 20 % increase of hit PMTs corresponding to unit deposited energy by electrons[1]; and (ii) ~ 10 % improvement of the energy resolution (now 19.5 % $/(E_e/10$ MeV$)^{0.5}$) for low energy electrons. The increase of the number of hit PMTs in an event improved event vertex resolution, and this led to reduction of poorly reconstructed events in the fiducial volume.

(2) The method of the spallation cuts was improved.

The criteria for spallation products of longer life time were improved. The improved criteria reduced the remaining background due to spallation products by a factor of about two.

Figure 5 compares the present final event rate with that before the improvements. The background rate was reduced by a factor of about three. Furthermore, this improvement enabled us to lower the energy threshold of the data analysis to ~ 7.5 MeV. The result of the recent data will be presented in near future.[9]

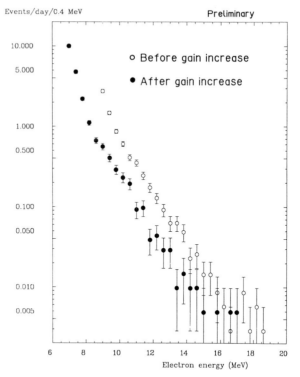

Figure 5. Differential energy distribution of low energy events after applying all cuts for the data before and after the improvements. The dead time due to the cuts is corrected.

---

[1] The PMT gain was increased without changing the discriminator threshold for each PMT. Thus, the collection efficiency of a single photoelectron level was increased and, consequently, the number of hit PMTs in an event was increased.

6. Conclusion

The observation of ^{8}B solar neutrinos in the Kamiokande-II detector was presented. The measured flux was 0.46 ± 0.13 (stat) ± 0.08 (sys) of the value predicted by the standard solar model, which was obtained based on 450 days of data in the time period of January 1987 through May 1988 with $E_e \geq 9.3$ MeV. The backgrounds to the solar neutrinos were reduced by a factor of about three after June 1988 by increasing the gain of PMTs and by improvements of cuts. This enabled us to lower the energy threshold of the data analysis to ~ 7.5 MeV. The statistical error of the recent data should be much smaller than before.

7. Note Added

After the conference, the recent solar neutrino data was analyzed. Based on 288 days of improved data taken in the time period of June 1988 through April 1989, the measured flux obtained from the data with $E_e \geq 7.5$ MeV was 0.39 ± 0.09 (stat) $\pm \sim 0.06$ (sys) (preliminary) of the value predicted by the standard solar model.[9]

8. Acknowledgements

We gratefully acknowledge the cooperation of the Kamioka Mining and Smelting Company. This work was supported by the Japanese Ministry of Education, Science and Culture, by the United States Department of Energy, and by the University of Pennsylvania Research Fund. Part of the analysis was carried out by FACOM M780 and M380 at the computer facilities of the Institute for Nuclear Study, University of Tokyo.

9. References

[1] J. K. Rowley, B.T. Cleveland and R. Davis, Jr., in "Solar Neutrinos and Neutrino Astronomy", ed. by M. L. Cherry, W. A. Fowler and K. Lande, AIP Conference Proc. No.126, p.1, 1985; R. Davis, Jr., in Proc. of the Seventh Workshop on Grand Unification, ICOBAN'86, ed. by J. Arafune, World Scientific, P.237, 1987.

[2] J. N. Bahcall and R. K. Ulrich, Rev. Mod. Phys. **60**, 297(1988). The ^{8}B flux adopted in this analysis is 5.8×10^6 /cm^2/sec.

[3] L. Wolfenstein, Phys. Rev. **D17**, 2369(1978); S. P. Mikheyev and A. Yu. Smilnov, Nuovo Cim. **9C**, 17(1986); H. A. Bethe, Phys. Rev. Lett. **56**, 1305(1986).

[4] M. B. Voloshin, M. I. Vysotskii, and L. B. Okun, Sov. J. Nuc. Phys. **44**, 440 (1986); M. B. Voloshin and M. I. Vysotskii, Sov. J. Nuc. Phys. **44**, 544(1986).

[5] S. M. Bilenky and B. Pontecorvo, Phys. Rep. **41**, 225(1978).

[6] R. Davis, Jr., in Proc. of the 13th Int. Conf. on Neutrino Physics and Astrophysics, Neutrino '88, ed by J. Schneps et al., World Scientific; R. Davis, Jr. in this proceedings.

[7] K. S. Hirata et al., Phys. Rev. **D38**, 448(1988).

[8] K. S. Hirata et al., Phys. Rev. Lett. **63**, 16(1989).

[9] K. S. Hirata et al., "Recent Solar Neutrino Data from the Kamiokande-II Detector", contributed paper to the XIV International Symposium of Lepton and Photon Interactions, Stanford, 1989; ICR Report 195-89-12.

THE GALLEX PROJECT

T.KIRSTEN
Max-Planck-Institut für Kernphysik
P.O.Box 103980
6900 Heidelberg

GALLEX COLLABORATION:
M.Breitenbach, W.Hampel, G.Heusser, J.Kiko, T.Kirsten, H.Lalla, A.Lenzing,
E.Pernicka, R.Plaga, B.Povh, C.Schlosser, H.Völk, R.Wink, K.Zuber
Max-Planck-Institut für Kernphysik,Heidelberg

R.v.Ammon, K.Ebert, E.Henrich, L.Stieglitz
Kernforschungszentrum Karlsruhe-KFK

E.Bellotti
Dip.di Fisica dell'Università,Milano and
Laboratori Nazionali del Gran Sasso-INFN
O.Cremonesi, E.Fiorini, C.Liguori, S.Ragazzi, L.Zanotti
Dip.di Fisica dell'Università, Milano

R.Mössbauer,A.Urban
Physik Dept. E15 , Technische Universität München

G.Berthomieu, E.Schatzman
Université de Nice-Observatoire

S.d'Angelo, C.Bacci, P.Belli, R.Bernabei, L.Paoluzi, R.Santonico
II-Universita di Roma -INFN

M.Cribier, G.Dupont, B.Pichard, J.Rich, M.Spiro, T.Stolarczyk, C.Tao, D.Vignaud
Centre d'Etudes Nucléaires de Saclay, Gif-sur Yvette

I.Dostrovsky
Weizmann Institute of Sciences,Rehovot

G.Friedlander,R.L.Hahn,J.K.Rowley,R.W.Stoenner,J.Weneser
Brookhaven National Laboratory, Upton,N.Y.

ABSTRACT. The **GALLEX** collaboration aims at the detection of solar neutrinos in a radiochemical experiment employing 30 tons of Gallium in form of concentrated aqueous Gallium-chloride solution.The detector is primarily sensitive to the otherwise inaccessible pp-neutrinos. Details of the experiment have been repeatedly described before [1-7]. Here we report the present status of implementation in the *Laboratori Nazionali del Gran Sasso* (Italy). So far, 12.2 tons of Gallium are at hand.The present status of development allows to start the first full scale run at the time when 30 tons of Gallium become available.This date is expected to be january,1990.

187

G. Berthomieu and M. Cribier (eds.), Inside the Sun, 187–199.
© 1990 *Kluwer Academic Publishers. Printed in the Netherlands.*

1.Introduction

The measurement of solar neutrinos allows to probe the state of the sun's interior in the most direct way.contrary to helioseismological investigations,the solar neutrino spectrum can provide detailed information on the ongoing fusion reactions,and,together with solar model calculations,on temperature and chemical composition[8,5]. This potential is partially lost if the neutrino fluxes leaving the solar core are modified by neutrino flavor oscillations(ν_e,ν_μ,ν_τ) on their way to the detector,yet the payoff could then be the confirmation of non-zero neutrino masses[9,10,11].

Should non-vanishing neutrino masses apply at all,then there is a very large probability that they can be unraveled through solar neutrino measurements.This is due to the MSW-effect [9] which effectively enlarges even initially small mixing angles once neutrinos pass through matter with high electron densities as in the case of the sun.This opens the possibility to probe neutrino masses $\sqrt{\Delta m_i^2}$ down to $\approx 10^{-6}$ev/c^2, a mass range highly interesting in Grand Unification schemes.

Measurements of ^{8}B-neutrinos with the radiochemical Homestake Chlorine detector[12] and with the Kamiokande Cerenkov detector[13] have firmly established a severe deficite relative to the standard solar model expectation.Whether this is due to neutrino oscillations or to deviations of the solar structure from the canonical model could be decided if the flux of the pp-neutrinos is measured,since the latter is directly tied to the solar luminosity and therefore hardly affected by not too unreasonable stellar structure modifications.

Apart from this principal decision,it could then also be possible to estimate from the pp-and ^{8}B-neutrino fluxes actual values for neutrino mass differences and/or mixing angles.This comes about since the reduction factors of the measurable ν_e-fluxes display distinct spectral response in the MSW-mechanism.

Experimentally,the low energy of pp-neutrinos (E_{max}=420keV) provides formidable problems.For many years to come,the only realistic possibility is a radiochemical Gallium-experiment,based on the inverse Beta-decay reaction

$$^{71}Ga\ (\ \nu_e,e^-\ )\ ^{71}Ge\ .$$

With an energy threshold of 233 keV,the neutrino capture rate on Ga is dominated by pp-neutrinos.The standard solar model prediction for Gallium has the highest reliability and the smallest error among all radiochemical neutrino detection schemes. The ^{71}Ge production rate is (132 + 20, -17) SNU[x] (3σ) [8],including 74 SNU from pp-and pep-neutrinos.

During 1979-1984, a pilot experiment employing 1.26 tons of Gallium has been carried out in collaboration between MPI Heidelberg and Brookhaven National Laboratory [14].The overall result was that the experiment is feasible if at least 30 tons of Gallium could be made available.This was not possible in the MPI/BNL-collaboration,but the GALLEX collaboration assembled in 1985 succeeded in assuring the required funds.Since that time work is beeing performed towards the implementation of the GALLEX-experiment at the INFN Gran Sasso underground laboratory L.N.G.S. (Laboratori Nazionali del Gran Sasso) and respective reports have been published[1-6].In this communication we do not intent to repeat previous reports but rather concentrate on a description of the actual implementation status,supported by numerous illustrations. As originally proposed,the starting date for data taking is planned for january,1990.

x) *SNU = solar neutrino unit,10^{-36} captures per atom and second.*

2.Status of the Experiment

The principal features of the experiment are summarized in Figure 1.In the following we describe the status of the various aspects of the experiment.

TARGET

30 t Ga as GaCl₃ in 1 tank

PRODUCTION RATE

1.09 capt./d (SSM)

.84 capt./d (Cons.model)

PRODUCTION BACKGROUND

muons: < 2% of signal

fast neutrons:< 1% of signal

target impurities: < 1% of signal

EXTRACTION

air purge → GeCl₄ → GeH₄ → purification (GC) → counter

as tested in 1.3 t BNL/MPI Pilot experiment

SEQUENCE

exposure~14 d ,extraction 1 d

:repeat:: 50 runs in 2 yrs

CALIBRATION

>800 kCi Cr⁵¹ source

Figure 1. Concept of the GALLEX experiment.

2.1. PROVISION OF GALLIUMCHLORIDE

We use an 8-normal aqueous Galliumchloride solution to facilitate Ge-extraction as volatile $GeCl_4$ by nitrogen purge of the target. 30 tons of Ga are contained in 100 t of solution.The solution is produced by Rhone Poulenc in Southern France and conditioned to a Cl/Ga-ratio of 3.25,the optimum for Ge-extractions according to our extraction test experiments.So far we have received,in three partial deliveries, 40.15 t of Galliumchloride solution.The Ga-content is 30.17 %,corresponding to 12.1 t of Ga. Legal contracts with our supplier call for the provision of the full 30 t of Ga till end of 1989.

All purity specifications are easily met by the product delivered so far (Table1).This was checked by neutron activation analysis (U,Th,Fe,As,Se,Ba) and by Radon-determinations using proportional counting.Additional Ge-extraction tests with spiked product solutions also gave satisfactory results. It is particularly remarkable that the critical ^{226}Ra-concentration is more than one order of magnitude below specification,making background-^{71}Ge production through actinide impurities completely negligible.We consider this to be a result of our intense guidance of the producer and our continuous analytical production control.

Altogether 14 teflon lined 1200-1 tanks are available for the transport of the solution from the producer to the LNGS (one "six-pack" in a 20" standard container per transport,containing ca. 4 t of gallium).The solution is received in a

Table 1. Key properties of the target solution.

		measured	specification
Ga	(g/l)	567.5	565 ± 10
Cl	(g/l)	938.5	
Cl/Ga	(mol/mol)	3.25	3.25
density	(g/ml)	1.8889	
^{226}Ra	(pCi/kg)	0.03	< 0.50 [x]
U	(ppb)	<0.05	< 25 [x]
Th	(ppb)	<0.1	< 2 [x]
Zn	(ppb)	≈ 20	
As	(ppb)	< 1	
Ba	(ppb)	≈ 20	

[x] These specifications would lead to a 1%-contribution to the standard solar model production rate for each individual background source.

Figure 2.
Technical Outside Facility (TOF) for $GaCl_3$-target conditioning.

Figure 3.
Interior of the TOF. Bridge balance(front), $10m^3$-tank(rear), storage tanks (large, side) and transport tanks (small, side).

Technical Outside Facility ("TOF") erected 1km from the tunnel entrance (Figure 2). Here it is transfered into storage tanks for later preparation in a 10 m³-mixing and transport tank equipped for fast truck transport into the tunnel (Figure 3). This allows to minimize the cosmic ray exposure time after the end of preextractions of cosmogenic ^{68,69,71}Ge in the 10 m³-tank within TOF. A bridge balance in the TOF serves to check the total quantities of solution during handling.

2.2. GALLEX UNDERGROUND LABORATORY FACILITIES

The Gallex facilities are located in the eastern wing of hall A in the Gran Sasso Underground Laboratory [15].Access is by 6.3 km highway from the tunnel entrance.The LNGS infrastructure includes a 40 t -hall crane, a number of auxiliary cranes as well as phone- and computer links to the outside institute.

The 3-story process building ("PB";Figure 4) accomodates the tank room (Figure 6) and the Ge-extraction facilities, such as absorber columns and control room. In smaller labs it houses the Ge-synthesis- and counter filling stations,the atomic absorption lab and the facilities for the Calciumnitrate neutron monitor.Plastic spill trays with the capacity to take up all the target solution plus diluting water in case of an emergency such as e.g. caused by an earthquake are installed below the side access road next to the tank room.

The 2-story counting building ("CB",Figure 5) contains the counting station in a Faraday cage (Ground Floor Lab,"GFL") and - connected via optical fiber- the First Floor Lab ("FFL"). The latter houses the electronics beyond digitalization, the computer facilities and a small auxiliary counting station.

Figure 4. Gallex target building.The counting building (Fig.5) is visible in the back.

Figure 5. Gallex counting building.

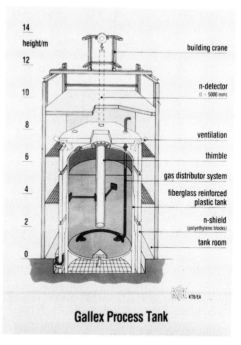

Gallex Process Tank

Figure 6. The two target tanks in their final position.Meanwhile the building is closed with panels.

Figure 7. Schematic cross section of the target tank.The neutron shield is an option only,at present it is not considered to be necessary.

Figure 8. The Karlsruhe pilot facility for testing Ge-absorption and defining the design of the actual Ge-extraction system.

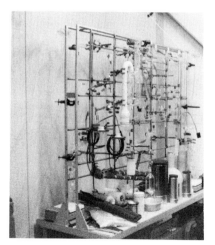 

Figure 9. Line for conversion of GeCl$_4$ in GeH$_4$ **Figure 10.** Counter filling station.

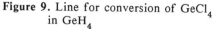

2.3 TARGET TANKS AND GERMANIUM EXTRACTION

The 54 m^3 of 8.2 molar Galliumchloride solution will be exposed in one single 70m^3 -tank.This facilitates extraction and a later artificial neutrino exposure with a ^{51}Cr - Megacurie neutrino source inserted into a central thimble. The latter can also take up the Ca(NO$_3$)$_2$-vessel for internal neutron monitoring (see 2.4.).

We have installed two equal tanks (Figure 6) for redundancy,safety,and eventual later tank inspections; yet only one tank is used at a time. The tanks are made by Plastilon Ltd (Finland) from PALATAL (brand name) unsaturated polyester reinforced by special low - U/Th- (Corning "S2") glass fiber.They are lined with PVDF. The same material is also used for the sparging system within the tanks (Figure 7).

Operating conditions and equippment design have been defined and optimized with the help of a pilot facility, scaled 1:9 , which is operative at Karlsruhe since 2 years (Figure 8).

The extraction will be performed by use of 3000 m^3 of nitrogen during 16 - 20 hours in the "once-through"-mode. Instead of unconvenient and noise generating compressors,we use evaporated liquid nitrogen. The volatile Germaniumchloride taken over by the sparging nitrogen is absorbed in 3 large absorber columns,(3.1m length, 25 cm diameter) filled with pyrex helices.

The GeH$_4$ preparation line (Figure 9) and counter filling line (Figure 10) are newly build from scratch for the Gran Sasso Lab as duplication of respective lines which have been operating at MPI since many years .

In preparing GeH$_4$ from GeCl$_4$ we use tritium free reagents and gaschromatographic purification. Conversion yields are 97+ %,for yield determinations we use an atomic absorption spectrophotometer installed on site. Extraction yields are,in addition, monitored by isotope dilution technique using off-site thermion mass-spectrometry at Karlsruhe. This implies spiking with separated Germanium isotopes.

2.4. CALCIUM NITRATE EXPERIMENT AND SIDE REACTIONS

Unwanted ^{71}Ge production is largely dominated by $^{71}Ga(p,n)^{71}Ge$ whereby the protons are secondaries from either natural radioactivity in the target or from the residual cosmic ray muon flux at the underground site($\approx 15/m^2$,d). A redetermination of the relevant ^{71}Ge-production rate in $GaCl_3$-solution at the CERN muon beam led to an estimate of the cosmic ray induced ^{71}Ge production rate in the target geometry of the experiment equivalent to 2.1 $\pm$ .7 SNU (1.6 % of the SSM production rate).

Contributions from the target impurities are completely negligible (see Table 1). The environemental fast neutron flux at the experiment site is low:
$(\Phi_n(>2.5$ MeV)= .23 x $10^{-6}/cm^2$,s [16]).

In order also to account for the neutrons originating from the tank walls and for the actual moderation conditions, in-situ measurements of the ^{71}Ge production rate via fast neutrons will be performed with the help of an elongated 470 1 vessel containing $Ca(NO_3)_2$ solution. It is inserted into the thimble of the target tank. From the ^{37}Ar produced via $^{40}Ca(n,\alpha)^{37}Ar$ in the $Ca(NO_3)_2$ solution,the ^{71}Ge production rate can be scaled . Figure 11 shows the $Ca(NO_3)_2$ vessel. The auxiliary equippment for purging the liquid, as well as for conditioning and counting the ^{37}Ar is fully operational, the ^{37}Ar recovery yield is 95 $\pm$ 2%. Experiments will be performed with as well as without target solution inside the tank.

Figure 11. $Ca(NO_3)_2$-vessel to monitor n-induced ^{71}Ge-production via ^{37}Ar from Ca.

2.5. LOW LEVEL COUNTING

After conversion of the extracted germanium into gaseous GeH_4 (German), low-level detection of ^{71}Ge is performed using miniaturized gas proportional counters located in a detector tank consisting of coincidence/anticoincidence devices (well-type NaI crystal pair and plastic scintillator) and passive shields (lead,copper,iron). ^{71}Ge (electron capture) decays result in ionizing events at ≈ 10.4 keV (K-peak) and ≈ 1.2 keV (L-peak) due to Auger electrons and/or stopped X-rays. Using a fast transient digitizer,the fully registered pulse shape serves to discriminate against backgrounds (especially Compton-electrons) .The fast wave form analyzer together with the superior properties of a newly developed preamplifier has led to a remarkable resolution power,up to the identification of individual features within a single primary ionization event (Figure 16).

2.5.1. Counters.

In five years of counter development we have continously refined and optimized the counter performance with respect to effiency and background. Our final design (counter type "HD2-Fe", Figure 12) is made from hyperpure Suprasil, has a solid Fe-cathode,and a directly sealed 13 μm - tungsten anode wire. Its key features are minimal dead volume (volume efficiency >90%), low capacity (< 1pF) , hence high signal/noise ratio, and low background rates (see below). 17 identical counters of this type alone are available. Using special quartz technology,their dimensions have been standardized,so any counter is alike the other.

2.5.2. Counter Environment.

Counters are supported in specially taylored Cu/low-activity-Pb housings, the preamplifier is also integrated (Figure 12). Their positioning within the NaI - well is shown in Figure 13.
The shield tank features:
- 8 counter positions within the pair spectrometer with the option to detect also ^{68}Ge (positron emitter) in the coincidence mode.
- 24 positions within a low radioactivity copper block (opposite end).
- high purity Cu-shield next to the NaI.
- outer steel vessel filled with low radioactivity lead.
- two sliding end-doors also filled with lead.
- air lock design with glove boxes to enable counter change without venting the low radon atmosphere inside.

The device is installed inside the Faraday cage in the GFL of the Gallex counting building (Figure 14). The analog part of the electronics installed in the GFL (Figure 15) is electrically decoupled from the MicroVax and its periphery installed above in the FFL via an optical fibre link. The data are supplied by the Camac oriented electronics to condensed disk with shared on-line access and permanent tape storage . Software for on-and off-line -data analysis is ready for use.The full system is presently being used for background measurements.

2.5.3. Counting Conditions.

Systematic studies of the counting properties of potential gas mixtures have revealed the general suitability of GeH_4-Xe mixtures for Xe-proportions from 0 - 95 % at pressures up to 2 atmospheres. These studies also gave guidance for optimizing the trade off among individual parameters such as slow drift velocity (low Xe, good rejection efficency), good energy resolution (high Xe,low gas pressure), K/L peak efficiency ratio ,amount of Ge- carrier which can be accomodated (order of 1mg), convenience of energy and rise time calibration, and overall background performance. Favored gas compositions are pure GeH_4 or 30 % GeH_4 / 70 % Xe - mixtures.

The neutrino exposure time of an individual run is planned to be ~20 days. Counting will be performed for periods of > 90 days, hence normally 5 or more multiplexed counters operate simultaneously.

2.5.4. Counter Backgrounds.

Counter background measurements on the various counter types under development had been performed before at the Heidelberg Low-Level Counting Laboratory,with full veto power to reject the frequent cosmic ray related events at shallow depth.Nevertheless, a reduction of the integral background rate (by a factor 2-3) was obtained after moving into the Gran Sasso tunnel. Measurements performed first in a

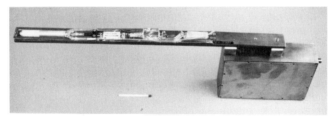

Figure 12. HD2(Fe)-proportional counter in its housing.Empty air space (Radon!) is minimized by a form fitted low activity-lead mould.A copper tube is slided over the counter (removed for this foto).The active volume is at the left end.

Figure 13. One counter box inserted for demonstration into one of eight counter positions within the well of the NaI-crystal.It is surrounded by ultra-pure copper.The outer shield is lead-filled steel.A fitting counterpiece (door),also lead in steel,is not seen in this opened position.At the opposite end of the tank,a low radioactivity copper block with 24 counter positions is mounted in an analagous manner.

Figure 14. The Faraday cage with the door opened to allow the view of the tank shield inside.

Figure 15. Electronics for pulse registration and pulse shape analysis, next to the tank shield.It also handles the signals from the peripheral detectors (NaI, Radon - detector). The computer, terminals and the rough periphery are decoupled from the Faraday cage via fiber optics leading to the electronics lab on top (FFL).

Figure 16. A (steep) genuine ^{71}Ge decay pulse and a background pulse of the same energy. The lower part of the figure magnifies the box insert in the upper part.

It demonstrates the power to resolve microscopic processes occuring within the counter for a given primary event.

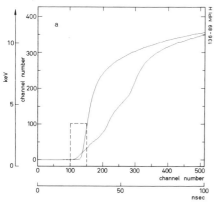

preliminary ("Bypass") station and later in the FFL gave good results. Employing pulse shape discrimination using a fast Tektronix transient digitizer (Figure 16) the results obtained for these counters are shown in Figure 17. The mean background rates are .15 cpd (L-peak) and .01 cpd (K-peak). These values allow to measure a 90 SNU - production rate of ^{71}Ge in a 4-year experiment with a relative error of 8 %. A further reduction might be expected from the operation of the Radon control system which so far was not in operation.

Background measured with counter type HDII—Fe
at the Gran Sasso

Figure 17. Background rates measured in the Gran Sasso.

Counter #	Location	Counting time [d]	Bkg count rate [counts/d]				
			>0.5 keV	L peak		K peak	
				Total	Fast	Total	Fast
37	Bypass	22.5	1.78	0.27	0.09	0.09	<0.04
59	FFL	56.3	0.94	0.21	0.12	0.12	0.02
70	FFL	90.7	1.06	0.41	0.20	0.09	0.01
All		179.5	1.05	0.31	0.15	0.09	0.01

2.6. ARTIFICIAL ^{51}Cr-NEUTRINO SOURCE

All technical provisions for inserting an artificial calibration source [17]into the target tank have been installed (thimble, source crane, mechanical structure adaption of the roof above the tank). The source experiment is presently planned in 1992, after about 18 months of observing solar neutrinos.

This work has been funded by the German Federal Minister for Research and Technology (BMFT) under the contract number 06HD554I.
This work has been funded by INFN (Italy) , CEA (France) , KFK Karlsruhe (Germany) , KRUPP Foundation (Germany), and others.

References

[1] Kirsten,T. (1986) The Gallex Project of Solar Neutrino Detection. Adv.Nucl.Astrophys. (Edition Frontieres), 85 -96.
[2] Kirsten,T. (1986) Progress Report on the "GALLEX" Solar Neutrino Project. 12th Int.Conf. Neutrino Physics and Astrophysics, World Scient. Publ. 317-329.
[3] Kirsten,T. (1987) Das GALLEX Sonnenneutrino-Experiment. Mitt. Astron.Ges. **68**, 59 - 70.
[4] Hampel,W. (1988) The present status of the Gallium solar neutrino detector GALLEX in: Neutrino Physics, Proc. Heidelberg Conf. Springer Publ. 230 - 238.
[5] Kirsten,T. (1988) Status of the solar neutrino problem. Proc.9th Workshop on Grand Unification, Aix-les-Bains, World Scientific. Publ. 221 - 241.
[6] Hampel,W. The Status of Gallex. Neutrino 88, Boston, World Scientific Publ. 311-316.
[7] Kirsten,T. (1989) Upcoming experiments and plans in low energy neutrino physics. Neutrino 88, Boston, World Scientific Publ.742-764.

[8] Bahcall,J. and Ulrich,R. (1988) Solar models, neutrino experiments and helioseismology. Rev.Mod.Phys. **60**, 297 - 372.

[9] Mikheyev,S. and Smirnov A. (1986) Resonance enhancement of oscillations in matter and solar neutrino spectroscopy. Sov.J.Nucl.Phys. **42**, 913-917.

[10] Bouchez,J.,Cribier,M.,Hampel,W.,Rich,J.,Spiro,M.,and Vignaud.D. (1986) Matter effects for solar neutrino oscillations. Z.Physik **C32**, 499 - 511.

[11] Hampel,W. (1986) The signal from the Gallium Solar Neutrino Detector:Implications for neutrino oscillations and solar models. Proc.Symp.Weak and Electromagnetic Interactions in Nuclei.Heidelberg,Springer Publ. 718 - 722.

[12] Davis,R. (1988) Solar Neutrinos. Neutrino 88, Boston, World Scientif.Publ. 518 - 525.

[13] Hirata,K. et al.   Observation of ^{8}B solar neutrinos in the Kamiokande II detector. Phys.Rev.Letters **63** , 16 - 19.

[14] Kirsten,T. (1984) The Gallium Solar Neutrino Experiment. Inst.Phys.Conf.Series **71**, 251 - 261.

[15] Paoluzi,L. (1989) The Gran Sasso National Laboratory. **A279**, 133 - 136.

[16] Belli,P., Bernabei,R., d'Angelo,S., dePascale,M., Paoluzi,L., Santonico,R., Taborgna,N., Iucci,N., and Villoresi,G. (1989) Deep underground neutron flux mesurements with large BF_3 counters. Report Rom2F/88/029, submitted to Nuovo Cimento.

[17]Cribier,M.,Pichard,B., Rich,J., Spiro,M., Vignaud,D., Besson,A., Bevilacqua,A., Caperan,F., Dupont,G., Sire,P., Gory,J., Hampel,W. and Kirsten,T. (1988) Study of a high intensity 746 keV neutrino source for the calibration of solar neutrino detectors. Nucl.Instr.Methods **A 265**, 574 - 586.

THE BAKSAN GALLIUM SOLAR NEUTRINO EXPERIMENT

V. N. Gavrin
Institute for Nuclear Research of the Academy of Sciences of the USSR
Moscow 117312, USSR

SAGE Collaboration:
INSTITUTE FOR NUCLEAR RESEARCH AS USSR: A. I. Abazov,
D. N. Abdurashitov, O. L. Anosov, O. V. Bychuk, S. N. Danshin,
L. A. Eroshkina, E. L. Faizov, V. N. Gavrin, V. I. Gayevsky,
S. V. Girin, A. V. Kalikhov, S. M. Kireyev, T. V. Knodel,
I. I. Knyshenko, V. N. Kornoukhov, S. A. Mezentseva, I. N. Mirmov,
A. V. Ostrinsky, V. V. Petukhov, A. M. Pshukov, N. E. Revzin,
A. A. Shikhin, E. D. Slyusareva, A. A. Tikhonov, P. V. Timofeyev,
E. P. Veretenkin, V. M. Vermul, V. E. Yantz, Yu. I. Zakharov,
G. T. Zatsepin, and V. L. Zhandarov;
LOS ALAMOS NATIONAL LABORATORY, USA: T. J. Bowles, B. T. Cleveland,
S. R. Elliott, H. A. O'Brien, D. L. Wark, and J. F. Wilkerson;
UNIVERSITY OF PENNSYLVANIA, USA: R. Davis, Jr. and K. Lande;
LOUISIANA STATE UNIVERSITY, USA: M. L. Cherry;
PRINCETON UNIVERSITY, USA: R. T. Kouzes.

ABSTRACT. A radiochemical ^{71}Ga-^{71}Ge experiment to determine the integral flux of neutrinos from the sun has been constructed at the Baksan Neutrino Observatory in the USSR. Measurements have begun with 30 tonnes of gallium. The experiment is being expanded with the addition of another 30 tonnes. The motivation, experimental procedures, and present status of this experiment are presented.

The inverse beta-decay of ^{37}Cl was proposed to measure the solar neutrino flux more than forty years ago [1,2]. Such a measurement was initiated in the 1960's in the Homestake gold mine, and the first result of the ^{37}Cl-^{37}Ar experiment was that the solar neutrino capture rate was less than 3 SNU [3] (1 SNU = 10^{-36} captures per target atom per second). This value disagreed with the then accepted theoretical prediction of 8-29 SNU [4]. Since then there have been many new measurements of the parameters used in calculating the solar neutrino flux, such as the nuclear reaction cross sections [5] and the composition of the solar surface [6]. These new measurements were incorporated in the Bahcall and Ulrich Standard Solar Model (SSM), which predicts 7.9 ± 2.6 (3σ) SNU for ^{37}Cl [7]. Another solar model calculation of Turck-Chieze et al. that uses somewhat different input data and physical assumptions gives 5.8 ± 1.3 SNU [8].

The overall average capture rate in ^{37}Cl measured at the Homestake installation from 1972-1988 was 2.2 ± 0.3 (1σ) SNU [9]. This disagreement between the theoretical calculation and experimental

G. Berthomieu and M. Cribier (eds.), Inside the Sun, 201–212.

measurements is often referred to as the solar neutrino problem (SNP). There are some time intervals when the average capture rate in the ^{37}Cl experiment has been different from this overall average. For example, from 1987-mid 1988 the rate was 4.2 ± 0.8 SNU [10]. Measurements during this same time period by the KAMIOKANDE II water Cherenkov detector [11] gave 0.46 ± 0.13(stat.) ± 0.08(syst.) of the value predicted by the SSM, consistent with the ^{37}Cl measurement.

The ^{37}Cl and KAMIOKANDE II experiments are sensitive primarily to the high energy ^{8}B solar neutrinos, which constitute less than 10^{-4} of the total estimated solar neutrino flux. The neutrinos produced in proton-proton fusion in the Sun (p-p neutrinos), which make up the bulk of this flux, are far below the energy thresholds of these experiments. The p-p neutrino production rate in the Sun is fundamentally linked to the observed solar luminosity, and is relatively insensitive to alterations in the Solar Model. The need for an experiment capable of detecting these low-energy neutrinos (endpoint energy 420 keV) has been apparent for many years. Such an experiment may be able to differentiate between solar model and particle physics explanations for the solar neutrino problem. This paper describes and gives the present status of the radiochemical solar neutrino experiment to detect the p-p neutrinos based on the ^{71}Ga-^{71}Ge transition that is being performed by a Soviet-American collaboration at the Baksan Neutrino Observatory of the Institute for Nuclear Research of the Academy of Sciences of the USSR.

1. Scientific Purposes

1.1 Solar Neutrinos

According to the Bahcall and Ulrich SSM [7] the dominant contribution to the total expected capture rate with ^{71}Ga (132 ± 20 SNU) arises from p-p neutrinos (71 SNU). Contributions made by ^{7}Be (34 SNU) and ^{8}B (14 SNU) neutrinos are also important. If the observed capture rate is significantly below 70 SNU, it would be very difficult to explain without invoking some phenomenon like Mikheyev-Smirnov-Wolfenstein (MSW) oscillations, a clear signal of physics beyond the standard model [12-16].

A capture rate in the vicinity of 70 SNU, i.e. the value predicted by the SSM for the p-p neutrinos alone, would most probably imply that both the ^{7}Be and ^{8}B neutrinos are suppressed. Such a suppression might be accounted for by new solar physics as well as by new weak interaction physics. For example, it could be understood if the solar interior were approximately $2 \cdot 10^6$ K cooler than predicted by the SSM or it could be the result of the MSW effect [13] with the parameters of oscillation $\delta m^2 = 10^{-5}$ eV^2 and $|\sin 2\theta| \geq 8 \cdot 10^{-2}$ (assuming predominant mixing of two neutrino states).

If the observed capture rate is greater than 70 SNU, the situation will be more complicated. If the rate is approximately 120 SNU, it will be most natural to attribute this to the combined effect of all neutrinos except those produced in the decay of ^{8}B. This could be understood either within the framework of the "cool Sun" model with the solar interior temperature approximately 10^6 K cooler than predicted by the SSM, or as a result of the MSW effect with $\delta m^2 = 10^{-4}$ eV^2, $\left|\sin2\theta\right| \geq 8\cdot10^{-2}$.

1.2 Supernova Neutrinos

The gallium neutrino experiment can also detect neutrinos from a supernova explosion. They could be distinguished from solar neutrinos by production of an appreciable number of ^{69}Ge atoms. According to recent experimental measurements of the isobaric analogue state in ^{69}Ge [17], 10-20 ^{69}Ge atoms are estimated to be produced in 60 tonnes of gallium by a supernova explosion at the center of the Galaxy.

1.3 Test of Electric Charge Conservation

The Ga-Ge experiment provides a unique opportunity to test electric charge conservation [18-21]. The mass of the nucleus ^{71}Ga is greater than that of the nucleus ^{71}Ge, but the conservation of electric charge prohibits the decay ^{71}Ga $\rightarrow$ ^{71}Ge + any neutral particle(s) (e.g., $\nu + \tilde{\nu}$, gamma-quanta, etc.). The best current limit on electric charge conservation with respect to nucleon decay was set with 300 kg of gallium in 1980 [22].

If the measured flux from the Sun turns out to be consistent with that predicted by the SSM minus ^{8}B neutrinos, one will be able to set the following limit:

$$T_{1/2}(^{71}\text{Ga} \rightarrow {}^{71}\text{Ge} + \text{any neutral particle(s)}) \geq 2\cdot10^{26} \text{ yr}.$$

Under some reasonable assumptions this may be expressed in terms of the element-independent ratio of branching probabilities for the elementary neutron decay [20]:

$$\epsilon^2 = \frac{\Gamma(n \rightarrow p + \nu + \tilde{\nu})}{\Gamma(n \rightarrow p + e^- + \tilde{\nu}_e)} \leq 10^{-26},$$

which would be a three order of magnitude improvement over the current result [22].

1.4 Test of ν_e - $\tilde{\nu}_e$ Non-Identity

One more result which may be obtained with the Ga-Ge experimental facilities is a new limit on ν_e - $\tilde{\nu}_e$ non-identity. This experiment would be analogous to a planned calibration experiment [23,24]. Experiments of this kind have been performed at a nuclear reactor [25] and at an accelerator [26]. One proposal is to place a gallium target into the antineutrino flux from a reactor [27]. An alternative experiment is to irradiate the gallium target with antineutrinos from a β^--active isotope with a halflife of a few tens of days. Possible sources [28] are ^{32}P, ^{170}Tm, and ^{90}Y (produced from the readily-available ^{90}Sr). If 10 tonnes of metallic gallium in the actual geometry of the gallium-germanium experiment is used [23] and if it is assumed that the solar neutrino flux (the main source of background) corresponds to the SSM [7,11], the following limit might be set on the ratio $\alpha^2 = \sigma_{exp}/\sigma_{theor}$ $(\nu_e = \tilde{\nu}_e) \leq 0.006$ with a ^{32}P source with an initial activity of 1 MCi. Such a result would be almost an order of magnitude improvement over the best current limit derived from [26] (see [27]).

2. History

The reaction $\nu_e + {}^{71}Ga \rightarrow e^- + {}^{71}Ge$ as a tool for detecting solar neutrinos was first proposed by V. A. Kuzmin [29] in 1964. The experiment seemed impractical at that time because of the very low worldwide production and the high cost of gallium. There also was no method developed for extracting and counting single ^{71}Ge atoms from the gallium target. The desirability of the Ga-Ge experiment became widely recognized in the 1970's, and intensive work began [30-36] on the chemical extraction of Ge from Ga, the counting of ^{71}Ge decays, and the production of the necessary amount of gallium in the Soviet Union.

Techniques of extracting minuscule amounts of Ge from Ga, essential for a gallium-germanium neutrino experiment, were developed for two detector materials: (1) gallium in the form of a concentrated hydrochloric acid solution [33] and (2) molten metallic gallium [30]. The experiment at Baksan uses a metallic gallium target. The most important advantages of the metal are reduced sensitivity to both internal and external backgrounds and a smaller volume. The smaller target volume results in a greater capture rate for electron neutrinos from a ^{51}Cr calibration source placed inside the gallium-containing vessel. The major disadvantages of the gallium metal target are the need for additional steps of chemical processing and somewhat greater quantities of fresh reagents for the extraction procedure.

3. Scheme of the Experiment

Figure 1 presents the overall scheme of the experiment.

3.1 Chemistry

The gallium is contained in chemical reactors that are lined with teflon and have an internal volume of 2 m^3. The reactors are provided with stirrers and with heaters that maintain the temperature just above the gallium melting point (29.8 C). Each reactor holds about 7 tonnes of gallium.

At the beginning of each run, a known amount of Ge carrier is added to the Ga in the form of a solid Ga-Ge alloy. The reactor contents are stirred so as to thoroughly disperse the carrier throughout the Ga metal. After a suitable exposure interval (3-6 weeks), the Ge carrier and any ^{71}Ge atoms that have been produced by neutrino capture are chemically extracted from the gallium.

It has been shown that contacting gallium metal with a weak acidic solution in the presence of an oxidizing agent results in the extraction of germanium into the aqueous phase. The extraction process thus begins by adding to each reactor an extraction solution containing 1 kg of HCl, 5.2 kg of H_2O_2, and 68.8 kg of H_2O. The total volume of the extraction solution is 70 l. To ensure that the starting reagents are free of germanium to an acceptable level, the HCl solution (concentration 7N) and the H_2O are purified before use. So as to minimize heating of the Ga, the HCl solution is cooled to -15 C and the water to 4 C. The mixture is intensively stirred and the Ga metal turns into a fine emulsion. The Ge dissolved in the Ga migrates to the surface of the emulsion droplets. In 7-9 min. the H_2O_2 is consumed; almost all of the emulsion spontaneously breaks down and the phases separate. The extraction procedure is then finished by adding 43 l of 7N HCl and stirring for 1 min. Less than 0.1% of the gallium has been oxidized and the gallium temperature has risen to about 50 C. The extraction solution is then siphoned away from each reactor and the reactors are washed by adding ~20 l of 0.5N HCl. This solution is vigorously stirred with the liquid gallium for about 1 min and is then siphoned away to be added to the previous extraction solution.

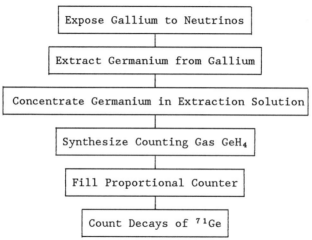

Figure 1: Scheme of the Experiment

All of the extracts from the separate reactors are combined and the Ge is then concentrated by vacuum distillation in glass apparatus. Since Ge is volatile from concentrated HCl solutions, the distillation is stopped when the volume has been reduced by a factor of four. The solution is then transferred to another glass vessel that is part of a sealed gas flow system. Purified 12N HCl is added to this solution to raise the HCl concentration to 9N and an argon purge is initiated. The argon flow (1.0 m^3/hour for 1.5 hours) sweeps the Ge as $GeCl_4$ from this acid solution into a volume of 1.0 l of H_2O. When this process is completed, a solvent extraction procedure is used to first extract the Ge into CCl_4 and then back-extract it into H_2O. This process is repeated three times and the residual CCl_4 is removed by heating the water to 90 C for 1.5 hours. To improve the efficiency of CCl_4 removal, a very small amount of hexane is added to the organic phase at the last step of the final back extraction. The next step of the procedure is to synthesize the gas GeH_4. The synthesis reaction that is used is optimized at a pH of 8-9, so NaOH is added to adjust the pH to this range. The resulting solution, whose volume is only 100 ml, is put into a small reactor flask in a sealed helium flow system and 50 ml of a 0.02N NaOH solution that contains about 2 g of $NaBH_4$ is added. GeH_4 is produced when this mixture is heated to about 70 C. The helium flow sweeps the GeH_4 into a gas chromatography system where it is purified. A measured quantity of xenon is added, and this mixture is put into a sealed proportional counter with a volume of about 0.75 cm^3. The GeH_4 sample is then counted for 1-6 months.

To prevent gallium oxidation, ~100 l of 1N HCl are added to each reactor when the extraction process is complete. This acid solution is discarded immediately before the next extraction.

3.2 Counting

^{71}Ge decays with an 11.4 day half life. The decay occurs solely by electron capture to the ground state of ^{71}Ga. The probabilities of K, L, and M-capture are 88%, 10.3%, and 1.7%, respectively. The only way to register such a decay is to detect the low-energy Auger electrons and X-rays produced during electron shell relaxation in the resulting ^{71}Ga atom. K-capture gives Auger electrons with an energy of 10.4 keV (41.5% of all decays), 9.2 keV X-rays accompanied by 1.2 keV Auger electrons from the subsequent M-L transition (41.2% of all decays), and 10.26 keV X-rays accompanied by 0.12 keV Auger electrons (5.3% of all decays). L- and M-capture give only Auger electrons with energies of 1.2 keV and 0.12 keV, respectively.

These low-energy electrons are detected in a small-volume proportional counter similar to that used in the Cl-Ar experiment. In these counters it is not practical to separate the weak signal arising from the 0.12 keV events from background and the X-rays almost always escape. If the electrons are detected with 100% efficiency, then the counter energy spectrum consists of events at 10.4 keV (41.5% of all decays) and 1.2 keV (51.5% of all decays).

The SSM predicts a production rate of 1.2 atoms of ^{71}Ge per day in 30 tonnes of Ga. At the end of a 4 week exposure period, 16 ^{71}Ge atoms will be present on the average. If it is assumed that there is a one day delay between the end of exposure and the beginning of counting, that Ge is extracted from the Ga and transferred to the proportional counter with 75% efficiency, and that the counting efficiency is 35%, then the mean number of detected ^{71}Ge atoms in each run is only 3.9.

The major source of background in the counters is local radioactivity. The counters are thus made from materials especially selected to have a low content of Ra, Th, and U (such as synthesized quartz and zone-refined iron) and are contained in large passive shields made from copper, lead, and tungsten. The counting is conducted deep underground where the nuclear component of the cosmic ray flux is negligible and the muon flux is very small. To further reject and to characterize background events, some counting channels have a NaI detector around the proportional counter that is used in coincidence/anticoincidence.

Background radioactivity primarily produces fast electrons in the counter. In contrast to the localized ionization produced by the Auger electrons from ^{71}Ge decay, these fast electrons give an extended ionization as they traverse the counter body. Since the risetime of the induced pulse increases as the radial extent of the ionization distance, it is possible to use pulse shape discrimination to separate the ^{71}Ge decays from background. This can be achieved by measuring the full pulse waveform for a few hundred nanoseconds or by differentiating the pulse with a time constant of about 10 nanoseconds and measuring the resulting amplitude (ADP method [37]).

Good rejection of background events occurs if the counting mixture contains 10% GeH$_4$ and 90% Xe at atmospheric pressure. This gas mixture gives a resolution of 18-21% at 5.9 keV. Some of the 9.2 keV X-rays are absorbed in this gas mixture and the measured total counting efficiency in a 2 FWHM window around 10.4 keV is 35%. This efficiency includes geometrical effects inside the counter and excludes events whose risetime is outside a 95% acceptance window.

3.3 ^{71}Ge Background

The main source of ^{71}Ge in the reactors other than from solar neutrinos is from protons arising as secondary particles produced by (1) external neutrons, (2) internal radioactivity, and (3) cosmic-ray muons. These protons can initiate the reaction ^{71}Ga(p,n)^{71}Ge. Extensive work has gone into measurements and calculations of these three background channels.

(1) Since the (n,p) cross sections on the Ga isotopes are small and the laboratory has been lined with low-background concrete, the fast neutron background is not significant. A calculation [38] has shown that the external neutron background in 30 tonnes of Ga metal produces no more than 0.01 atoms of ^{71}Ge per day.

(2) The background from internal radioactivity is mainly determined by the concentrations of U, Th, and Ra in the gallium. Measurements [39] of these concentrations, combined with measured yields of ^{71}Ge from alpha particles [33], indicate that less than 0.01 atoms of ^{71}Ge will be produced per day in 30 tonnes of Ga metal.

(3) Based on the measured muon flux in the laboratory, the production rates of the germanium isotopes from cosmic-ray muons have been calculated [40] to be 0.005 ^{71}Ge, 0.013 ^{69}Ge, and 0.009 ^{68}Ge atoms per day in 30 tonnes of Ga metal.

A valuable feature of the gallium experiment is that the ratios of the production rates for the isotopes ^{68}Ge, ^{69}Ge, and ^{71}Ge that arise from these background processes differ sharply from the corresponding ratios for production by solar neutrinos. This fact can be used to identify the source of the observed signal in the Ga-Ge detector. These background processes are all sufficiently small that the Ga experiment has the capability of observing a significant depression of the solar neutrino flux below the SSM predicted value.

Since these sources of ^{71}Ge background have all been made small, the major difficulty for the experiment that they pose is the need to remove from the gallium the large quantities of long-lived ^{68}Ge (half life = 271 days) that was produced by cosmic rays while the gallium was on the surface. ^{68}Ge decays only by electron capture, so its decays cannot be differentiated from those of ^{71}Ge. The subsequent decay of ^{68}Ga (half life = 1.14 hours) is by positron emission in 90% of the cases. In a proportional counter with 5 mm cathode diameter filled with 90% Xe and 10% GeH$_4$, the ^{68}Ga decay gives an energy spectrum with a broad peak whose maximum is at about 1.0 keV. These ^{68}Ga decays can generally be identified by rise-time analysis of the counter pulse and by detecting a coincidence pulse in the surrounding NaI crystal.

Another type of background that arises only during counting can come from tritium in the counting gas GeH$_4$. So as to eliminate this source of counter background, special methods for synthesizing NaBH$_4$ have been developed [41] that use starting ingredients that can be selected to have a low tritium content.

4. Present Experiment Status

The gallium-germanium neutrino experiment is situated in an underground laboratory specially built in the Baksan Valley of the Northern Caucasus, USSR. The laboratory is 60 m long, 10 m wide, and 12 m high. It is located 3.5 km from the entrance of a horizontal adit driven into the side of Mount Andyrchi, and has an overhead shielding of 4700 mwe. The muon flux has been measured to be $(2.4 \pm 0.3) \cdot 10^{-9}$ cm^{-2}sec^{-1}. To reduce the background from fast neutrons, the laboratory has been fully lined with 60 cm of low-radioactivity concrete and a 6 mm inner steel shell. Reactor filling began in April 1988 and the experiment now contains 30 tonnes of gallium in four reactors. Each reactor has undergone at least 15 extractions to remove the germanium isotopes that were produced by cosmic-ray interactions while the Ga was on the surface.

The efficiency of extraction of germanium from the reactors is measured at several stages of the extraction procedure. When the efficiency is to be determined, inert Ge carrier (approximately 120 micrograms alloyed into 50 g. of Ga) is added to each reactor. As the process of germanium concentration is carried out, several samples are routinely taken and the Ge content in each is determined by atomic absorption analysis (AAS). The first samples are taken from the extraction solutions from each reactor. Other samples are removed after the germanium has been swept into 1.0 l of H$_2$O and after the back-extraction from CCl$_4$ into H$_2$O. A final determination of the quantity of germanium is made by measuring the volume of synthesized GeH$_4$. Table 1 gives typical results of germanium content in some of these samples. The major uncertainty in these measurements is in the amount of Ge carrier added to the reactors. The Ge concentration in the carrier slugs is determined by exhaustive extraction of Ge from several representative slugs using the same procedure as for Ge extraction from the large reactors. The Ge in the extraction solutions is then measured by AAS. The error on the quantity of Ge in the carrier is estimated to be $\pm$ 5%.

Table 1: Measurements of Germanium Yields During Extraction Process

Results are corrected for losses due to the samples
taken for analysis and are expressed in micrograms of Ge.

Germanium Carrier Initially Added to Four Reactors	468 ± 23
Germanium in Extraction Solution before GeH$_4$ Synthesis	366 ± 12
Germanium in Synthesized GeH$_4$	342 ± 10

The overall extraction efficiency from the Ge added to the
reactors to the synthesized GeH$_4$ is approximately 75%. The standard
procedure is to conduct three extractions in series within a period of 5
days without adding additional carrier to the reactors. The GeH$_4$
samples from each of these three extractions are usually counted
separately.

Two low-background counting systems, each with four channels, have
been used to count the GeH$_4$ samples. Four counting channels have a NaI
crystal for background suppression and recognition of ^{68}Ge-decay events.
Both of these systems use the ADP-method for pulse shape discrimination.
The energy, ADP, time, and NaI energy are recorded for each event.

The total background rate of selected counters filled with 90% Xe,
10% GeH$_4$ has been measured in the energy interval of 0.7-13.0 keV to be
approximately 1.5 events per day. In the region of the Ge K-peak
energy, the counter background is 1 event per month.

Counting of the germanium samples began during the extraction of
the germanium isotopes produced by cosmic-ray interactions. After the
gallium had been stored underground for 3 months, the initial activity
of the Ge in the 30 tonnes of Ga was 7700 events per day in the energy
region of the Ge K-peak. For recent runs, the activity in the Ge K-peak
(35% counting efficiency) is less than 1 count per day.

Some residual radioactivity is still present that produces events
in the energy range of 1-15 keV. These events are predominantly below 6
keV; they have both slow and fast risetime components with some events
in coincidence with a NaI detector. In recent extractions this activity
does not appear to be decreasing at the same rate as in the earlier
extractions. Many measurements of the activity obtained in various
isolated steps of the extraction procedure have been conducted, but the
statistics available are quite low and the source of this activity has
not yet been definitely identified.

The counting rate after the latest extractions enables one to set
a limit on the half life of ^{71}Ga for non-charge conserving decay of
$T_{1/2}$ (^{71}Ga $\rightarrow$ ^{71}Ge + ...) > 3·10^{25} years. This is a two order of
magnitude improvement over the best previously existing limit [22], and
gives the value $\epsilon^2 < 7·10^{-26}$.

5. Future Plans

Monthly extractions from the 30 tonnes of gallium will continue.
Simultaneous with this work, further experiments to understand the
source of the remaining background will be conducted and the detector

will be expanded to contain 60 tonnes of gallium. The additional 30
tonnes is now stored underground and will be put into reactors in the
next few months. Almost all of the glassware for the 60-tonne
extraction system is installed and tests of the full-scale extraction
system are now being conducted. At the background levels presently
achieved, the full 60-tonne detector should yield a statistical accuracy
of better than 25% after one year of operation if the production rate is
132 SNU.

The number of channels in the present counting systems will very
soon be increased to 12. An additional 12 counting channels with full
pulse shape digitization and storage are expected to be operational this
summer. Significant quantities of ^{72}Ge, ^{73}Ge, and ^{76}Ge are available
that will be used sequentially for the Ge carrier. Mass spectrometric
analysis of the isotopes in the germanium extracted from the reactors
should yield an improved understanding of the chemical process.

A calibration of the gallium detector is planned using a reactor
produced ^{51}Cr neutrino source. It is expected that an activity of 0.8
MCi will be obtained by irradiating about 200 g of enriched chromium
(86% ^{50}Cr) in Soviet reactor SM-2 (thermal flux = $3.2 \cdot 10^{15}$ neutrons/cm^2-
sec [42,43]). Approximately 400 decays from the ^{71}Ge atoms produced by
this source are expected to be detected, yielding a statistical accuracy
in the calibration of 5%.

ACKNOWLEDGMENTS

The authors wish to thank A. E. Chudakov, G. T. Garvey, M. A. Markov,
V. A. Matveev, J. M. Moss, V. A. Rubakov, and A. N. Tavkhelidze for
their continued interest in our work and for stimulating discussions.
We are also grateful to J. K. Rowley and R. W. Stoenner for chemical
advice and to E. N. Alekseyev and A. A. Pomansky for their vital help in
our work. The U. S. participants wish to acknowledge the support of the
U. S. Department of Energy and the National Science Foundation.

REFERENCES

1. B. Pontecorvo, 'Inverse Beta Decay', Chalk River report PD-205
    (1946); Usp. Fiz. Nauk. 141 (1983) 675 [Sov. Phys. Usp. 26 1087].
2. L. W. Alvarez, 'A Proposed Experimental Test of the Neutrino
    Theory', Univ. of California Radiation Laboratory Report UCRL-328,
    Berkeley, California (1949).
3. R. Davis, Jr., D. S. Harmer, and K. C. Hoffman, Phys. Rev. Lett. 20
    (1966) 1205.
4. J. N. Bahcall and G. Shaviv, Astrophys. J. 153 (1968) 113.
5. P. D. Parker, in Physics of the Sun, ed. by P. A. Sturrock,
    D. Reidel Pub. Co., New York, vol. I, pp. 15 (1986).
6. N. Greveese, Physica Scripta T8 (1984) 49; L. H. Aller in
    Spectroscopy of Astrophysical Plasmas, ed. by A. Dalgarno and
    D. Layzer, Cambridge Univ. Press, Cambridge, pp. 89 (1986).
7. J. N. Bahcall and R. K. Ulrich, Rev. Mod. Phys. 60 (1988) 297.
8. S. Turck-Chieze, S. Cahen, M. Casse, and C. Doom, Astroph. J. 335
    (1988) 415.

9. R. Davis, Jr., B. T. Cleveland, and J. K. Rowley, in *Proceedings of Underground Physics '87 Int'l. Symposium*, Baksan Valley, USSR, 17-19 August 1987, ed. by G. V. Domogatsky et al., Nauka Pub. Co., Moscow, p. 6 (1988).

10. R. Davis, Jr., K. Lande, B. T. Cleveland, J. Ullman, and J. K. Rowley in *Proceedings of Neutrino '88 Conference*, Boston 5-11 June 1988, ed. by J. Schneps, World Scientific, Singapore (1989).

11. K. S. Hirata, T. Kajita, K. Kifune, K. Kihara, M. Nakahata, K. Nakamura, S. Ohara, Y. Oyama, N. Sato, M. Takita, Y. Totsuka, Y. Yaginuma, M. Mori, A. Suzuki, K. Takahashi, T. Tanimori, M. Yamada, M. Koshiba, T. Suda, K. Miyano, H. Miyata, H. Takei, K. Kaneyuki, Y. Nagashima, Y. Suzuki, E. W. Beier, L. R. Feldscher, E. D. Frank, W. Frati, S. B. Kim, A. K. Mann, F. M. Newcomer, R. Van Berg, and W. Zhang, Phys. Rev. Lett. $\underline{63}$ (1989) 16.

12. S. P. Mikheyev and A. Yu. Smirnov, Yadernaya Fizika $\underline{42}$ (1985) 1441; Nuovo Cimento $\underline{9C}$ (1986) 17; L. Wolfenstein, Phys. Rev. $\underline{D17}$ (1978) 2369.

13. S. P Mikheyev and A. Yu. Smirnov, Usp. Fiz. Nauk $\underline{153}$ (1987) 3.

14. M. B. Voloshin, M. I. Vysotsky, and L. B. Okun, ZhETF $\underline{91}$ (1986) 754.

15. E. Kh. Akhmedov and O. V. Bychuk, ZhETF $\underline{95}$ (1989) 442.

16. Z. G. Berezhiani and M. I. Vysotsky, Phys. Lett. $\underline{199B}$ (1987) 281.

17. A. E. Champagne, R. T. Kouzes, M. M. Lowry, A. B. McDonald, and Z. Q. Mao, Phys. Rev. $\underline{C38}$ (1988) 2430.

18. G. Feinberg and M. Goldhaber, Proc. Natl. Acad. Sci. USA $\underline{45}$ (1959) 1301.

19. R. I. Steinberg, Bull. Am. Phys. Soc. $\underline{21}$ (1976) 528.

20. J. N. Bahcall, Rev. Mod. Phys. $\underline{50}$ (1978) 881.

21. O. V. Bychuk and V. N. Gavrin, Pis'ma v ZhETF $\underline{48}$ (1988) 116.

22. I. R. Barabanov, E. P. Veretenkin, V. N. Gavrin, Yu. I. Zakharov, G. T. Zatsepin, G. Ya. Novikova, I. V. Orekhov, and M. I. Churmayeva, Pis'ma v ZhETF $\underline{32}$ (1980) 384; Sov. Phys. JETP Lett. $\underline{32}$ (1981) 359.

23. V. N. Gavrin, S. N. Danshin, G. T. Zatsepin, and A. V. Kopylov, 'Consideration of Possibilities for Calibration of a Gallium-Germanium Neutrino Telescope with an Artificial Neutrino Source', Preprint INR AS USSR P-0335 (1984).

24. E. P. Veretenkin, V. M. Vermul, and P. V. Timofeyev, 'Determination of Impurities in Metallic Chromium for a Neutrino Source', Preprint INR AS USSR P-0565, Moscow (1987).

25. R. J. Davis, Jr., in *Radioisotopes in Scientific Research, Proc. of the First UNESCO Int'l. Conf.*, Paris, ed. by R. C. Extermann, Pergamon Press (1957), vol. 1, pp. 728; R. Davis, Jr., Bull. Amer. Phys. Soc. $\underline{4}$ (1959) 217; see also J. N. Bahcall and H. Primakoff, Phys. Rev. $\underline{18D}$ (1978) 3463.

26. A. M. Cooper, Phys. Lett. $\underline{112B}$ (1982) 97.

27. I. R. Barabanov, A. A. Borovoi, V. N. Gavrin, G. T. Zatsepin, A. Yu. Smirnov, and A. N. Kheruvimov, 'A Proposed Experiment to Investigate the Difference Between ν_e and $\bar{\nu}_e$', Preprint INR AS USSR P-0466 (1986).

28. O. V. Bychuk and V. N. Gavrin, Pis'ma v ZhETF $\underline{49}$ (1989) 244.

212

29. A. Kuzmin, Lebedev Physical Institute Preprint A-62 (1964); Sov. Phys. JETP **49** (1965) 1532.
30. I. R. Barabanov, E. P. Veretenkin, V. N. Gavrin, S. N. Danshin, L. A. Eroshkina, G. T. Zatsepin, Yu. I. Zakharov, S. A. Klimova, Yu. V. Klimov, T. V. Knodel, A. V. Kopylov, I. V. Orekhov, A. A. Tikhonov, and M. I. Churmayeva, *Proc. of Conf. on Solar Neutrinos and Neutrino Astronomy*, Homestake, 1984, ed. by M. L. Cherry, K. Lande, and W. A. Fowler, Am. Inst. of Phys. Conf. Proc. **126** (1985) 175; *Proc. of the Soviet-American Workshop on the Problem of Detecting Solar Neutrinos*, Izvestiya Akademii Nauk SSSR **51** (1987) 1225.
31. W. Hampel, *Proc. of Conf. on Solar Neutrinos and Neutrino Astronomy*, Homestake, 1984, ed. by M. L. Cherry, K. Lande, and W. A. Fowler, Am. Inst. of Phys. Conf. Proc. **126** (1985) 162.
32. A. A. Pomansky, *Kosmicheskiye Luchi*, Nauka, Moscow **10** (1963) 95.
33. J. N. Bahcall, B. T. Cleveland, R. Davis, Jr., I. Dostrovsky, J. C. Evans, W. Frati, G. Friedlander, K. Lande, J. K. Rowley, R. W. Stoenner, and J. Weneser, Phys. Rev. Lett. **40** (1978) 1351.
34. I. R. Barabanov, V. N. Gavrin, Yu. I. Zakharov, and G. T. Zatsepin, *Proc. 5th Intl. Conf. Neutrino '75*, Balatonfured, Hungary **2** (1975) 385.
35. I. Dostrovsky, *Proc. Informal Conf. on Status and Future of Solar Neutrino Research*, Brookhaven, New York, BNL Report 50879 **1** (1978) 231.
36. W. Hampel and Gallium Solar Neutrino Experiment Collaboration, *Proc. of Science Underground Workshop*, Los Alamos, New Mexico, ed. by M. M. Nieto et al., Am. Inst. of Phys. Conf. Proc. **96** (1983) 88.
37. E. F. Bennet, Rev. Sci. Inst. **33** (1962) 1153; R. Davis, Jr., J. C. Evans, V. Radeka, and L. C. Rogers, *Proc. of European Physics Conference 'Neutrino 72'*, Balatonfured, Hungary **1** (1972) 5.
38. I. R. Barabanov, V. N. Gavrin, L. P. Prokopyeva, and V. E. Yants, 'Low Internal Neutron-Activity Concretes for Lining Chambers of Radiochemical Solar Neutrino Detectors', Preprint INR AS USSR P-0559 (1987).
39. V. N. Gavrin, S. N. Danshin, A. V. Kopylov, and V. I. Cherkhovsky, 'Low-Background Semiconductor Gamma Spectrometer for the Measurement of Ultralow Concentrations of ^{238}U, ^{226}Ra, and ^{232}Th' Preprint INR AS USSR P-0494 (1987).
40. V. N. Gavrin and Yu. I. Zacharov, 'Production of Radioactive Isotopes in Metallic Gallium by Cosmic Rays and the Background of a Gallium-Germanium Detector of Solar Neutrinos', Preprint INR AS USSR P-0335 (1984).
41. D. P. Alexandrov, T. N. Dimova, N. G. Eliseeva, and T. V. Knodel, 'Preparation of Low-Tritium Borohydrides from Molten Alkali Metals', Report at the Fourth All-Union Meeting on the Chemistry of Hydrides, Dushanbe, USSR, Nov. 1987, page 37.
42. A. M. Petrosyants, *Problems of Atomic Science and Technology*, Atomizdat, Moscow (1979).
43. S. M. Fainberg et al., 'Physical and Operational Characteristics of Reactor SM-2', *Proc. of the Third Int'l. UN Conference on the Peaceful Uses of Atomic Energy*, (1969).

NEUTRINO MASSES, NEUTRINO OSCILLATIONS AND DARK MATTER *

HAIM HARARI
Weizmann Institute of Science
76100 Rehovot
Israel.

ABSTRACT. We discuss bounds on neutrino masses using an analysis based on direct measurements, cosmological bounds, oscillation experiments, the solar neutrino puzzle and theoretical considerations on neutrino decays. We present four possible solutions for the mass range of the three neutrino flavors. We outline experiments which can distinguish among these solutions and discuss their implications for the cosmological dark matter problem.

1. Introduction

In this report we discuss several issues related to neutrino masses, neutrino oscillations, neutrino decays and neutrinos as dark matter candidates. None of these have ever been convincingly observed. But neutrino physics is largely an art of learning a great deal by observing nothing. It can give us extremely useful information on physics beyond the standard model and can help us probe, indirectly, higher energy scales and cosmological problems.

Much of the literature on neutrino masses, oscillations and decays is based on very specific theoretical models (e.g. GUTS, left-right symmetry, string inspired models, majoron schemes, etc.). Within the framework of a well defined model it is sometimes possible to make specific predictions concerning various physical quantities. However, in the absence of direct observations of masses, mixing or decays, such predictions are often less than useful. At the other extreme we find general "theorems" or arguments based on "the most general case". These are often useless and lead to very weak predictions.

Our approach here will be different. We start by defining a simple reasonable theoretical framework, motivated by plausibility arguments. It will be much more general than any specific detailed model, but not general enough to become useless. We than review all the information that we can collect on the interrelated problems of neutrino masses, oscillations and decays. We use information

* Supported in part by the Israel Commission for Basic Research, the U.S.–Israel Binational Science Foundation, and the Minerva Foundation

G. Berthomieu and M. Cribier (eds.), Inside the Sun, 213–230.

from direct mass measurements, cosmology, oscillation experiments, theoretical arguments, comparison with non-neutrino experiments and solar neutrino rates. When we combine all of this information and use our plausible theoretical framework, we reach surprisingly strong conclusions concerning neutrino properties. We are also led to related interesting speculations concerning the cosmological dark matter problem and we discuss crucial experiments which can answer some of the open questions. Of particular interest to us is a $\nu_\mu - \nu_\tau$ oscillation experiment which may prove or disprove the hypothesis that ν_τ is responsible for the cosmological dark matter.

Much of the work discussed here is based on collaboration[1] with Y. Nir, to whom I am indebted for many helpful discussions. An earlier version of this report was included in the Proceedings of the Neutrino 88 conference[2], and the last section is based on a recent proposal[3] for a $\nu_\mu - \nu_\tau$ oscillation experiments.

2. The See-Saw Mechanism for Neutrino Masses and a Speculation on Neutrino Mass Ratios

Neutrinos are either exactly massless or extremely light. They are lighter than the corresponding charged leptons at least by *several* orders of magnitude and, as we will show below, probably by *seven or more* orders of magnitude.

There is no convincing explanation for exactly massless neutrinos. In the minimal standard model one simply declares that there is no right handed neutrino. In that case, the neutrino mass is precluded by chiral symmetry. However, this really amounts to assuming the answer, rather than predicting it. Much worse – even if we accept an exactly massless neutrino in the standard model, practically any physics beyond the standard model, will reintroduce the neutrino mass. The only known exception is *minimal* SU(5) which both disagrees with experiment and leaves crucial issues (hierarchy problems, generation puzzle) totally unexplained.

An *exactly* massless neutrino requires an absolute new symmetry which remains unbroken to all orders both within the standard model and in any new physics beyond it. No one has suggested such a symmetry. There is no good reason to believe that neutrinos are indeed *exactly* massless.

If neutrinos are *approximately* massless, i.e. they have a negligible mass with respect to other leptons and quarks, we must again insist on a simple explanation. It should tell us how certain fermions can be much lighter than others. Whatever that explanation is, it should single out the neutrinos and clarify why it is them, and no other fermions, which are so light. In that respect, the suggestion that neutrinos have Dirac masses and that these masses just "happen to be small" (with a Yukawa coupling smaller than 10^{-10} for ν_e!), is totally unacceptable.

Fortunately, there is one simple and elegant theoretical framework[4] which explains why neutrinos, *and only neutrinos*, are much lighter than all other fermions. This framework was first proposed in the context of GUTs, but it has no necessary relation to GUTs. It is much more general, and may appear naturally in practically any theory beyond the standard model.

The idea is simple: neutrinos, like any other fermion, have Dirac masses which are assumed to be of the usual order of magnitude. Neutrinos, *unlike* any other fermion in the standard model, may also have a Majorana mass. The Majorana mass of a *left-handed* neutrino may only come from a Higgs *triplet* (because ν_L is an $I_3 = \frac{1}{2}$ member of a weak isodoublet and a $\nu_L \nu_L$ term must therefore transform like an $I_3 = 1$ component of a triplet, coupling only to Higgs isotriplet). Higgs triplets are unwanted, unnecessary and, if they existed, they would have spoiled the mass relation $M_W = M_Z cos\theta_W$. In contrast, the Majorana mass of a *right-handed* neutrino may come only from a Higgs *singlet* (because ν_R is an $I = 0$ singlet, hence $\nu_R \nu_R$ can only couple to a Higgs isosinglet). Higgs singlets are harmless to the standard model. They do not contribute to the W or Z mass and do not influence $M_W = M_Z cos\theta_W$. But Higgs singlets are actually present and are practically necessary in any theory beyond the standard model. If such a theory is based on any symmetry which is larger than $SU(3) \times SU(2) \times U(1)$ (e.g. GUTs, left-right symmetry, technicolor, horizontal symmetry, most composite models), the higher symmetry will be broken down to the standard model symmetry by a Higgs field which is a non-singlet under the higher symmetry (and therefore breaks it) but is a singlet under the standard model gauge group (and therefore preserves it). Such a singlet Higgs will have a vacuum expectation value of order Λ, where Λ is some new energy scale of the new "beyond standard" physics. We know that Λ can be anywhere between, say, O(TeV) and M_{planck}. Neutrinos, unlike charged leptons and unlike quarks, could then have a "normal-size" Dirac mass *and* a large right-handed Majorana mass. This situation is logical, natural and almost inescapable in many models beyond the standard model. It is a situation which is unique for neutrinos and does not apply to other leptons or quarks. It leads to a simple explanation for the small mass of the observed left-handed neutrinos.

The above scenario leads to 2×2 mass matrix[4] for each generation of neutrinos. Ignoring generation mixing (we will return to it in a minute) we obtain:

$$\begin{pmatrix} 0 & m_D \\ m_D & M \end{pmatrix}$$

where m_D is a Dirac mass assumed to be comparable to the mass of the corresponding charged lepton and M is a Majorana mass due to a Higgs singlet and driven by a new energy scale Λ. This is the famous "see-saw" matrix.

In the quark case, quarks in the same generation have mass ratios $\frac{m_u}{m_d} \sim 0.55$; $\frac{m_c}{m_s} \sim 9$; $15 \lesssim \frac{m_t}{m_b} \lesssim 40$. We therefore guess that m_D is roughly within one order of magnitude of m_ℓ, the mass of the corresponding charged lepton.

The two eigenvalues of the mass matrix are:

$$m_1 \sim \frac{m_D^2}{M} \; ; \; m_2 \sim M$$

The first eigenstate (with mass m_1) is almost purely left-handed with a small right-handed admixture, determined by the small ratio $\frac{m_D}{M}$. Similarly, the heavy eigenstate is almost purely right-handed. The ratio $\frac{m_D}{M}$ is at most 10^{-3} (for $\frac{m_\tau}{1\,TeV}$) but could be much smaller (e.g. $\frac{m_e}{M_{GUT}} \sim 10^{-18}$).

Assuming $m_D \sim O(m_\ell)$ and $M \sim O(\Lambda)$, the light eigenstates is *much* lighter than an ordinary lepton:

$$m_1 \sim \frac{m_D^2}{M} \sim \frac{m_\ell^2}{\Lambda} \sim m_\ell \left(\frac{m_\ell}{\Lambda}\right) \ll m_\ell$$

That is the "see-saw" explanation[4] for the lightness of the neutrino. The heavy eigenstate is much heavier than an ordinary lepton and its mass could be anywhere between $O(TeV)$ and the Planck scale.

The approximate relation $m_\nu \sim \frac{m_\ell^2}{\Lambda}$ tells us that for a higher energy scale of the new physics, we obtain smaller neutrino masses. For instance, for $m_\ell \sim m_e \sim O\,(MeV)$ we obtain $m(\nu_e) \sim O\,(eV)$ for $\Lambda \sim O\,(TeV)$; but $m(\nu_e) \sim 10^{-12}$ eV for $\Lambda \sim M_{GUT}$. It is interesting that $\Lambda \sim O\,(TeV)$ is the next achievable frontier of high energy physics and, at the same time, $m(\nu_e) \sim O(eV)$ is one order of magnitude away from the direct mass measurements, neutrinoless double beta decay and the limit derived from SN1987A. A peculiar coincidence!

If the Λ-scale is identical for all three generations and if all generation mixing angles are small, the light neutrino masses are proportional to the squared masses of the corresponding leptons. Thus, regardless of the value of Λ, *but as long as Λ is generation independent*, we obtain:

$$m(\nu_e) : m(\nu_\mu) : m(\nu_\tau) \sim m_e^2 : m_\mu^2 : m_\tau^2$$

This result could be modified by many factors. The relation $M \sim O(\Lambda)$ may be corrected by unknown Yukawa couplings. The relation $m_\ell \sim m_D(\nu)$ could easily be wrong by an order of magnitude.

The Λ values or the Yukawa couplings which determine the Majorana masses of different generations can be quite different from each other. For instance: we may have $\Lambda_e : \Lambda_\mu : \Lambda_\tau \sim m_e : m_\mu : m_\tau$, in which case:

$$m(\nu_e) : m(\nu_\mu) : m(\nu_\tau) \sim m_e : m_\mu : m_\tau$$

The nondiagonal elements of the 3×3 Dirac mass matrix and/or the 3×3 Majorana mass matrix may also change the situation. We therefore consider the above pattern as very qualitative. We actually need to assume only:

(i)   The see-saw mechanism operates.

(ii)   The light neutrino mass is within a few orders of magnitude from $\frac{m_D^2}{\Lambda}$.

(iii)   The neutrino masses obey ratios:

$$m(\nu_e) : m(\nu_\mu) : m(\nu_\tau) \sim m_e^P : m_\mu^P : m_\tau^P$$

where the power P obeys $1 \lesssim P \lesssim 2$ (we may even relax this inequality to $\frac{1}{2} \lesssim P \lesssim 3$, without changing too many conclusions).

These are very weak and reasonable assumptions. We will refer to them as the "reasonable see-saw" scenario[1].

Note that we have not assumed anything about the explicit theoretical model leading to the "see-saw". Neither GUTS nor right-handed currents are absolutely necessary.

Our "reasonable see-saw" assumption leads, among other things, to the rough estimate

$$O(10) \lesssim \frac{m(\nu_\tau)}{m(\nu_\mu)} \lesssim O(10^3)$$

We will see that this relation, if true, leads to far-reaching consequences.

3. Experimental Bounds on Neutrino Masses and Oscillations

Direct measurements have yielded the following bounds on neutrino masses[5]:

$$m(\nu_e) < 18\,eV; \ m(\nu_\mu) < 250\,keV; \ m(\nu_\tau) < 35\,MeV$$

To these we may add the bound derived from SN1987A, an almost direct bound. Depending on various assumptions, values between 10 and 20 eV are obtained as the upper bound on $m(\nu_e)$ from SN1987A.

A fourth generation of neutrino may exist, although the well-known argument on nucleosynthesis indicates that a fifth light neutrino is ruled out and a fourth one is not very likely[6].

It is interesting that the present direct experimental bounds obey ratios which are not too different from those of the squared masses of the corresponding charged leptons. For instance, if $m(\nu_e) : m(\nu_\mu) : m(\nu_\tau) = m_e^2 : m_\mu^2 : m_\tau^2$ and if $m(\nu_\tau)$ were found to be *exactly* at its present experimental bound of 35 MeV, we would expect $m(\nu_\mu) \sim 120$ KeV, $m(\nu_e) \sim 3\,eV$. Hence, the unlikely possibility that *all* three neutrino masses are not too far from their present direct mass limits[7] is compatible with the reasonable see-saw, provided that M is actually around 100 GeV or so. Of course, there is no good reason to expect nature to choose such a possibility.

Most of the information concerning leptonic mixing angles is obtained from neutrino oscillations experiments. There have been several unconfirmed claims of two-standard-deviation effects in various oscillation experiments, but all of them were contradicted by later experiments. A given negative oscillation experiment can only rule out a region in the $\Delta_{ij} - sin^2 2\theta_{ij}$ plane, where $\Delta_{ij} = m_i^2 - m_j^2$, m_i and m_j are the masses of the mixed neutrinos and θ_{ij} is their leptonic mixing angle. Consequently, no value of θ_{ij} can be completely excluded. It can be excluded only for a specific range of Δ_{ij}-values.

We have no theoretical information concerning the values of the leptonic mixing angles. The three quark mixing angles are

$$\theta_{12} = 0.22, \ \theta_{23} = 0.043 \pm 0.008, \ \theta_{13} \leq 0.01$$

or:

$$sin^2 2\theta_{12} = 0.18, \ sin^2 2\theta_{23} \sim 0.005 - 0.01, \ sin^2 2\theta_{13} \leq 4 \times 10^{-4} \ .$$

One possible guess is that the leptonic mixing angles are roughly comparable to the quark angles. It is possible to support this guess by the following hand waving argument: It is very likely that there are relations between fermion masses and mixing angles. Many models actually indicate that a given mixing angle between two generations is somehow related to the mass ratio of the fermions in the same two generations. The detailed formulae vary from one model to another. The angles may be proportional to mass ratios or to square roots of mass ratios, etc. However, given the mass ratios, the angles are determined in all such models.

The mass pattern of the three charged leptons is not very different from that of the quarks. If fermion mass ratios indeed determine the mixing angles, one might therefore expect that the leptonic mixing angles are within, say, an order of magnitude of the quark angles.

If we explicitly assume that $\theta_{e\mu} = \theta_{12}^q \sim 0.22$, we conclude from oscillation experiments that $\Delta_{12} \lesssim 0.1$ eV^2. If we assume $\theta_{\mu\tau} = \theta_{23}^q \sim 0.043$, we conclude from experiment that $\Delta_{23} \lesssim 1$ eV^2. However, if $\theta_{\mu\tau}$ is slightly smaller (say 0.03) we have no bound at all on Δ_{23}. We will return to this subject towards the end of our discussion.

The well-known MSW effect[8] may explain the deficiency of solar neutrinos in the Davis experiment[9]. The Bethe solution requires[10] $\Delta_{ij} \sim 10^{-4}$ eV^2, i.e. $m(\nu_\mu)$ or $m(\nu_\tau)$ of order 10^{-2} eV. The Rosen-Gelb solution[11] allows Δ_{ij}-values ranging between $10^{-7} - 10^{-4}$ eV^2, depending on θ_{ij}. Consequently, in all solutions, the heavier neutrino mass (ν_μ or ν_τ) is between 10^{-4} eV and 10^{-2} eV. All of these values are well below the bounds obtained from direct neutrino oscillation experiments in accelerators or reactors. If $m(\nu_\mu) \sim O(10^{-2}$ eV$)$, the see-saw mechanism yields $\Lambda \sim O$ (10^9 GeV). For $m(\nu_\tau) \sim O$ (10^{-2} eV), we obtain $\Lambda \sim O$ (10^{12} GeV).

4. Cosmological Bounds

A well known cosmological argument tells us that[12]

$$\sum_i N_o(\nu_i) m(\nu_i) < \rho_0 = \Omega \rho_c = \Omega \cdot \frac{3H_o^2}{8\pi G} = \Omega h^2 \cdot 11 \frac{KeV}{cm^3}$$

where ν_i is a neutrino flavor, N_o is the number density of ν_i, ρ_0 is the present density of the universe, ρ_c is the critical density corresponding to a flat universe, H_o is the Hubble parameter, G is Newton's constant and $h=H_o/100 \, km/sec/Mpc$.

For the accepted values of $\Omega \leq 2$; $0.5 \leq h \leq 1$; $10^{10} \leq t_o \leq 2 \cdot 10^{10}$ yrs (t_o=present age of the universe) we obtain $\Omega h^2 \leq 0.65$.

The neutrino decoupling temperature is O(MeV). For $m(\nu) < O(MeV)$, all neutrino species have a number density $\frac{3}{11} N_o(\gamma)$ where $N_o(\gamma) \sim 400 cm^{-1}$ is the number density of photons. We then obtain:

$$m(\nu_i) < 100 \, \Omega h^2 \leq 65 \, eV$$

For $m(\nu) > O(MeV)$, each number density is suppressed by an appropriate Boltzman factor $e^{\frac{-m(\nu_i)}{kT}}$. A more complicated treatment is needed, leading to a *lower* limit[13] (for $\Omega h^2 \leq 0.65$):

$$m(\nu_i) \geq 4.2 \, GeV$$

All mass values between 65 eV and 4.2 GeV are thus excluded, under very general assumptions which essentially depend only on the gross features of Big Bang Cosmology. The only strong assumption hidden in the argument is the assumption that neutrinos are stable. We return to this issue below. However, if we ignore the possibility of neutrino decays for a brief moment, we must conclude that ν_μ and ν_τ are lighter than 65 eV. If we add the "reasonable see-saw" assumption we find:

$$m(\nu_\tau) < 65 \, eV; \; m(\nu_\mu) < few \, eV; \; m(\nu_e) < 10^{-2} eV$$

This is an extremely strong result. It means that ν_τ, ν_μ and ν_e are lighter than their present *direct* mass limits by approximately six, five and three orders of magnitude, respectively. If it holds, it means that all *direct* experiments of neutrino mass measurements are absolutely hopeless for the foreseeable future. This far-reaching conclusion can be avoided only if neutrinos decay. We now turn to this possibility.

The cosmological bounds $m(\nu_i) < 65$ eV or $m(\nu_i) > 4.2$ GeV hold only for stable neutrinos. They are based on the assumption that the energy density due to neutrinos today is related to the density at the decoupling temperature by the simple proportionality

$$\rho \propto \frac{1}{R^3}$$

i.e. as the universe expands the density is inversely proportional to the volume.

If neutrinos are unstable, the density behaves differently. After the decay, the universe becomes "radiation dominated", obeying:

$$\rho \propto \frac{1}{R^4}$$

Consequently, the density at the decay time may be larger than the density allowed for stable neutrinos. The unstable neutrinos may therefore have a higher mass. For shorter neutrino lifetime, the effect becomes larger.

A straightforward analysis[14] yields the following bounds on the masses of unstable neutrinos[1]:

$$m_\nu^2 \tau_\nu \leq 2 \times 10^{20} \; eV^2 \cdot sec \text{ for } m_\nu < O \text{ (MeV)}$$
$$m_\nu^{-4} \tau_\nu \leq 1.5 \times 10^{-22} \; eV^{-4} \text{ for } m_\nu > O \text{ (MeV)}$$

These bounds mean that a neutrino may be heavier than 65 eV provided that its lifetime is short enough. In particular, $m(\nu_\tau)$ may be close to 35 MeV if $\tau(\nu_\tau) \leq 8$ yrs and $m(\nu_\mu)$ may be around 250 keV if $\tau(\nu_\mu) \leq 100$ yrs.

There is no reason to expect neutrinos to be stable. In fact, if the neutrinos have masses, it is almost certain that ν_τ and ν_μ decay into lighter neutrinos. If that is the case, it would appear that the 65 eV cosmological limits is not valid and our strong conclusion of the previous section is unapplicable.

However, before we jump into such a conclusion, we must consider the possible rates for different neutrino decay modes. Only if these rates are sufficiently high, the 65 eV bound is evaded. If neutrinos are unstable but decay slowly, the 65 eV bound remains valid. We must therefore now turn to estimate neutrino decay rates.

5. Neutrino Decays and their Implications for the Cosmological Bounds

A complete analysis of all possible neutrino decays in all possible models[1] (beyond the standard model) is beyond the scope of this report. Here we briefly summarize the main results.

There are three major classes of neutrino decay modes:
(i) *Radiative decays* such as $\nu_i \to \nu_j + \gamma$, $\nu_i \to \nu_j + \gamma + \gamma$, $\nu_\tau \to \nu_i + e^+ e^-$ where i, j are neutrino flavors.
(ii) *Decays into neutrinos* such as $\nu_i \to \nu_j + \nu_k + \nu_\ell$ where the decay products may be neutrinos and/or antineutrinos of the same or of different flavors.
(iii) *Decays into Goldstone bosons* such as $\nu_i \to \bar{\nu}_j$ + majoron or $\nu_i \to \nu_j$ + familon.

Radiative neutrino decays (class (i)) have been analyzed both within the standard model[15] and in "beyond standard" schemes. In all cases one can derive bounds on $\tau(\nu)$ which depend on the mass of the decaying neutrino. An investigation of the $m(\nu) - \tau(\nu)$ plane shows that below $m(\nu) \sim O(\text{MeV})$ there is no overlap between the region allowed by the cosmological constraints on unstable neutrinos and the region allowd by the decay rate for radiative neutrino decay. Thus, if radiative decays were the only neutrino decays we would have concluded:

$$m(\nu_\tau) < 65\,eV \ \text{ or } \ O(MeV) < m(\nu_\tau) < 35\,MeV$$
$$m(\nu_\mu) < 65\,eV$$
$$m(\nu_e) < 65\,eV$$

The next class of decays $(\nu \rightarrow 3\nu)$ can be mediated within the standard model by Z° exchange. It can also occur in "beyond standard" theories by the exchange of a horizontal gauge boson, by the exchange of a Higgs triplet Δ_L (with two units of lepton number) or by some interaction among subleptons in a composite model. The Z exchange contribution is too slow[16]. In all other cases we cannot calculate the decay rate directly but we can relate it to the observed experimental bound on the decay of a charged lepton into three lighter charged leptons (e.g. $\tau^- \rightarrow e^- e^+ e^-$ etc.). We again obtain[1] an allowed region in the $m(\nu) - \tau(\nu)$ plane, leading to the conclusion that $m(\nu)$ is either below 65 eV or above O(MeV).

In all of the above decay modes $m(\nu_\mu)$ must be lighter than 65 eV. From our "reasonable see-saw" we then conclude:

$$m(\nu_\tau) \leq 10^3 m(\nu_\mu) \ll MeV$$

Hence, the possibility that $m(\nu_\tau) > O(MeV)$ which is allowed by the estimated ν_τ decay rates and by the cosmological bound on unstable neutrinos, is incompatible with $m(\nu_\mu) < 65eV$ and the "reasonable see-saws".

The combined result of cosmology, neutrino decays and the reasonable see-saw is, therefore:

$$m(\nu_\tau) < 65\,eV$$
$$m(\nu_\mu) < few\,eV$$
$$m(\nu_e) < 10^{-2}\,eV$$

We are back to the extremely strong result obtained earlier for the case of stable neutrinos!

There is, however, one class of decay modes which, under extremely unlikely circumstances, may allow larger neutrino masses. The decay

$$\nu_\mu \rightarrow \bar{\nu}_e + majoron$$

may be consistent with $m(\nu_\mu)$ around 50-250 keV. For that to happen[7], the Majorana mass term must be driven by a Λ-scale around 100 GeV or less, a very unlikely proposition. In addition, the age of the universe must be at most 12 or 13 Gyr. If both condition are obeyed, one cannot exclude the weird possibility that all three neutrino flavors have masses near their present direct bounds, i.e.:

$$m(\nu_\tau) \sim O\,(10\,MeV)$$
$$m(\nu_\mu) \sim O\,(100\,keV)$$
$$m(\nu_e) \sim O\,(1\,eV)\,.$$

This small window (or "peephole") requires too many miracles to be taken seriously. However, it can be tested and excluded experimentally if the direct bound on $m(\nu_\mu)$ is improved by less than one order of magnitude.

We conclude: It is almost certain that:

$$m(\nu_\tau) < 65\,eV$$
$$m(\nu_\mu) < few\ eV$$
$$m(\nu_e) < 10^{-2}\,eV$$

But the remote possibility of "neutrinos at the limit" cannot be excluded.

6. Four Interesting Possibilities For The Neutrino Mass Values, and their Implications for the Cosmological Dark Matter Problem

Dark matter seems to exist at several different scales in the universe.[17] Here we only discuss the cosmological dark matter, i.e. the matter responsible for an $\Omega=1$ flat universe or for Ω being not too different from one. Other forms of dark matter may or may not coincide with the cosmological dark matter. The leading particle physics candidates for the cosmological dark matter are:

(i) *"Light Neutrinos"*. These must have a mass in the range 15-65 eV. For the most likely values of $\Omega=1$ and $t_o=15$ Gyr we need $m(\nu) \sim 20$ eV. If we take note of our combined bound from cosmology, neutrino decay estimates and the "reasonable see-saw", we must conclude that, among the known neutrinos, only ν_τ can provide us with such a mass. Hence – the first dark matter candidate is ν_τ with $m \sim O$ (20 eV).

(ii) *WIMPS*. These are particles with masses of order few GeV. One such candidate is an ordinary fourth generation neutrino ν_σ with mass near 4.2 GeV. We know that a hypothetical fourth charged lepton (say, σ) must be lighter than 350 GeV or else it will disturb $M_W = M_Z cos\theta_W$ by too much. Hence, $\frac{m(\sigma)}{m(\tau)} \leq 200$. However, if $m(\nu_\tau) < 65$ eV, $\frac{m(\sigma)}{m(\tau)} \leq 200$ and $\frac{m(\nu_\sigma)}{m(\nu_\tau)} \leq \left[\frac{m(\sigma)}{m(t)}\right]^2$, we obtain $m(\nu_\sigma) < 3$ MeV and the 4.2 GeV mass is excluded. On the other hand, if ν_τ, ν_μ and ν_e are all "at the limit", ν_σ can easily be around 4.2 GeV. Other likely WIMP's are the photino as well as a variety of other types of heavy neutrino-like objects.

(iii) *Axions*. Here we need a new physics scale which is around 10^{12} GeV or perhaps a bit (but not much) lower.

As we see below, all of these possibilities may be related to our discussion of neutrino masses.

On the basis of everything we said so far we find that there are four possible solutions corresponding to four interesting mass ranges for the neutrinos. For each of these solutions we can offer one or several crucial experimental tests. Each one of them has important implications for the cosmological dark matter problem. All our conclusions are based on our fairly general framework. We combine experimental data, simple cosmological arguments, rough estimates of neutrino decay rates and the "reasonable see-saw" assumption. Based on these assumptions we now review the four solutions:

Solution I: Neutrinos at the limit (the Glashow solution[7]). In this case $m(\nu_\tau) \sim O$ (10 MeV); $m(\nu_\mu) \sim O$ (100 keV); $m(\nu_e) \sim O$ (1 eV). A possible fourth neutrino ν_σ might be (if it exists) around 5-100 GeV, depending on $m(\sigma)$. A 5-GeV ν_σ may form the cosmological dark matter. Crucial experiments are direct mass measurements of ν_τ, ν_μ and ν_e and neutrinoless double beta decay. All of these experiments probe the relevant mass regions. All mixing angles must be extremely small (based on present data). The new physics scale Λ is extremely low and is uncomfortably close to M_W. Neutrino lifetimes are also "at the limit" and the dominant decay mode is into antineutrino + majoron. We consider this solution extremely unlikely and almost perverse.

Solution II: Our favorite solution is the possibility that ν_τ forms the cosmological dark matter. Note that if solution I is excluded (i.e. if $m(\nu_\mu) < O$ (100 keV)), the next highest allowed mass range is:

$$m(\nu_\tau) \sim 15 - 65\,eV$$
$$m(\nu_\mu) \sim O\,(1\,eV)$$
$$m(\nu_e) \lesssim O\,(10^{-2}\,eV)\;.$$

In this case all direct mass measurements are useless. Neutrinoless beta decay is hopeless. The new scale Λ is O(100 PeV). There is no fourth generation neutrino. The crucial experiment[3] is a search for $\nu_\tau - \nu_\mu$ oscillations for $\Delta_{\mu\tau} \sim O(10^3 \text{ eV}^2)$ but small $\theta_{\mu\tau}$ values (below $\theta_{\mu\tau} \sim 0.05$) which is the present limit for the relevant $\Delta_{\mu\tau}$ value. We discuss this experiment in detail below.

Solution III: The Bethe solution[10] for the solar neutrino puzzle. If both solution I and solution II are wrong, the next range is:

$$m(\nu_\tau) \sim few\,eV$$
$$m(\nu_\mu) \sim O\,(10^{-2}\,eV)$$
$$m(\nu_e) < O\,(10^{-4}\,eV)$$

The value of Λ is around 10^9 GeV. A possible fourth neutrino ν_σ may have a mass around 65 eV and form (if it exists) the cosmological dark matter. Direct mass measurements, and neutrinoless double beta decay experiments are hopeless. The crucial experiments are the Gallium solar neutrino experiments and, possibly, $\nu_\tau - \nu_\mu$ oscillations.

Solution IV: The Bethe-Rosen-Gelb solution[10,11]. If solutions I, II and III are all excluded, we reach the next interesting range in which $m(\nu_\tau) \sim O(10^{-2} \text{ eV})$ obeying the Bethe solution[10] for the solar neutrino puzzle. However, in that case, $m(\nu_\mu)$ may well be around $10^{-4} - 10^{-3}$ eV, consistent with the Rosen-Gelb solution[11]. Either or both of these neutrinos might have resonant oscillations with ν_e (whose mass would be below 10^{-6} eV). The only doable experiments in this case are the solar neutrino experiments. No other experiment mentioned in this report has a chance. The value of Λ is $O(10^{12} \text{ GeV})$, consistent with the range required by axions providing the cosmological dark matter.

It is, of course, possible that all neutrino masses are even lower than the values of solution IV. This would be very sad indeed and no known experiment can probe such a range.

Based on our reasonable and general framework we showed that there are four possible solutions for the mass range of the light neutrinos. Each of these solutions can be tested by one or more crucial doable experiments. Our prejudice (which, like all prejudices, is unfounded) it to favor solution II in which the cosmological dark matter is ν_τ. Solution III is probably a close second. Time and experiment will tell.

7. A Crucial Dark Matter Experiment

The crucial experiment which can "clinch" solution II is a very special neutrino oscillation experiments, probing ν_τ masses at the level of 15-65 eV.

We are discussing here a ν_τ mass value which is six orders of magnitude below the best direct limit[5] $m(\nu_\tau) < 35 \, MeV$. The only way to probe this mass region are neutrino oscillations involving ν_τ. Since we assumed that $m(\nu_\tau) \gg m(\nu_\mu) \gg m(\nu_e)$, and we are interested in the range $15 \, eV \leq m(\nu_\tau) \leq 65 \, eV$, we must consider only $\nu_\tau - \nu_\mu$ and $\nu_\tau - \nu_e$ oscillations and we know that, to a good approximation, $\Delta m^2 \approx [m(\nu_\tau)]^2 \approx (200 - 4500) \, eV^2$.

What can we say about the $\nu_\tau - \nu_e$ and the $\nu_\tau - \nu_\mu$ mixing angles $\theta_{\tau e}$ and $\theta_{\tau \mu}$?

In section 3 we already speculated that quark and lepton mixing angles might be of the same order of magnitude. The angle $\theta_{\tau e}$ mixes non-adjacent generations. It is analogous to $\theta_{13}^{(q)}$ in the quark sector, which is known to be smaller (but probably not *much* smaller) than 10^{-2}. If $\theta_{\tau e} \approx \theta_{13}^{(q)}$ we expect $\sin^2 2\theta_{\tau e} \leq 4 \cdot 10^{-4}$. The best $\nu_\tau - \nu_e$ oscillation data[18] (as well as the best ν_e "disappearance" data) reach only much larger values of $\sin^2 2\theta_{\tau e}$ and therefore tell us nothing about $m(\nu_\tau)$.

This leaves us with $\nu_\tau - \nu_\mu$ oscillations as the last resort. The angle $\theta_{\tau \mu}$ mixes *adjacent* generations. It is analogous to $\theta_{23}^{(q)}$ in the quark sector. Experimentally, $\sin \theta_{23}^{(q)} = 0.043 \pm 0.008$. If we had $\theta_{\tau \mu} = \theta_{23}^{(q)}$ we would expect $\sin^2 2\theta_{\tau \mu} \approx 0.005 - 0.010$. In the quark sector, we have another mixing angle which connects neighbouring generations: the original Cabibbo angle, obeying $\sin\theta_{12}^{(q)} = 0.22$ or $\sin^2 2\theta_{12}^{(q)} = 0.18$. We do not really know why $\theta_{12}^{(q)} \gg \theta_{23}^{(q)}$. We also do not know the actual value of $\theta_{\tau\mu}$, but on the basis of the above analogy to the quark sector, it might be anywhere, say, between 0.03 and 0.22. the pattern of the charged lepton mass ratios is not very different from that of the quark mass ratios. Most theoretical models expect mixing angles to be somehow related to fermion mass ratios. We may therefore "guess" that the $\theta_{\tau\mu}$ is not far from the above range, possibly below it, but not too far below. Since $\theta_{13}^{(q)}$ is probably near 0.01, and the mixing of "distant" generations is expected to be smaller, we propose a very conservative lower bound $\theta_{\tau\mu} \geq 0.01$. This would mean $\sin^2 2\theta_{\tau\mu} \geq 4\cdot 10^{-4}$. This bound seems safe although, in principle, arbitrarily small values of $\theta_{\tau\mu}$ cannot be excluded. What we need is, therefore, a $\nu_\tau - \nu_\mu$ oscillation experiment probing the region of Δm^2 between 200 and 4500 eV^2 *and reaching* $\sin^2 2\theta_{\tau\mu}$ *values which are at least as low as* $4 \cdot 10^{-4}$, *preferrably even lower.*

The relevant range in Δm^2 is easily accessible. How far can we go in the other crucial variable, $\sin^2 2\theta_{\tau\mu}$? The best ν_μ "disappearance" experiments reach only[8] $\sin^2 2\theta_{\mu x} \approx 0.05$, far above the required range. By far the best $\nu_\mu - \nu_\tau$ data comes from[9] Fermilab experiment E531, using a hybrid combination of an emulsion and a spectrometer. This experiment, at the 90% confidence level, reached $\sin^2 2\theta_{\tau\mu} \approx 4\cdot 10^{-3}$, just enough to exclude $\theta_{\tau\mu} = \theta_{23}^{(q)}$. What we now need is an improved experiment that can reach *at least* down to $\sin^2 2\theta_{\tau\mu} \approx 4 \cdot 10^{-4}$, hopefully below it. Such an experiment will provide us with an excellent probe of the possibility that the cosmological dark matter is due to tau-neutrinos.

The E531 experiment[19] was not originally designed to search for ν_τ oscillations. It was a by-product of a charm lifetime experiment. It still achieved, by far, the best $\nu_\mu - \nu_\tau$ oscillation data. In that experiment, approximately 4000 neutrino interactions were detected. A τ candidate was defined as an event with a kink (having $p_T > 125\ MeV$) or a three-prong secondary vertex, no prompt muon (to eliminate standard $\nu_\mu \to \mu$ events), a negative charged track (to eliminate charm events) and a minimum momentum for the τ ($p_\tau > 2.5\ GeV$, to avoid confusion with other background). With these cuts, most τ events should survive, but no candidate events were found. The experiment, with these cuts, had no background at all. On the basis of zero τ candidates and 1870 ordinary charged current events with an identified μ, the range of $\sin^2 2\theta_{\tau\mu} \leq 4 \cdot 10^{-3}$ was obtained.

Improving the bound by at least an order of magnitude would require a new dedicated experiment using similar techniques. The emulsion seems necessary in order to observe τ tracks with a typical length of a few hundred microns. The spectrometer is needed in order to point towards the suspected vertex. Conceptually, the simplest method would be to repeat the essential features of experiment E531 with a larger number of events. One needs *at least* 20,000 charged current neutrino interactions with identified muons, preferrably more. Depending on the efficiency and the acceptance for muon identification, this would require a total of at least 30,000 and probably 40,000 neutrino interactions.

This can be achieved by any combination of more emulsion, higher beam intensity and longer running time. Assuming that the transverse size of the detector covers most of the width of the neutrino beam, the number of neutrino interactions can be roughly estimated by the following crude formula:

$$\left[\frac{N_{\nu-events}}{1000}\right] = \eta \cdot \left[\frac{E_p}{100\ GeV}\right]\left[\frac{n_p}{10^{18}}\right]\left[\frac{M_{target}}{1\ ton}\right]$$

where E_p and n_p are, respectively, the energy and the number of protons on target and M_{target} is the active target mass. The coefficient η is always of order one and it contains all the details of the beam, detector, etc. In a sample of CERN and Fermilab experiments over the last few years, η-values between 0.6 and 3.5 are obtained. For our purposes, we need to generate a factor of 40 on the left hand side of our equation.

For a single realistic run at Fermilab with 800 GeV protons and 10^{18} protons on target, we therefore have:

$$\left[\frac{N_{\nu-events}}{1000}\right] = 8\eta \cdot \left[\frac{M_{target}}{1\ ton}\right].$$

For $\eta = 1$ we therefore need, say, two runs with at least 2.5 *tons* of emulsion. The situation for the CERN SPS is somewhat better. Because of the higher beam intensity and the higher repetition rate of the machine, and in spite of the lower energy, one obtains for a typical realistic run $E_p = 400\ GeV$, $n_p = 6 \cdot 10^{18}$, yielding:

$$\left[\frac{N_{\nu-events}}{1000}\right] = 24\eta \cdot \left[\frac{M_{target}}{1\ ton}\right].$$

With $\eta = 1$, two such runs with *800 kg* (or 200 liters) of emulsion would do the job. Some of the above numbers could be modified by factors of two, depending on the quality of the neutrino beam, the length of the run, the percentage of machine protons dedicated to the experiment, the distance of the detector, the acceptance and efficiency, etc. In fact, we believe that by optimizing all of these parameters, it may be possible to obtain the required sensitivity with a somewhat smaller amount of emulsion, possibly below 100 liters. For $\eta \approx 3$ (a value which have been achieved in past experiments), one needs approximately 70 liters.

With so many events, scanning the emulsion becomes a difficult and lengthy procedure. Almost all scanned events would involve a muon which is detected by the spectrometer and traced back to a primary vertex in the emulsion. Rejecting these events is a fairly rapid procedure. Selecting the serious candidates and scanning them is the heart of the experiment. A dedicated ν_τ experiment which is not a by-product of something else, may allow a more efficient procedure of selecting candidate events before the cuts.

It may be worthwhile to concentrate on specific decay modes of τ (*e.g.* single hadron or three prongs or electron) and in this way considerably reduce the necessary amount of scanning. The price paid would, of course, be the necessity of having a higher total number of events and therefore a proportionately larger amount of emulsion.

It seems that the best method would be to concentrate on events containing an energetic negative electron and no muon. Such events would include 17% of all τ-leptons, necessitating a total number of events which is six times larger, *i.e.* a total of 250,000 neutrino interactions. However, such a procedure would eliminate all normal charged current events and almost all neutral current events. The main *physics* background here would come from ν_e contamination in the neutrino beam, usually estimated at 1%. This would yield approximately 1,500 ν_e-initiated charged current events. Most of the scanned events would be of this type. If the electron comes from the primary vertex in the emulsion, the event should be rejected. If a kink is observed for an e^-, it is a τ^- candidate. In spite of the sixfold increase in the total number of neutrino interactions, the absolute number of scanned events will be reduced by more than an order of magnitude, relative to the case in which one searches for all τ decay modes.

The total amount of emulsion needed for performing this version of the experiment at CERN will have to be of the order of 500 liters (assuming $\eta \approx 3$). The typical effective transverse area of the neutrino beam at a distance of 1 km is a few squared meters (say, $3m^2$), leading to a total emulsion thickness of the order of 15 cm or five radiation lengths. In order to overcome showers, conversions and other facts of life, it would be advantageous to use several layers of emulsion (say, each with a depth of 1 cm) separated by tracking chambers which can help identify the electrons and distinguish them from various types of background. The combined electronic information from the detector behind the emulsion and the chambers between the emulsion plates could help identify true electron events, reducing the total number of scanned events to a few thousands, a number similar to that of experiment E531. Scanning will consist of searching for the relatively simple signature of a kink involving a short track of a few hundred microns followed by a single negative electron.

It is conceivable that the experiment can also be performed with other detectors containing a track-sensitive target. It might be interesting to pursue this possibility. However, the requirement of hundreds of kilograms of active target and the necessity of observing τ-tracks of a few hundred microns are not easily reconciled in other methods. A particularly attractive possibility along these lines is the idea of using scintillating optical fibres in order to detect τ-tracks in a neutrino beam[20].

It is, in principle, also possible to detect τ leptons without explicitly observing their tracks, using much larger active targets and higher event rates. However, at the level of sensitivity required here, background becomes an extremely serious problem in such experiments.

If τ events are discovered, we must be certain that they come from $\nu'_\mu s$ which oscillated into $\nu'_\tau s$ rather than from a ν_τ-contamination which exists in the neutrino beam as a result of direct hadronic decays. The prime candidate for such decays is the $c\bar{s}$ meson, known as F or D_s. The decay of F is the dominant mechanism for producing ν_τ in beam dump experiments. However, for the type of experiment discussed here, at a distance of, say, 1 km, the number of τ events originating from F-decay is expected to be negligible. It may become the limiting factor if the $\nu_\mu - \nu_\tau$ oscillation experiment is ever pushed to even lower values of $\sin^2 2\theta_{\tau\mu}$. The background due to "direct ν_τ" can, in principle, be measured by turning down, removing or diverting the focused neutrino beam. At lower energies (such as at CERN), the F background is smaller than at higher energies (such as at Fermilab).

We conclude that the proposed experiment is difficult, but not impossible. The potential reward is, in our opinion, extremely significant.

If the experiment is performed and oscillations are found, it will provide us with information on $m(\nu_\tau)$. A $precise$ determination of $m(\nu_\tau)$ may require additional, more complicated, experiments at different distances and/or energies. However, the existence of any $\nu_\mu - \nu_\tau$ oscillations in an experiment of the type discussed here, would indicate that $m(\nu_\tau)$ is at $least$ a few $eV's$, making it a very likely candidate for the dark matter. If $m(\nu_\tau)$ is found to be in the appropriate mass range, it is probably the cosmological dark matter of the universe and it becomes the dominant contributor to its energy! This would correspond to our solution II in section 6.

If the result is negative down to $\sin^2 2\theta_{\tau\mu} \approx 4 \cdot 10^{-4}$ and if, like E531, the experiment is sensitive to $m(\nu_\tau)$-values as low as a few eV, we face two possibilities: The most likely one is that $m(\nu_\tau)$ is at, or below, few eV and it does not form the cosmological dark matter of the universe. In that case, $m(\nu_\mu)$ is most likely to be at, or below, 10^{-2} eV, just the range required for explaining the solar neutrino puzzle by $\nu_\mu - \nu_e$ oscillations[15−16]. This would correspond to solution III, and, of course, solution IV may also be possible.

The second possibility (in the case of a negative result) is that ν_τ is still around $15 - 65$ eV, but for some peculiar reason $\theta_{\tau\mu} < 0.01$, well below the analogous quark angles and possibly even below the angle $\theta_{13}^{(q)}$. This would be a very small angle and it is not suggested by any known model. However, such a situation cannot be ruled out and the only way to cope with it would be to push the experiment even further, to lower values of $\sin^2 2\theta_{\tau\mu}$.

If $m(\nu_\tau)$ is in the $15 - 65$ eV range, $m(\nu_\mu)$ is likely to be approximately around 0.1 eV. In such a case, $\nu_\mu - \nu_e$ oscillations at $\Delta m^2 \approx 10^{-2}$ eV^2 become relevant. Such experiments are being now contemplated. However, even if $\nu_\mu - \nu_e$ oscillations are discovered at $m(\nu_\mu) \approx 0.1$ eV, we still cannot be sure that ν_τ is the cosmological dark matter. Only a direct observation of $\nu_\tau - \nu_\mu$ oscillations will be convincing.

We summarize: the experiment described in detail in this section is designed to test solution II (as presented in section 6). We believe that it is a crucial experiment. We hope it is done in the near future.

8. References

1. Harari, H. and Nir, Y., Nucl. Phys. B292, 251 (1987).
2. Harari, H., Proceedings of the "Neutrino 88" Conference, Medford, Mass., U.S.A., June, 1988.
3. Harari, H., Phys. Lett., 216B, 413 (1989).
4. Yanagida, T., in Proc. Workshop on Unified Theory and Baryon Number in the Universe, eds. O. Sawada and A. Sugamoto, (KEK, 1979): Gell-Mann, M., Ramond, P. and Slansky, R, in Supergravity, eds. P. van Nieuwenhuizen and D. Freedman (North-Holland, 1980).
5. Fritsch, M. et al., Phys. Lett. 173B, 485 (1986); Abela, R. et al., Phys. Lett. 146B, 431 (1984); Albrecht, H. et al., Phys. Lett. 202B, 149 (1988).
6. Steigman, G., Schramm, D.N. and Gunn, J.E., Phys. Lett. B66, 202 (1973).
7. Glashow, S.L., Proc. Rencontres de Physique de la Vallee d'Aosta, La Thuile, Italy, 1987. See also reference 1.
8. Wolfenstein, L., Phys. Rev. D17, 2369 (1978); Mikheyev, S.P. and Smirnov, A.Yu., in Proc. 10th Int. Workshop on Weak Interactions, Savonlinna, Finland (1985).
9. Davis, R., Proceedings of the "Neutrino 88" Conference.
10. Bethe, H.A. Phys. Rev. Lett. 56, 1305 (1986).
11. Rosen, S.P. and Gelb, J.M., Phys. Rev. D34, 969 (1986).
12. For various reviews of dark matter see e.g. Proc. of the 1986 Jerusalem Winter School on Dark Matter, eds. J.N. Bahcall, T. Piran and S. Weinberg.
13. Gerstein, S.S. and Zeldovich, Ya.B., JETP Lett. 4, 120 (1966); Cowsik, R. and McClelland, J., Phys. Rev. Lett. 29, 669 (1972).
14. Lee, B.W. and Weinberg, S., Phys. Rev. Lett. 39, 165 (1977); Hut, P. Phys. Lett. 69B, 85 (1977).
15. Dicus, D.A., Kolb, E.W., and Teplitz, V.L., Phys. Rev. Lett. 39, 168 (1977); (E)39, 973 (1977).
16. Pal, P.B. and Wolfenstein, L., Phys. Rev. D23, 766 (1982); Nieves, J.F., Phys. Rev. D28, 1664 (1983).
17. Hosotani, Y., Nucl. Phys. 191, 411 (1981).
18. For a recent review of neutrino oscillations see e.g. R.A. Eichler, Proc. of the International Symposium on Lepton Photon Interactions at High Energies, Hamburg, 1987.

19. E531 Collaboratio, N. Ushida et al., Phys. Rev. Lett. $\underline{57}$, 289 (1986).
20. See V. Zacek, Proceedings of the Rencontre de Moriond, January 1988, to be published.

MATTER EFFECTS IN NEUTRINO PROPAGATION

A. YU. SMIRNOV
Institute for Nuclear Research
Academy of Sciences of the USSR
60th October Anniversary prospect 7a
117312 Moscou
USSR

ABSTRACT. Conditions and dynamics of the resonant neutrino conversion are described. We discuss the applications of the effect to the solar neutrinos as well as present status of the conversion inside the Sun. The influence of different matter density perturbations on the conversion is considered, and in this connection the possible effects of parametric and stochastic enhancement of the influence are remarked.

1. Introduction

Resonant neutrino conversion is the process of nonreversible transformation of one neutrino species into another one (in a system of mixed neutrinos) during the propagation through the matter with monotonously changing density. The transformation occurs continuously according to density change, and mainly, in the resonance layer (the layer in which the density varies in definite limits about the resonance value) [1,2].

According to definition the key points are : 1) mixing of neutrinos - the transitions occur between mixed neutrino components; 2) interactions of neutrinos with matter - the elastic forward scattering is essential [3]; 3) resonance (crossing of the resonance layer); 4) change of density, moreover the density should change slowly to satisfy the adiabaticity condition. In the first part of review we consider these items and then describe dynamics of conversion (for more details see reviews [4-6]).

Conditions of resonance conversion are fulfilled for the Sun in a wide region of neutrino parameters (mass difference $\Delta m^2 = m_2^2 - m_1^2$ and mixing angle, θ) [1]. The conversion (ν_e - ν_μ, ν_e - ν_τ) diminishes signals in the solar neutrino experiments : ν_μ or ν_τ are sterile at the low energies with respect to charged currents interactions. The effect depends on neutrino energy, and consequently the conversion distorts the energy

G. Berthomieu and M. Cribier (eds.), Inside the Sun, 231–250.

spectrum of the electron neutrinos. Conversion may solve the solar neutrino problem [1]. It can ensure up to 15-fold suppression of the argon production rate in Cl - Ar -experiment, and rather independent suppression of ν e-scattering signal (Kamiokande-II). In a view of existing data and forthcoming Ga - Ge -experiment - results the present status of resonant conversion will be discussed.

In the third part of the review we consider the influence of different matter density perturbations on resonant conversion. Such a consideration has two aspects. Firstly, it enables to understand the sensitivity of the conversion effects to the model of the Sun and to evaluate the possibility of measuring the solar parameters with neutrino data. Secondly, if the perturbations depend on time then variations of v_e -flux are induced via the resonant conversion. The parametric and stochastic enhancement of perturbation effects will be remarked.

2. Resonant neutrino conversion

2.1. CONDITIONS OF RESONANT CONVERSION

2.1.1. *Mixing*. The necessary condition of resonant conversion in a system of two neutrinos, for example v_e - v_μ, is the mixing of these neutrinos. Mixing implies the interaction, which transforms v_e into v_μ. In the simplest case the mixing (so called vacuum mixing) is induced by the nondiagonal mass terms of the Hamiltonian. Now v_e and v_μ which are called the flavor states or eigenstates of the weak interactions turn out to be the coherent mixtures of v_1 and v_2 -- the states with definite masses m_1 and m_2 (eigenstates of the mass matrix) :

$$v_e = \cos\theta\, v_1 + \sin\theta\, v_2$$
$$v_\mu = -\sin\theta\, v_1 + \cos\theta\, v_2 \tag{1}$$

Here θ is the mixing angle. Flavor states themselves have no definite masses.

2.1.2. *Refraction*. Conversion is stipulated by the interactions of neutrinos with matter, and precisely -- by the elastic forward scattering, which is reduced to the appearance of the refraction indexes, n_e and n_μ, for neutrino waves [3] :

$$(n\text{-}1) \propto G_F N/k$$

Here G_F is the Fermi constant, N is the concentration of particles in medium, k is the momentum of neutrino. Matter influences on the evolution of mixed neutrinos if the indexes are different $n_e \neq n_\mu$. So, the conversion takes place in the transparent for neutrinos mediums, being nonsymetric with respect to the mixed components. The influence is described by the refraction length, l_0, defined as the distance over which an additional phase difference between ν_e an ν_μ waves due to interactions becomes 2π [3]. The l_0 determines spatial scale of matter effects. For ν_e - ν_μ -system one has

$$l_0 = \frac{2\pi}{k \left| n_e - n_\mu \right|} = \frac{2\pi}{\sqrt{2} \, G_F \, N_e} = \frac{2\pi m_N}{\sqrt{2} \, G_F \, \rho^{eff}} \qquad (2)$$

where $N_e = \rho . Y_e / m_N$ is the electron concentration, $\rho^{eff} = \rho . Y_e$ is the effective density, m_N is the nucleon mass. The corresponding width of matter :

$$d_0 = l_0 \cdot \rho = \frac{2\pi m_N}{\sqrt{2} \, G_F} = 3.5 \; 10^9 \; g/cm^2$$

is the universal constant, which in fact fixes the field of applications of the phenomena : the width d $\geq$ d_0 is needed.

Matter effects can be described equivalently in terms of potentials, V_e and V_μ, in which ν_e and ν_μ are moving : $V \propto (n-1)k \propto G_F N$ [7]

2.1.3. *Neutrino eigenstates in matter. Effective mixing in matter.* Effective (dynamical) mixing is used for the description of neutrino evolution [3]. In matter this mixing is introduced with respect to the neutrino eigenstates in matter, ν_{1m} and ν_{2m}, which are determined as the eigenstates of the total Hamiltonian, including the neutrino interactions (i.e. potentials V_e and V_μ) :

$$\nu_e = \cos \theta_m \, \nu_{1m} + \sin \theta_m \, \nu_{2m}$$
$$\nu_\mu = -\sin \theta_m \, \nu_{1m} + \cos \theta_m \, \nu_{2m} \qquad (3)$$

θ_m is the mixing angle in matter. The eigenstates ν_{im} have definite energies (eigenvalues) as well as definite phase and group velocities. In vacuum ν_{im} coincide with ν_i and $\theta_m = \theta$. Matter changes mixing, and moreover θ_m depends on matter density. As it can be shown at $\rho^{eff} \neq 0$ the transitions $\nu_1 \div \nu_2$ takes place and hence ν_i are no more the eigenstates of the Hamiltonian : $\nu_{im} \neq \nu_i$.

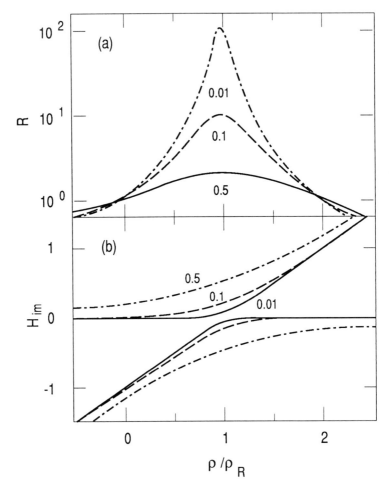

Figure 1. The dependence of (a) the resonance factor $R = sin^2 2\theta_m/sin^2 2\theta$ and (b) level energies in units $\Delta m^2 . (2k\ cos2\theta)^{-1}$ on the effective density for different value of $sin^2 2\theta$ (numbers of the curves).

2.1.4. Resonance. The dependance of mixing parameter, $sin^2 2\theta_m$, on the density has a resonance character [1] (fig. 1). At

$$\rho_R = \frac{m_N \cos 2\theta}{2\sqrt{2}\ G_F} \cdot \frac{\Delta m^2}{E} \qquad (4)$$

where $\Delta m^2 = m_2^2 - m_1^2$ and E is the energy of neutrino, this parameter reaches the maximum : $sin^2 2\theta_m = 1$ for arbitrary small θ. The ρ_R is called the resonant density. Half width of resonance at half height is proportional to the vacuum mixing :

$$\Delta\rho_R = \rho_R \cdot tan\ 2\theta \qquad (5)$$

The width of resonance layer, $2\Delta\rho_R$ fixes the scale on which resonant conversion occurs.

Parameter $sin^2 2\theta_m$ defines (as $sin^2 2\theta$ does in vacuum) the depth of neutrino oscillations. Therefore in matter with resonant density the depth is maximal.

Resonance condition (4) can be written as

$$l_v = l_0 \cdot cos\ 2\theta \qquad (6)$$

where $l_v = 4\pi E/\Delta m^2$ is the vacuum oscillation length. According to (6) in resonance (at small θ) period of system (l_v) itself coincides with period of external medium (l_0).

The value $\rho = \rho_R$ is the specific point in the dependance of θ_m on ρ. When ρ varies from $\rho \gg \rho_R$ to zero, the angle θ_m diminishes from $\pi/2$ to θ. At $\rho = \rho_R$ the $\theta_m = \pi/4$.

2.1.5. *Flavor changing of neutrino eigenstates*. Mixing angle θ_m determines according to (3) a flavor, or v_e-, v_μ -content, of the neutrino eigenstates, v_{im}. Propagating through a matter with varying density the v_{im} change their flavors. If θ is small and ρ varies from $\rho \gg \rho_R$ to $\rho \ll \rho_R$, the flavors of v_{im} change almost completely. For example, v_{2m} coincides practically with v_e at $\rho \gg \rho_R$ and with v_μ at $\rho \ll \rho_R$. Mainly a flavor changes in the resonance layer : $\rho_R - \Delta\rho_R < \rho < \rho_R + \Delta\rho_R$.

2.1.6 *Level crossing*. Cabibbo [8] and Bethe [7] had given the interpretation of the resonance in terms of eigenvalues H_{im} i.e. the level energies corresponding to v_{im} (fig. 1b). Potentials, V_e and V_μ, are proportional to a density, and consequently the total energies of v_e and v_μ, H_e and H_μ, are the linear functions of ρ. At resonance they cross :

$$H_e(\rho_R) = H_\mu(\rho_R) \qquad (7)$$

Mixing of v_e and v_μ rejects level crossing : the eigenvalues $H_{im}(\rho)$, are not equal but their splitting is minimal in resonance (fig. 1b). In the resonance layer when ρ increases the curve $H_{1m}(\rho)$ goes from the line $H_e(\rho)$ to the line $H_\mu(\rho)$, the $H_{2m}(\rho)$ vice versa -- from $H_\mu(\rho)$ to $H_e(\rho)$.

In resonance the difference of the potentials $V = V_\mu - V_e$, compensates the difference of level energies related to masses. As the V for neutrinos and antineutrinos have an opposite sign the resonance as well as resonant conversion in a given medium takes place for neutrinos or antineutrinos only, depending on the respective signs of Δm^2, V and $\cos 2\theta$.

2.1.7. *Adiabaticity*. [1,2,9,10]. Adiabaticity implies a slowness of density change. If ρ changes adiabatically (slowly), then the transitions between the eigenstates ν_{1m} - ν_{2m} can be neglected; the admixtures of ν_{im} in a given ν -state conserve; the system has a time to adjust itself to external condition (density) variations. The condition of adiabaticity can be written as

$$\frac{d\theta_m}{dr} \ll \frac{d\varphi}{dr} \equiv |H_{2m} - H_{1m}| \tag{8}$$

where φ is the phase difference between ν_{1m} and ν_{2m} waves $(d\theta_m(\rho)/dr \propto d\rho/dr)$. The condition is the most crucial in resonance, where the level splitting is minimal [1,2] :

$$2\Delta r_R \geq l_m^R \tag{9}$$

Here $\Delta r_R = l_0 . \tan 2\theta$, $[l_\rho \equiv (d\rho/dr)^{-1} . \rho]$ is the spatial half width of the resonance layer; l_m^R is the oscillation length in resonance : $l_m^R = l_\nu/\sin 2\theta$. According to (9) at least one oscillation length should be obtained in resonance layer.

The degree of adiabaticity violation is determined by the adiabaticity parameter [4] :

$$\text{æ}_R = \frac{\Delta r_R}{l_m^R} = \frac{\sin^2 2\theta}{4\pi \cos 2\theta} \cdot \frac{\Delta m^2}{E} \cdot l_\rho \tag{10}$$

At $\text{æ}_R \leq 0.1$ the violation is strong and the transitions $\nu_{1m} \rightarrow \nu_{2m}$ become essential.

So far we have considered main conditions of resonance conversion : 1) mixing, 2) resonance (level crossing), 3) adiabaticity (see also the general discussion in [1]).

2.2. DYNAMICS OF RESONANT CONVERSION.

Resonant neutrino conversion is, in fact, the change of the flavor of neutrino state on the adiabatic (or weakly nonadiabatic) crossing of resonance layer. The decomposition of given ν -state over ν - eigenstates, ν_{im}, and the change the flavors of ν_{im} themselves are all we need to trace the conversion. The dynamics is reduced to the change of the admixtures of ν_{im} in $\nu(t)$.

To describe the process of conversion the survival probability, $P(t)$, i.e. the probability to find the neutrino of initial type in a given moment t, is used.

One can single out three different regimes of conversion 1) nonoscillatory, 2) oscillatory adiabatic, and 3) nonadiabatic.

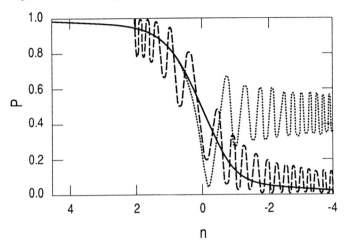

Figure 2. Spatial picture of resonant neutrino conversion. The dependence of survival probability on the distance from the resonance layer : $n = (\rho - \rho_R)/\Delta\rho_R$ for nonoscillatory transformation (solid line), oscillatory adiabatic transition (dashed line), adiabaticity violation regime (dotted line).

2.2.1. *Nonoscillatory transition* [1,2]. Let neutrinos, for definiteness ν_e, are produced at $\rho \gg \rho_R$ [more precisely at $(\rho - \rho_R)/\Delta\rho_R \gg 1$], and the adiabaticity is satisfied during the propagation. Then in the initial moment according to the first condition ($\rho \gg \rho_R$) the mixing angle equals $\pi/2$ and hence the neutrino state coincides practically with ν_{2m} : $\nu(t_0) = \nu_e \cong \nu_{2m}$. According to the second condition (adiabaticity) $\nu(t)$ will coincide with ν_{2m} throughout the propagation. Consequently, $\nu(t)$ changes its flavor together with ν_{2m}; the flavor of $\nu(t)$ follows density variations, like the flavor of ν_{2m} does (fig.2). If the final density is zero (or $\ll \rho_R$), then at the exit one has $\nu(t_f) = \nu_{2m}(\rho = 0) = \nu_2 = \nu_e sin\theta + \nu_\mu cos\theta$ and then the survival probability is [1] :

$$P = P(\nu_e \to \nu_e) = sin^2\theta \qquad (11)$$

The smaller vacuum mixing, the stronger transformation of the neutrino type into another one can be achieved in contrast with vacuum oscillations.

2.2.2. *Adiabatic oscillatory transition* [2]. let the adiabaticity condition is fulfilled but the neutrinos are produced not far from thr resonance layer : $|\rho_0 - \rho_R| <$ several $\Delta\rho_R$. Then $\nu(t)$ contains the comparable admixtures of both eigenstates and moreover these admixtures will conserve. As $\nu(t)$ does not coincide with one of the eigenstates the flavor of $\nu(t)$ oscillates. Oscillations superpose on the conversion (fig.2), but the averaged probability changes according to density variations [2,9,10] :

$$\bar{P} = \frac{1}{2} - \frac{1}{2} \cos 2\theta_m(\rho_0) \, \cos 2\theta_m(\rho) \tag{12}$$

when ρ_0 approaches ρ_R, the transformation becomes weaker : P increases. At $\rho_0 = \rho_R$: $P = 1/2$. (In the adiabatic regime not only P but also the depth of oscillations, $Ap = \sin2\theta_m(\rho_0).\sin2\theta_m(\rho)$ depends on the initial and final densities only and does not depend on density distribution in the intermediate points.

2.2.3. *Adiabaticity violation regime.* If the adiabaticity is violated, the admixtures of neutrino eigenstates in a given $\nu(t)$ change. Let neutrinos are produced far from the resonance layer ($\rho_0 \gg \rho_R$), then $\nu(t_0) = \nu_e$ consists of ν_{2m} predominatingly. But the admixture of ν_{1m} appears in $\nu(t)$ in the course of propagation. Consequently the probability $P(t)$ begins to oscillate and the transition becomes weaker, than in nonoscillatory case (fig.2). The averaged probability can be written now as

$$\bar{P} = \frac{1}{2} + (\frac{1}{2} - P_{21}) \cos 2\theta_m(\rho_0) \, \cos 2\theta_m(\rho) \tag{13}$$

where P_{21} is the probability of ν_{2m} - ν_{1m} transition At small θ and at the values of ρ_0 and ρ_f beyond the resonance layer, the P_{21} can be evaluated using the well known Landau-Zener result [12,13]. It gives the probability of transition between two levels under linear with time perturbation, which mixes these levels and induces their crossing :

$$P_{21} = exp(-\pi^2 æ_R) \tag{14}$$

There is a lot of improvements and generalizations of formula (14) in literature [14].

2.2.4. *Analogies.* There are many analogies of resonant conversion in different fields of physics. Let remark two of them.

The system of mixed neutrinos is similar to the weakly coupled oscillators (for example, two pendulums) [15]. Matter effect is equivalent to a change of pendulum eigenfrequences ω_1 and ω_2. Resonance corresponds to the equality $\omega_1 = \omega_2$. Now the conversion is nonreversible transmission of oscillations from one pendulum to another one under slow changing of frequences from $\omega_1 \ll \omega_2$ to $\omega_1 \gg \omega_2$.

Another analogy is the electron spin flip in the rotating magnetic field [16]. The states with the projections $1/2$ and $-1/2$ correspond to ν_e and ν_m.

2.3. DIFFERENT REALIZATIONS OF RESONANT NEUTRINO CONVERSION.

The general conditions : mixing, level crossing, adiabaticity may have many realizations even for neutrino systems [11]. Depending on properties of neutrino states are mixed one can single out three types of conversion.

2.3.1. *Flavor conversion* (considered above). Neutrino states with different flavors but the same helicities are involved, for example n_{eL} - $n_{\mu L}$. In the course of conversion the flavor changes, and the helicity conserves. In general all three known neutrino species, ν_e, ν_μ, ν_t, are mixed. Such a system has three resonances but in usual medium (electrons, nuclei) in the lowest order of the perturbation theory the ν_μ and ν_τ interact equally and only two level crossings are possible. In $(\nu_e - \nu_\mu)$ and $(\nu_e - \nu_\tau)$ -chanels [17]. If $\Delta m^2_{21} = m^2_2 - m^2_1$ differs from $\Delta m^2_{31} = m^2_3 - m^2_1$ sufficiently (which realizes for mass hierarchy, $m_1 \ll m_2 \ll m_3$), then the corresponding resonances are splitted on the density scale. In this case crossing of resonances can be described independently -- three neutrino task is reduced to two neutrino task (see reviews [4-6]).

2.3.2. *Spin conversion* [18]. Conversion takes place in a system of left and right components composed the same dirac neutrino : ν_{eL} and ν_{eR}. Mixing is induced by the interaction of neutrino magnetic moment, μ_ν, with magnetic field, B [the exchange (mixing) energy is $2\mu_\nu B$]. Both level splitting and level crossing are stipulated by the refraction in inhomogenious matter. Now helicity is changed and flavor conserves.

2.3.3. *Spin-flavor conversion*. Both flavor and helicity of neutrino state are changed, for example ν_{eL} - $\nu_{\mu R}$ [19]. Mixing is induced by the interaction of the nondiagonal (transition) magnetic moment, μ_ν, with magnetic field. Now the mixing parameter is

$$"sin \ 2\theta" = (2\mu_\nu B) \cdot E/\Delta m^2 \tag{15}$$

Mass difference and refraction (potentials) give the contributions to level splitting. Level crossing is related to the change of density.

Considered conversion types differ by the dependences of the adiabaticity parameter, $æ_R$, on the neutrino energy. One has $æ_R \sim 1/E$ for the flavor conversion (10), $æ_R$ = constant for the spin precession and

$$æ_R = (2\mu_\nu B)^2 \cdot E/(\pi \Delta m^2 l_\rho)$$ (16)

for the spin flavor conversion. In the last case at B = const, for example, $æ_R$ is proportional to E.

3. Resonant conversion of solar neutrinos. Present status

3.1. EFFECTS OF FLAVOR CONVERSION

Neutrinos, ν_e, produced in the center of Sun, cross the layer of matter with monotonously decreasing density. Resonant flavor conversion results in suppression of the initial ν_e -flux, which depends on Δm^2, $sin^2 2\theta$ and the energy of neutrinos :

$$F(E, \Delta m^2, sin^2 2\theta) = P(E, \Delta m^2, sin^2 2\theta) \cdot F_0(E)$$

Here F_0 is the flux without conversion. The suppression factor p (averaged survival probability) as the function of $E/\Delta m^2$ has a bath-like shape [1] (fig. 3).

Moreover there are two specific energies in $P(E/\Delta m^2)$ dependence : E_c and E_a. 1). The energy of resonance turning on, E_c, is fixed via the resonance condition (4) by the central density ρ_c :

$$E_c/\Delta m^2 = m_N/(2\sqrt{2} G_F \rho_c)$$ (17)

For neutrinos with E < E_c there is no resonance (level) crossing. 2). The energy of adiabaticity violation, E_a, is determined by adiabaticity condition $æ_R(E_a) = 0.5$:

$$E_a/\Delta m^2 = l_\rho \cdot sin^2 2\theta/(2\pi cos 2\theta)$$ (18)

(see (9,10)). When E increases the adiabaticity parameter diminishes : $æ_R \sim 1/E(l_0$ = const. in a wide region of ρ), the adiabaticity violates stronger and P increases (E > E_a). The energy E_a is proportional to $sin^2 2\theta$ -- with decreasing vacuum mixing the baths become narrower.

If E rises, the resonance layer shifts to the surface of the Sun. In the region of resonance turning on ($E = E_c \pm$ several $.\Delta E_R$) the oscillatory adiabatic transition takes place , here $\Delta E_R = E_c.\tan 2\theta$, ΔE_R is the width of resonance in energy scale. At (E_c + several. ΔE_R) ÷ E_a the conditions for the nonoscillatory transformation are fulfilled and $P = \sin^2\theta$. At $E > E_a$ neutrinos convert in adiabaticity violation regime (see sect.2.2.3., (13, 14). The mutual position of the suppression bath and the neutrino spectrum depends on Δm^2, and if the spectrum falls on the edges of the bath, it distorts.

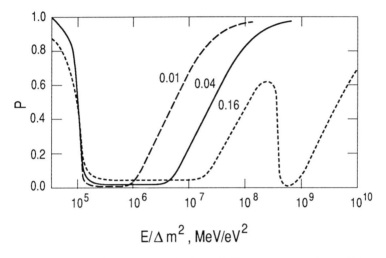

Figure 3. Suppression baths of flavor conversion. The dependences of suppression factors on $E/\Delta m^2$ for different values of $\sin^2 2\theta$ (numbers on the curves). The suppression pit for 3v-mixing (an example, dashed line).

Two remarks are in order. 1). The averaging of P over the v - production region results in the smoothing of the left edges of baths. 2) For three neutrino mixing the suppression factor P(E) is the superposition of two baths, which are shifted on the axis E one with respect to another by factor of $\Delta m_{13}^2 / \Delta m_{12}^2$ (fig.3).

3.2. SPIN-FLAVOR CONVERSION AND SPIN CONVERSION

In contrast with flavor conversion for spin flavor case [19] (1) mixing parameter "$\sin^2 2\theta$ is not constant : it depends on the magnetic field and the neutrino energy (15), (2) effective density is $\rho^{eff} = \rho (Y_e - Y_n)$ for v_e-$\bar{v}_\mu$, where Y_n is the number of neutrons per nucleon, (3) the dependence of the adiabaticity parameter on the neutrino energy differs from that of flavor case. If $B(r) \sim \rho^\alpha$ and $\alpha < 0.5$, then æ$_R$ rises with increasing E and the adiabaticity becomes better (or restores). This results in rather complicated shape of suppression pits, which in general

242

differs from flavor conversion baths (especially on the bottom and nonadiabatic edge) fig.4 [20]. P (E) depends on the magnetic field profile, which is essentially unknown. In some cases (B(r)) the shape of pit may coincide with that of the flavor conversion and the problem appears to distinguish these two cases.

For spin conversion the energy splitting is $\Delta H = G_F(\rho(Y_e - Y_n/2)/m_N$ and there is no level crossing inside the Sun. Moreover the effect does not depend on neutrino energy. It can be evaluated from the spin-flavor results by the transition $\Delta m^2/E \to 0$ (P at the largest values of $E/\Delta m^2$ in fig.4) [20].

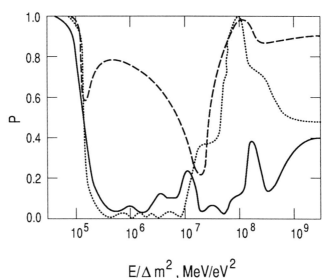

Figure 4. The suppression pits for spin-flavor conversion. The dependences of suppression factor on $E/\Delta m^2$ for different distributions of the magnetic field. Maximal strengh of field $B_1 = 10^6 - 10^7$ Gauss in the center of the Sun and $B_0 = 10^4$ Gauss in the convective zone were used (for more details see [20]. $\mu_v = 10^{-11} \mu_B$.

3.3 PRESENT STATUS OF RESONANT NEUTRINO CONVERSION

The suppression factor for ^{37}Ar-production rate in Cl-Ar-experiment due to flavor conversion is

$$R_{Ar}(\Delta m^2, sin^2 2\theta) = \frac{1}{Q_{Ar}^{SSM}} \int dE \sigma(E) F^0(E) P(E, \Delta m^2, sin^2 2\theta)$$

(19)

where $Q^{SSM}_{AR} = \int dE\sigma(E)F^0(E)$, $\sigma(E)$ is the cross-section, F^0 (E) is the neutrino flux on the Earth in the standard solar model (SSM) [21]. The equation R_{Ar} (Δm^2, $sin^2 2\theta$) = c = const determines lines of $_2$f equal suppression [1,2,4-6] (or ISOSNU lines [22] on Δm^2-$\sin^2 2\theta$-plot (fig.5a). Consequently, Davis's data [23], $R^{exp}_{AR} = (0.2 \div 0.5)$ can be explaned in terms of resonant flavor conversion if the values of Δm^2 and $\sin^2 2\theta$ find between ISOSNU lines $c_{min} = R^{exp}_{min} = 0.2$ and $c_{max} = 0.5$. The system of three neutrinos is described by two points on Δm^2 - $\sin^2 2\theta$ - plot and as can be shown [24], to achieve the observed suppression of Q_{Ar} at least one point should be placed between ISOSNU lines $c_{min} = R^{exp}_{min} = 0.2$ and $c_{max} = \sqrt{R^{exp}_{max}} = 0.7$.

The suppression of ν e-signal in Kamiokande-II experiment [25] which is sensitive to the high energy part of boron neutrino spectrum only, is rather independent on the suppression of Q_{Ar} (fig.5b) [24]. Preliminary Kamiokande-II result with 1σ errors [25] excludes some part of loops in fig.5b. i.e. definite values of Δm^2 and $\sin^2 2\theta$.

The suppression factor of Ge-production rate in Ga-Ge experiment, R_{Ge}, is introduced similarly to (19). According to fig.5c. at $R_{Ar} = 0.2 \div 0.5$ any result for R_{Ge} from 0.04 to 0.98 is possible and moreover 1σ Kamiokande-II limit restricts this region of prediction very weakly.

The above consideration (fig.5) corresponds to fixed parameters of standard solar model. In fact, one should consider the central temperature, T_c, abundances, X,Y,Z and others as free parameters and carry out the multidimensional analysis, which includes both solar and neutrino parameters. In that time the crucial signature of resonant conversion, which enables to distinguish its effects from the astrophysical ones, is the distortion of continuous energy spectra from individual reactions (pp-, ^{8}B-decay). The distortion is described by suppression bath (fig.3) and depends on Δm^2 and $\sin^2 2\theta$.

Spin-flavor conversion may result in strong suppression of both ^{37}Ar-production rate and number of ν e-scattering events as well as in the variety of the energy spectrum distortion, if the transition magnetic moment $\mu_\nu > 10^{-11} \mu_B$ (μ_B is the Bohr magneton).

Possible anticorrelations of solar neutrino flux and 11-years solar activity [23] can be explained immediately by the magnetic field variations via the spin-flavor conversion. But it is not excluded that the change of ν -flux is related to the variations of density distribution which influences on flavor conversion (see sect.4).

244

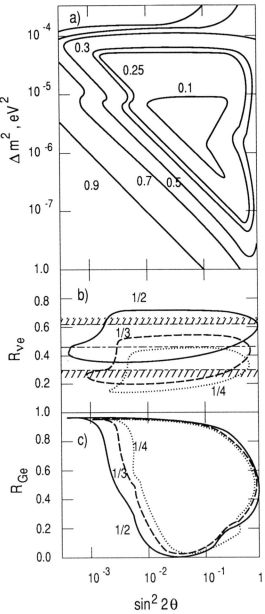

Figure 5.a). Lines of equal suppression (ISOSNU lines) of ^{37}Ar production rate in Cl-Ar experiment. Numbers on the curves are the suppression factors, R_{Ar}.

b). The suppression factor of number of $\nu e - \nu e$ events with the energies of final electrons $E_e > 9.5$ MeV as the function of $\sin^2 2\theta$ at fixed suppression of Ar-production rate (numbers on the curves). The points of loops correspond to definite points of ISOSNU lines. The horizontal lines are the Kamiokande-II result with 1σ errors.

c). The same as in fig. b) for the Ge-production rate in Ga-Ge experiment.

4. Matter density perturbations and resonance conversion.

The influence of density perturbations on conversion depends on relation between the oscillation length, I_m, and the spatial scale of perturbation, I_{fI} (several model calculations were performed in [28, 29, 11]).

The smooth variations of density profile with typical scale $I_{fI} \cong R_0$ result in rather small change of suppression baths, which is controled by the energies $E_c \sim 1/\rho$ and $E_a \sim d\log\rho/dr$ (17,18). Such a change reduces mainly to snift of the edges of baths and is equivalent to renormalization of Δm^2 and $\sin^2 2\theta$. Futher on we will concentrate on the case $I_{fI} \cong I_m << R_0$.

4.1. CONDITIONS FOR MAXIMAL EFFECT OF DENSITY PERTURBATIONS

As the scales, I_{fI}, and amplitudes $A_\rho = \Delta\rho/\rho$ of possible density perturbations in the Sun are essentialy unknown and moreover A_ρ are considered to be small (no more than several percents), it is worthwhile to evaluate the minimal values of A_ρ and I_{fI} which can be observed with solar neutrino data. We begin with conditions under which the influence of ρ −perturbation on resonance conversion is maximal.

1). The perturbations should be placed in the resonance layer (at $\rho = \rho_R$ (E)). Indeed, density fluctuation, $\Delta\rho_{fI}$, can be considered as the perturbation of potential : $\Delta V = \sqrt{2}. \, G_F.\Delta\rho_{fI}/m_N$. The perturbation induces the transition between levels with probability $P = 1$ if ΔV is of order of level splitting, ΔH. The splitting is minimal in resonance (fig. 1b) $\Delta H_R = \sqrt{2} \, G_F.\rho_R.\tan2\theta/m_N$, and from the inequality $\Delta V > \Delta H$ one obtains $\Delta\rho_{fI} > \Delta\rho_R$ or $A_\rho = \Delta\rho_{fI}/\rho > \tan 2\theta$. The smaller R_θ, the smaller A_ρ is sufficient for a large effect.

2). To avoid the averaging, which suppresses the perturbation effect , neutrino should be in definite level. Maximal effect takes place in the region of nonoscillatory transformation -- on the bottom of the bath.

3). The effect can be additionally enhanced if there are several density fluctuations, N_{fI}, on the way of neutrinos, and the parametric condition is fulfilled [30,31] :

$$l_{fI} \cong l_m(\rho) \qquad (20)$$

In this case the effects of individual fluctuations add constructively. Now the same change of conversion probability can be obtained with smaller value of A_ρ : $A_\rho \sim 1/N_{fI}$. If the perturbations are distributed randomly in space, then a stochastic enhancement takes place and $A_\rho \sim 1/\sqrt{N}_{fI}$.

4). The effect depends on the shape of density perturbation but in any cases perturbations should violate adiabaticity. Conditions 1). and 2). for a given ρ-fluctuations can be satisfied at definite neutrino energy only. Moreover they are compatible if the relation between the l_{fl} and the position of fluctuation inside the Sun, r_0 is fulfilled :

$$l_{fl} = \frac{2\pi m_N}{\rho(r_0)\sqrt{2} G_F \tan 2\theta} \tag{21}$$

The effect of density fluctuation looks like a peak on the bottom of the bath (under the considered conditions). The height of the peak equals

$$P^{max} \approx \left(\frac{A_\rho N_{fl}}{2\tan 2\theta}\right)^2$$

and the width is iversely proportional to number of structures $\Delta E \sim \tan 2\theta/N_{fl}$. If the condition (21) is not fulfilled, then two peaks appear corresponding to the resonance at $E = E_R$ and to parametric condition at $E = E_P$ (parametric resonance). If neutrino state contains the admixtures of both levels (which is realized on the nonadiabatic edge), then the ρ-perturbation results in that the enhancement of P will be altered with energy by suppression.

4.2. EFFECTS OF ρ-PERTURBATION. NUMERICAL RESULTS

To illustrate the results discussed above Krastev and myself have calculated the probabilities of resonant conversion (averaged over oscillations in final state) for

$$\rho(r) = \rho^{ssm}(r) \cdot \left[1 + A_\rho \theta(r-r_0) \cdot \theta(r_1-r) \cdot \sin 2\pi r/l_{fl}\right]$$

where ρ^{ssm} is the profile of standard solar model, $\theta(r)$ is the step-function (the perturbations are in the interval $r \div r_1$), A_ρ = const. In fig.6 the effects on the bottom of the bath (a-c) and on the nonadiabatic edge (d) are shown.

4.3. DENSITY VARIATIONS AND SOLAR NEUTRINO FLUX

Let us adopt that variations of boron neutrinos can be observed if they take place in the energy region $\Delta E/E \geq 0.2$ and if the amplitude of variations is $\Delta P/P \geq 0.3$. For beryllium neutrinos line the region of variations can be much smaller $\Delta E/E > 0.02$ and even very narrow peaks of fig.6 can induce an appreciable effect. Using the results of sect. 4.2,3. the parameters of smallest density fluctuations ($l_{fl}A_\rho$) are evaluated, which give an observable variations : ($\rho 10^{-3}R_{\odot}$, 3.10^{-3}) at r_0 = $0.2R_{\odot}$, ($10^{-2}R_{\odot}$, 0.03) at $r_0 = 0.5R_{\odot}$ ($0.1R_{\odot}$, 0.1) at $r_0 = 0.7R_{\odot}$.

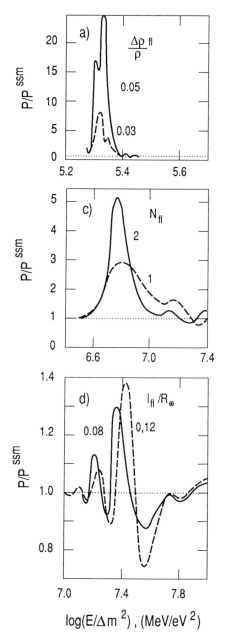

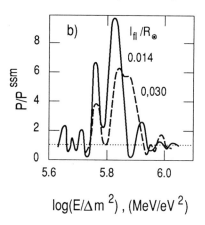

$\log(E/\Delta m^2)$, (MeV/eV^2)

Figure 6. The respective change of the survival probability $P(E)/P^{ssm}(E)$ for different parameters of density perturbations : r_0 - the position inside the Sun, A_ρ the amplitude, I_{fl}-the spatial scale in radial direction, N_{fl} - the number of periodic structures.

$\log(E/\Delta m^2)$, (MeV/eV^2)

a). r_0 = $0.18R_{\odot}$, $\sin^2 2\theta$ = 0.01, N_{fl} = 1, I_{fl} = $0.015R_{\odot}$.
b). r_0 = $0.30R_{\odot}$, $\sin^2 2\theta$ = 0.01, N_{fl} = 1, A_ρ = 0.05.
c). r_0 = $0.51R_{\odot}$, $\sin^2 2\theta$ = 0.25, A_ρ = 0.2, I_{fl} = $0.035R_{\odot}$
d). r_0 = $0.74R_{\odot}$, $\sin^2 2\theta$ = 0.25, A_ρ = 0.2, N_{fl} = 2.

248

The most strong respective change of P (E) occurs on the bottom of the baths, where in absence of ρ-perturbation the suppression of v -flux is strong. Then a scenario can be considered in which without conversion the v -flux is large, so that $Q^0{}_{Ar} = (13 - 15)$ SNU but due to flavor conversion and ρ−variations the Q_{Ar} change from 1 to 4-5 SNU [31].

Density variations may be related to the magnetic fields [32], to the instabilities induced by rotation or nuclear energy release. Semiannual variations of signals on the Earth may be stipulated by nonsphericity of the Sun and to the annual change of Earth position. Another possibility is the ρ-variations due to g-mode of oscillations of the Sun; the parametric and/or stochastic enhancement of perturbation effect may take place. The change of the oscillation activity (power of vibrations) with time can induce the corresponding change of v -signals.

5. Conclusion

1). Resonant neutrino conversion -- the change of flavor of neutrino state according to matter density variations -- is the analog of well known and established phenomena from different fields of physics.

2). For conversion the massiveness and mixing of neutrinos are needed only. Conversion takes place inside the Sun if $\Delta m^2 = (10^{-8} - 10^{-4})$ eV^2 and $\sin^2 2\theta > 10^{-4}$. Spin-flavor precession demands sufficiently large magnetic moment of neutrino : $> 10^{-11}\mu_B$.

3). Resonant conversion allows to explain the suppressions of signals in Cl-Ar and Kamiokande-II experiments. Moreover it can explain the possible variations of signals via magnetic field variations (in the spin-flavor case) or via density profile variations with $\Delta\rho/\rho > 3.10^{-3}$ and $I_{fI} > 10^{-3}R_\odot$ (in flavor as well as spin-flavor cases). Inverseiy, the search for time variations of solar neutrino signals is sensitive to $\Delta\rho/\rho > 3.10^{-3}$ and $I_{fI} > 10^{-3}R_\odot$

4). Resonant conversion gives the unique method of measuring the neutrino masses and mixing in very plausible region of magnitudes, which is not achieved by usual experiments. If the conversion effects will not be found then large region of Δm^2 and $\sin^2 2\theta$ as well as $\mu_v > 10^{-11}\mu_B$ can be excluded.

Acknowledgements

I am gratefull to J. Boratav, M. Cribier, E. Schatzman and D. Vignaud for the discussions and hospitality during my visit in France.

References

1. S.P. Mikheyev and A. Yu. Smirnov, Sov.J. Nucl.Phys., 42 (1985) 913-919; Nuovo Cimento, C9 (1986) 17-26.

2. S.P. Mikheyev and A. Yu. Smirnov, Sov.Phys.JETP, 64 (1986) 4-7.

3. L. Wolfenstein, Phys.Rev.D17 (1978) 2369-2374; ibidem D20 (1979) 2634-2635.

4. S.P. Mikheyev and A.Yu.Smirnov,Usp.Fiz.Nauk, 153 (1987) 3-58.

5. S.P. Mikheyev and A.Yu. Smirnov, Prog.in Particle and Nucl.Phys., Pergamon Press (to be published) (1989)

6. T.K. Kuo and J. Pantaleone, Rev.Mod.Phys., (1989) to be published.

7. H. Bethe, Phys.Rev.Lett., 56 (1986) 1305-1308.

8. N. Cabibbo, 10th Int.Workshop on weak interactions and neutrinos (Savonlinna, Finland 1985) unpublished.

9. A. Messiah Proc. of the 6th Moriond Workshop on Massive Neutrinos in Astrophysics and Particle Physics (Tignes, France) eds. O. Fackler and J. Tran Thahn Van, 373-389.

10. V. Barger et al, Phys.Rev., D34 (1986) 980-983.

11. A.Yu. Smirnov Proc.of the Int.Conf. "Neutrino-88", and Preprint MPI-PAE/PTh 48/88.

12. S.J. Parke, Phys.Rev.Lett, 57 (1986) 1275-1278.

13. W.C. Haxton, Phys.Rev.Lett., 57 (1986) 1271-1274.

14. P.I. Krastev and S.T. Petcov, Phys. Lett.B205 (1988) 84-92 and references therein.

15. S.P. Mikheyev and A.Yu. Smirnov Proc.of the 12th Int.Conf. Neutrino-86, (Sendai, Japan), eds. T. Kitagaki and H. Yuta 177-193.

16. J. Bouches et al, Z. Phys., C32 (1986) 499-511.

V.K. Ermilova et al, Sov.JETP Letters, 43 (1986) 353-355.

17. T.K. Kuo and J. Pantaleone, Phys.Rev.Lett., 57 (1986) 1805-1808.

18. M.B. Voloshin et al, Sov.JETP, 91 (1986) 754-765, and E.Kh. Akhmedov, see in [19].

19. E.Kh. Akhmedov, Phys.Lett., 213B (1988) 64-68.

20. E.Kh. Akhmedov and O.V. Bychuk, ZhETF, 95 (1989) 442-457

21. J.N. Bahcall and R.K. Ulrich, Rev.Mod.Phys., 60 (1988) 297-372.

22. S.J. Parke and T.P. Walker, Phys.Rev.lett., 57 (1986) 2352-2355.

23. R. Davis, in"Inside the Sun" eds. G. Berthomieu and M. Cribier, Kluwer Academic Publishers, Dordrecht.

24. S.P. Mikheyev and A.Yu. Smirnov, Phys.Lett., 200B (1988) 560-564.

25. M. Nakahata in "Inside the Sun", eds. G. Berthomieu and M. Cribier, Kluwer Academic Plublishers, Dordrecht.

26. V. Gavrin in "Inside the Sun" (see ref.25).

27. T. Kirsten in "Inside the Sun" (see ref.25).

28. A. Schafer and S.E. Koonin, Phys.Lett., B185 (1987) 417-420.

29. M. Cribier and D. Vignaud, see in ref.[32].

30. V.K. Ermilova et al, Kr.Soob.Fiz. Lebedev Institute, 5, (1986) 26.

31. P.I. Krastev and A.Yu. Smirnov, Phys.Lett. (1989) to be published.

32. E. Schatzman and E. Ribes, Proc. of the 7th Moriond Workshop "New and Exotic phenomena", ed. by O. Fackler and J. Tran Thanh Van, (1987) 368.

PART 3

HELIOSEISMOLOGY

A REVIEW OF OBSERVATIONAL HELIOSEISMOLOGY

Thomas L. Duvall Jr.
Laboratory for Astronomy and Solar Physics
NASA/Goddard Space Flight Center
Greenbelt, MD 20771 USA

Introduction

There have been several excellent reviews of observational helioseismology in recent years. These include the reviews by Harvey (1988), Libbrecht (1988), and van der Raay (1988) presented at a recent conference in Tenerife. The present effort will concentrate on the progress made on solar rotation recently.

Basic Helioseismology

The Sun is a resonant cavity that supports many ($\sim 10^7$) modes of oscillation. The modes that are most easily observed are the acoustic or p-modes. The eigenfunctions for these modes are:

$$E = f_{nl}(r)Y_{lm}(\theta,\phi)e^{i2\pi\nu_{nlm}t}$$

$f_{nl}(r)$ is the radial part of the separable eigenfunction where r is the radial coordinate measured from the center of the star. n is the number of radial nodes in the eigenfunction. $Y_{lm}(\theta,\phi)$ is the spherical harmonic function, where θ is the colatitude and ϕ is the longitude. The spherical harmonic degree l is the number of nodes of the spherical harmonic measured along a great circle that makes an angle $\cos^{-1}(\frac{m}{l(l+1)})$ with the equator. The azimuthal order m is the number of nodes around the equator. The frequency of the eigenmode, ν_{nlm}, depends on the mode. Much of our information derived about the solar interior from helioseismology comes from the measurement of these frequencies.

Some examples of the spherical harmonics for $l = 40$ and different m values are shown in Fig. 1. The limiting of these functions in latitude is not obvious at the lowest degrees. There is a term in the spherical harmonic function that multiplies the overall function that is $\sin^m(\theta)$. This term causes the latitudinal falloff at high m values. So for the sectoral modes ($m = l$), we are observing equatorially concentrated quantities, while for zonal modes ($m = 0$), we have an approximately equal weighting in latitude. This is true both in latitude and in radius: we observe integral quantities with the integral extending over different ranges of the independent variable. To learn about a solar parameter versus depth or latitude requires an equivalent differentiation of the data, a somewhat noisy process. If the eigenfunction of a mode does not extend into a certain region, we cannot learn anything directly about the region from this mode. This is why it is difficult to learn much about the deep interior from p-modes as not many of the

G. Berthomieu and M. Cribier (eds.), Inside the Sun, 253–264.

modes extend into that region.

The p-modes satisfy a dispersion relation that is best seen as an observational power spectrum (Fig. 2). The m variation of power has been suppressed. The dispersion relation, which shows the frequency ν_{nl} versus l would consist of closely spaced dots along the "ridges" of power in Fig. 2. Each ridge corresponds to a constant value of n or radial harmonic.

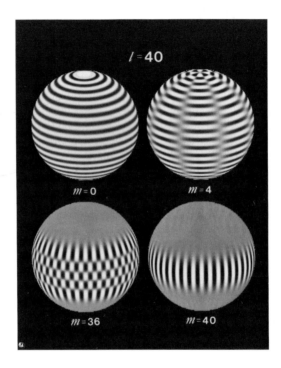

Fig. 1. Some examples of the spherical harmonic functions for degree $l = 40$. The white and black areas would correspond to receding and approaching areas in velocity observations. Note the increasing equatorial concentration as the azimuthal order m approaches its maximum value at $l = m$.

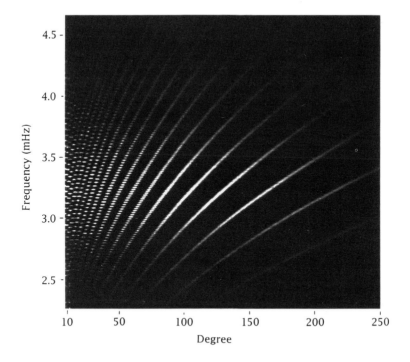

Fig. 2. This is a portion of the space-time spectrum of intensity oscillations of 50 hours of images from Duvall *et al.* (1987).

Global modes

Our use of the frequencies to infer properties about the solar interior depends critically on the assumption that the waves are giving a true global average of solar properties. This may not be true for all the waves that we observe. Which waves are global modes? If waves can travel a circumference coherently to interfere with themselves, then they are global. So, the lifetime must be greater than the travel time for a circumference. The energy from a wave will travel at a velocity given by the group velocity: $\frac{d\omega}{dk} = 2\pi R_{\odot}\frac{d\nu}{dl}$. So the time to travel a circumference is $T = 2\pi\frac{R_{\odot}}{d\omega/dk} = \frac{1}{d\nu/dl}$. If the frequency width of a feature is $< \frac{1}{T}$, then it will be global. Or, equivalently width $< \frac{d\nu}{dl}$. $\frac{d\nu}{dl}$ is the frequency spacing between modes of the same n and adjacent degree l. In our observational spectra, we always have the modes from several adjacent l values in a single frequency spectrum because of our inability to see the back side of the Sun. The spherical harmonic functions are a complete set over the whole sphere. An example of a well-separated set of modes is shown in Fig. 3a. So the condition for waves to be global reduces observationally to the condition of being able to separate the adjacent l "sidebands". There are areas of the $k-\omega$ diagram in which the modes are global as *e.g.* Fig. 3a. In addition, there are currently observed areas where this is not the case. An example is shown in Fig. 3b, where we see the normal phenomenon of the mode width increasing with frequency. At lower frequency we are separating the adjacent l-values while at higher frequencies we are not.

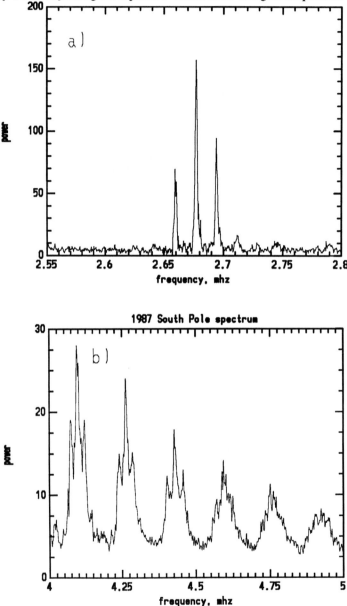

Fig. 3. Some examples of high resolution spectra at degree 50 from the data described by Jefferies *et al.* (1988). In a) we see a low frequency sprectral area where the spatial sidebands are well resolved. In b) we see the transition from resolving the spatial sidebands to not

resolving them as frequency increases. Both spectra have been averaged over the azimuthal order with rotation removed. In addition, the high frequency spectrum has been smoothed by convolution with a Gaussian of full width at half maximum of $5\mu Hz$.

Some care needs to be exercised in using results derived from such an area of the spectrum. The waves are not really a global average of solar properties but more of a local average over the observing aperture. Also for frequency estimation if the system response to adjacent degree modes is asymmetric (which it often is), then frequency estimates will be biased by the unequal weighting of the unresolved waves. This is a serious source of systematic error at high degrees, where the frequency separation of adjacent degree modes is small and unresolved in short observation sequences.

To how high in degree are any waves global? This is a question for which we do not currently have a good answer, with a lower limit of $l = 150$ for the Antarctic data of 1987. This question is currently under active investigation by several observational groups. It is an important question to answer for designing analysis techniques. If waves are not global, we may not need to go to the expense of computing a spherical harmonic decomposition at these degrees.

Frequency estimation

A problem that has not received enough attention from solar oscillation observers is the estimation of mode frequencies from the spectra that we observe. The problem has some subtle difficulties that are not always appreciated. The most important of these is probably that the statistics of the spectrum are not Gaussian. The standard deviation of the power at a certain frequency is equal to the power at that frequency. This means that points with high signal should not have a very high weight as they are sometimes given. This problem of the statistics is sometimes ignored and a standard unweighted nonlinear leastsquares fit is made to a region of the spectrum to a line profile (e.g. Libbrecht (1986), Lazrek et al. (1988)). There should not be any systematic errors associated with this procedure as long as lines are symmetric (which is generally assumed anyway). The incorrect weighting will lead to random frequency errors that are larger than those given by an optimum technique. Also, some fitting algorithms will tend to settle on one sharp spike as being the profile in question as reported by Sorensen (1988).

An advance was made in our understanding of the properties of the spectrum with the doctoral thesis of Woodard (1984). He showed that for the case of a harmonic oscillator excited by random noise that the power spectrum will be distributed as chi-square with two degrees of freedom, or that the distribution function at a given frequency of the power as measured in a large number of independent trials would be given by

$$\gamma(p_0) = \frac{1}{p_0} e^{-p/p_0}.$$

p_0 is the mean or expectation value of the power that one would obtain from doing a large number of experiments and averaging the power spectra. p is the power in a given realization. He then showed that the observed spectra were consistent with this distribution. One consequence of this is the innate uncertainty of mode frequencies. Even if there is no instrumental noise and no solar background at the frequency of the mode in question, there is still an uncertainty in measuring the mode frequency because of the stochastic nature of the excitation process. An approximate expression for this uncertainty is

$$\sigma = \sqrt{\frac{w}{4\pi T}} \, ,$$

where w is the fwhm of the mode and T is the length of the observing sequence. This leads to nonnegligible errors. As a concrete example, consider $w = 1\mu Hz$, $T = 90 days$. The result is $\sigma = 0.1\mu Hz$. For long observing sequences, this noise source, which we might call realization noise, dominates in many situations according to the simulations of Duvall and Harvey (1986). p_0 is a function of frequency in our spectra with Lorentzian profiles representing the modes on top of an underlying smooth background.

The Lorentz profile is the one expected for a harmonic oscillator excited by random noise. To date there has not been a good observational demonstration that this is the correct profile to use. A logical consequence of the random oscillator model is that the power and phase are random from point to point in the observed spectrum (Jenkins and Watts, 1968) at frequencies separated by at least $1/T$ where T is the length of the time series. This will not be exactly true for gapped data sets. A way to simulate an observed power spectrum is then to assume a mean or expected spectrum and then at each "observed" frequency to pick a random number consistent with the above distribution. An example of a simulated realization and its associated expectation value are shown in Fig. 4 from Woodard (1984).

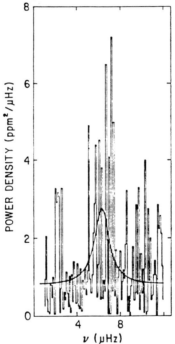

Fig. 4. A simulated mode (the histogram) along with its limit spectrum (the smooth curve). The power is independent from point to point with only a statistical relation to the limit spectrum. If one did a large number of realizations and averaged the power spectra, the limit spectrum would result.

The problem we would normally like to solve is to estimate the parameters of the underlying mean spectrum given a realization. The maximum likelihood technique described by Duvall and Harvey (1986) is a good way to do this. If we consider the mean spectrum as a function of frequency $p_0(v)$ to be a sum of Lorenztians plus background of unknown parameters, we can then construct the joint probability density that we would observe the spectrum that we did as the product of individual probabilities of the form shown above:

$$L = \prod_i \gamma(p_0(v_i)) = e^{-\sum_i [\frac{p(v_i)}{p_0(v_i)} + \ln p_0(v_i)]}$$

The method then consists of maximizing this probability distribution or likelihood function as a function of the parameters describing the spectrum. Maximizing the likelihood function is equivalent to minimizing the negative of the argument of the exponential:

$$-\ln L = \sum_i [\frac{p(v_i)}{p_0(v_i)} + \ln p_0(v_i)]$$

The above function is minimized using standard techniques.

Solar Rotation

If the sun were spherically symmetric and nonrotating, its mode frequencies would be independent of the azimuthal order m. Fortunately for helioseismologists this is not the case as it permits some of the most interesting inferences about the solar interior. The solar rotation is the largest departure from sphericity in its effect on the mode frequencies. The largest part of the frequency shift is due to the advection of the mode. That is, the mode is fixed to the rotating sun and the observer sees the pattern moving. This causes a Doppler shift of the mode's frequency which varies linearly with the azimuthal order m. The mode is advected at a rate which depends on the rotation rate in the region in which it is concentrated. By examining modes with different radial and latitudinal regions of concentration, we can learn about the rotation versus depth and latitude in the solar interior.

Observationally we express the variation of frequency in a multiplet (fixed n,l, varying m) as a Legendre polynomial series:

$$v_{nlm} - v_{nl} = L \sum_i a_i P_i(-m/L),$$

where P_i is the Legendre polynomial and $L = \sqrt{l(l+1)}$. Some observers use l instead of L in this relation with the result being small differences in the $a_i's$. The odd terms in this sum yield the direct effect of solar rotation while the even terms contain information about latitudinal variation of the mean structure and about internal magnetic fields.

The coefficient of the first term, a_1, is by far the largest coefficient, being about a factor of 20 larger than the next largest, a_3. It signifies roughly a latitudinal average of the rotation. If the rotation were only a function of depth, it would be the only nonzero coefficient. The next odd coefficient, a_3, is a measure of latitudinal differential rotation. a_5 is similarly a measure of latitudinal differential rotation but is somewhat smaller than a_3 and is not well determined as yet.

The a_i coefficients have been measured by a number of observers in the intermediate degree range of $l = 10-60$ (e.g. Libbrecht (1988) and Brown and Morrow (1987)). The results have led to a consistent picture of the internal rotation versus depth and latitude for the outer half by radius of the sun. In this picture, the rotation is constant with depth and latitude throughout the convection zone and the decrease of rotation rate between equator and pole of 20% that we see at the surface persists throughout the convection zone. Below the convection zone is a transition zone of depth at most $0.1R_{\odot}$ (it has not been resolved yet; Christensen-Dalsgaard and Schou (1988)). Below this depth the sun rotates as a solid body: no differential rotation in latitude or depth.

To see how well (or not) this model compares to the results, it is useful to consider the "forward" problem. That is, given a model of the interior rotation, what values of the a_i will we observe? Morrow (1988) has considered several interesting cases which I will show here. In all of these figures the calculations are compared with the data of Brown and Morrow (1987). The variation of the a_i is shown versus the degree l of the mode which is a proxy for depth, lower l corresponding to larger depths.

A model of the rotation in the solar interior that has received much attention is the fluid dynamic calculation of Gilman (1977), Gilman and Miller (1986), and Glatzmaier (1987). This model suggests that the rotation of the convection zone should be constant on cylinders. In Fig. 5, we show (following Morrow (1988)) a comparison of the a_1, a_3 and a_5 coefficients for a model with constant rotation on cylinders (the bottom curve). The top curve in this figure is a model with rotation constant with depth in the convection zone but having the surface latitudinal differential rotation. The model with rotation constant with depth but with normal surface latitudinal differential rotation obviously fits the data pretty well while the rotation constant on cylinders model is inadequate. It is on the basis of this figure that rotation constant on cylinders is considered to be excluded by the helioseismic data.

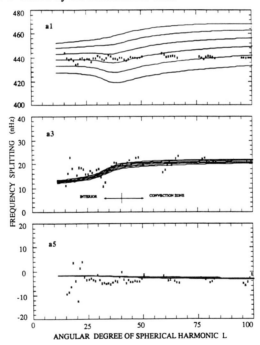

Fig. 5. The odd rotation coefficients a_i for models with rotation constant on cylinders (bottom curve for a_1), rotation constant with radius throughout the convection zone (top curve for a_1), and intermediate models from Morrow (1988).

The approximate constancy with the degree of the mode l of the a_1 coefficient suggests that the rotation averaged over latitude is approximately constant. This point is brought home clearly in Fig. 6 from Morrow (1988) which shows several models compared that all have differential rotation only in latitude in the convection zone and rotation constant in latitude and depth below this level but at a rate that varies from model to model. The center rate is near the value derived for the correct latitudinal averaging to get a constant a_1.

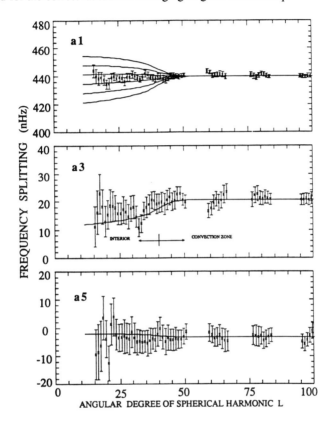

Fig. 6. The odd rotation coefficients a_i for models in which the rotation in the convection zone is independent of radius and has the same latitudinal structure as the surface but the interior constant rotation rate is varied for the different models. The rate is varied by 11% of the mean value between the bottom and top curves for a_1.

The depth at which the rotation switches from surface latitudinal differential rotation to constant rotation in latitude is investigated in Fig. 7 again from Morrow (1988). a_3 is seen to be the sensitive parameter in this case. As the depth of the rotation transition is varied over a total range of 0.25 $R_\odot$, a_3 varies by an amount that is distinguishable by the data. The top curve corresponds to a deeper region of latitudinal differential rotation.

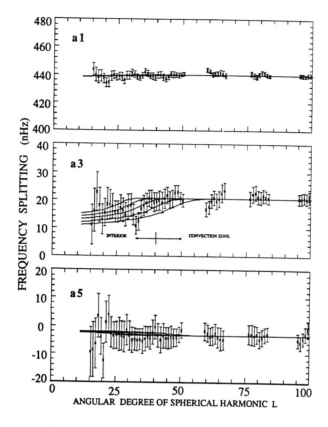

Fig. 7. Rotation coefficients for models in which the depth of the convection zone is varied. The bottom curve for a_3 is for the shallowest convection zone.

These curves show that we can make some pretty strong statements about the solar interior rotation, at least over the outer half of the sun by radius. The convection zone does not have rotation constant on cylinders but the rotation looks much as it is at the surface. The rotation immediately below the differentially rotating layer is constant with latitude at a rate that corresponds to about 30 degrees latitude. The actual depth at which the transition occurs between differential and rigid rotation is somewhat uncertain because of slight differences in the results of different observers. The current results should provide significant input to workers studying the solar dynamo.

References

"Study of Solar Structure Based om P-mode Helioseismology", J. Christensen-Dalsgaard, in *Seismology of the Sun and Sun-like Stars*, ESA SP-286, 431-450, 1988.

"On the Measurement of Solar Rotation Using High-degree P-mode Oscillations", M. F. Woodard and K. G. Libbrecht, in *Seismology of the Sun and Sun-like Stars*, ESA SP-286, 67-71, 1988.

"Theory of Solar Oscillations", D. O. Gough, in *Future Missions in Solar, Heliospheric, and Space Plasma Physics*, ed. E. Rolfe, ESA SP-235, European Space Agency, Noordwijk, 1985.

"Inverse Problem: Acoustic Potential vs. Acoustic Length", H. Shibahashi, in *Advances in Helio- and Asteroseismology*, ed. J. Christensen-Dalsgaard and S. Frandsen, D. Reidel, 133-136, 1988.

"On Taking Observers Seriously", D. O. Gough, in *Seismology of the Sun and Sun-like Stars*, ESA SP-286, 679-683, 1988.

"EBK Quantization of Stellar Waves", D. O. Gough, in *Hydrodynamic and Magnetohydrodynamic Problems in the Sun and Stars*, ed. Y. Osaki, University of Tokyo, Tokyo, 117-143, 1986.

"A New Picture for the Internal Rotation of the Sun", C. A. Morrow, Ph. D. thesis, Univ. of Colorado, Boulder; available as NCAR Cooperative Thesis No. 116, 1988.

"Solar Internal Rotation from Helioseismology", J. W. Harvey, in *Seismology of the Sun and Sun-like Stars*, ESA SP-286, 55-66, 1988.

"The Excitation and Damping of Solar P-modes", K. G. Libbrecht, in *Seismology of the Sun and Sun-like Stars*, ESA SP-286, 3-10, 1988.

"Long Period Solar Oscillations", H. B. van der Raay, in *Seismology of the Sun and Sun-like Stars*, ESA SP-286, 339-352, 1988.

"Is There an Unusual Solar Core?", K. G. Libbrecht, *Nature* **319**, 753, 1986.

"Modelling the Solar Oscillation Time Series as a Randomly Excited Oscillator", M. Lazrek, Ph. Delache, and E. Fossat, in *Seismology of the Sun and Sun-like Stars*, ESA SP-286, 673-676, 1988.

"Measurements of Oscillation Parameters from Synthetic Time Series", J. M. Sorensen, in *Seismology of the Sun and Sun-like Stars*, ESA SP-286, 47-54, 1988.

"Short Period Oscillations in the Total Solar Irradiance", M. F. Woodard, Ph. D. thesis, University of California, San Diego, 1984.

"Solar Doppler Shifts: Sources of Continuous Spectra," T. L. Duvall, Jr. and J. W. Harvey, in *Seismology of the Sun and the Distant Stars*, (ed. D. Gough), NATO ASI Ser., Reidel, Dordrecht, p.105, 1986.

Spectral Analysis and its Applications, G. M. Jenkins and D. G. Watts, Holden-Day, San

Francisco, 1968.

"Solar P-mode Frequency Splittings", K. G. Libbrecht, in *Seismology of the Sun and Sun-like Stars*, ESA SP-286, 131-136, 1988.

"Depth and Latitude Dependence of Solar Rotation", T. M. Brown and C. A. Morrow, *Ap. J.* **314**, L21-L26, 1987.

"Differential Rotation in the Solar Interior", J. Christensen-Dalsgaard and J. Schou, in *Seismology of the Sun and Sun-like Stars*, ESA SP-286, 149-156, 1988.

"Nonlinear Dynamics of Boussinesq Convection in a Deep Rotating Spherical Shell", P. A. Gilman, *Geophys. Astrophys. Fluid Dynamics,* **8**, 93, 1977.

"Nonlinear Convection of a Compressible Fluid in a Rotating Spherical Shell", P. A. Gilman and J. Miller, *Ap. J. Supp.,* **61**, 211, 1986.

"A Review of What Numerical Simulations Tell Us about the Internal Rotation of the Sun", G. A. Glatzmaier, in *The Solar Internal Angular Velocity: Theory, Observations, and Relationship to Solar Magnetic Fields*, eds. B. R. Durney and S. Sofia, Reidel, Dordrecht, 263, 1987.

"Latitude and Depth Variation of Solar Rotation," T. L. Duvall, Jr., J. W. Harvey and M. A. Pomerantz, in *The Internal Solar Angular Velocity*, ed. B. R. Durney and S. Sofia, pub. D. Reidel, Dordrecht, 19, 1987.

"Helioseismology from the South Pole: Comparison of 1987 and 1981 Results", S. M. Jefferies, M. A. Pomerantz, T. L. Duvall Jr., J. W. Harvey and D. B. Jaksha, in *Seismology of the Sun and Sun-like Stars*, ESA SP-286, 279-284, 1988.

NETWORKS FOR HELIOSEISMIC OBSERVATIONS

Frank Hill
National Solar Observatory
National Optical Astronomy Observatories[*]
Tucson, Arizona 85726 USA

ABSTRACT - Helioseismology from a single ground-based observatory is severely compromised by the diurnal rising and setting of the Sun. This causes sidelobes to appear in the helioseismic power spectrum at multiples of $\pm$ 11.57 µHz from each solar line, contaminating the spectrum and rendering mode identification and frequency measurement extremely difficult. The difficulty can be overcome in three ways — observing from a fully sunlit orbit in space, observing from the Polar regions, or observing with a network of stations placed around the Earth. This paper discusses the networks that are either currently in operation or being planned. These include the Global Oscillation Network Group (GONG) project, the Birmingham network, the IRIS network of the University of Nice, and the SCLERA network of the University of Arizona. The scientific objectives and instrumentation of these networks are briefly described. Theoretical predictions for network performance are compared with actual results. The problem of merging simultaneous data from multiple instruments is discussed, as well as the relationship of the networks with the helioseismology experiments on the SOHO space mission.

1. The Need For Continuous Data — The Scientific Issues

Helioseismology, the study of solar oscillations, is a powerful tool for probing the solar interior. The oscillations are global and permeate the entire Sun, thus their properties are sensitive to the Sun's internal structure. Careful measurements of the frequencies, amplitudes, and lifetimes of the oscillations provide information on the internal sound speed, temperature, rotation rate, composition, opacity, convective flow patterns, and magnetic field. Comparison of the observed frequencies with theoretical predictions derived from solar models helps to test potential explanations for the observed deficiency of solar neutrinos such as the cosmion or WIMP model (*e.g.* Faulkner, Gough and Vahia 1986; Bahcall and Ulrich 1988; see also several papers in these proceedings and contributed posters in the *Solar Physics* volume).

The most useful measurements are the frequencies of the oscillations as a function of their spherical harmonic "quantum numbers" — ℓ, the spherical harmonic degree, m, the azimuthal degree, and n, the radial order. It is relatively simple to compute these frequencies from theoretical solar models, and the observed frequencies can also be directly combined via inverse techniques to infer the depth dependence of physical conditions (see Gough 1985 for a review of inverse techniques). Frequencies are also the most easily measurable property of the oscillations. The temporal length of the sequence of observations, T, and the precision with which the frequencies ν can be determined, $\Delta\nu$, are simply related: $\Delta\nu = 1/T$. Thus longer time strings are always the goal of the observational helioseismologist. Unfortunately for him or her, the Earth rotates and the Sun sets every day, placing periodic gaps in the sequence of

---

[*]Operated by the Association of Universities for Research in Astronomy, Inc. under contract with the National Science Foundation.

G. Berthomieu and M. Cribier (eds.), Inside the Sun, 265–278.

observations.

This is especially unfortunate because observational time strings of many days, months, or even years are essential for helioseismic discrimination between different models of internal solar structure. As an example, the solar surface rotation period varies from about 28 days at the equator to 32 days at the pole. The rotation changes the frequency of the modes by an amount proportional to their m-value (the number of wavelengths around the equator of the coordinate system); the constant of proportionality is approximately the inverse of the rotation period. In order to resolve these shifts, one must have a frequency resolution of twice the shift, implying a minimum observational duration of two rotation periods or about 60 days. Even longer time strings are needed to investigate solar cycle changes that may occur over an 11-year period.

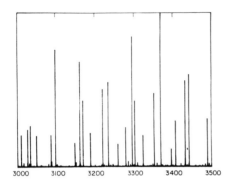 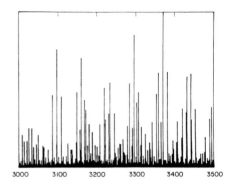

FIGURE 1. Left: The 3000 to 3500 µHz band of a simulated solar spectrum as might be obtained from continuous observations covering one year. Right: The same portion of the simulated spectrum as might be obtained at a single mid-latitude site experiencing the day-night cycle and typical weather patterns.

Fourier analysis shows that when two functions are multiplied together in the temporal domain, their transforms are convolved in the frequency domain. For helioseismology, the two functions are the time series of solar data and a time series of ones or zeroes representing the times of data (ones) or no data (zeroes) called the window function. Thus, in the frequency domain we obtain the real solar spectrum convolved with the spectrum of the window. The window function has a regular diurnal periodic structure, thus its spectrum consists of a strong peak at a frequency of 1/day, or 11.57 µHz, along with its harmonics at integral multiples of the fundamental 1/day frequency. These peaks are called diurnal sidelobes; the fundamental diurnal sidelobe typically has an amplitude that is 60% of the amplitude of the zero frequency component of the window spectrum. Since the solar spectrum and the window spectrum are both nearly sets of discrete delta functions, the convolution of the window spectrum with the solar spectrum results in a replica of the window spectrum appearing symmetrically around every solar spectral line, as shown in Figure 1. As can be seen in Figure 1, there are many instances where a solar line overlies a diurnal sidelobe, greatly compromising mode identification and frequency measurements. An example of a window function is shown in Figure 2, where the window as actually observed at Tucson is displayed on a grid representing one hour increments, with black representing times of no observations due to either weather or nighttime. Nighttime is clearly visible as the prominent black swaths across the diagram. The first 60 µHz of the power spectrum of the window in Figure 2 is shown in Figure 3. The fundamental and the first four harmonics of the diurnal sidelobes are clearly visible, as is a noise background.

Several data processing schemes to eliminate the effects of the diurnal sidelobes have been investigated. Radio astronomers have long faced the problem of unwanted sidelobes, and have developed an iterative peak subtraction algorithm known as CLEAN. It has been used to search for solar gravity modes (Delache and Scherrer 1983; Scherrer 1984), and to remove the sidelobes from p-mode spectra

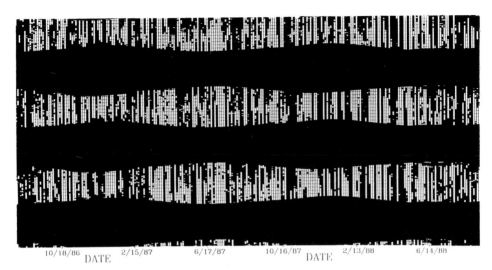

FIGURE 2. An example of an actual window function observed at Tucson, Arizona. Each box in the grid represents one hour of observations, with the hours arranged in columns of 72 hour or 3 day intervals. Times of no sunlight are indicated by black areas. Night and its seasonal variation is clearly visible as the three prominent black swaths across the figure.

(Duvall and Harvey 1984; Henning and Scherrer 1986). In the context of helioseismology, the CLEAN algorithm can misidentify a sidelobe as a solar frequency. Another method rearranges the data to fill in the daily gaps, but this reduces the frequency resolution of the spectrum (Kuhn 1982, 1984). Linear iterative deconvolution techniques tend to increase the noise in the spectra, while nonlinear spectral deconvolution (Connes and Connes 1984) assumes that the actual signal is band limited and that power outside the band comes from the gaps. This assumption is not met by the solar oscillations (Scherrer 1986). Maximum entropy has been used to extrapolate data into gaps (Fahlman and Ulrych 1982), and this method has been studied in the context of helioseismology (Brown and Christensen-Dalsgaard 1985, unpublished). They found that solar spectra can be reliably recovered if the filling factor, or duty cycle, is greater than 0.8, and the signal-to-noise ratio is larger than 100. Finally, Bayesian probability techniques have been used to combine all available *a priori* information along with the observed power spectra to extract and indentify the modes (Morrow and Brown 1988; Brown 1988). This method has also proven to be capable of misidentifying modes. One must draw the conclusion that, so far, no technique is available that will allow the reliable elimination of the diurnal sidelobes from spectra obtained at a single ground-based site. We must look to other observing strategies to eliminate the daily gaps as much as possible during the observing process itself.

2. Strategies For Obtaining Continuous Data

There are three basic strategies that enable helioseismologists to obtain nearly continuous data. These strategies are observing at the polar regions, observing from space, or observing from a network of ground-based stations. The strategies are summarized in Table 1. All of the strategies face environmental and logistical challenges, and produce large data flows if they involve observations with moderate to high spatial resolution.

The Sun stays above the horizon for six months at a time at the polar regions, thus observations from these regions are capable of providing more than the 14 continuous hours available from lower

268

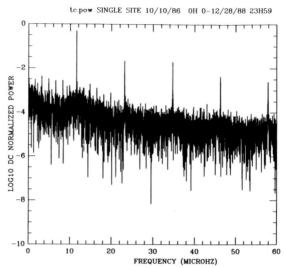

FIGURE 3. The first 60 μHz of the power spectrum of the window function in Figure 2. The power has been normalized by the power in the zero-frequency or DC component and the logarithm taken. The fundamental and first four harmonics of the diurnal sidelobes are clearly visible. It is the reduction of these sidelobes that is the goal of the network concept.

latitude single sites. In practice, the presence of a continental land mass at the South pole and the associated infrastructure makes it the pole of choice for observational astronomy. Several helioseismology experiments have been carried out at the South pole, including non-imaging observations (Grec, Fossat and Pomerantz 1983); diameter measurements (Stebbins and Wilson 1983), and imaged observations (Jefferies *et al.* 1988b). There are three advantages to a South pole experiment. First, since only a single instrument is involved, there is no need to merge simultaneous data sets. Second, the cost of the experiment is substantially lower than that of the other two strategies. Finally, it is possible to recover from instrumental failure, although it is rather more difficult than at sites at lower latitudes. The disadvantages are twofold: First, the weather patterns at the South pole place a limit on the uninterrupted duration of the observations of about seven days, far short of the sixty needed to cleanly separate modes with adjacent values of m. Secondly, the presence of the Earth's atmosphere degrades the spatial response of the solar power spectra, especially at higher values of ℓ (F. Hill *et al.* 1984). The logistical challenge consists of the remoteness of the site and the harshness of the environment.

A space platform can be placed in a fully sunlit orbit about the Earth or around a Lagrangian point. Various studies have been made of space-borne helioseismic observations (*e.g.* Newkirk 1980, Noyes and Rhodes 1984), and the European Space Agency will be launching the SOHO mission in the mid 1990s (Domingo 1988; see also the paper by Bonnet in these *Proceedings*). There are two advantages to the space-borne observations: First, there is naturally no degradation of the observations from atmospheric seeing. Secondly, no data merging is required for a single instrument. Three disadvantages exist: First, there is little chance for recovery from an instrumental failure, especially for a spacecraft in orbit around the L_1 point. Second, telemetry bandwidth limitations render it impossible to transmit the entire data set from a high-resolution imaging experiment, thus requiring on-board data processing and forcing a selection of data parameters. Telemetry ground stations are typically three-site networks which are vulnerable to single-site instrumental failures. Furthermore, they may not provide 24-hour coverage in all seasons. Finally, a space experiment is monetarily the most expensive method of obtaining nearly continuous data. The lifetime and hence the frequency resolution of a space experiment is limited by the expendable supplies such as propellant, and the logistical challenges consist of placing the vehicle in orbit, and coping

Table 1
A Comparison of Observing Strategies for
Obtaining Continuous Helioseismic Data

Strategy	Advantages	Disadvantages
South Pole Sun always above horizon during Austral summer	• No data merging required • Inexpensive • Failure recovery possible	• Limited frequency resolution • Terrestrial atmospheric degradation
Space platform Can be placed in fully sunlit orbit (*e.g.* L_1)	• No atmospheric degradation • No data merging required	• No failure recovery • Telemetry limitations • Expensive
Network Longitude coverage and redundancy reduces diurnal cycle and impact of weather and failures	• Unlimited frequency resolution • Failure recovery possible	• Requires data merging • Terrestrial atmospheric degradation

with the radiation and thermal environment.

A network of identical observing stations redundantly distributed in longitude reduces the diurnal sidelobes as well as the impact of weather and instrumental failures. The optimal number of sites for a network is six, distributed in longitude so that at least two stations are potentially observing at all times. Models predict that such a network should provide observations with a duty cycle of about 93% (F. Hill and Newkirk 1985); as discussed in section 4, this agrees well with actual results (*e.g.* F. Hill and the GONG Site Survey Team 1988). With a duty cycle this high, and assuming that the instrument is well-designed and built with a signal-to-noise ratio greater than 100, then the remaining gaps should be easily filled using the maximum entropy method. The advantages of a network are twofold: First, unlimited frequency resolution is available in principle. Secondly, it is possible to recover from instrumental failures. There are two disadvantages: The existence of multiple sites observing simultaneously requires that the data be merged, a problem for which there is currently little previous experience. Secondly, the observations are degraded by seeing and transparency gradients in the Earth's atmosphere. The logistical challenge consists of manufacturing a number of identical instruments and engineering them to withstand almost the entire range of climates found on the Earth. In addition, the international aspect provides the opportunity to sample the customs of many nations. The expense of an imaging network is moderate, being substantially less than a space experiment, but more than a South pole experiment.

3. Summary Of Current And Future Networks

Four major networks are either currently in partial operation, or being planned. None of these networks have yet reached full operational status of six or more sites. In addition, there have been a number of two-site networks with limited lifetimes.

Historically, the original helioseismology network is the Birmingham Solar Seismology Network, with headquarters at the University of Birmingham in the United Kingdom (Claverie *et al.* 1984; Aindow *et al.* 1988; Elsworth *et al.* 1988b). It was proposed in 1975, and funded for a full six stations in 1987. The first station was installed at Izaña, Tenerife in the Spanish Canary Islands off the western coast of Africa in 1975. A second station was installed in 1981 at Haleakala, Hawaii, and a third automatic station began operations in 1985 at Carnarvon, Western Australia. A fourth station will be installed in 1989 at Sutherland, South Africa. Two additional sites will be selected and installed at a future time. The Birmingham Network instrumentation consists of a non-imaging potassium optical resonance spectrometer. Since the experiment is non-imaging, the observations are sensitive only to oscillations with $0 \leq \ell \leq 3$. Scientific objectives of the experiment have included the measurement of line widths (Isaak 1986; Elsworth *et al.* 1988a); solar internal structure (Claverie *et al.* 1979); rotational splitting (Jefferies *et al.* 1988a); g modes (Isaak *et al.* 1984; van der Raay 1988; Garcia, Pallé and Roca Cortés 1988); solar cycle frequency changes (Isaak *et al.* 1988; Jiménez *et al.* 1988a, Pallé, Régulo and Roca Cortés 1989); and the identification of modes in the low-ℓ portion of the solar spectrum (Pallé *et al.* 1986; Anguera *et al.* 1989).

A second non-imaging network is being developed and administered at the Université de Nice in France. This network is known as IRIS (Installation d'un Réseau International de Sismologie Solaire or International Research on the Interior of the Sun) (Fossat 1988). This network was proposed in 1983 and funded for seven stations in 1984. A prototype instrument was operated at La Silla, Chile from May 1986 to the Spring of 1987. Three instruments are currently operating at Stanford, California (operations began in 1987), Kumbel, Uzbekistan, U. S. S. R. (1988), and l'Oukaimeden, Morocco (1988). Two more instruments are scheduled to come on-line in 1989 at La Silla, and Izaña. The final two instruments will be installed in 1990 at Learmonth, Western Australia, and Haleakala. The operation of the IRIS network is scheduled to continue for one full solar cycle of 11 years, ending in 2001. The IRIS instrument is similar to the Birmingham device, but uses sodium instead of potassium (Grec, Fossat and Vernin 1976). The full-disk observations are sensitive to oscillations with $0 \leq \ell \leq 3$. The Nice group has been active with a single instrument for some time. Scientific goals of the IRIS project include solar cycle frequency changes (Fossat *et al.* 1987; Gelly, Fossat and Grec 1988); g modes (Fossat *et al.* 1988; Provost and Berthomieu 1988); frequency and amplitude measurements (Grec, Fossat and Pomerantz 1983); and asteroseismology (Gelly, Grec and Fossat 1986; Gelly *et al.* 1988b; Schmider *et al.* 1988). The Nice group has also observed at the South pole (Gelly *et al.* 1988a).

Measurements of the solar diameter have also been used to search for solar oscillations (H. Hill 1984). Although the results of such studies are controversial (*e.g.* Yerle 1988), a network of instruments to obtain the measurements is being planned. The SCLERA (Santa Catalina Laboratory for Experimental Relativity and Astrometry) network is being developed at the University of Arizona in Tucson. Two sites at Tucson, Arizona and Yunnan Observatory, Kunming, China, have been selected, and negotiations are underway for a third site in the U.S.S.R. Other sites will be determined at a future time. The instrumentation will comprise a combination of diameter measurements and broad-band photometry of the central portion of the disk. The scientific objectives of the group have included the gravitational quadrupole moment (H. Hill, Bos and Goode 1982); the 160-minute oscillation (H. Hill, Tash and Padin 1986); g modes (H. Hill, Gao and Rosenwald 1988); internal rotation (H. Hill *et al.* 1986); and the neutrino problem (H. Hill 1986).

The final major network is perhaps the most ambitious. The Global Oscillation Network Group (GONG) project is planning to place six imaging instruments around the world in 1992 (Harvey, Kennedy and Leibacher 1987; Harvey *et al.* 1988a). The network was proposed in 1984 and funded in 1985. Guided by models (F. Hill and Newkirk 1985), a site survey is underway to select the six sites from fourteen candidates (Fischer *et al.* 1986; F. Hill and the GONG Site Survey Team 1988). The candidate sites are Cerro Tololo and Las Campanas, Chile; Izaña; l'Oukaimeden; Riyadh, Saudi Arabia; Udaipur, India; Urumqi, China; Learmonth; Haleakala, Mauna Kea and Mauna Loa, Hawaii; and Big Bear, Yuma and Tucson, U.S.A. The site survey has been underway since 1985, and site selection is currently planned for 1991. Deployment of the network is now scheduled for 1992, with observations starting in 1993 and ending in 1996. The instrument is under development and comprises a Fourier tachometer design using a

Michelson interferometer and a Lyöt prefilter (Harvey and the GONG Instrument Development Team 1988). The detector nominally will be a 256×256 CCD, and the instrument will feature laser calibration, automatic operation, and Exabyte cartridge data storage. The amount of data produced by the experiment will total approximately three terabytes, necessitating careful planning of the data reduction system (Kennedy and Pintar 1988; Pintar and the GONG Data Reduction Team 1988). The project is open to all helioseismologists, and so its scientific objectives cover the entire range of helioseismology. With the current detector configuration, oscillations with $0 \leq \ell \leq 150$ will be observable. Scientific interests of the project staff include the internal solar rotation (Leibacher 1984; Duvall et al. 1984; Duvall, Harvey and Pomerantz 1986; Harvey 1988; F. Hill 1987; F. Hill et al. 1988); solar cycle frequency changes (Jefferies et al. 1988b); and large-scale flow mapping (F. Hill 1988, 1989, see also the contributed posters from this conference).

In addition to these four major helioseismology networks, at least three limited-duration two-site networks have either existed or are planned. The most successful has been the ESTEC-sponsored Solar Luminosity Oscillation Telescope (SLOT) project, which installed an instrument at Izãna in 1984, and another at Observatorio de San Pedro Mártir, Baja California, Mexico in 1987 (Jiménez et al. 1988b). This two-site network used a four-channel photometer to observe the oscillations in intensity and correlate them with Doppler measurements. This provides information on the adiabatic behavior of the solar atmosphere. Observations with this network have recently ended. Another two-site network has recently begun to obtain non-imaged solar observations with a magneto-optical filter (Cacciani et al. 1988). The two sites are at Mt. Wilson, California, U.S.A., and at the University "La Sapienza", Rome, Italy. This network began coordinated observations in September of 1988. Finally, a two-site network of imaged observations with a video-magnetograph was attempted by workers at the California Institute of Technology. The two sites were at Big Bear, California, and at Tel Aviv, Israel. Instrumental differences greatly complicated the data merging problem and the attempt was abandoned.

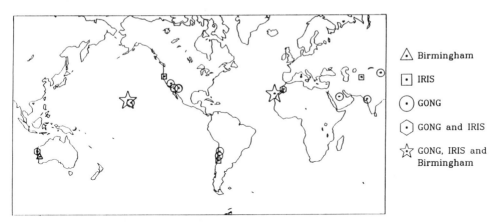

FIGURE 4. A world map showing the actual, planned, or candidate sites for the Birmingham, IRIS, and GONG networks. Note the considerable overlap of the networks, especially in Hawaii and the Canary Islands. This overlap is imposed by the distribution of land on the Earth.

4. Network Design And Performance — Predicted And Actual

The main advantage of a network over a single site consists of the reduction of the day/night cycle and the impact of both weather and instrumental failures. In order to maximize this advantage, the location of the sites must be arranged such that at least two and preferably more than two stations are able to potentially observe at any time. If this condition is not met, then the loss of the single station covering a given band of Universal Time will result in the immediate reappearance of the diurnal sidelobes in the window power spectrum. The multiple-coverage requirement constrains the choice of locations of the sites, but an even stronger constraint is imposed by the distribution of land on the Earth. The multiple-coverage requirement, along with an average observing day at a single site of eight hours, results in a pattern of six sites, spaced as evenly as possible around the globe. In order to take advantage of the seasonal weather patterns, the six sites should ideally alternate between Northern and Southern hemisphere locations. Perusal of a globe shows that there is virtually no reasonable site in the mid-Atlantic Ocean, forcing one location to be in the mid-Pacific, namely, Hawaii. Once this is realized, then all other longitude bands are set by geography: One must have sites in Australia, mid-Asia, Western Africa, South America, and the Western Coast of North America. These constraints are demonstrated by Figure 4, which shows the locations of the chosen or candidate sites for three of the major networks (Birmingham, IRIS, and GONG) on a map of the world. There is considerable overlap in the choice of sites. Further constraints are imposed scientifically and logistically. Existing, developed astronomical observatories are clearly advantageous, especially if they are solar and have established helioseismology programs. The efficiency of communication and shipping channels are also of concern.

A model of network performance was developed by F. Hill and Newkirk (1985). This model used a simple four-parameter probabilistic description of the temporal distribution of cloud cover at a given site. The most important parameters of the model are the fraction of cloudy time at a site in the summer and winter. For the model, these parameters were estimated from small-scale climatological maps. Such maps cannot reflect micro-meteorological conditions at individual sites that may alter the parameter estimates. The results of the model predicted that a well-chosen six-site network would achieve a duty cycle of about 93.5%.

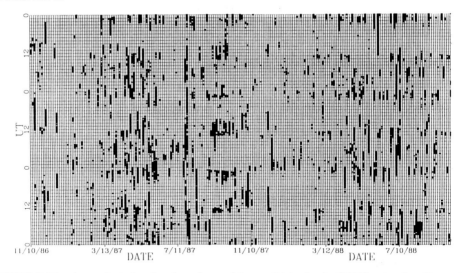

FIGURE 5. The observed window function of one of the candidate six-site GONG networks. Comparison of this window with that in Figure 2 shows the effectiveness of the network in eliminating the diurnal cycle.

The GONG project has been running a network of normal-incidence pyrheliometers since 1985 to measure the parameters and test the model, and to establish communication, shipping, and working relationships with the candidate sites (F. Hill et al. 1988a). Figure 5 shows the actual window function obtained by one of several possible six-site networks in the site survey. Comparison with Figure 2 shows that the diurnal cycle has been nearly completely obliterated. The duty cycle obtained by this network is 93.91%, in good agreement with the model prediction. Figure 6 displays the power spectrum of the network window function in Figure 5, and shows that the diurnal sidelobes have been greatly attenuated. Comparison of Figure 6 with Figure 3 shows that the amplitude of the fundamental daily sidelobe has been reduced by a factor of 400, while the combined amplitudes of the first five sidelobes has dropped by a factor of 200, and the overall background noise power in the first 60 μHz is reduced by a factor of 50. The longest segment of uninterrupted observations so far achieved by the GONG site survey network is about 418 hours. Thus, the network has fulfilled its design goals, and is capable of producing helioseismic data that is virtually free of contamination by the diurnal sidelobes.

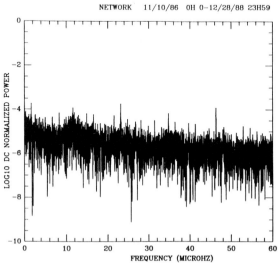

FIGURE 6. The first 60 μHz of the power spectrum of the window function in Figure 5. Comparison of this spectrum with that in Figure 3 shows that the amplitude of the fundamental sidelobe has been reduced by a factor of 400, the combined amplitudes of the fundamental and first four harmonic sidelobes has been reduced by a factor of 200, and the overall noise background has been reduced by a factor of 50.

Figure 7 presents the expected duty cycle of a well-chosen network as a function of the number of sites in the network. The plot shows that the duty cycle rises with the number of sites in the network. It also shows that the dispersion in duty cycle between different site choices decreases with the number of sites. This decrease is also borne out by the results of the GONG site survey. Twelve possible well-chosen networks have so far been examined; they all have duty cycles between 93 and 94%. This is another demonstration of the ability of the network strategy to reduce the effects of weather and instrumental downtime. Figure 7 also plots the actual duty cycles obtained by the GONG site survey for six sites, and the actual duty cycles obtained by the Birmingham network for a few realizations of two and three site networks. The Birmingham results are somewhat lower for the two-site case, and substantially lower for the three-site case. However, there have been logistical problems with the third site that have prevented this version of the Birmingham network from realizing its true potential. Further testing of the model can be done by using the GONG site survey data to form networks with varying numbers of sites; this is in progress.

5. The Data Merging Problem

A well-chosen network attempts to have two or more sites observing at all times. The GONG site survey shows that this goal is achieved typically 65% of the time. Thus, at every temporal step in the observations, two to four simultaneous images will frequently be available. It is clearly desirable to combine

274

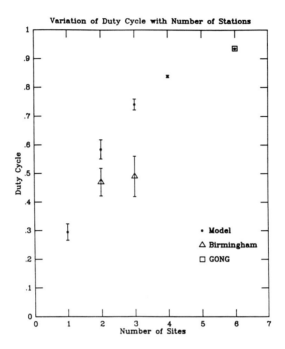

FIGURE 7. The expected duty cycle of a well-chosen network as a function of the number of stations in the network. The theoretical predictions are compared with the actual results from the Birmingham network and the GONG site survey.

these simultaneous images, but a strategy for doing so does not yet exist because of a lack of previous experience. The problem has been faced by the Birmingham network, but it is much simpler to merge a single non-imaged pixel, rather than a set of images. There are many possible strategies, and the final choice will most likely be a combination of approaches matched to the scientific questions being addressed.

The first step is most likely to develop a set of quality assessment measures. Four basic classes of measures can be readily discussed: The first class of global measures concerns the entire image — the rms velocity, intensity, instrumental characteristics, shape of the limb, seeing, scattering, site location, season, zenith angle and transparency variations. The transparency variations are especially important for non-imaged observations because they can easily cause apparent velocity gradients across the entire image in conjunction with the solar rotation gradient (Belmonte et al. 1988). The second class of quality assessment parameters apply to a single pixel — detector characteristics such as hot pixels, differential stretching distortions, and pixel-sized transparency gradients. The third class of parameters apply to temporal sequences of images, and basically comprise deviations from running means. The fourth class concerns the power spectra of the oscillations, and comprise the amplitudes of the spherical harmonics in various bands of ν and ℓ.

Once these quality measures are determined, it seems intuitively clear that the general answer to the merging problem is a weighted average, but again there are several choices. At what point in the data reduction should the merging be done — before or after the spherical harmonic transform? For instance, seeing effects such as blurring are ℓ-dependent. How should the weights be chosen — declare one ℓ-range, pixel, or image better than the others, thereby setting the weights to either 1 or 0 only; or perform a general weighted average of all available images with the weights continuously variable between 1 and 0? In the latter case, one then has the problem of calculating the weights.

There are issues in the temporal domain as well. What are the merging effects on the temporal power spectra, *e.g.* since there are six sites in the network, is there any evidence of a four-hour periodicity? There is no strong evidence of enhanced power at this period in the GONG site survey data, but it may be present at a level high enough to affect subtle searches for long-period oscillations such as *g* modes. Further, discontinuities in the temporal sequence will introduce spurious high frequencies into the spectra.

There are clearly many ideas to explore. Development of the techniques is going on in three directions. The experience of the Birmingham and IRIS networks will provide insight into the non-imaged problem. The GONG site survey data has been used for a preliminary look at the temporal domain effects. The most valuable route for the GONG project involves the use of artificial solar velocity data that has been created and degraded by a set of programs, rather than the Sun, the Earth's atmosphere, and an observing instrument (Hathaway 1988; Global Oscillation Network Group 1987). The advantage of this approach is that the input answer is available to assess the quality of the techniques.

The question of efficient implementation must also be addressed. The GONG data flow will be substantial, and the merging of the data must be done as efficiently as possible. It is thus highly desirable that the process be done automatically, but the nature of the problem may require the application of human judgement, a process that is relatively slow. Artificial intelligence, expert systems, and parallel processing may have to be incorporated into this phase of the GONG data reduction.

6. Networks And Space Projects

The combination of both networks and space projects for helioseismology will result in more useful knowledge than either approach alone. The techniques are complementary in some respects, and similar in others. Both aspects are useful: similar observations enable calibration and validation of results, while dissimilar observations can be used to extend the range of data.

The SOHO mission will carry both imaging and non-imaging instrumentation. Non-imaging helioseismology data will be provided by the Global Oscillations at Low Frequencies (GOLF) (Damé 1988) and Variability of solar IRradiance and Gravity Oscillations (VIRGO) (Fröhlich *et al.* 1988a) instruments on board SOHO, and have already been obtained by the SMM/ACRIM experiment (Woodard and Hudson 1983) and the IPHIR experiment (Fröhlich *et al.* 1988b). These non-imaging space experiments provide the opportunity for comparison with results from the Birmingham and IRIS networks.

The imaging experiment on SOHO is known as the Solar Oscillations Imager (SOI) (Scherrer *et al.* 1988), and comprises another variation on the Fourier tachometer technique but using two Michelson interferometers instead of one (Hoeksema *et al.* 1988). The SOI experiment most closely corresponds to the GONG network. The SOI data with a 1024×1024 detector will have much higher spatial resolution than the GONG data, and will also be free of atmospheric distortion. This will allow the measurement of the impact of terrestrial atmospheric effects on the GONG data. Care will have to be taken with the telemetry of the SOI data. The bandwidth limits do not allow complete recovery of the entire SOI data stream, forcing the development of on-board processing and data selection. In addition, the use of the Deep Space Network for the telemetry means that the SOI experiment is also a network, but comprising only three sites. While weather is not as great a factor for radio communication as for optical astronomy, the DSN is vulnerable to instrumental and scheduling problems. In addition, the DSN has an uncovered gap during the Northern Hemisphere winter season. The current scheduling of the two projects (SOHO launch in 1995 with a six-year lifetime; GONG observations covering 1993 to 1996) will provide spatially-resolved helioseismology data over a large fraction of the solar cycle. The data for the two experiments will be analyzed and archived together. Thus, the ground-based networks and the space-borne experiments form a complementary and synergetic approach to helioseismology.

7. Summary And Conclusion

The development of ground-based networks is a fundamental aspect of current work in helioseismology. The networks are an efficient and practical method of greatly reducing the impact of the diurnal rising and setting of the Sun on the power spectra of solar oscillations. The use of networks, in concert with space experiments, holds the promise of greatly increasing our understanding of the inside of the Sun through order of magnitude improvements in the quality of solar oscillation spectra. While this increase is of importance for solar physics, it also bears on aspects of both particle physics and cosmology through the solar neutrino problem. Thus, investment in helioseismology networks will pay handsome scientific dividends.

Acknowledgements

This paper has benefited from critical readings by John Leibacher, Jack Harvey, and Jim Kennedy. Information on the networks was contributed by Eric Fossat, George Isaak, and Henry Hill.

References

Many of the references in this list appeared in a few conference proceedings. For compactness, these proceedings will be identified by the location of the conference in the list. The full references are:

Aarhus: *Advances in Helio- and Asteroseismology, IAU Symp. 123*, ed. J. Christensen-Dalsgaard and S. Frandsen, Dordrecht: Reidel, 1988.

Cambridge: *Seismology of the Sun and the Distant Stars*, ed. D. O. Gough, Dordrecht: Reidel, 1986.

Snowmass: *Solar Seismology From Space*, ed. R. K. Ulrich, J. Harvey, E. J. Rhodes, Jr., and J. Toomre, NASA JPL 84-84, 1984.

Tenerife: *Seismology of the Sun and the Sun-Like Stars*, ed. E. J. Rolfe, ESA SP-286, 1988.

Aindow, A., Elsworth, Y. P., Isaak, G. R., McLeod, C. P., New, R., and van der Raay, H. B. 1988, Tenerife, p. 157.

Anguera Gubau, M., Pallé, P. L., Pérez Hernandez, F., and Roca Cortés, T. 1989, *Contributed Papers to Inside the Sun, submitted to Solar Phys.*

Bahcall, J. N., and Ulrich, R. K. 1988, *Rev. Mod. Phys.* **60**, 297.

Belmonte, J. A., Elsworth, Y., Isaak, G. R., New, R., Pallé, P. L., and Roca Cortés, T. 1988, Tenerife, p. 177.

Brown, T. M. 1988, Aarhus, p. 491.

Cacciani, A., Rosati, P., Ricci, D., Marquedant, R., and Smith, E. 1988, Tenerife, p. 181.

Claverie, A., Isaak, G. R., McLeod, C. P., van der Raay, H. B., and Roca Cortés, T. 1979, *Nature* **282**, 591.

Claverie, A., Isaak, G. R., McLeod, C. P., van der Raay, H. B., Pallé, P. L., and Roca Cortés, T. 1984, *Mem. Soc. Astron. Ital.* **55**, 63.

Connes, J., and Connes, P. 1984, in *Space Research Prospects in Stellar Activity and Variability*, ed. A. Mangeney and F. Praderie, Meudon: Obs. de Paris, p. 135.

Damé, L. 1988, Tenerife, p. 367.

Delache, P., and Scherrer, P. H. 1983, *Nature* **306**, 651.

Domingo, V. 1988, Tenerife, p. 363.

Duvall, T. L., Jr., Dziembowski, W., Goode, P. R., Gough, D. O., Harvey, J. W., and Leibacher, J. W. 1984, *Nature* **310**, 22.

Duvall, T. L., Jr., and Harvey, J. W. 1984, *Nature* **310**, 19.

Duvall, T. L., Jr., Harvey, J. W., and Pomerantz, M. A. 1986, *Nature* **321**, 500.

Elsworth, Y. P., Isaak, G. R., Jefferies, S. M., McLeod, C. P., New, R., Pallé, P. L., Régulo, C., and Roca Cortés, T. 1988a, Tenerife, p. 27.

Elsworth, Y. P., Isaak, G. R., Jefferies, S. M., McLeod, C. P., New, R., van der Raay, H. B., Pallé, P. L., Régulo, C., and Roca Cortés, T. 1988b, Tenerife, p. 535.

Fahlman, G. G., and Ulrych, T. J. 1982, *Mon. Not. Roy. Astron. Soc.* **199**, 53.

Faulkner, J., Gough, D. O., and Vahia, M. N. 1986, *Nature* **321**, 226.

Fischer, G., Hill, F., Jones, W., Leibacher, J. W., McCurnin, W., Stebbins, R. T., and Wagner, J. 1986, *Solar Phys.* **103**, 33.

Fossat, E. 1988, Tenerife, p. 161.

Fossat, E., Gelly, B., Grec, G., and Pomerantz, M. A. 1987, *Astron. Astrophys.* **177**, L47.

Fossat, E., Grec, G., Gavrjusev, V., and Gavrjuseva, E. 1988, Tenerife, p. 393.

Fröhlich, C., Andersen, B. N., Berthomieu, G., Crommelynck, D., Delache, P., Domingo, V., Jiménez, A., Jones, A. R., Roca Cortés, T., and Wehrli, C. 1988a, Tenerife, p. 371.

Fröhlich, C., Bonnet, R. M., Bruns, A. V., Delaboudinière, J. P., Domingo, V., Kotov, V. A., Kollath, Z., Rashkovsky, D. N., Toutain, T., Vial, J. C., and Wehrli, C. 1988b, Tenerife, p. 359.

Garcia, C., Pallé, P. L., and Roca Cortés, T. 1988, Tenerife, p. 353.

Gelly, B., Fossat, E., and Grec, G. 1988, *Astron. Astrophys.* **200**, L29.

Gelly, B., Fossat, E., Grec, G., and Pomerantz, M. 1988a, Aarhus, p. 21.

Gelly, B., Fossat, E., Grec, G., and Schmider, F.-X. 1988b, *Astron. Astrophys.* **200**, 207.

Gelly, B., Grec, G., and Fossat, E. 1986, *Astron. Astrophys.* **164**, 383.

Global Oscillation Network Group 1987, *Report #5: The 1987 Artificial Data Workshop.*

Gough, D. O. 1985, *Solar Phys.* **100**, 65.

Grec, G., Fossat, E., and Pomerantz, M. A. 1983, *Solar Phys.* **82**, 55.

Grec, G., Fossat, E., and Vernin, J. 1976, *Astron. Astrophys.* **50**, 221.

Harvey, J. W. 1988, Tenerife, p. 55.

Harvey, J. W., and the GONG Instrument Development Team 1988, Tenerife, p. 203.

Harvey, J. W., Hill, F., Kennedy, J. R., Leibacher, J. W., and Livingston, W. C. 1988, *Adv. Space Res.* **8**, (11)117.

Harvey, J. W., Kennedy, J. R., and Leibacher, J. W. 1987, *Sky & Telescope* **74**, 470.

Hathaway, D. H. 1988, *Solar Phys.* **117**, 329.

Henning, H. M., and Scherrer, P. H. 1986, Cambridge, p. 55.

Hill, F. 1987, in *The Internal Solar Angular Velocity*, ed. B. R. Durney and S. Sofia, Dordrecht: Reidel, p. 45.

Hill, F. 1988, *Astrophys. J.* **333**, 996.

Hill, F. 1989, *Astrophys. J.* **343**, in press.

Hill, F., and the GONG Site Survey Team 1988, Tenerife, p. 209.

Hill, F., Gough, D., Merryfield, W. J., and Toomre, J. 1984, in *Probing the Depths of a Star: The Study of Solar Oscillations From Space*, ed. R. W. Noyes and E. J. Rhodes, Jr., NASA JPL 400-237, p. 37.

Hill, F., Gough, D. O., Toomre, J., and Haber, D. A. 1988, Aarhus, p. 45.

Hill, F., and Newkirk, G. A. 1985, *Solar Phys.* **95**, 201.

Hill, H. A. 1984, *Astrophys. J.* **290**, 765.

Hill, H. A. 1986, in *Neutrino '86: The Twelfth International Conference on Neutrino Physics and Astrophysics*, ed. T. Kitagaki and H. Yuta, Singapore: Kim Hup Lee Publishers, p. 221.

Hill, H. A., Bos, R. J., and Goode, P. R. 1982, *Phys. Rev. Lett.* **49**, 1794.

Hill, H. A., Gao, Q., and Rosenwald, R. D. 1988, Tenerife, p. 403.

Hill, H. A., Rabaey, G. F., Yakowitz, D. S., and Rosenwald, R. D. 1986, *Astrophys. J.* **310**, 444.

Hill, H. A., Tash, J., and Padin, C. 1986, *Astrophys. J.* **304**, 560.

Hoeksema, J. T., Scherrer, P. H., Title, A. M., and Tarbell, T. D. 1988, Tenerife, p. 407.

Isaak, G. R. 1986, Cambridge, p. 223.

Isaak, G. R., Jefferies, S. M., McLeod, C. P., New, R., van der Raay, H. B., Pallé, P. L., Régulo, C., and Roca Cortés, T. 1988, Aarhus, p. 201.

Isaak, G. R., van der Raay, H. B., Pallé, P. L., Roca Cortés, T., and Delache, P. 1984, *Mem. Soc. Astron. Ital.* **55**, 91.

Jefferies, S. M., McLeod, C. P., van der Raay, H. B., Pallé, P. L., and Roca Cortés, T. 1988a, Aarhus, p. 25.

Jefferies, S. M., Pomerantz, M. A., Duvall, T. L., Jr., Harvey, J. W., and Jaksha, D. B. 1988b, Tenerife, p. 279.

Jiménez, A., Pallé, P. L., Pérez, J. C., Régulo, C., Roca Cort'es, T., Isaak, G. R., McLeod, C. P., and van der Raay, H. B. 1988a, Aarhus, p. 205.

Jiménez, A., Pallé, P. L., Roca Cortés, T., Andersen, N. B., Domingo, V., Jones, A., Alvarez, M., and Ledezma, E. 1988b, Tenerife, p. 163.

Kennedy, J. R., and Pintar, J. A. 1988, in *Astronomy From Large Databases*, ed. F. Murtagh and A. Heck, Garching: ESO, p. 367.

Kuhn, J. R. 1982, *Astron. J.* **87**, 196.

Kuhn, J. R. 1984, Snowmass, p. 293.

Leibacher, J. W. 1984, in *Theoretical Problems in Stellar Stability and Oscillations*, ed. A. Noels and M. Gabriel, Liège: Institut d'Astrophysique, p. 298.

Morrow, C. A., and Brown, T. M. 1988, Aarhus, p. 485.

Newkirk, G. A. 1980, editor, *Solar Cycle and Dynamics Mission, Final Report*, NASA Goddard Space Flight Center.

Noyes, R. W., and Rhodes, E. J., Jr. 1984, editors, *Probing the Depths of a Star: The Study of Solar Oscillations From Space*, NASA JPL 400-237.

Pallé, P. L., Perez, J. C., Régulo, C. Roca Cortés, T., Isaak, G. R., McLeod, C. P., and van der Raay, H. B. 1986, *Astron. Astrophys.* **170**, 114.

Pallé, P. L., Régulo, C., and Roca Cortés, T. 1989, *Contributed Papers to Inside the Sun, submitted to Solar Phys.*

Pintar, J. A., and the GONG Data Team 1988, Tenerife, p. 217.

Provost, J., and Berthomieu, G. 1988, Tenerife, p. 387.

van der Raay, H. B. 1988, Tenerife, p. 339.

Scherrer, P. H. 1984, Snowmass, p. 173.

Scherrer, P. H. 1986, Cambridge, p. 117.

Scherrer, P. H., Hoeksema, J. T., Bogart, R. S., and the SOI Co-Investigator Team 1988, Tenerife, p. 375.

Schmider, F.-X., Fossat, E., Grec, G., and Gelly, B. 1988, Tenerife, p. 605.

Stebbins, R. T., and Wilson, C. 1983, *Solar Phys.* **82**, 43.

Woodard, M. F. and Hudson, H. S. 1983, *Nature* **305**, 589.

Yerle, R. 1988, Aarhus, p. 87.

Recent Helioseismological Results from Space:

Solar p-mode Oscillations from IPHIR on the PHOBOS Mission

C.Fröhlich[1], T.Toutain[2], R.M.Bonnet[3], A.V.Bruns[4],
J.P.Delaboudinière[2], V.Domingo[5], V.A.Kotov[4], Z.Kollath[6],
D.N.Rashkovsky[4], J.C.Vial[2], Ch.Wehrli[1]

[1] Physikalisch-Meteorologisches Observatorium Davos
   World Radiation Center
   CH-7260 Davos-Dorf
[2] Laboratoire de Physique Stellaire et Planetaire
   F-91370 Verrières-le-Buisson
[3] Science Directorate of ESA
   F-75738 Paris Cedex 15
[4] Crimean Astrophysical Observatory
   Nauchny, Crimea 334413, USSR
[5] Space Science Department of ESA
   NL-2200 AG Noordwijk
[6] Central Research Institute for Physics
   H-1525 Budapest

ABSTRACT. IPHIR (InterPlanetary Helioseismology by IRradiance measurements) is a solar irradiance experiment on the USSR planetary mission PHOBOS to Mars and its satellite Phobos. The experiment was built by an international consortium including PMOD/WRC, LPSP, SSD/ESA, KrAO and CRIP. The sensor is a three channel sunphotometer (SPM) which measures the solar spectral irradiance at 335, 500 and 865 nm with a precision of better than 1 part-per-million (ppm). It is the first experiment dedicated to the investigation of solar oscillations from space. The results presented here are from a first evaluation of data gathered during 160 days of the cruise phase of PHOBOS II, launched on July, 12th 1988. The long uninterrupted observation produces a spectrum of the solar p-mode oscillations in the 5-minute range with a very high signal-to-noise ratio, which allows an accurate determination of frequencies and line shapes of these modes.

1. Introduction

Solar pressure mode oscillations can be observed by measuring either velocity variations or changes in the brightness temperature. By looking at the Sun as a star (no imaging) only low degree modes ($l \leq 4$) can be seen. The velocity signals have amplitudes of a few cm/s and the brightness signals of a few parts-per-million (ppm) of the irradiance. From the earth's surface p-mode oscillations can be observed with velocity measurements, whereas observations in brightness are marginal due to noise produced by transmittance changes of the earth' atmo-

279

G. Berthomieu and M. Cribier (eds.), Inside the Sun, 279–288.

sphere. A major drawback of ground observations is the daily window function producing aliases at ±11.57 µHz (1 day period) and multiples thereof. This problem can be partly overcome with global multi-station networks as e.g. by the Birmingham group (Aindow et al.,1988), IRIS (Fossat, 1988) and GONG (e.g. F.Hill and Newkirk, 1985). Their operation requires quite important logistic efforts and the fitting of the data of two adjacent stations can still produce aliases. An other possibility are observations from the geographic south pole, where uninterrupted time series of up to 10 days can be gathered (e.g. Grec et al., 1983). Up to now observations from space have been limited to brightness measurements: by ACRIM on the Solar Maximum Mission (SMM) of NASA (Woodard and Hudson, 1983) and the IPHIR experiment on the USSR mission to Phobos (Fröhlich et al., 1988). Irradiance measurements are advantageous because they are not influenced by variable radial velocities of space-crafts relative to the Sun.

The best orbit for space observations over long periods of time is the Lagrange point L1 where SOHO, the Solar and Heliospheric Observatory of ESA/NASA, will be placed (1995 launch). SOHO will carry helioseismology experiments with brightness and velocity observations of the global and resolved Sun (experiments VIRGO, GOLF and SOI, see e.g. Bonnet, 1989). The next best orbits are the cruise phases of planetary missions and geostationary orbits, which allow uninterrupted observations of the Sun during at least a few months. The circular orbits around the Earth have durations of about 95 minutes, as e.g. SMM. This yields aliases at about ±180 µHz and multiples thereof, which deteriorate the signal-to-noise-ratio, but are still less disturbing the 5-minutes p-modes than the daily aliases of the ground observations.

The opportunity for a helioseismology experiment on a planetary mission came up in 1985 when the final payload of the USSR mission to the martian satellite Phobos was planned. The cruise phase of this mission offered the unique chance to gather for the first time uninterrupted time series of several months duration. The excellent results from ACRIM (Woodard and Hudson, 1983) suggested solar irradiance measurements for this experiment, and more specifically, the good experience with sunphotometers lead to a proposal of spectral measurements with a three channel instrument. This kind of sunphotometer (SPM) was already developed for other investigations (Brusa et al., 1983) and had proven its ability to observe solar oscillations with high resolution (Fröhlich and van der Raay, 1984) during stratospheric balloon experiments.

2. Instrument Description

The sensor part consists of a three channel SPM and a two axis sun sensor (TASS) with perpendicular linear arrays which monitor the pointing towards the Sun. The data processing unit with a microprocessor in a separate box controls the sensor, handles the data and ensures the interface with the spacecraft. A detailed description of IPHIR is given by Fröhlich et al.(1988). The SPM has three independent channels at 335, 500 and 865 nm (half-power-bandwidth of 5 nm) each consisting of a Silicon-diode interference-filter combination in a sealed housing. The channels have no image forming optics, but measure irradiance (W/m^2 within the bandpass of the filter) with a field-of-view of ±2.5° determined by two apertures, one (2 mm diam.) in front of the detector and the other (9 mm diam.) at a distance of 104 mm. The noise of the detector-amplifier circuit is less than 1 ppm of the full-scale signal for an integration time of the order of 10 s. The analog signals of the three channels are measured simultaneously by three dedicated voltage-to-frequency converters and counter circuits. The basic sampling period is 8.22 s, out of which 64 ms are used to perform electrical calibrations. Five such observation are summed before transmission to reduce the data rate; thus the sampling interval amounts to 41.1 s.

3. Data Handling and Evaluation

The data from the experiment are received every 4 days at the Space Research Institute (IKI) in Moscow, and are distributed to the team either directly by tape or through the Centre National d'Etudes Spatiales (CNES) in Toulouse (France) over a permanent telephone line. The data coverage is excellent with virtually no gaps due to transmission or operations. Two problems have been encountered with the sensors: degradation of the sensitivity and an unexpected influence of the offset pointing on the signal of all three channels. The degradation depends strongly on wavelength and is maximal for the blue channel with a 1/e time constant for the sensitivity decay of about 30 days. As we are interested in relative changes and as the degradation of the green and red channel (1/e: 300 and 600 days at the beginning of the mission) is moderate this smooth decrease in sensitivity is not really harming the results of the latter two channels. The effect of the pointing, however, is a more serious problem. The pointing of the S/C was specified to be within ± 1° and indeed the axis of the spacecraft oscillates slowly between ±0.6° in the ecliptic plane around the axis to the Sun with constant speed and fast turns. These movements have periods between 40 and 80 minutes and are further modulated by movements in the plane perpendicular to the ecliptic. During offset pointing the signal increases due to straylight, produced within the instrument by the light reflected at a small angle from the highly reflecting detector aperture into the baffle and back to the detector. The movement of the spacecraft translates into a roughly quadratic change between very pointed apexes. The amplitude of these peaks can reach several tenths of a percent of the signal. As the SPM have up to now always been used on well pointed platforms, this unexpected kind of behavior was detected only during the mission and was later examined in detail by ground tests with the spare instrument. It is obvious that the small amplitudes of the solar oscillations of only a few ppm are strongly masked by this effect.

An efficient method to reduce the influence of this adverse pointing effect for on the frequency spectrum of the p-mode oscillations in 5-minute range is to use a filter of the form $y_i - \Sigma f(y_{i+k})$ with $-6 \leq k \leq +6$ and f being e.g. a Hanning taper. At the pointed apexes, however, such a filter fails to smooth the data and a more sophisticated method has to be used: in the range of ±13 points around an apex both sides are fitted with a least square method to concave parabolas with their minima to the left and right of the apex reproducing the peak at the apex. The data are detrended with these two parabola and the center of the filter run over the range of ±6 points from the apex. By this method the time series can be smoothed to such a degree, that the noise in power spectrum from 1.5 to 6 mHz remains well below the ppm range and is mostly of solar origin.

4. Results

The power spectrum for the 5-minutes range is shown in Figure 1. The signal-to-noise ratio is 25:1 for the strongest p-mode, that is very high compared to the one of the ACRIM data of 7:1 (Woodard, 1984). It has to be noted that ACRIM was not designed to be used for such studies and the instrument operation is indeed not really suited for it. The signal-to-noise ratio improvement in IPHIR is due to the low instrumental noise (inherent and by proper sampling) and due to the continuity of the data (constant window function compared to the 35-40 minutes gaps every 90 minutes in ACRIM due to the orbit of SMM).

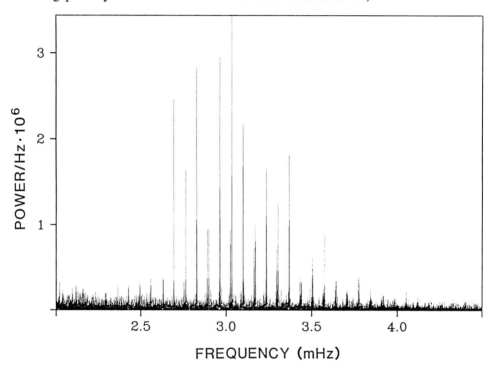

Figure 1: Power spectrum of the red channel in the 5-minutes range showing very accurately the solar p-mode oscillations with degrees l=0-2 and orders n=16-27. The time series spans 160 days from 15. July to 22. December 1988.

For low degree and high order modes (l<<n) the frequencies can be analyzed using the asymptotic approximation (e.g. Christensen-Dalsgaard, 1989):

$$\nu_{l,n} = \nu_0 \cdot (l/2 + n + \varepsilon) + \delta_{l,n}$$

ν_0, the spacing between different orders of the same degree, corresponds to the travel time of sound wave from surface to surface through the center and is a mainly determined by the sound speed near the surface. ε is usually taken as 1/2. The correction term $\delta_{l,n}$ manifested by the small frequency difference between the modes with degrees 0 and 2 and n different by one, depends on the properties in the core. Thus it can be used to distinguish between different solar models, e.g. standard, mixed and WIMP (Däppen et al., 1986, Faulkner et al., 1986). Although the formula is only an approximation and parameters may depend on frequency this is a very useful way to characterize the p-modes. By comparing the observed frequencies with the asymptotic formula the modes can be identified and some evaluation of the structure of the Sun (helioseismology) performed.

The frequency centroids of the strongest lines have been determined by either fitting Lorentz-lines or by searching the bary-center of the line within a range of ± 2.5 μHz using the spectrum of the whole period (Figure 1). The differences between the two methods are normally smaller than 0.1 μHz but can reach 0.5 μHz (l=1, n=18). This difference indicates an asymmetry of the lines which could be due to the statistics of the excitation giving rise to a biased sample although the observation time is thirty to fifty times the 1/e lifetime of the modes. More likely, the asymmetry arises from a combination of frequency shift during the observation and a changes in the amplitudes. The rapid increase of solar activity should increase the frequency by about 0.1 μHz from July to the end of 1988. This value is estimated from the frequency data of the declining cycle from 1980 towards the minimum (Woodard and Noyes, 1985, Gelly et al., 1988; Pallé et al., 1988) using the solar irradiance data for scaling the values to the 1988 period (Kuhn et al., 1988, Kuhn, 1989; Fröhlich et al., 1990). As to the amplitude variation, Figure 3 shows indeed a quite unequal amplitude repartition during the period of observation. Until a more detailed analysis is available, best estimates for the frequency centroids are the means of the two values as presented in Table 1. The uncertainty is probably of the order of 0.1 μHz.

Table 1: Frequency centroids for l=0-2 and n=17-23

n	l=0	l=1	l=2
17		2559.40	2620.39
18	2629.84	2693.43	2754.02
19	2764.30	2828.45	2889.86
20	2898.07	2963.51	3024.82
21	3034.17	3098.65	3159.60
22	3169.16	3233.63	3295.19
23	3304.18		

For the determination of the asymptotic parameters and their frequency dependence, quadratic fits of the form

$$\nu_{l,n} = a \cdot (n + l/2 - 21.5)^2 + b \cdot (n + l/2 - 21.5) + c$$

are performed. For l=0,1 and 2 the following constants and the 1σ standard deviation of the fit in μHz are listed in Table 2.

Table 2: Coefficients of quadratic fits of mode frequencies

	a	b	c	σ
l=0:	0.0665	135.029	3101.55	0.09
l=1:	0.0947	135.194	3098.48	0.20
l=2:	0.0815	135.269	3092.43	0.47

The constant b corresponds to v_0 and depends in a systematic way on the degree, what had already been observed in the ACRIM data of 1980 (Woodard, 1984). The ACRIM values, however, are lower by about 0.05 μHz. On the other hand, the constants c of IPHIR are lower than the ACRIM ones by $\leq$0.05 μHz. Differences between the a values are probably not significant.

The $\delta_{l,n}$ is calculated from the l=0 and l=2 values of Table 1 by fitting a straight line as function of frequency. The standard deviation of the fit is substantially decreased if the first value is dropped. This may indicate that the straight line is not an appropriate approximation. From the five remaining values the $\delta_{l,n}$ amounts to 9.35 μHz at 3.0 mHz and shows a slope of -2.4 μHz/mHz. This compares well with the results from ACRIM (9.30; Woodard, 1984) and is somewhat higher than the one calculated from the observations 1977-1985 at Tenerife (9.04; Jimenes et al., 1988).

In order to obtain more detailed information about the line-shapes, a superimposed analysis has been made for the modes listed in Table 1. For the superposition the centers of the l=0 and 1 lines are assumed to be at the frequencies calculated from the quadratic fit. The result is shown in Figure 2. The mean line width of the l=0 lines (n=18-23) equals 1.45 μHz; the real value is probably lower as the lines are superimposed at mean frequencies which adds the deviations from the fit to the line width. The value is lower than the one of Woodard (1984) for l=0 and n=20-23 but within his error bars. It is higher than the one (0.98 μHz) reported by the Birmingham group for the same l=0 and n=18-23 modes (Elsworth et al., 1988). The width of l=1 is 2.3 μHz; the difference to the l=0 width yields an estimate of the unresolved splitting of the l=1 mode of 0.85 μHz for a difference in m=2; the deduced rotational speed would amount to 0.42 μHz or 27.4 days. The superposition of the l=2 on the plot l=0/2 is not very accurate and depends on the difference $\delta_{l,n}$ which is a function of frequency. An interesting feature of is the broad pedestal of the lines. This is probably due to modulation of the intensity with periods longer than about 5 days from e.g. active regions.

The evolution in time of the amplitudes of p-modes can give information about the excitation and damping of these modes. Figure 3 shows this variation for the modes l=0-2 and n=17-27. The analysis is done for 5-day periods and it shows large variations as already known from ground observations (e.g. Jefferies et al, 1986). The continuity of the data of IPHIR, however, renders the results more reliable. They show a more or less random distribution with a suggestion of an more or less steady increase with time for l=0 and two maxima for l=1 in first and last third and no obvious correlation between the different degress seems to be present. It is obvious that the temporal behavior of p-modes contains a great deal of information and a detailed study on this subject is under way.

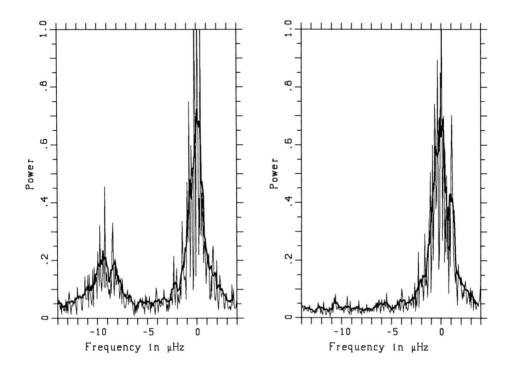

Figure 2: Epoch analysis of l=0 and l=2 (left) and l=1 (right) for orders 17 to 23. The difference in line-width of the l=0 and l=1 is due to unresolved rotational splitting. All lines show a broad pedestal due to modulation by low frequency irradiance variation.

5. Conclusions

The results presented here are the first outcome of the evaluation of the vast information contained in these data. The continuity of the observations with IPHIR are the major advantage of this experiment compared to ground based and Earth orbiting space observations. The results indicate that the achievable accuracy for the determination of absolute frequencies is limited, possibly due to the stability of the mode themselves and their changes during a solar cycle. On the other hand, these data allow an un-biased study of p-mode amplitudes and their evolution in time.

286

ℓ   n

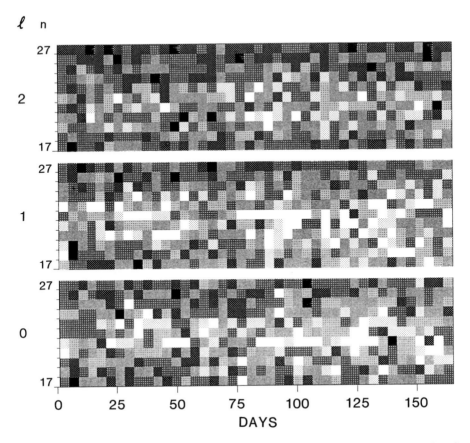

Figure 3: Time variation of the amplitudes of $l=0..2$ and n=17..27 modes from 31 time series of 5 days length each. The 7 levels of gray scale span relative amplitudes from 0.3 (black) to 1.3 (white). The highly variable amplitudes demonstrate the importance of long uninterrupted time series for the determination of unbiased mean characteristics of the p-modes.

6. Acknowledgements

The completion of this experiment would not have been possible without the continuous engagement of the engineering and technical groups at the Space Research Institute (IKI) at Moscow, and at the institutes building the hardware under the leadership of B.Cougrand (LPSP), and H.J.Roth (PMOD/ WRC), which is gratefully acknowledged. Thanks are extended for the support by the Swiss National Science Foundation, the Centre National d'Etudes Spatiales and by funds of the Space Science Department of ESA, the Crimean Astrophysical Observatory and the Central Research Institute for Physics.

7. References

Aindow,A., Elsworth,G.R., Isaak,G.R., McLeod,C.P., New,R. and van der Raay, H.B., 1988, 'The current Status of the Birmingham Solar Seismology Network', in *Seismology of the sun and sun-like stars,* 157-160, ESA SP-286, Noordwijk.

Bonnet,R., 1989, 'Future Prospects of Helioseismology from Space', this issue.

Brusa,R.W., Fröhlich,C. & Wehrli, Ch., 1983, 'Solar Radiometry from High-Altitude Balloons', in *Sixth ESA Symposium on European rocket & balloon Programmes,* 429-433, ESA SP-183, Noordwijk.

Christensen-Dalsgaard,J., 1989, 'Data on the Internal Structure derived from Helioseismic Observations, this issue.

Däppen,W., Gilliland,R.L. & Christensen-Dalsgaard,J., 1986, 'Weakly Interacting Massive Particles, Solar Neutrinos, and Solar Oscillations', *Nature,* **321,** 229-231.

Elsworth,Y., Isaak,G.R., Jefferies,S.M., McLeod,C.P., New,R., Pallé,P.L., Régulo,C. & Roca-Cortes,T., 1988, 'Line Width of Low Degree Acoustic Modes of the Sun', in *Seismology of the sun and sun-like stars,* 27-29, ESA SP-286, Noordwijk.

Faulkner,J., Gough,D.O. & Vahia,M.N., 1986, 'Weakly Interacting Massive Particles and Solar Oscillations', *Nature,* **321,** 226-229.

Fröhlich,C., Bonnet,R.M., Bruns,A.V., Delaboudinière,J.P., Domingo,V., Kotov,V.A., Kollath,Z., Rashkovsky,T., Toutain,T., Vial,J.C. & Wehrli, Ch., 1988, 'IPHIR: The Helioseismology Experiment on the Phobos Mission', in *Seismology of the sun and sun-like stars,* 359-362, ESA SP-286, Noordwijk.

Fröhlich,C. & van der Raay,H.B., 1984, 'Global Solar Oscillations in Irradiance and Velocity: a Comparison', in *The Hydrodynamics of the Sun,* 17-20, ESA SP-220, Noordwijk.

Fröhlich,C., Foukal,P.V., Hickey,J.R., Hudson,H.S. & Willson,R.C., 1990, 'Solar Irradiance Variability from Modern Measurements', in *The Sun in Time (in preparation),* University of Arizona Press, Tucson.

Fossat,E., 1988, 'IRIS: a Network for Full-Disc Helioseismology', in *Seismology of the sun and sun-like stars,* 161-162, ESA SP-286, Noordwijk.

Gelly,B., Fossat,E. & Grec,G., 1988, 'Solar p-mode Frequency Variations between 1980 and 1986', in *Seismology of the sun and sun-like stars,* 275-277, ESA SP-286, Noordwijk.

Grec,G., Fossat,E. & Pomerantz,M., 1983, 'Full Disk Observations of Solar Oscillations from the Geographic South Pole: Latest Results', *Sol.Phys.,* **82,** 55-66.

Hill,F, & Newkirk Jr.,G., 1985, 'On the expected Performance of a Solar Oscillation Network', *Sol.Phys.,* **95**, 201-219.

Jiménes,A., Pallé,P.L., Péres,J.C., Régulo,C., Roca-Cortes,T., Isaak,G.R., McLeod,C.P., van der Raay,H.B., 1986, 'The Solar Oscillation Spectrum and the Solar Cycle', in *Advances in Helio- and Asteroseismology, J.Christensen-Dalsgaard & S.Frandsen, eds.,* 205-209, Reidel Publ., Dordrecht.

Jefferies,S.M., McLeod,C.P., van der Raay,H.B., Pallé,P.L. & Roca-Cortes,T., 1986, 'Splitting of Low l Solar P-Modes', in *Advances in Helio- and Asteroseismology, J.Christensen-Dalsgaard & S.Frandsen, eds.,* 25-28, Reidel Publ., Dordrecht.

Kuhn,J.R., Libbrecht,K.G. & Dicke,R.H., 1988, 'The Surface Temperature of the Sun and Changes in the Solar Constant', *Science*, **242**, 908-911.

Kuhn,J.R., 1989, 'Helioseismic Observations of the Solar Cycle', *Astrophys.J.*, **339**, L45-L47.

Pallé,P.L., Régulo,C. & Roca Cortes, T., 1988, 'Frequency Shift of Solar p-modes as seen by Cross-Correlation Analysis', in *Seismology of the sun and sun-like stars,* 285-289, ESA SP-286, Noordwijk.

Woodard,M., 1984, 'Sort-Period Oscillation in the Solar Irradiance', Ph.D. Thesis, Univ.of San Diego, San Diego.

Woodard,M. & Hudson,H.S., 1983, 'Frequencies, Amplitudes and Linewidths of Solar Oscillation from Total Irradiance Observations', *Nature,* **305**, 589-593.

Woodard,M. & Noyes,R.W., 1985, 'Change of Solar Oscillation Eigenfrequencies with the Solar Cycle', *Nature*, **318**, 449-450.

FUTURE PROSPECTS OF HELIOSEISMOLOGY FROM SPACE

R.M. BONNET
European Space Agency
8-10 rue Mario Nikis
75738 Paris Cedex 15 - France

ABSTRACT. In view of their costs, space-borne instruments should be considered only for their exclusive capabilities in helio-and asteroseismology. Space-borne high resolution spectrometers and photometers operate free of atmospheric pertubations, can be put on special orbits offering continuous (uninterrupted) observations and therefore offer the best opportunity for high signal-over-noise ratio. The recent data obtained on board the Phobos-2 mission clearly evidences this fact. The ESA-NASA SOHO observatory will be the first mission of its kind carrying a comprehensive set of instruments to analyse the gravity and acoustic modes of solar oscillations over an uninterrupted period of at least 2 years. Projects also exist to observe oscillations of the solar diameter. Long term observation of the solar constant may provide a clue to the understanding of the origins of the solar cycle. Simultaneous out of eliptic measurements may nicely complement our data set and offer unambiguous views on the asymmetries of the solar interior. Space observations are probably the only means to get access to the deep solar interior through the detections of g modes. They offer the only prospect in the exploitation of asteroseismology over a larger number of stars.

1.   INTRODUCTION

Like most branches of astronomy, helioseismology started in ground based observatories and has up to now flourished and provided very impressive results without the contribution of space techniques. In the past 10 years our understanding and our knowledge of the interior of the Sun has been confronted to helioseismological data obtained with the observations made from the ground of either the global or resolved motions of the solar surface as induced by the acoustic modes of intrinsic solar oscillations. The detailed analysis of their fre-quencies and the use of complex inversion techniques made possible through the availability of modern and powerful computers has already allowed to determine the variation along the solar radius of the sound velocity and to infer from it the temperature throughout the

G. Berthomieu and M. Cribier (eds.), Inside the Sun, 289–304.
© 1990 Kluwer Academic Publishers. Printed in the Netherlands.

convection zone down to a depth of approximately 0.5 solar radii (Gough (1988)). The set of reliable data already obtained which spans more than half a solar cycle, yields already some clue as to the relationships between the cycle and the convection zone, and the variation of the asphericity of the velocity of sound.

Furthermore, differential rotation has been inferred which already gives some indication on the dynamics of the solar interior down to 0.5 solar radii.

Helioseismology is such a rich field and so much has been obtained from the ground that one may, rightfully, question the necessity to start exploiting this field from space. This being said, the advent of space techniques and the initiation of several space missions or experiments devoted to helioseismology lead us to discuss briefly the role that these techniques might play in the future and what progress can be expected from them.

2.   WHY SHOULD WE GO TO SPACE ?

We should not go to space when our goals can be fulfilled from the ground. This statement is of a general nature and applies particularly well to the case of helioseismology. This is why we will not speak here of measurements which can be made, and very well, from the ground such as those of zonal velocity and magnetic fields or zonal temperatures (Kuhn et al. (1988)).

On the contrary, space offers intrinsic absolute advantages but these should always be judged against the necessarily expensive nature of space borne experiments. The advantages for helioseismology can be summarized as follows :

2.1   The elimination of atmospheric pertubations.

The presence of the earth atmosphere limits the quality and the nature of helioseismological data obtained from the ground. The degradation by atmospheric turbulence of the surface patterns induced by p-modes with degrees larger than a few hundreds is very substantial as shown in the excellent report published by NASA (Noyes and Rhodes (1984)). For example, the precise determination of the abundance of helium in the subphotospheric layers require observations with l ranging between 400 and 1200 which are not possible from the ground.

The measurement of the variation in the dimension of the solar diameter is definitely very difficult if not impossible from the ground as was clearly demonstrated by R. Sofia (1989). Even a perfectly stable atmosphere would not be free of refraction effects. Such effects have been invoked in order to explain the 160 min oscillation whose existence has, for a long time, been a very controversial issue (Elsworth et al. (1989)).

2.2   The possibility of photometric measurements

Photometric data play a very important role in helioseismology. Due to their very small amplitude (intrinsic solar radiance or irradiance variations are usually measured in fractions of millionths) they cannot easily be disentangled from atmospheric pertubations and require quasi essentially the use of space techniques.

Photometric measurements are preferred when one wants to look very close to the solar limb due to the rapid fall-off of velocity oscillations there, although the rms amplitude of the modes over the intrinsic solar noise make them less easy to use. This is particularly serious for low degree and g modes whose amplitude is well known to be very small. However, the relative simplicity of the instrumentation (a photometer) may easily overcome this drawback and in addition one can select the best spectral band for the optimum detection of surface patterns. This possibility indeed couples the advantages of 2.1 and 2.2 : from space there is no more wavelength limitation since atmospheric absorption has vanished. The ultra-violet spectrum is especially suited to the study of oscillations near the region of the temperature minimum.

2.3   The possibility of uninterrupted observations

From space we can have access to unique observing sites, such as full-sun orbits. On Earth this possibility exists but only at the poles and for a limited number of days per year, weather permitting, or using networks of stations like those in the GONG (Harvey (1988)) and IRIS (Fossat (1988)) projects. In space no such problems remain. Uninterrupted observations are essential to get access to the highest frequency resolution which is a key parameter for the detection of the low amplitude, long lived g modes, not yet achieved from the ground. For example, with 2 years of uninterrupted data one can achieve a resolution in frequency of 0.02 micro-hertz which is essential for separating the various g modes whose frequencies are increasing like $1/\sqrt{}$. This unique advantage derives from the fact that discontinuous data sets introduce side lobes in the frequency spectrum which broaden the observed frequency of individual modes.

Furthermore, orbits can be selected which minimize the line of sight velocity thereby increasing the S/N ratio for velocity oscillations. From space, stereoscopic observations are also achievable opening a completely new prospect for helioseismology.

3.   WHAT PARAMETERS ARE BEST SUITED FOR SPACE-BORNE OBSERVATIONS ?

The term "helioseismology" is used here in the brodest sense, encompassing observations which carry information on both the physical conditions and the dynamics of the interior of the Sun even though they are not exactly concerned with the so-called gravity or acoustic vibration modes.

3.1   The Solar constant

The total solar irradiance has been monitored nearly continuously from space since 1980 with the Active Cavity Radiometer Irradiance Monitor (ACRIM) on the Solar Maximum Mission and the NIMBUS-7 (Willson and Hudson (1988)) satellite and there is clear evidence that it varies in phase with the solar cycle by at least 0.1% between maximum and minimum. Kuhn et al. (1988) and Gough (1988) have discussed this variation in terms of an asphericity of the sound velocity inside the Sun. It can be shown that latitudinal temperature variations can be made responsible for part of the observed variation which indicate that the source of the cycle is rooted at least as deep as the bottom of the convection zone. Future progress will come from a continuation of the measurements from either SMM or any future mission. A few years ago COSPAR issued a resolution recommending that there should always be at least one radiometer at work in space. Such devices as ACRIM are small enough to be carried on any sun-pointed spacecraft or part of a spacecraft like the solar arrays for example.

3.2   Intensity (radiance) oscillations

Intensity oscillations measured with a photometric accuracy of at least 10^{-7} may be our best way to study g modes in the frequency range between 1 and 150 micro-hertz. It has been shown that for low frequencies, which characterize g mdoes, the relative contribution of perturbations (noise) due to supergranulation is much smaller for intensity than for velocity measurements.

The new results presented by Fröhlich (1990) at this colloquium prove without ambiguity the very strong interest in having long term, continuous intensity oscillation measurements and may give an indication of what amplitude might be expected for the observation of g modes. In addition, spectral irradiance measurements at selected wavelengths may provide some information of their propagation properties, through phase and amplitude differences between the various modes.

It is also important to have some spatial resolution on the disc in order to be able to identify and to separate the modes with different l numbers.

3.3   Velocity oscillations

Velocity oscillations, either global or resolved on the disc, provide the best means of studying the high frequency portion of the power spectrum which is filled uniquely with acoustic modes.

High resolution observations give access to modes which are trapped just underneath the photosphere and which, up to now, have not been observed from the ground due to viewing limitations. On the contrary, low degree modes but of high order represent good tools for the probing of the deep solar interior.

3.4   Solar diameter measurements

The variations in luminosity might well be associated/correlated with variations in the solar diameter. An accuracy of a few 10^{-3} arcsec is now within reach and these measurements may offer the best way to determine the low frequency modes (the g modes) and to probe the deep interior.

4.   NEAR TERM PROSPECTS

We now focus on three main space projects which in the near term should cast some new light on the overall properties of solar oscillatory modes. At the time of writing this paper the brilliant data obtained with IPHIR have not yet been fully processed and analysed and are still our first hope of detecting the gravity modes or of fixing upper limits to their amplitude relative to the noise. The next major step will come when the SOHO mission and its comprehensive set of instruments will fly.

4.1   The SOHO spacecraft

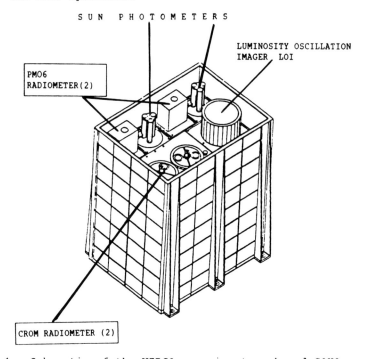

Fig. 1 - Schematic of the VIRGO experiment on-board SOHO, consisting

Table 1 - The main characteristics of the Virgo Experiment

Principal investigator :	C. Fröhlich Davos, Switzerland
Main scientific objectives :	. low degree (ℓ = 0-7) irradiance oscillations
	. precision $<$ 10^{-6} (for 10 s integration)
	. solar constant measurements : absolute accuracy : 0.15% precision : $10\text{-}50.10^{-6}$
Technique :	. active cavity radiometers
	. sun photometers operating at 335,500 and 865 nm
	. luminosity oscillation imager (55 mm telescope, f=1.3 m, solar image formed on 16 separate pixels)
Mass :	14.6 kg
Power :	16.6 W
Bit rate :	0.1 Kbps

2 PMO6 a of d 2 CROM radiometers, 2 sets of 3 sunphotometers each
operating at 335, 500 and 865 nm ($\Delta\lambda$ = 5 nm). The LOI consists of a 55
mm Cassegrain telescope with a servo controlled secondary mirror. The
detector is a 16 element diode array offering the capability to
resolve modes with l numbers from zero to seven.

SOHO is the first ever mission dedicated to helioseismology. A
complete description can be found in Domingo and Poland (1989). The
project is part of the Solar Terrestrial Science Programme of ESA
which also includes the four Cluster spacecraft and is conducted in
cooperation with NASA. ESA will build and integrate the spacecraft,
and NASA will launch its experiments and operate it during its nominal
two years lifetime. ESA and NASA share the payload. The launch of the
mission could be envisaged for 1995 on board an American expendable
vehicle. SOHO will be placed on a special 180 days halo orbit around
the L_1, Lagrangian point located at 1.5 million km from the Earth,
providing the required uninterrupted full Sun capability and
minimizing the spacecraft radial velocity with respect to the Sun, an
important requirement for the measurement of velocity oscillations.
This velocity will be known with an accuracy better than 2 cm/sec. The
spacecraft is 3-axis stabilized and provides an absolute pointing
accuracy of 10 arcsec and a stability of 1 arcsec over 15 minutes of
time. The spacecraft weight is about 1.3 metric tons including
approximately 150 kg of propellant and the telemetry bit rate is 40
Kps with no interruption in the data coverage. For the Solar
Oscillation Imager, MDI, the bit rate can reach 200 Kbps for real time
transmission or for tape dump.

All experiments have been selected and are in the process of
development in the various institutes. Not all of them are devoted to
helioseismology but only 18% of the payload mass i.e. $\simeq$ 90 Kg and 30
% of the power, i.e. $\simeq$ 102 W. Three instruments will share the
payload, VIRGO (Fröhlich et al. (1989) the Michelson Doppler Imager or
MDI (Scherrer et al. 1989) and GOLF (Gabriel et al. 1989)) which will
investigate, irradiance oscillations, highly resolved velocity (high
l) and possibly solar diameter oscillations, and global oscillations
respectively. Tables 1, 2 and 3 summarize the main characteristics of
these instruments. More details can be found in Domingo and Poland
(1989)). It should be noticed that VIRGO will indeed have the
capability of resolving modes of l values between zero and 7 (fig. 1).

4.2   The Solar Disc Sextant

The SDS is an instrument dedicated to the measurement of the
oscillations of the Solar diameter. This instrument (Sofia (1989)) is
already operating onboard a balloon and is proposed to NASA in the
framework of their small explorers series. The required accuracy for
the measurement is 0.003 arc seconds which is impossible to achieve
directly at the focus of an imaging telescope. However, relative
measurements with this accuracy are achievable once we can compare two
fields of view separated by a broad angle.

Table 2 - Main characteristics of the Michelson Doppler Imager

Principal investigator :

P. Scherrer
Stanford University

Main scientific objectives :

. High degree (ℓ / 4500)
velocity oscillations, with a
. precision of 0.002 cm/s
over 2 years ($\frac{1}{16 \text{ nhz}}$), Solar
noise limited

. Solar limb oscillations :
0.04 arc sec /pixel or
7.10^{-4} arc sec for oblateness

Technique :

Fourier tachometer : phase of
the line profile gives doppler
shift.
Resolution : 4 and 1.5 arc sec

Mass :

43.4 kg

Power :

55 W

Bit rate :

5 (+160) Kbps

This is the case for the SDS as shown in fig. 2. The key element in the design is a glass wedge defining a very stable angle of $\simeq$ 1000 arcsec which, in combination with a classical Cassegrain telescope image two opposite limbs of the Sun along the same diameter. A set of five linear array detectors are used to measure the light fluctuations incurred by variations in the solar diameter.

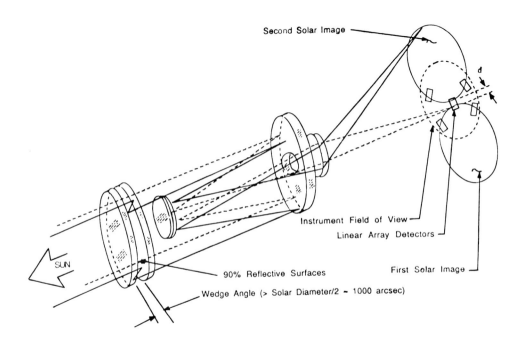

Fig. 2 - The Solar Disc Sextant optical principle offering a 0.003 arcsec accuracy for the measurement of variations of the solar diameter

Table 3 - Main characteristics of GOLF

Principal investigator :	A. Gabriel LPSP/IAS, Verrieres-le-Buisson
Main scientific objectives :	. Low degree velocity oscillations g modes, low order p modes . Precision $\simeq$ 0.1 mm/s . Solar noise limited
Technique :	Na vapour resonant scattering cell using Zeeman splitting in $\vec{B}$ = 4700 G Global measurement on the Θ
Mass :	31.2 kg
Power :	30 W
Bit rate :	0.128 Kbps

The SDS has already flown twice on board a balloon. It is not yet clear when it will fly on board a spacecraft.

4.3   Measurements of the Solar Constant

Because of their relative simplicity, several Solar Irradiance variability instruments are foreseen to fly in space in the near future.

A Solar Irradiance Variability instrument called SOVA placed under the responsibility of P. Crommelynck will be carried into space for at least 6 months on board the European Retrievable Carrier (EURECA) of ESA, scheduled for a launch onboard the space Shuttle in 1991. A second ACRIM instrument will measure the solar constant during 15 minutes every orbit on board NASA's UARS mission to be launched also in 1991. This instrument, together with the first ACRIM onboard SMM, and SOVA, will probably yield the most accurate evaluation ever of the solar constant and of its variation with time.

According to Noyes (1988) it is intended that successors to ACRIM may fly on the US series of Geophysical Orbiting Environmental Satellites (GOES) although this has not been confirmed. It is also very likely that similar instruments will be proposed on the polar platforms which are part of the set of platforms built by the US, ESA and Japan and which are connected to the International Space Station.

Finally, the Soviets have in their planning a solar project called KORONASS that will include a solar irradiance oscillations instrument. The implementation of this project is foreseen in the 1991-1993 time scale.

5.   LONGER TIME PROSPECTS

Present limitations in either the developed instrumentation or the duration of observations may affect or even hamper the detection of g modes which are our ultimate hope of getting access to physical conditions at the deep core of the Sun. Future solar probes such as those foreseen in ESA, NASA and even the Soviet programme may provide a means to detect these modes through their effect on the trajectory of the probe, if carefully monitored. Fig. 3 shows the ballistic profile of the Vulcan mission which is under study at ESA (Bertotti et al. 1988). None of these probes is foreseen to fly before the end of this century.

VULCAN - BALLISTIC MISSION PROFILE

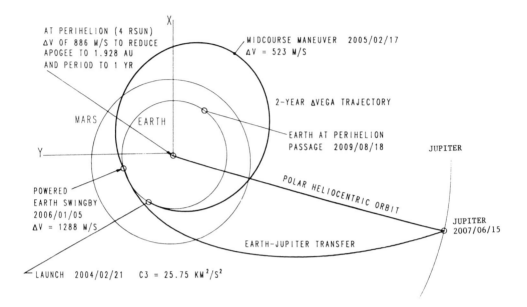

AT PERIHELION (4 RSUN)
ΔV OF 886 M/S TO REDUCE
APOGEE TO 1.928 AU
AND PERIOD TO 1 YR

MIDCOURSE MANEUVER   2005/02/17
ΔV = 523 M/S

2-YEAR ΔVEGA TRAJECTORY

MARS   EARTH

EARTH AT PERIHELION
PASSAGE   2009/08/18

JUPITER

POLAR HELIOCENTRIC ORBIT

POWERED
EARTH SWINGBY
2006/01/05
ΔV = 1288 M/S

JUPITER
2007/06/15

EARTH-JUPITER TRANSFER

LAUNCH   2004/02/21   C3 = 25.75 KM^2/S^2

Fig. 3 - Ballistic profile of the Vulcan mission presently under study at ESA

In the even longer term we may witness future progress coming from the great potentiality of two possible, although not yet even proposed, missions. After the launch of Ulysses (presently foreseen on 5 October 1990) and its hopefully successful operation, it may well be envisaged that a successor to what is presently the first out of the ecliptic mission appears as a next logical step in the understanding of the interplanetary medium and of the Sun itself. Such a mission should logically include helioseismology instruments and certainly an irradiance monitor like ACRIM, or a VIRGO type instrument. Latitudinal effects such as those reported by Kuhn in the distribution of zonal temperatures may henceforth be unambiguously interpreted. ESA in its Horizon 2000 plan (Bonnet and Bleeker (1984)) and also the National Academy of Science (Donahue (1988)) have such missions in their very long term plans.

Due to orbit maintenance and the necessarily limited capacity of the gas tanks onboard SOHO the mission is limited in time and may, at most, span half a solar cycle. We are just starting now to see the influence of the solar cycle on the various helioseismology parameters and we may wish to see more of this effect in particular if it is proven that the source of the cycle is deeply rooted inside the sun (Gough (1988)) and in that respect repeated observations with the same instrumentation over several cycles might lead to important progress. What follows rests solely on the author's imagination at this stage but we may imagine that when the astronomical and scientific capabilities of a lunar site will be exploited, in other words when we will have in the course of next century an operational lunar base, the capability thus developed may include a helioseismology facility.

We may think for example of a network of small stations located near the lunar pole. At a latitude of 75 degrees, given the fact that the rotation axis of the Moon is inclined by 1.5 degrees on the plane of the ecliptic, a network of 4 stations regularly positionned on a circonference of only 2730 km will serve all the purposes required for uninterrupted observation of intensity and velocity oscillations. If we are clever enough, we could even conceive these stations so that one may look at some stars during the lunar night. Solar energy is plentiful on the Moon at the pole (there is no atmospheric absorption) and can easily be used to power each station of the network on a continuous basis. It is at times good to dream !

6.   ASTEROSEISMOLOGY

Although the study of stellar interiors is not the main purpose of this meeting, we cannot talk about future prospects without mentioning the future prospects of asteroseismology from space. This topic was discussed in quite some extent by Noyes (1988) in terms of stellar magnitude, size of the telescope and integration time. For example, a star of V=12 would require 100 days of observations with a 1 m aperture telescope using broad band visible radiation. Noyes also clearly shows that the detection of velocity oscillations would require the use of very large telescopes, a solution which is not realistic if we want to sample a reasonable number of stars, although this might be envisaged for a small number of them. We list below a few projects which in the near future will perform stellar seismology observations.

6.1   The Hubble Space Telescope

An observation is planned on α cen A using the high speed photometer on HST which has the required sensitivity. This star is a prime target to compare with the Sun because of its well known mass, radius, and effective temperature, and also because it is possible to observe from the HST orbit for long uninterrupted periods. The launch of the Hubble Space Telescope is now foreseen for March 1990.

6.2   EVRIS

EVRIS is a project envisaged in the framework of the Franco-Soviet space programme and is foreseen to be placed on-board the Mars 94 mission, to be operated during the cruise phase.

brighter than V=4. During the mission, 30 such stars can be observed with integration times as high as 25 days. This modest instrument of 3.3 kg will be developed in the framework of a Consortium led by Dr. A.Baglin (Meudon Observatory) and G. Bisnovatyi-Kogan (IKI, URSS) and several co-investigators from Austria, Denmark, Spain, Switzerland and the United-Kingdom.

6.3 Asteroseismology from Cassini

Cassini is a NASA-ESA mission to Saturn and Titan, to be launched at the earliest in April 1996. (Lebreton and Scoon (1988)). During the cruise phase there are several opportunities to observe stars. Although the scientific package has not yet been selected, it is likely that there will be an asteroseismology instrument proposed on board.

7. CONCLUSIONS

This presentation and the discussion of future prospects of helioseismology from space is probably not complete. It should be seen in the context of the new results obtained with IPHIR, the only dedicated helioseismology experiment to have been operated in space so far. Because of their relatively high costs space experiments should be envisaged only when ground based observations bump on intrinsic physical limitations.

Clearly, space-borne measurements may provide in the future the best chance to detect if ever possible the long period gravity modes, which have so far resisted unambiguous determination from the ground. As usual, long uninterrupted sequences of observations are required. Observations for at least a complete solar cycle may provide a powerful means for detecting the source and inferring the mechanism(s) of the cycle.

Space techniques also offer unique advantages for the detection of stellar oscillation modes, essentially through broad band photometry. Several small scale projects envisaged on forthcoming interplanetary missions may open bright new perspectives in the not too distant future.

REFERENCES

Bonnet, R.M., and Bleeker, J. (1984), 'Space Science Horizon 2000' - ESA SP-1070.

Bertotti, B., Hoyng, P., Richter, A.K., Roxburgh, I.W., Schüssler, M., and Stenflo, J.O. (1988), 'The Vulcan Mission', Interim Report of ESA Science and Technology definition team, ESA/SCI(88)7, 1-21.

Domingo, V., and Poland, A.I. (1989), 'SOHO - An Observatory to study the solar interior and the solar atmosphere, 'The SOHO Mission' - ESA SP-1104, 7-12.

Donahue, T.M. (1988), 'Space Science in the twenty-first century. - Imperatives for the decades 1995 to 2015', National Academic Press, Washington, D.C.

Elsworth, Y.P., Jefferies, S.M., Mc Leod, C.P., New, R., Pallé, P.L., Van der Raay, H.B., Regulo, C., and Roca-Cortès, T. (1989), Astrop. J. 338, 557-562.

Fossat, E. (1988), 'IRIS : A network for full disk helioseismology', 'Seismology of the Sun and Sun like stars', - ESA SP-286, 161-162.

Fröhlich, C., Andersen, B.N., Berthomieu, G., Crommelynck, D., Delache, P., Domingo, V., V. Jimenez, A., Jones, A.R., Roca-Cortès, T and, Wehrli, C. (1989), 'VIRGO - The Solar monitor experiment on SOHO', 'The SOHO Mission', ESA SP-1104, 19-23.

Fröhlich, C. (1990) 'Recent helioseismological results from space', 'Inside the Sun', These Proceedings.

Gabriel, A.H., Bocchia, R., Bonnet, R.M., Cesarsky, C., Christensen-Dalsgaard, J., Damé, L., Delache, P., Deubner, F.L., Foing, B., Fossat, E., Fröhlich, C., Gorisse, M., Gough, D., Grec, G., Hoyng, P., Pallé, P., Paul, J., Robillot, J.P., Roca-Cortès, T., Stenflo, J.L., Ulrich, R.K., and van der Raay, H.B. (1989), 'GOLF-Global oscillations at low frequencies for SOHO mission', 'The SOHO Mission', ESA SP-1104, 13-17.

Gough, D.O. (1988), 'On taking Observers seriously', 'Seismology of the Sun and Sun like Stars', ESA SP 286, 679-682.

Harvey, J., and the GONG Instrument Team (1988), 'The GONG instrument', 'Seismology of the Sun and Sun like Stars, ESA SP-286 , 203-208.

Kuhn, J.R., Librecht, K.G., and Dicke, R.H. (1988), 'The Surface temperature of the Sun and changes in the solar constant', Science 242, 908-910.

Lebreton, J.P., and Scoon, G. (1988), 'Cassini - A mission to Saturn and Titan', ESA Bulletin 55, 24-30.

Noyes, R.W. (1988), 'Space Observations of Solar and Stellar oscillations', 'Advances in Helio and Asteroseismology', Christensen-Dalsgaard and S. Frandsen, Editors, Reidel Publishing Company, 527-534.

Noyes, R.W., and Rhodes, E.J. Jr. (1984), 'Probing the depths of a star : the study of solar oscillations from space', 'Report of the NASA Science Working Group on the study of Solar Oscillations from Space', NASA Report, July 1984.

Scherrer, P.H., Hoeksema, J.T., Bogart, R.S., Walker A.B.C. Jr., Title, A.M., Tarbell, T.D., Wolfson, C.J., Brown, T.M., Christensen-Dalsgaard, C., Gough, D.O., Kuhn, J.R., Leibacher, J.W., Libbrecht, K.G., Noyes, R.W., Rhodes, E.J., Toomre, J., Zweibel, E.G., and Ulrich, R.K. Jr. (1989), 'SOI - The Solar Oscillation imager for SOHO', 'The SOHO Mission', ESA SP-1104, 25-30.

Sofia, S. (1989), 'The Solar Disk sextant', Private communication

Willson, A.C., and Hudson, H.S. (1988), 'Solar luminosity variations in solar cycle 21', Nature 322, 810-814.

HELIOSEISMIC INVESTIGATION OF SOLAR INTERNAL STRUCTURE.

Jørgen Christensen-Dalsgaard
Astronomisk Institut, Aarhus Universitet
DK-8000 Aarhus C
Denmark

ABSTRACT. The solar oscillation frequencies provide our only means of obtaining detailed information about conditions inside the Sun. Here I give a brief overview of the relevant properties of solar models and solar oscillations, and present examples of the dependence of the oscillation frequencies on the structure of the model. Furthermore I discuss some results obtained so far from analysis of observed frequencies.

1. Introduction.

Observations of solar oscillations have given us a large amount of very precise data on the properties of the solar interior. Recent compilations of observed frequencies (Duvall *et al.* 1988; Libbrecht & Kaufman 1988) list over 2000 frequencies, with estimated errors that are in some cases less than 0.01 per cent. This must be compared with the other observational data that is, or may be, relevant to tests of solar models: the mass, radius and luminosity, all of which are known with comparable precision, and the neutrino flux, which, as is evident from other contributions in this volume, is subject to considerable observational and theoretical uncertainties.

The physical nature and behaviour of the oscillations are in general well understood. The observed modes correspond to standing acoustic waves, or *p modes*. Given a solar model it is relatively straightforward to compute its oscillation frequencies; the details of the behaviour of the oscillations in the uppermost part of the convection zone and the atmosphere are still somewhat uncertain, but the effects of this region can to a large extent be eliminated through suitable analysis of the observations. Apart from this difficulty, the frequencies provide a clean diagnostics of conditions inside the Sun.

What can we hope to learn from these data? An immediate goal is to determine empirically the variation of the relevant properties, in particular the sound speed and perhaps the density, throughout the Sun. Aside from satisfying our curiosity about conditions in the solar interior, this may provide constraints on conditions in the solar core, and hence on the rate of neutrino emission, or lead to determination of the depth of the solar convection zone, which is important to understanding the generation of the solar

G. Berthomieu and M. Cribier (eds.), Inside the Sun, 305–326.
© 1990 *Kluwer Academic Publishers. Printed in the Netherlands.*

magnetic field (Rädler, these proceedings) and the evolution of the solar surface abundance of, *e.g.* lithium (Baglin & Lebreton, these proceedings).

A more fundamental purpose, however, is to study the basic processes that determine the structure of the solar interior. Computations of stellar evolution are based on assumptions of perhaps questionable validity, and require information, which is often uncertain, about the properties of matter under the conditions in stellar interiors. Analysis of the solar frequencies provides a detailed test of computations of solar models, and may therefore uncover weaknesses in the assumptions that could affect other stellar models. Furthermore, the frequencies are sensitive at a significant level to even quite subtle details of the equation of state or the opacity. Thus it is possible to use the observations to study properties of plasmas under conditions so extreme that they cannot be reproduced in the laboratory.

2. Properties of the solar interior.

It is useful to review very briefly normal calculations of solar models, and their possible shortcomings (see also Bahcall, these proceedings; Turck-Chièze *et al.* 1988; Turck-Chièze 1990). It is assumed that the model is in hydrostatic and thermal equilibrium. Evolution is controlled by the gradual fusion of hydrogen into helium; it is assumed that there is no mixing in the solar interior, so that the composition in any given mass-shell is determined solely by the local nuclear burning. With these assumptions the structure is largely determined by the *microphysics* of the solar interior, *i.e.*

- the equation of state
- the opacity
- the nuclear energy generation rates.

In addition, the computation requires that the solar mass is known, as well as the initial chemical composition, which is assumed to be uniform. The goal is to compute a model at the age of the present Sun, which is also assumed to be known, with the observed radius and surface luminosity.

In practice, the initial helium abundance Y_0 cannot be determined independently and must be regarded as a free parameter of the calculation, as must the "mixing-length" parameter α which measures the efficiency of convective energy transport near the solar surface. Y_0 and α are adjusted until the model of the present Sun has the correct radius and luminosity. In this way one obtains what is sometimes called a "standard solar model". It is evidently dependent on the uncertainties in the assumed microphysics, but is otherwise well-defined.

The equation of state is discussed in these proceedings by Ebeling; furthermore the opacity is discussed in separate papers by Cox and Iglesias. Cox also considers the effects of the opacities on the solar models and their frequencies and predicted neutrino flux. The equation of state must obviously take into account the transition from very little ionization in the solar atmosphere to essentially full ionization in the solar interior. Also thermodynamic consistency must be ensured. However, the implementations used in actual calculations of solar models differ widely in complexity. Among the simplest is the one, in the following referred to as EFF, proposed by

Eggleton, Faulkner & Flannery (1973); this assumes the atoms to be in their ground states and ignores essentially all interactions between the constituents of the gas, but does include a thermodynamically consistent, if largely arbitrary, transition to full ionization in the solar core. At the opposite extreme is the socalled MHD equation of state developed by B. &. D Mihalas, Hummer & Däppen (Hummer & Mihalas 1988; Mihalas, Däppen & Hummer 1988; Däppen et al. 1988; see also Däppen 1988). Here the excitation of the atoms is included in considerable detail, and the interactions are described by assigning to each level an occupation probability which depends on the perturbations from other particles. Christensen-Dalsgaard, Däppen & Lebreton (1988) compared models and frequencies computed with these two formulations. I return to this comparison in section 5. Note also that Däppen, Lebreton & Rogers (1990) made a comparison between the MHD equation of state and a conceptually very different formulation (e.g. Rogers 1981, 1986).

The opacity is in general obtained from interpolation in tables. Commonly used have been the tables by Cox & Stewart (1970), Cox & Tabor (1976) and, more recently, tables computed with the Los Alamos Opacity Library (Huebner et al. 1977). The computation of these tables is complicated by the need to take into account the ionization states and level populations of the atoms responsible for the absorption or scattering of radiation, and to include the effect of large numbers of absorption lines. Differences in the treatment of such effects lead to substantial differences in the computed opacities (Iglesias, Rogers & Wilson 1987; Rozsnyai 1989; Courtaud et al. 1990). Thus the opacity is, at least as far as the microphysics is concerned, probably the major source of uncertainty in solar model computations. Particular difficulties may be associated with the opacity in the solar atmosphere. Here recent Los Alamos calculations (cf. Cox, Guzik & Kidman 1989) found opacities up to a factor 2 higher than previous values. Below I consider opacities from the Los Alamos Opacity Library supplemented with the new low-temperature opacities (in the following LAOL) and compare with results obtained with the Cox & Tabor tables (CT).

The computation of "standard" solar models ignores, or grossly simplifies, a number of processes that might be labelled the macrophysics of the Sun. These include

- energy transport
- dynamics of convection
- convective overshoot
- molecular diffusion
- core mixing
- magnetic fields

Energy transport by radiation is treated adequately in the solar interior in the diffusion approximation; on the other hand energy transport by convection is treated in a rather crude way, with furthermore depends on the a priori unknown parameter α. Near the surface convection is probably sufficiently vigorous to have dynamic effects on the average hydrostatic equilibrium, yet such effects are often ignored. At the lower boundary of the convection zone motion is normally supposed to stop at the point where convective instability ceases; there is no doubt, however, that motion extends into the convectively stable region, through convective overshoot, although the extent of the overshoot is uncertain. Molecular diffusion is likely to have some effect on the composition profile in the convectively stable region, yet with a few exceptions has been ignored. Instabilities in the deep interior could lead to material mixing, affecting the composition profile and hence solar evolution. (Note that mixing in the solar interior is reviewed in the papers by Spruit and by Zahn, these proceedings). Finally, magnetic fields dominate the structure of the upper solar atmosphere and may have some effect at the photospheric level. The nature or strength of the subphotospheric field is unknown, but one probably cannot totally exclude a field of sufficient magnitude to have an effect on the overall structure of the Sun.

Despite the complications it introduces, convection in a certain sense simplifies the structure of the outer parts of the Sun. Regardless of the uncertain details of convective energy transport, there is no doubt that except in a thin boundary layer near its top the convection zone is very nearly adiabatically stratified, so that gradient of density ρ is given by

$$\frac{d\ln\rho}{dr} \approx \frac{1}{\Gamma_1} \frac{d\ln p}{dr}, \tag{2.1}$$

where r is distance from the centre, p is pressure, and $\Gamma_1 = (\partial\ln p/\partial\ln\rho)_s$, the derivative being at constant specific entropy s. The structure of the adiabatic part of the convection zone is determined by this relation, together with the equation of hydrostatic support. Hence it only depends on the equation of state, the composition and the constant value of the specific entropy, which in turn is essentially fixed by the value of α; in particular, the convection zone structure is insensitive to the opacity.

It should also be noted that much of the uncertain macrophysics is concentrated very near the surface. This is true of the dynamical effects of convection, since convective velocities are likely to be very small elsewhere, of the details of convective energy transport, and of the effects of the visible magnetic field. Apart from convective overshoot and a hypothetical strong internal magnetic field, the remaining difficulties listed are concerned with the composition profile in the radiative interior of the model. Although the list of problems is not exhaustive, this argument gives some support to the simplified view of solar structure shown in Figure 1.

Quite apart from the uncertainties in the physics, it is important to consider with sufficient care the *numerical accuracy* in the computation of solar models. To utilize fully the precision of the observed frequencies to study the properties of the solar interior, we must require that the error in the calculation, given the physics, is no greater than the observational error. This is a far more stringent requirement than is normally imposed on stellar evolution calculations. In an attempt to meet it, a collaboration has been set up under the GONG project (Hill, these proceedings) to compare independently computed solar models with precisely defined physics. Some initial results were reported by Morel, Provost & Berthomieu (1990). In one case, the differences between pressure, density and sound speed in two independent models of the present Sun have been reduced to below 0.06 per cent. This is encouraging, although it still does not quite meet the precision of the most accurately determined observed frequencies.

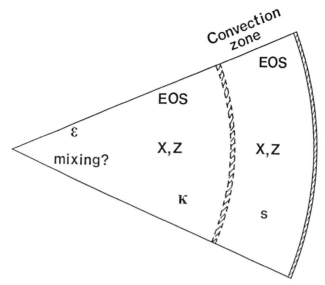

Figure 1. Schematic representation of solar structure. The thin hashed area near the surface indicates the region where the physics is uncertain, because of effects of convection, nonadiabaticity, *etc.* At the base of the convection zone, convective overshoot introduces additional uncertainty. The structure of the adiabatic part of the convection zone is determined by the equation of state (EOS), and the constant values of specific entropy *s*, and composition (given by the abundances *X* and *Z* of hydrogen and heavy elements). Beneath the convection zone the structure also depends on opacity κ and the energy generation rate ε.

3. Properties of solar oscillations.

A mode of oscillation of the Sun is characterized by three wave numbers: the *radial order n* which, roughly, gives the number of zeros in the eigenfunction in the radial direction; the *degree ℓ*; and the *azimuthal order m*, ranging between $-\ell$ and ℓ, which measures the number of zeros in longitude. The degree is related to the horizontal wavenumber k_h and wavelength λ of the mode at radius r by

$$k_h = \frac{2\pi}{\lambda} = \frac{L}{r},\tag{3.1}$$

where $L = \sqrt{\ell(\ell+1)}$.

Apart from damping or excitation, the time dependence of a single mode is harmonic, as $\cos(\omega t)$. In general the *angular frequency* $\omega = \omega_{n\ell m}$ depends on all three wave numbers. However, if rotation or other departures from spherical symmetry are ignored, $\omega_{n\ell m}$ does not depend on m. This follows from the fact that in this case there is no preferred axis in the star; since m depends on the choice of coordinate axis, the physics of the oscillations, and hence their frequencies, must be independent of m. I shall adopt this approximation here; it should be noted in passing, however, that the m-

dependence of the frequencies permit studies of solar internal rotation (*e.g.* Duvall *et al.* 1984; Brown *et al.* 1989). - In addition to ω, the *cyclic frequency* $\nu = \omega/(2\pi) = 1/P$, is commonly used, particularly in discussions of observed frequencies; here P is the oscillation period.

In calculations of solar oscillation frequencies it is common to ignore a number of complicating features that are so far badly understood, such as

- nonadiabaticity
- excitation, more generally
- dynamical effects of convection
- detailed atmospheric behaviour
- magnetic fields.

These approximations are in some sense similar to those underlying the computation of standard solar models. Calculations that do take into account some of the features (*e.g.* Christensen-Dalsgaard & Frandsen 1983; Kidman & Cox 1984; Balmforth & Gough 1988, 1990) show that they may change the frequencies by several μHz. Thus they have a substantial effect on comparisons between observed and computed frequencies. On the other hand, the complications are all (again with the possible exception of a very strong deep-seated magnetic field) located near the solar surface. Thus they add to the uncertainty of the surface region indicated in Figure 1 but do not directly affect the properties of the oscillations in the deeper solar interior.

With this simplification the computation of the oscillation frequencies is a straightforward numerical problem. Nevertheless, some care is evidently needed to obtain sufficient precision, particularly in view of the fact that the radial order of some of the observed modes is high.

As an aid to understanding the results of the numerical calculations, and to interpret the observations, asymptotic theory has been very useful. The p modes can be approximated locally by plane sound waves, with the dispersion relation $k^2 \equiv k_r^2 + k_h^2 = \omega^2/c^2$. Here k_r and k_h are the radial and horizontal components of the wave vector, and c is the adiabatic sound speed. For a mode of oscillation, k_h is given by equation (3.1), to that

$$k_r^2 = \frac{\omega^2}{c^2} - \frac{L^2}{r^2}. \tag{3.2}$$

Close to the surface, c is small and hence k_r is large. Here the modes propagate almost vertically. With increasing depth, c increases and k_r decreases, until the point is reached where $k_r = 0$ and the wave propagates horizontally. The location $r = r_t$ of this *turning point* is determined by

$$\frac{c(r_t)}{r_t} = \frac{\omega}{L}. \tag{3.3}$$

It corresponds to a point of total internal reflection; for $r < r_t$, $k_r^2 < 0$, and the mode decays exponentially. The behaviour at the surface requires a more careful analysis, which shows that below a critical cut-off frequency (which in the solar atmosphere corresponds to a cyclic frequency of about 5200 μHz) the wave is reflected by the steep density gradient. Thus the wave propagates in a series of "bounces" between the surface and the turning point. A mode of oscillation is a standing wave, formed as an interference pattern between such bouncing waves. It is trapped between the surface and r_t, and hence its frequency depends largely on conditions in this region.

Due to the rapid decrease of the sound speed with increasing radius, the first term on the right hand side of equation (3.2) is substantially larger than the second except near or below the turning point. Thus except near their turning points modes of the same frequency but different degree have essentially the same k_r; thus the properties of the modes, and their response to solar structure, are similar.

Figure 2 illustrates r_t as calculated from equation (3.3). Modes at highest observed values of ℓ are confined to the outermost fraction of a percent of the solar radius, whereas the lowest-degree modes penetrate essentially to the centre.

This simple description of the p modes may be extended to give an asymptotic relation for their frequencies (Gough 1984; Christensen-Dalsgaard et al. 1985): The condition for a standing wave is that the change in phase in the radial direction is an integral multiple of π, apart from a contribution which takes into account the phase change at the inner turning point and at the surface. This condition may be expressed as

$$\int_{r_t}^{R} k_r \, dr \approx (n + \alpha)\pi, \quad \text{or} \quad \int_{r_t}^{R} \left[1 - \left[\frac{Lc}{\omega r} \right]^2 \right]^{1/2} \frac{dr}{c} \approx \frac{\pi(n + \alpha)}{\omega}, \quad (3.4)$$

where we used equation (3.2); here α is the quantity which takes into

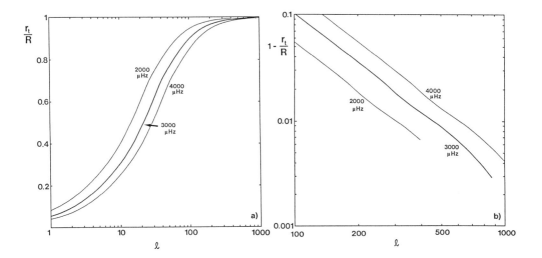

Figure 2. The turning point radius r_t (a) and the penetration depth $R - r_t$ (b), in units of the solar radius R, as a function of degree ℓ for three values of the frequency ν. This has been calculated from equation (3.3) for a model of the present Sun.

account the phase change at the reflection points and hence in particular depends on conditions near the surface. From equation (3.3) follows that the left hand side of equation (3.4) is a function $F(\omega/L)$ of ω/L. Thus this equation establishes a very particular relation among the p-mode frequencies. That solar oscillations satisfy such a relation was first found by Duvall (1982) from observed frequencies.

For small ℓ, equation (3.4), with L replaced by $\ell + \tfrac{1}{2}$, reduces to

$$\nu \sim \left(n + \frac{\ell}{2} + \frac{1}{4} + \alpha\right)\Delta\nu \tag{3.5}$$

where

$$\Delta\nu = \left[2\int_0^R \frac{dr}{c}\right]^{-1} \tag{3.6}$$

is the inverse of twice the sound travel time between the centre and the surface (e.g. Tassoul 1980). Thus there is approximately a uniform spacing $\Delta\nu$ between modes of same degree, but different order. Equation (3.5) also predicts the approximate equality $\nu_{n\ell} \approx \nu_{n-1,\ell+2}$. This frequency pattern has been observed for the solar 5 min modes of low degree and may be used in the search for stellar oscillations of solar type.

It is of great interest to consider the *deviations* from this simple relation. The separation $\delta\nu_{n\ell} = \nu_{n\ell} - \nu_{n-1,\ell+2}$ is predominantly determined by conditions in the solar core (e.g. Provost 1984; Gough 1986a), since, as argued above, only here does the behaviour of the modes depend substantially on ℓ. A more careful analysis shows that the average separation satisfies

$$\langle\delta\nu_{n\ell}\rangle \approx (4\ell + 6)D_0. \tag{3.7}$$

In section 5 I consider examples of the dependence of the constant D_0 on the structure of the solar core.

By linearizing equation (3.4) one may obtain an expression for the frequency change $\delta\omega$ caused by a change in the solar model, with resulting changes δc and $\delta\alpha$ in c and α (Christensen-Dalsgaard, Gough & Pérez Hernández 1988). Since c/r decreases quite rapidly with increasing r, $L^2c^2/r^2\omega^2 \ll 1$ except near the turning point r_t, and as a first approximation may be neglected in the resulting expression. If furthermore the term in $\delta\alpha$ can be neglected, the result is the very simple relation between the changes in sound speed and frequency:

$$\frac{\delta\omega}{\omega} \approx \frac{\int_{r_t}^R \frac{\delta c}{c}\frac{dr}{c}}{\int_{r_t}^R \frac{dr}{c}} \tag{3.8}$$

This shows that the change in sound speed in a region of the Sun affects the frequency with a weight determined by the time spent by the mode, regarded as a superposition of traveling waves, in that region. Thus changes near the surface, where the sound speed is low, have relatively large effects on the frequencies. Although this expression is only a rough approximation, it is a useful guide in attempts to interpret frequency differences between models, or between observed and computed frequencies. Note that according to equation (3.8) $\delta\omega/\omega$ depends on the properties of the mode only through

r_t, which in turn is determined by ω/L (*cf.* equation (3.3)).

From a physical point of view, the denominator in equation (3.8) corrects for the fact that with increasing r_t the modes extend over a smaller fraction of the solar mass, and hence their frequencies are easier to perturb. A similar result is obtained from a perturbation analysis, based on the exact oscillation equations, of the effects of modifications to the model or the physics of the oscillations (*e.g.* Christensen-Dalsgaard 1988a; Christensen-Dalsgaard & Berthomieu 1990). In analyses of frequency differences between models, or between observations and theory, this effect may be eliminated by considering scaled frequency differences $Q_{n\ell}\delta\omega_{n\ell}$. Here

$$Q_{n\ell} = \frac{E_{n\ell}}{\overline{E}_0(\omega_{n\ell})}, \tag{3.9}$$

where $E_{n\ell}$ is a measure of the inertia in the mode, integrated over the volume of the Sun, and $\overline{E}_0(\omega)$ is the value of $E_{n\ell}$ for $\ell = 0$, interpolated to the frequency ω. Roughly speaking, $Q_{n\ell}\delta\omega_{n\ell}$ measures the effect of the part of the modification which is confined to the region where the actual mode is trapped, on a radial mode of the same frequency.

As argued above, near the surface the behaviour of the oscillations depends on frequency but not on ℓ. Thus, if the modification is confined close to the surface its effect on the frequency, when corrected for the ℓ-dependence of the mode inertia, is a function of frequency alone; so therefore is $Q_{n\ell}\delta\omega_{n\ell}$. The condition for this to be true is that the extent of the region over which the modification is significant is much smaller than the depth of penetration of the modes considered. It follows that if $Q_{n\ell}\delta\omega_{n\ell}$ *does* depend on ℓ for a set of modes, the change in the model extends at least to the lower turning point of those modes.

The upper reflection of the modes occurs at increasing depth with decreasing frequency; thus the mode amplitude very near the surface, relative to the amplitude in the interior, decreases (Libbrecht 1988; Christensen-Dalsgaard 1988b). It follows that low-frequency modes are insensitive to modifications that are confined to the superficial layers of the model.

These properties of the oscillations are particularly important in the light of the currently unavoidable errors near the surface of the model. These errors may be expected to lead to scaled frequency errors that are essentially independent of ℓ and small at low frequency. Frequency errors that do not have these properties therefore indicate errors in the bulk of the model.

4. Sensitivity of the frequencies to changes in solar structure.

To provide an illustration of the principles discussed in the previous section, it is interesting to consider specific examples of how a solar model and its frequencies respond to changes in the physics. For p modes, which are the only modes considered here, it follows from section 3 that the change in the sound speed is particularly relevant. A detailed discussion of the effects of various modifications was presented by Christensen-Dalsgaard (1988a). Cox *et al.* (1989) studied the effects of molecular diffusion, Christensen-

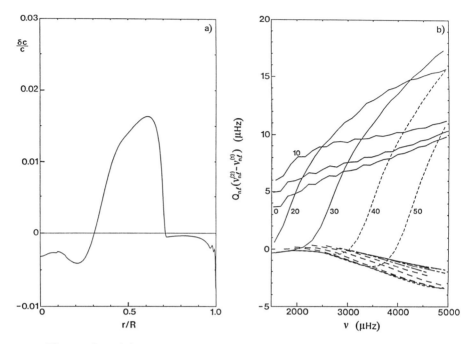

Figure 3. (a) Sound-speed difference, at fixed fractional radius r/R, between the model with modified opacity in the interior and the reference model, in the sense (modified model) - (reference model). (b) Frequency differences between the same two models. The differences have been scaled by the inertia ratio $Q_{n\ell}$ (cf. equation (3.9)). Points corresponding to a given value of ℓ have been connected, according to the following convention: ℓ = 0, 5, 10, 20, 30 (————); ℓ = 40, 50, 70, 100 (--------); ℓ = 150, 200, 300, 400 (— — — —); and ℓ = 500, 600, 700, 800, 900, 1000 (— - — - —). In addition a few values of ℓ have been indicated in the figure.

Dalsgaard, Däppen & Lebreton (1988) compared frequencies computed with different assumptions about the equation of state, and Gough & Novotny (1990) studied the effect of the assumed solar age. An interesting analysis of the effects of the opacity was presented by Korzennik & Ulrich (1989).

Here I concentrate on effects of modifications to the opacity. Indeed, it was argued in section 2 that apart from the very uncertain region near the solar surface, the opacity is probably the least well-determined of the physical properties required to compute a model of the Sun. The results presented here were discussed in considerably more detail by Christensen-Dalsgaard & Berthomieu (1990).

All models were calibrated to have the solar radius and luminosity. The reference model was similar to model 1 of Christensen-Dalsgaard (1982), although it was computed with considerably better numerical precision. To obtain the modified models, the opacity κ was changed by adding to $\log\kappa$ a function of temperature which was only different from zero in a restricted temperature interval. I first consider the effect of a modification in the opacity near the base of the convection zone, the maximum change in opacity at fixed T and ρ being about 26 per cent. In Figure 3a is shown the resulting change in the sound speed. A striking feature is the comparatively small change in much of the convection zone. In fact it is easy to show that except in the outer part of the convection zone the sound speed is approximately determined by the total mass and surface radius of the model, which are the same in the two cases (e.g. Christensen-Dalsgaard 1986). The opacity increase caused an increase in the depth of the convection zone by about 0.02 R. As a result, the temperature gradient is higher (being adiabatic) in the modified model than in the reference model just beneath the bottom of the convection zone of the latter. This is the reason for the increase in $\delta c/c$ with increasing depth just beneath the convection zone.

Frequency differences between the modified and the reference model are shown in Figure 3b. In accordance with the discussion in section 3, the differences have been scaled by the ratio $Q_{n\ell}$ of mode inertias (cf. equation (3.9)); for the highest-degree modes the raw differences are larger by a factor of about 5 than those shown. These differences can be understood relatively simply in terms of the differences in sound speed shown in Figure 3a and the behaviour of the turning point illustrated in Figure 2. At very low degree the modes penetrate essentially to the centre, and the frequency change is given by the weighted average in equation (3.8), which is dominated by the region of positive δc beneath the convection zone. As ℓ increases to 10 the turning point moves out through the region of slightly negative δc near the core, and $Q_{n\ell}\delta\nu_{n\ell}$ increases. At higher degrees,

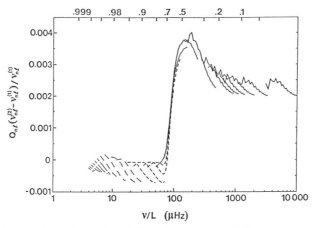

Figure 4. Scaled relative frequency differences corresponding to the differences in Figure 3b, but plotted against $\nu/(\ell + \tfrac{1}{2})$. The upper axis is labelled in terms of the corresponding turning point position r_t/R. The tick marks are at r_t/R = 0.05, 0.1, 0.2, 0.3, 0.5, 0.7, 0.8, 0.9, 0.95, 0.98, 0.99, 0.995 and 0.999. The same coding of the lines is used as in Figure 3b.

316

beginning at low frequency for $\ell = 20$, and at increasing ν when ℓ increases to 50, the modes become largely confined within the convection zone, where δc is negative; thus the frequency differences are negative.   It should be noticed that the negative δc near the surface has a substantial effect on the frequencies, despite its insignificant appearance in Figure 3a.   The reason is the weighting with c^{-1} (*cf.* equation (3.8)) which makes the frequencies very sensitive to changes in the model near the surface.

It follows from this description that the behaviour of the frequency differences can to a large extent be described in terms of the turning point position r_t.   This is seen more clearly in Figure 4, where $Q_{n\ell}\delta\nu_{n\ell}/\nu_{n\ell}$ is plotted against $\nu/(\ell + \tfrac{1}{2})$ which according the equation (3.3) determines r_t (it follows from a more careful asymptotic analysis that L in equation (3.2) should be replaced by $\ell + \tfrac{1}{2}$).   Particularly striking is the transition near $\nu/(\ell + \tfrac{1}{2}) = 100$, where the turning point moves from beneath to above the base of the convection zone, and the modes therefore no longer penetrate into the region of positive $\delta c/c$.

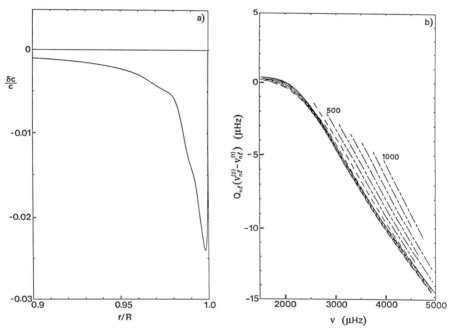

Figure 5.   (a) Sound-speed difference, at fixed fractional radius r/R, between a model of the present Sun where the opacity has been artificially increased by up to about a factor 2 near the surface and a normal model, in the sense (modified model) - (reference model).   (b) Scaled frequency differences between the same two models.   Points corresponding to a given value of ℓ have been connected, according to the same convention as in Figure 3b.

To investigate the effects of the recently suggested opacity increase in the solar photosphere (*cf.* section 2), I computed a model where $\log_{10}\kappa$ was increased by 0.3 in the atmosphere and the upper part of the convection zone. Figure 5a shows the resulting sound-speed difference between the models, in the outer part of the convection zone. The change in the deeper parts of the model is essentially negligible. Corresponding scaled frequency differences are illustrated in Figure 5b. Modes of degree less than about 500 all penetrate well beyond the region where the sound speed is affected (*cf.* Figure 2), and for these the scaled frequency change is mainly a function of frequency, but depends little on the depth of penetration of the mode, and hence on ℓ. At higher degree the modes sample only part of the negative sound speed difference, and the frequency change is smaller.

Quite apart from the specific modification considered, this example illustrates the important point, discussed in section 3, that changes near the surface cause scaled frequency changes that depend mainly on frequency and are small at low frequency. Qualitatively similar changes result from modifications to the treatment of the upper, significantly superadiabatic part of the convection zone (Christensen-Dalsgaard 1986). More generally, it seems likely that the uncertainties in the treatment of the surface layers (nonadiabaticity of the oscillations; the treatment of convection; possible effects of magnetic fields; *etc.*) would have a similar effect on the frequencies. Thus in analyzing observed frequencies it is possible to absorb these uncertainties by allowing an undetermined frequency-dependent part of the scaled differences between observed and computed frequencies.

5. Comparison with observed frequencies.

It is of obvious interest to compare observed frequencies of solar oscillations with frequencies of representative models. Here I consider 4 such models, differing in the treatment of the equation of state or the opacity, as discussed in section 2. The equation of state was obtained either from the EFF or the MHD formulation. The opacity was obtained from interpolation in either the CT or the LAOL tables. Finally the nuclear reaction parameters were essentially as in Bahcall & Ulrich (1988). In the following the models will be labelled as, for example, (EFF, CT) for the model with the Eggleton *et al.* equation of state and the Cox & Tabor opacity table.

Figure 6 shows differences between selected observed frequencies from Duvall *et al.* (1988) and Libbrecht & Kaufman (1988), and computed frequencies for the (EFF, CT) model. The dominant feature in this plot is the increase in the magnitude of the differences with increasing degree. However, this is largely caused by the variation in the mode inertia due to the decrease of the extent of the region where the modes are trapped. If instead one considers the scaled frequency differences $Q_{n\ell}\delta\nu_{n\ell}$, shown in Figure 7a, most of the ℓ-dependence is eliminated. As argued in section 3, this suggests that the dominant errors in the model or the frequency calculation is located close to the solar surface.

It is interesting to analyze in more detail these scaled differences, as well as those for the other three combinations of equation of state and opacity which are also shown in Figure 7. A closer look at Figure 7a shows

318

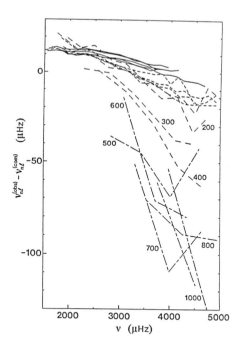

Figure 6. Differences between observed frequencies and frequencies for a model computed with the EFF equation of state and the CT opacity tables. Points corresponding to a given value of ℓ have been connected, according to the same convention as in Figure 3b.

Figure 7 (following page). Scaled differences between observed frequencies and frequencies computed for models of the present Sun. The observed data are from Duvall *et al.* (1988) and Libbrecht and Kaufman (1988). As described in the text, the models differ in the choice of equation of state and opacity tables as follows: a) EFF equation of state, CT opacities; b) MHD equation of state, CT opacities; c) EFF equation of state, LAOL opacities; d) MHD equation of state, LAOL opacities. Points corresponding to a given value of ℓ have been connected, according to the same convention as in Figure 3b.

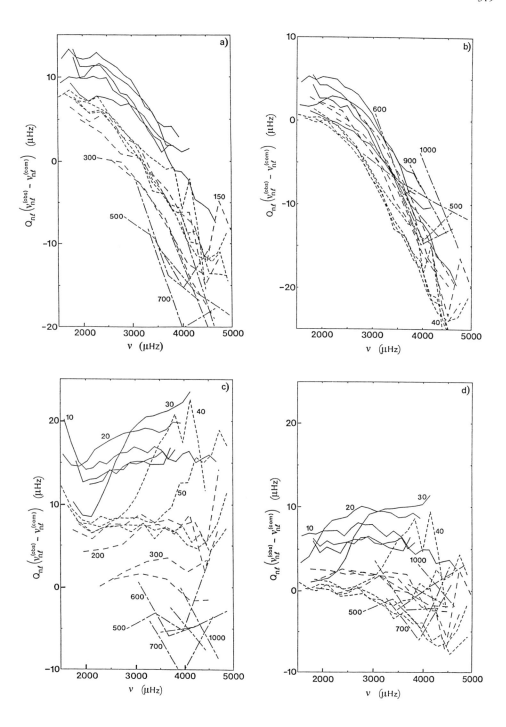

that there is still a systematic variation with ℓ in $Q_{n\ell}\delta\nu_{n\ell}$. This is visible as a shift between $\ell = 20$ and 40, and a gradual increase in the magnitude of the differences at higher ℓ. Also, there are substantial differences at low frequency. Both these features indicate that there are significant errors in the interior of the model. If the EFF equation of state is replaced by the MHD formulation (*cf.* Figure 7b), the differences at low frequency are reduced substantially, as is the ℓ-dependence. This strongly indicates that the error in the interior of the model, particularly in the convection zone where modes of degree higher than about 50 are trapped, has been reduced by the introduction of the MHD equation of state (Christensen-Dalsgaard, Däppen & Lebreton 1988). The dominant remaining trend is the strong frequency dependence of the differences, as well as a barely visible shift around $\ell = 40$. The latter feature must be associated with errors in the model around the turning point of modes of degree 40, *i.e.* at or below the base of the convection zone (*e.g.* Christensen-Dalsgaard & Gough 1984). The presence of a frequency-dependent part of the differences could have been expected, given the approximations in the calculation. The neglect of nonadiabaticity and the effects of convection could well introduce scaled frequency errors of this magnitude, and, as argued in section 3, they would be expected to be largely independent of ℓ.

In the (EFF, LAOL) model (Figure 7c) the variation in $Q_{n\ell}\delta\nu_{n\ell}$ with ℓ is increased substantially over the (EFF, CT) model. Again the use of the MHD equation of state (Figure 7d) causes a marked decrease in the variation for modes trapped in the convection zone, confirming the earlier conclusion that MHD gives the better representation of the equation of state. For modes penetrating beneath the convection zone, however, there remains a substantial ℓ-dependence, indicating that the use of the LAOL opacities increases the errors in the radiative interior of the model. Indeed, although the LAOL opacities are generally somewhat larger than the CT values, near the base of the convection zone they are about 10 per cent smaller, causing the depth of the convection zone to be smaller in the LAOL models. Thus, for example, the depth is 0.283 R in the (MHD, CT) model and 0.267 R in the (MHD, LAOL) model. A more careful analysis of the oscillation frequencies shows that in the Sun the depth of the convection zone is close to the former value (Christensen-Dalsgaard, Gough & Thompson 1990) The difference in convection zone depth causes a pattern of frequency differences, when going from modes penetrating beneath to modes trapped within the convection zone that is qualitatively similar to what was observed in Figure 3b. That the LAOL opacities may be too low in this region was also noticed by Cox *et al.* (1989) and Korzennik & Ulrich (1989).

The most striking effect of using the LAOL opacities, however, is that the frequency-dependent part of the differences has been reduced substantially. This is most evident in Figure 7d, for the (MHD, LAOL) model. For this model the errors, for modes of degree exceeding 100, are comparable with the estimated observational errors. The effect of using the LAOL opacity on this part of the differences is caused by the opacity increase at low temperature. The effect of such an increase was presented in Figure 5b and is comparable with the differences in the (MHD, CT) case shown in Figure 7b. Some care is required, however, when interpreting this result. It might be tempting to take the agreement between observation and theory at face value, as an indication that the LAOL opacities are to be preferred at low temperature, and that there is then little remaining error in the description of the model and the oscillations near the surface. Such a conclusion would be premature, given the known inadequacies in the frequency computation. It is likely that the agreement in Figure 7d is fortuitous, resulting from a partial cancellation of several sources of error.

It was argued in section 3 that for low-degree modes the quantity D_0, which is related to the average frequency difference $\langle \nu_{n\ell} - \nu_{n-1\ell\cdot2} \rangle$ by equation (3.7), is a measure of conditions in the solar core. The average separation has been measured with considerable precision (e.g. Jiménez et al. 1988, Gelly et al. 1988). However, since the asymptotic relation (3.7) is not exact, the value of D_0 depends on how it is obtained. Here I use a least squares fit (e.g. Scherrer et al. 1983) to the frequencies of modes of degree 0 - 3, including those for which $17 \le n + \frac{1}{2}\ell \le 29$, corresponding in frequency to the range between approximately 2500 and 4100 μHz. This leads to a determination of D_0, as well as an average value $\Delta\nu_0$ of the overall frequency spacing $\Delta\nu$ (cf. equation (3.6)).

I have applied this analysis to the observed frequencies, and to the frequencies compared with the observations in Figure 7. In addition I have considered a model whose core has been partially mixed; the hydrogen abundance profile $X(m)$, as a function of the mass m, was obtained from that of the Re* = 100 model of Schatzman et al. (1981) by scaling with a constant factor, chosen to obtain the correct luminosity (cf. Christensen-Dalsgaard 1986). The results are shown in Table I. The most striking feature is the difference between the results for the normal and the partially mixed model. The increase in D_0 in the mixed model is caused by the increase in the sound speed c in the core. For an approximately ideal gas c is given by

$$c^2 \approx \frac{\Gamma_1 k_B T}{\mu m_u},$$

(5.1)

where T is temperature, μ is the mean molecular weight, k_B is Boltzman's constant and m_u is the atomic mass unit. In the mixed model the central hydrogen abundance is higher, and μ is consequently smaller; on the other hand there is little difference in the temperature. As a result c is higher. For the remaining models there is some scatter in the values of D_0, the general tendency being that the computed values are close to, but slightly higher than the value obtained from the observations. On the other hand D_0 for the mixed model is evidently not consistent with the observed value. Thus mixing as severe as that proposed by Schatzman et al. appears to be ruled out by the observed frequencies (see also Cox & Kidman 1984; Provost 1984; Christensen-Dalsgaard 1986).

It has been pointed out (Spergel, these proceedings; Faulkner 1990) that the presence in the solar core of a very small population of hypothetical "weakly interacting massive particles" (WIMPs) could contribute to the energy transport in the core and hence lower the central temperature and consequently the neutrino flux. The reduction in the central temperature, which occurs without substantial modifications in the composition profile, would lead to a reduction of the sound speed in the core, and hence to a reduction in D_0 (Faulkner, Gough & Vahia 1986, Däppen, Gilliland & Christensen-Dalsgaard 1986, Gilliland & Däppen 1988). If parameters are

Table I.

Frequencies	$\Delta\nu_0$	D_0
Duvall *et al.* observations	135.15 μHz	1.487 μHz
EFF, CT model	136.52 μHz	1.551 μHz
MHD, CT model	136.78 μHz	1.533 μHz
EFF, LAOL model	135.30 μHz	1.557 μHz
MHD, LAOL model	135.57 μHz	1.500 μHz
Schatzman *et al.* mixed model	136.68 μHz	1.974 μHz

Average frequency separations (*cf.* equations (3.5) and (3.7)) for the compilation of observed frequencies by Duvall *et al.* (1988), as well as for a number of solar models. The first four models are "standard" solar models, differing in the equation of state or the opacity, whereas in the last model the hydrogen profile simulates partial mixing by "turbulent diffusion" (Schatzman *et al.* 1981).

chosen for the WIMPs such that the model has the observed neutrino capture rate, D_0 is typically reduced by 8 - 15 per cent relative to the corresponding normal models. The earlier calculations indicated that this improved the agreement between the computed and the observed values of D_0. However, the values presented in Table I are somewhat smaller than those obtained previously for the normal models, the difference being probably due to an improvement in the numerical precision. Thus it appears likely that for corresponding models with WIMPs D_0 would be significantly *lower* than the observations. Certainly there is no evidence in the present results that modifications beyond the "standard" model is required to bring theory and observations into agreement on D_0 (see also Cox *et al.* 1989). Indeed, Gough & Kosovichev (1988) found that the sound speed inferred from inverting more extensive sets of oscillation frequencies appeared to be inconsistent with models including WIMPs (see also Gough, these proceedings).

6. Discussion.

The principal result of the present paper is that solar oscillation frequencies are sensitive to the physics of the solar interior, and that by suitable analysis of the observations it is possible to to some extent to separate the effects of various aspects of the physics. The separation is aided by the properties of the oscillations, in that the observations contain modes that penetrate to very different depths. Furthermore, the structure of the convection zone is largely independent of opacity; by studying modes that are entirely trapped in the convection zone, one therefore gets information about the equation of state that is depends little on the uncertainties in the opacities. On the other hand, beneath the convection zone the gas is essentially fully ionized and the equation of state is relatively simple; here the principal uncertainty is therefore in the opacity. Thus, by judicious choice of data there is some hope that we may learn both about the equation of state and the opacity of matter under stellar condition.

A special problem is encountered near the solar surface, where the physics of both the structure of the Sun and the oscillations is uncertain. However, by suitable scaling of frequency differences between the observations and the model the effects of this region can to a large extent be eliminated. Thus it is possible to study the properties of the solar interior separately from the uncertainties of the surface region.

These principles were illustrated by studying the effects on the model and the frequencies of artificial modifications to the opacity, and by comparing with the observations frequencies of models differing in the equation of state or the opacity. The differences in the the physics were within the range of currently used formulations. These differences caused frequency changes that far exceeded the observational errors. Furthermore it appeared that the use of a more sophisticated equation of state lead to a considerable improvement between observed and computed frequencies. The situation with regards to the opacity is less clear. The use of the newer opacities in the solar interior apparently increased the discrepancy between theory and observation; on the other hand a recently proposed major increase in the opacity in the solar atmosphere seemed to cause a significant improvement in the agreement, reducing the differences between theory and observation to a level close to the observational error for those modes that are trapped in the convection zone. Given the remaining uncertainties affecting this part of the model, however, this agreement is of doubtful significance.

I have only considered the simplest use of the observed oscillation frequencies. Much more detailed information can be obtained by applying *inverse analyses* (*e.g.* Gough 1985, 1986b; Gough & Kosovichev 1988; Gough & Thompson 1990). The data are in fact of sufficient quality to permit the determination of the run of sound speed as a function of position in much of the Sun (*e.g.* Christensen-Dalsgaard *et al.* 1985; Brodsky & Vorontsov 1987; Vorontsov 1988; Kosovichev 1988; Christensen-Dalsgaard, Gough & Thompson 1988, 1989; Sekii & Shibahashi 1989), or for other properties of the solar interior. Gough & Kosovichev (1988) and Kosovichev (1990) suggested that the data indicate slight mixing of material in the solar core, although to a far smaller extent than in the Schatzman *et al.* (1981) model discussed in section 5. Dziembowski, Pamyatnykh & Sienkiewicz (1990) found that the minimum neutrino flux consistent with results of an inversion were considerably *higher* than the flux predicted by standard solar models; the sensitivity of this conclusion to random and systematic errors in the data has still to be tested, however.

Helioseismology has already contributed significantly to our knowledge about the interior of the Sun. It is perhaps surprising (and to some possibly even disappointing) that so far no major departure from standard evolution theory has been revealed by the results. Certainly they have offered no solution to the neutrino problem; in contrast there is a tendency for models with a low neutrino flux (*e.g.* with substantial core mixing or energy transport by WIMPs) to be inconsistent with the seismic data. We are only the at the beginning of seismic investigations of the Sun, however. In the coming decade we shall witness a major increase in the amount and quality of observations of solar oscillations, as ground-based networks of oscillation observatories (Hill, these proceedings) and space-based facilities (Bonnet, these proceedings) become operational. Such data will allow us to look for more subtle failures of standard models. We can hope to constrain conditions in the solar core to the extent that a reliable estimate can be made of the neutrino spectrum which is produced by nuclear reactions; the detailed observations of the neutrino spectrum which will become available in the same period could then be used to investigate the properties of the neutrino. Finally, it should be possible to separate the uncertainty in the "macrophysics" of the solar interior from the effects of the microphysics, and hence to investigate the latter in considerable detail; we would then be in a position to use the Sun as a laboratory for the study of basic properties of plasmas.

Acknowledgements: I am grateful Y. Lebreton for her assistance in installing the LAOL opacity tables and updating the nuclear reaction rates in my evolution code. The Danish Natural Science Research Council is thanked for supporting the computations reported here and my participation in the meeting.

References.

Bahcall, J. N. & Ulrich, R. K., 1988. *Rev. Mod. Phys.*, **60**, 297.

Balmforth, N. J. & Gough, D. O., 1988. *Seismology of the Sun & Sun-like Stars*, p. 47, eds Domingo, V. & Rolfe, E. J., ESA SP-286.

Balmforth, N. J. & Gough, D. O., 1990. *Solar Phys.* (special issue, IAU Coll. 121), submitted.

Brodsky, M. A. & Vorontsov, S. V., 1987. *Pis'ma Astron. Zh.*, **13**, 438 (English translation: *Sov. Astron. Lett.*, **13**, 179).

Brown, T. M., Christensen-Dalsgaard, J., Dziembowski, W. A., Goode, P., Gough, D. O., Morrow, C. A., 1989. *Astrophys. J.*, **343**, 526.

Christensen-Dalsgaard, J., 1982. *Mon. Not. R. astr. Soc.*, **199**, 735.

Christensen-Dalsgaard, J., 1986. *Seismology of the Sun and the distant Stars*, p. 23, ed. Gough, D. O., Dordrecht, D. Reidel Publ. Co..

Christensen-Dalsgaard, J., 1988a. *Seismology of the Sun & Sun-like Stars*, p. 431, eds Domingo, V. & Rolfe, E. J., ESA SP-286.

Christensen-Dalsgaard, J., 1988b. *Multimode Stellar Pulsations* p. 153, eds Kovacs, G., Szabados, L. & Szeidl, B.; Kultura Press and Konkoly Observatory, Budapest.

Christensen-Dalsgaard, J. & Berthomieu, B., 1990 In *Solar interior and atmosphere*, eds Cox, A. N., Livingston, W. C. & Matthews, M., Space Science Series, University of Arizona Press, submitted.

Christensen-Dalsgaard, J. & Frandsen, S., 1983. *Solar Phys.*, **82**, 165.

Christensen-Dalsgaard, J. & Gough, D. O., 1984. *Solar seismology from space*, p. 79, eds Ulrich, R. K., Harvey, J., Rhodes, E. J. & Toomre, J., NASA, JPL Publ. 84-84.

Christensen-Dalsgaard, J., Duvall, T. L., Gough, D. O., Harvey, J. W. & Rhodes, E. J., 1985. *Nature*, **315**, 378.

Christensen-Dalsgaard, J., Däppen, W. & Lebreton, Y., 1988. *Nature*, **336**, 634.

Christensen-Dalsgaard, J., Gough, D. O. & Pérez Hernández, F., 1988. *Mon.*

Not. R. astr. Soc., **235**, 875.

Christensen-Dalsgaard, J., Gough, D. O. & Thompson, M. J., 1988. *Seismology of the Sun & Sun-like Stars*, p. 493, eds V. Domingo & E. J. Rolfe, ESA SP-286, Noordwijk, Holland.

Christensen-Dalsgaard, J., Gough, D. O. & Thompson, M. J., 1989. *Mon. Not. R. astr. Soc.*, **238**, 481.

Christensen-Dalsgaard, J., Gough, D. O. & Thompson, M. J., 1990. *Mon. Not. R. astr. Soc.*, submitted.

Courtaud, D., Damamme, G., Genot, E., Vuillemin, M. & Turck-Chièze, 1990. *Solar Phys.* (special issue, IAU Coll. 121), submitted.

Cox, A. N. & Kidman, R. B., 1984. *Theoretical Problems in Stellar Stability and Oscillations*, p. 259, (Institute d'Astrophysique, Liège)

Cox, A. N. & Stewart, J. N., 1970. *Astrophys. J. Suppl.*, **19**, 243.

Cox, A. N. & Tabor, J. E., 1976. *Astrophys. J. Suppl.*, **31**, 271.

Cox, A. N., Guzik, J. A. & Kidman, R. B., 1989. *Astrophys. J.*, **342**, 1187.

Duvall, T. L., 1982. *Nature*, **300**, 242.

Duvall, T. L., Dziembowski, W. A., Goode, P. R., Gough, D. O., Harvey, J. W. & Leibacher, J. W., 1984. *Nature*, **310**, 22.

Duvall, T. L., Harvey, J. W., Libbrecht, K. G., Popp, B. D. & Pomerantz, M. A., 1988. *Astrophys. J.*, **324**, 1158.

Dziembowski, W. A., Pamyatnykh, A. A. & Sienkiewicz, R., 1990. *Mon. Not. R. astr. Soc.*, submitted.

Däppen, W., 1988. *Seismology of the Sun & Sun-like Stars*, p. 451, eds Domingo, V. & Rolfe, E. J., ESA SP-286.

Däppen, W., Gilliland, R. L. & Christensen-Dalsgaard, J., 1986. *Nature*, 321, 229.

Däppen, W., Lebreton, Y. & Rogers, F., 1990. *Solar Phys.* (special issue, IAU Coll. 121), submitted.

Däppen, W., Mihalas, D., Hummer, D. G. & Mihalas, B. W., 1988. *Astrophys. J.*, **332**, 261.

Eggleton, P. P., Faulkner, J. & Flannery, B. P., 1973. *Astron. Astrophys.*, **23**, 325.

Faulkner, J., 1990. *Solar Phys.* (special issue, IAU Coll. 121), submitted.

Faulkner, J., Gough, D. O. & Vahia, M. N., 1986. *Nature*, 321, 226.

Gelly, B., Fossat, E., Grec, G. & Schmider, F.-X., 1988. *Astron. Astrophys.*, **200**, 207.

Gilliland, R. L. & Däppen, W., 1988. *Astrophys. J.*, **324**, 1153.

Gough, D. O., 1984. *Phil. Trans. R. Soc. London*, A **313**, 27.

Gough, D. O., 1985. *Solar Phys.*, **100**, 65.

Gough, D. O., 1986a. *Hydrodynamic and magnetohydrodynamic problems in the sun and stars*, p. 117, ed Osaki, Y., University of Tokyo Press.

Gough, D. O., 1986b. *Seismology of the Sun and the distant Stars*, p. 125, ed. Gough, D. O., Reidel, Dordrecht.

Gough, D. O. & Kosovichev, A. G., 1988. *Seismology of the Sun & Sun-like Stars*, p. 195, eds Domingo, V. & Rolfe, E. J., ESA SP-286.

Gough, D. O. & Novotny, E., 1990. *Solar Phys.* (special issue, IAU Coll. 121), submitted.

Gough, D. O. & Thompson, M. J., 1990. In *Solar interior and atmosphere*, eds Cox, A. N., Livingston, W. C. & Matthews, M., Space Science Series, University of Arizona Press, submitted.

Huebner, W. F., Merts, A. L., Magee, N. H. & Argo, M. F., 1977.

Astrophysical Opacity Library, Los Alamos Scientific Laboratory report LA-6760-M.

Hummer, D. G. & Mihalas, D., 1988. *Astrophys. J.*, **331**, 794.

Iglesias, C. A., Rogers, F. J. & Wilson, B. G., 1987. *Astrophys. J.*, **322**, L45.

Jiménez, A., Pallé, P. L., Pérez, J. C., Régulo, C., Roca Cortés, T., Isaak, G. R., McLeod, C. P. & van der Raay, H. B., 1988. *Proc. IAU Symposium No 123*, p. 205, eds Christensen-Dalsgaard, J. & Frandsen, S., Reidel, Dordrecht.

Kidman, R. B. & Cox, A. N., 1984. *Solar seismology from space*, p. 335, eds Ulrich, R. K., Harvey, J., Rhodes, E. J. & Toomre, J., NASA, JPL Publ. 84-84.

Korzennik, S. G. & Ulrich, R. K., 1989. *Astrophys. J.*, **339**, 1144.

Kosovichev, A. G., 1988. *Seismology of the Sun & Sun-like Stars*, p. 533, eds Domingo, V. & Rolfe, E. J., ESA SP-286.

Kosovichev, A. G., 1990. *Solar Phys.* (special issue, IAU Coll. 121), submitted.

Libbrecht, K. G., 1988. *Seismology of the Sun & Sun-like Stars*, p. 3, eds Domingo, V. & Rolfe, E. J., ESA SP-286.

Libbrecht, K. G. & Kaufman, J. M., 1988. *Astrophys. J.*, **324**, 1172.

Mihalas, D., Däppen, W. & Hummer, D. G., 1988. *Astrophys. J.*, **331**, 815.

Morel, P., Provost, J. & Berthomieu, G., 1990. *Solar Phys.* (special issue, IAU Coll. 121), submitted.

Provost, J., 1984. *Proc. IAU Symposium No 105: "Observational Tests of Stellar Evolution Theory"*, p. 47, eds Maeder, A. & Renzini, A., Reidel.

Rogers, F., 1981. *Phys. Rev.*, **A24**, 1531.

Rogers, F., 1986. *Astrophys. J.* **310** 723.

Rozsnyai, B. F., 1989. *Astrophys. J.*, **341**, 414.

Schatzman, E. & Maeder, A., Angrand, F. & Glowinski, R., 1981. *Astron. Astrophys.*, **96**, 1.

Scherrer, P. H., Wilcox, J. M., Christensen-Dalsgaard, J. & Gough, D. O., 1983. *Solar Phys.*, **82**, 75.

Sekii, T. & Shibahashi, H., 1989. *Publ. Astron. Soc. Japan*, **41**, 311.

Tassoul, M., 1980. *Astrophys. J. Suppl.*, **43**, 469.

Turck-Chièze, S., 1990. *Solar Phys.* (special issue, IAU Coll. 121), submitted.

Turck-Chièze, S., Cahen, S., Cassé, M. & Doom, C., 1988. *Astrophys. J.*, **335**, 415.

Vorontsov, S. V., 1988. *Seismology of the Sun & Sun-like Stars*, p. 475, eds Domingo, V. & Rolfe, E. J., ESA SP-286.

USING HELIOSEISMIC DATA TO PROBE THE HYDROGEN ABUNDANCE
IN THE SOLAR CORE

D. O. GOUGH
Institute of Astronomy and Department of Applied Mathematics
and Theoretical Physics, University of Cambridge, UK; and
Joint Institute for Laboratory Astrophysics, University of
Colorado and National Institute of Standards and Technology,
Boulder, Colorado, USA

A. G. KOSOVICHEV
Crimea Astrophysical Observatory
Nauchny, Crimea, USSR

ABSTRACT. A procedure for inverting helioseismic data to determine
the hydrogen abundance in the radiative interior of the sun is briefly
described. Using Backus-Gilbert optimal averaging, the variation of
sound speed, density and hydrogen abundance in the energy-generating
core is estimated from low-degree p-mode frequencies. The result
provides some evidence for there having been some redistribution of
material during the sun's main-sequence evolution. The inversion also
suggests that the evolutionary age of the sun is perhaps some 10 per
cent greater than the generally accepted value, and that the solar
neutrino flux, based on standard nuclear and particle physics, is
about 75 per cent of the standard-model value.

1. INTRODUCTION

Helioseismology provides a unique tool for investigating the chemical
composition of the interior of the sun. Information can be obtained
either indirectly, by comparing the observed frequencies with the
eigenfrequencies of theoretical solar models (e.g. Christensen-
Dalsgaard and Gough, 1980, 1981), or directly, by measuring the influence on the sound speed in the regions of ionization of abundant
elements (Gough, 1984). Frequency comparisons have been carried out,
for example, for models with varying initial chemical abundances, and
with turbulent diffusive mixing or enhanced energy transport by weakly
interacting massive particles in the core [see reviews by Gough (1983,
1985) and Christensen-Dalsgaard (1988)]. In principle, the direct
method can be used to determine the helium abundance in the convection
zone, though attempts to do so have not yet yielded a reliable value
(Däppen, Gough and Thompson, 1988).
    It should be noted that the eigenfrequencies of a spherically
symmetrical hydrostatic solar model depend only on the pressure p,
density ρ and adiabatic exponent γ of the equilibrium state.
Therefore, any attempt to relate them to chemical composition $\underset{\sim}{X}$ must
require a knowledge of the equation of state: $\gamma = \gamma(p,\rho,\underset{\sim}{X})$. In the

327

G. Berthomieu and M. Cribier (eds.), Inside the Sun, 327–340.

radiative interior of the sun the most abundant elements are almost totally ionized, and γ ($\approx 5/3$) is extremely insensitive to $\underset{\sim}{X}$. Therefore, only indirect methods are available to study $\underset{\sim}{X}$ in the core, requiring the seismic analysis to be supplemented with additional assumptions.

In this paper we report on a helioseismic inversion in which it is assumed that the sun is in thermal balance. By equating the generation of thermonuclear energy in the core with radiative transfer, and assuming the total heavy-element abundance Z to be given, an estimate of the deviation δX of the hydrogen abundance X in the core of the sun from that in a standard solar model is obtained. With the use of the equation of state it is then possible to estimate the temperature deviation, and thence revise the theoretical computation of the neutrino flux.

2. HELIOSEISMIC INVERSION UNDER THE CONSTRAINT OF THERMAL BALANCE

Direct inversions, based on iterative linearization of the relation between frequency and structure differences between the sun and a theoretical model, depend on satisfying linear integral constraints such as

$$\frac{\delta\omega_{n,\ell}^2}{\omega_{n,\ell}^2} = \int_0^R \left(K_{\rho,\gamma}^{(n,\ell)} \frac{\delta\rho}{\rho} + K_{\gamma,\rho}^{(n,\ell)} \frac{\delta\gamma}{\gamma} \right) dr \quad , \tag{2.1}$$

or

$$\frac{\delta\omega_{n,\ell}^2}{\omega_{n,\ell}^2} = \int_0^R \left(K_{c^2,\gamma}^{(n,\ell)} \frac{\delta c^2}{c^2} + K_{\gamma,c}^{(n,\ell)} \frac{\delta\gamma}{\gamma} \right) dr \quad , \tag{2.2}$$

(e.g. Gough, 1985; Kosovichev, 1986; Gough and Kosovichev, 1988), where $\delta\omega_{n,\ell}$ is the perturbation to the frequency $\omega_{n,\ell}$ of order n and degree ℓ produced by small deviations $\delta\rho$, $\delta\gamma$, or equivalently δc, $\delta\gamma$, in ρ, γ and sound speed c. Spherical symmetry has been assumed, so that degeneracy splitting in azimuthal order m is ignored; that also reduces integrals over the volume of the sun to integrals with respect to the single radial coordinate r. Only two of the three independent thermodynamic state variables ρ, c and γ appear in each constraint, because in deriving those constraints the equations of hydrostatic support

$$\frac{dp}{dr} = - \frac{Gm\rho}{r^2} \quad , \tag{2.3}$$

$$\frac{dm}{dr} = 4\pi\rho r^2 \quad , \tag{2.4}$$

(together with a knowledge of the mass M and radius R of the sun) have been imposed, as indeed they have been also in determining the displacement oscillation eigenfunctions $\xi_{n,\ell}$ of the theoretical model, upon which the kernels $K_{\alpha,\beta}^{(n,\ell)}$ depend.

The kernels in equations (2.1) and (2.2) characterize the sensitivity of the oscillation frequencies to deviations in the structure

at different radii in the star. Using the constraints (2.1) and
(2.2), those deviations can be inferred from the data $\delta\omega^2_{n,\ell}$ by a
standard inversion procedure.

Our interest here is in the core of the sun, where $\gamma \approx 5/3$.
Therefore, notwithstanding possible contributions to $\delta\omega^2_{n,\ell}/\omega^2_{n,\ell}$ from
γ deviations in the ionization zones, we have set $\delta\gamma = 0$ in equations
(2.1) and (2.2); our hope is that from the data set with which we are
working it is possible to construct averaging kernels for linear com-
binations $\Omega_k = \sum \alpha_{k,n\ell}(\delta\omega^2_{n,\ell}/\omega^2_{n,\ell})$ of data that are sufficiently small
in the outer regions of the sun such that any contribution from $\delta\gamma$
to Ω_k is small, even though the contribution to $\delta\omega^2_{n,\ell}/\omega^2_{n,\ell}$ may not
be. Thus we approximate equations (2.1) and (2.2) by

$$\frac{\delta\omega^2}{\omega^2} = \langle K_\rho, \frac{\delta\rho}{\rho}\rangle \quad , \tag{2.5}$$

and

$$\frac{\delta\omega^2}{\omega^2} = \langle K_{c^2}, \frac{\delta c^2}{c^2}\rangle \quad , \tag{2.6}$$

where the angular brackets denote integration with respect to r. For
clarity, we have now suppressed the labels n,ℓ identifying the modes,
and also the second subscript on the kernels.

The frequency conditions (2.5) or (2.6) do not adequately con-
strain the possible deviations $\delta\ln\rho$ or $\delta\ln c^2$ from solar models, since
they do not demand the condition m=M at r=R. This can be achieved
simply by imposing the additional constraint:

$$0 = \langle 4\pi\rho r^2, \frac{\delta\rho}{\rho}\rangle \quad , \tag{2.7}$$

which is of the same form as the other conditions (2.5). The density
kernel $4\pi\rho r^2$ can be transformed into a corresponding c^2 kernel by the
same procedure as is used for transforming the other density kernels
(cf. Kosovichev, 1988; Gough, 1989). In the remainder of the discussion it is assumed that this constraint is included amongst the frequency constraints (2.5) and (2.6).

Equations (2.5) or (2.6) can be transformed into a relation
between $\delta\ln\omega^2$ and composition deviations by additionally imposing the
constraint of thermal balance through the equations:

$$\frac{dL}{dr} = 4\pi\rho r^2 \epsilon \quad , \tag{2.8}$$

$$\frac{dT}{dr} = \begin{cases} -\dfrac{3\kappa\rho L}{64\sigma r^2 T^3} \, , & \text{in radiative zones} \\[2ex] \left(\dfrac{dT}{dr}\right)_c \, , & \text{in the convection zone} \end{cases} \tag{2.9}$$

where L is luminosity, T is temperature, ϵ is the energy-generation
rate per unit mass, κ is opacity and σ is the Stefan-Boltzmann constant. The function denoted by $(dT/dr)_c$ is provided by the (mixing-

length) formalism relating temperature gradient to energy transport in a convection zone. In addition one requires a means of determining the equilibrium values of $\varepsilon(\rho,T,X,Z)$ and $\kappa(\rho,T,X,Z)$ and their partial derivatives. The transformation then yields a new constraint, which may be written

$$\frac{\delta\omega^2}{\omega^2} = \langle K_X, \frac{\delta X}{X}\rangle + \langle K_Z, \frac{\delta Z}{Z}\rangle \quad . \tag{2.10}$$

It is a straightforward matter to compute the kernels in equation (2.10) from K_ρ using the adjoint of the linearized structure equations (e.g. Marchuk, 1977; cf. Masters, 1979). For simplicity, and for want of an obviously more suitable procedure, we have set

$$\delta Z = 0 \quad , \tag{2.11}$$

(we could equally well have chosen $\delta(Z/X) = 0$ without substantially complicating the analysis), under which circumstances the linearized structure equations (2.3), (2.4), (2.8) and (2.9) formally become, for example,

$$A\frac{\delta\rho}{\rho} = \frac{\delta X}{X} \quad , \tag{2.12}$$

where A is a linear differential operator. This equation must be supplemented with appropriate boundary conditions, which are derived from demanding regularity of the deviation at the centre of the sun and from the requirement that conditions in the photosphere are unchanged, which we adopt in the form $\delta\ln\rho = 0$, $\delta\ln L = 0$ at $r=R$. Substituting equations (2.11) and (2.12) into equation (2.10) yields

$$\frac{\delta\omega^2}{\omega^2} = \langle K_X, A\frac{\delta\rho}{\rho}\rangle = \langle A^* K_X, \frac{\delta\rho}{\rho}\rangle \quad , \tag{2.13}$$

where A^* is the adjoint of A. By demanding that the expressions on the right-hand sides of equations (2.5) and (2.13) are identical, one thus obtains

$$A^* K_X = K_\rho \quad . \tag{2.14}$$

Thus, the hydrogen-abundance kernels K_X are obtained as solutions of the adjoint linearized structure equations, subject to adjoint boundary conditions which are determined by requiring that the integrated parts arising from the partial integrations necessary to derive the second of equations (2.13) from the first vanish for all functions $\delta\rho/\rho$ that satisfy the boundary conditions supplementing equation (2.12).

Kernels K_Z are computed similarly. Examples of kernels K_X, K_Z, K_ρ and K_{c^2} are presented by Gough and Kosovichev (1988).

3. INVERSION OF FREQUENCIES

Inversions using the constraints (2.5), (2.6), and the constraints (2.10) with the condition (2.11), have been separately carried out to determine $\delta\ln\rho$, $\delta\ln c^2$ and $\delta\ln X$ using the optimal averaging procedure of Backus and Gilbert (1968; see also Gough (1985) for a discussion of

its application to the solar interior). Only the 119 frequencies of
modes with $\ell \leq 5$ reported by Jiménez et al. (1988) and Henning and
Scherrer (1986) were used; these are the modes that are most important
in diagnosing the structure of the core. Christensen-Dalsgaard's
(1982) solar model 1 was adopted as the reference.

To test the procedure we first inverted the frequencies of the
chemically homogeneous model used by Christensen-Dalsgaard and Gough
(1980) and Christensen-Dalsgaard et al. (1989). The results are
compared with the exact deviations $\delta\ln\rho$, $\delta\ln c^2$ and $\delta\ln X$ in Figures 1,
2 and 3. It can be judged from those figures how accurate the pro-
cedure might be when optimal averages are identified with point
values. It should be borne in mind, however, that this test is not
realistic: the deviations are much larger than those of the sun, which
must have a deleterious effect on the accuracy of the linearized frequency constraints; on the other hand, the theoretical frequencies of
the homogeneous model were computed (without deliberately introducing
errors) in precisely the same manner as those of the reference model,
and the values of Z are known to be the same for the two models, both
of which tend to enhance the apparent reliability of the inversions.

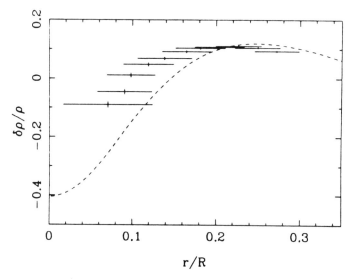

Figure 1. The dashed line is the relative difference between the
density of a homogeneous model of the sun and that of a standard
model. The crosses represent optimal averages deduced from the
differences between eigenfrequencies of the two models corresponding
to those modes that were used for the inversion of the solar data.
The horizontal components of the crosses indicate the widths of the
optimal averaging kernels and represent the resolution of the inversion; the vertical components indicate the standard errors. The
latter were computed assuming the standard errors for the corresponding observed frequencies.

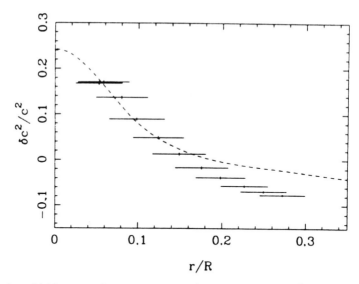

Figure 2. Difference between squared sound speeds of the homogeneous and the standard models, and the inversion results. The notation is otherwise the same as in Figure 1.

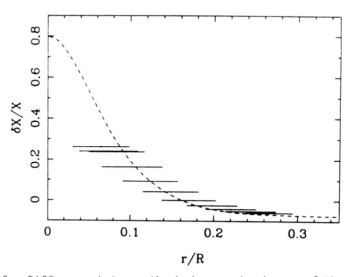

Figure 3. Difference between the hydrogen abundances of the homogeneous and the standard models, and the inversion results. The notation is otherwise the same as in Figure 1.

Notice, however, that because the two models have different hydrogen
abundances there is a substantial difference in the value of γ in the
ionization zones, possibly of greater magnitude than for the sun.
Therefore the inversion does test our ignoring the γ deviations in
equations (2.5) and (2.6) for the mode set we have used. It is evident from the inversions that the structure of the small inner region
r ≲ 0.05 R cannot be resolved by the p-mode data.

Inversions of the solar data have been reported previously (Gough
and Kosovichev, 1988), and are reproduced in Figures 4, 5 and 6. The
sun appears to be some 10 per cent denser at the centre than the standard solar model. Moreover, the inversion for $\delta \ln X$ suggests a flatter
hydrogen-abundance profile than that of the model. This is suggestive
of there having been some degree of material redistribution in the
core.

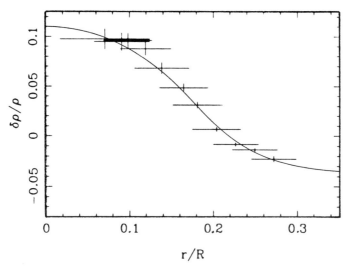

Figure 4. Optimal averages of the relative density deviation of the
sun from the reference solar model, obtained by inversion of the low-
degree p-mode frequencies, represented as crosses in the manner of
Figures 1-3. The continuous curve was drawn freehand through the
averages, and was used for estimating the neutrino flux.

4. MAIN-SEQUENCE AGE AND NEUTRINO FLUX

If one were to accept the continuous line drawn through the points in
Figure 6 to be representative of the deviation of the solar hydrogen
abundance from that of the standard model, one would conclude that
more hydrogen has been consumed in the sun than is predicted theoretically. Since the main-sequence variation of L with time is only
weakly dependent on the fine details of conditions in the cores of
calibrated solar models, this would appear to imply that the sun is

334

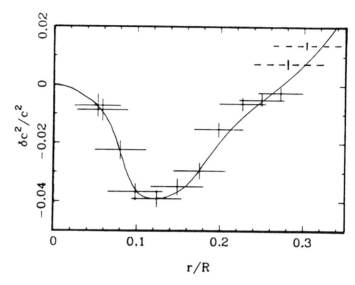

Figure 5. Optimal averages of the relative deviation of the square of
the sound speed in the sun from that of the reference theoretical
model. The continuous crosses denote averages obtained by inverting
the constraints (2.6) imposed by the low-degree modes alone; the
dashed crosses denote optimal averages, presented by Gough and
Kosovichev (1988), of the modes with $\ell \leq 3$ provided by Jiménez et al.
(1988) together with all the modes listed by Duvall et al. (1988)
whose lower turning points are below r = 0.7 R. The continuous curve
was drawn, essentially freehand, through the averages, taking account
of the inversions for r > 0.35 R which lie outside the bound of this
figure.

older than is generally believed, by an amount δt given approximately
by

$$\delta t \simeq E\bar{L}^{-1} \int_0^M (\delta X_0 - \delta X) dm \quad , \tag{4.1}$$

where δX_0 is the deviation of the primordial hydrogen abundance from
that of the model, $E = 6.3 \times 10^{18}$ erg g^{-1} is the energy released per
unit mass in the conversion of hydrogen to helium, and $\bar{L} \simeq 0.85$ $L_\odot$ is
the mean main-sequence surface luminosity of the sun (averaged over
time since the start of main-sequence evolution). If one sets δX_0
equal to the constant value attained for r $\gtrsim$ 0.25 R in Figure 6,
then $\delta t \simeq 6 \times 10^8$y. However, if $\delta X_0 = -0.005$, which is hardly inconsistent with the inversion, the increment δt is reduced to zero.

One can also estimate the solar neutrino flux. For this purpose
it is adequate to use the perfect gas law for fully ionized material
in the form

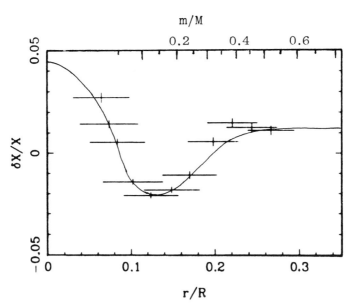

Figure 6. Optimal averages of the relative deviation of the hydrogen-abundance in the sun from that in the reference solar model, obtained by inversion of the low-degree p-mode frequencies. The continuous curve was drawn freehand, and was used together with the curve plotted in Figure 5 to estimate the temperature which is plotted in Figure 7. The upper abscissa scale is the relative mass variable for the sun, inferred from the continuous curve plotted in Figure 4. The notation is otherwise the same as in Figure 4.

$$T = \frac{\mu c^2}{\gamma R} \quad , \qquad\qquad\qquad (4.2)$$

to estimate the temperature T, where $\mu \simeq 4/(5X+3)$ is the mean molecular weight and R is the gas constant. From the continuous lines drawn in Figures 5 and 6, the temperature deviation δT can be obtained by linearizing equation (4.2) in the deviations. The resulting temperature distribution we thus infer for the central regions of the sun is shown in Figure 7.

The major contribution, $F_{\nu 8}$, to the neutrino flux F_ν comes from the decay of ^8B. It can be estimated from the formula

$$F_{\nu 8} = \frac{\lambda_8}{\lambda_{8m}} F_{\lambda 8m} \quad , \qquad\qquad (4.3)$$

where the subscript m denotes the value obtained from the theoretical model and

$$\lambda_i = \int_0^M \phi_i dm \quad , \qquad\qquad\qquad (4.4)$$

with

336

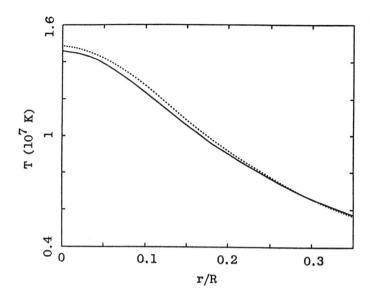

Figure 7. The continuous curve is the temperature in the core of the sun, inferred from the sound-speed and hydrogen-abundance deviations plotted as continuous curves in Figures 5 and 6. For comparison, the temperature of the reference theoretical model of Christensen-Dalsgaard (1982) is plotted as a dotted curve.

$$\phi_8 = \frac{1-X}{1+X} \, X^2 \rho T^{24.5} \quad , \tag{4.5}$$

(cf. Gough, 1988). The fluxes $F_{\nu 7}$ and $F_{\nu CNO}$ from the electron capture by ^{7}Be in the proton-proton chain and from the decay of ^{13}N and ^{15}O in the CNO cycle were scaled similarly, with $\phi_7 = X(1-X)\rho T^{11}$ and $\phi_{CNO} = XZ\rho T^{20}$. We thus found $\lambda_8/\lambda_{8m} = 0.67$, $\lambda_7/\lambda_{7m} = 0.84$ and $\lambda_{CNO}/\lambda_{CNOm} = 0.73$. The remaining contribution, principally from the p-p reaction of the proton-proton chain, is small and is hardly changed and was taken to be invariant.

Theoretical neutrino fluxes are not available for Christensen-Dalsgaard's model 1. We therefore adopted the model values quoted by Bahcall and Ulrich (1988), namely $F_{\nu pepm} = 0.2$ snu, $F_{\nu 7m} = 1.1$ snu, $F_{\nu 8m} = 6.1$ snu and $F_{\nu CNOm} = 0.4$ snu. The resulting total solar neutrino flux is thus found to be $F_\nu \simeq 5.5$ snu.

5. DISCUSSION

The aim of the analysis summarized in this paper has been to make a first estimate of the implications of helioseismic data regarding the structure of the energy-generating core of the sun. We have carried out separate inversions for the deviations $\delta\rho$ and δc^2 of the density and the square of the sound speed in the sun from the corresponding quantities in a standard solar model, the results of which are illustrated in Figures 4 and 5. The continuous lines drawn through the results are our preliminary estimates of those deviations in the inner 35 per cent of the solar radius. They were drawn essentially freehand, and, notwithstanding any systematic errors that may have corrupted our inversions, are particularly uncertain for $r \lesssim 0.05$ R, where the structure of the star has little influence on the p-mode frequencies, and for $r \gtrsim 0.25$ R. To make inferences outside the energy-generating core would require supplementing the seismic data we have used with observed frequencies of modes of higher degree.

Our inversions have not been carried out in a wholly consistent manner. One of the most obvious approximations we have made is to neglect the deviation of γ in the ionization zones. The degree to which this spoils our inferences can be judged from Figures 1 and 2, which display the results of using theoretical eigenfrequencies to infer the structure of a chemically homogeneous solar model. In carrying out those inversions, we used the same mode set as that for the solar inversions. The homogeneous model has the same value of Z, but a different hydrogen abundance. Therefore the ionization zones are different. Since γ varies substantially only in and above the ionization zones, in the outer layers of the sun, we conclude that this experiment actually gives an indication of our ability to eliminate the influence of not solely our neglect of $\delta\gamma$, but also the effect of all the other uncertainties in the complicated physics of the upper layers of the convection zone. Notice that the relative deviations $\delta\rho/\rho$ and $\delta c^2/c^2$ plotted in Figures 1 and 2 are not extremely small compared with unity. Therefore the linearized constraints (2.5)-(2.7) used for the inversions may not be good approximations, which no doubt accounts in part for the differences between the actual deviations and those inferred. We have made no attempt to iterate the procedure.

Of course the inversions for $\delta\rho$ and δc^2 are not independent: the kernels K_ρ and K_{c^2} are related via the hydrostatic constraint, which has already been assumed to hold in the derivation of the linearized oscillation equations. Therefore, in principle, just one of the inversions should suffice; from a knowledge of $\delta\rho$ and the equilibrium state of the reference theoretical solar model one can calculate δc^2, and vice versa. However, we have carried out both inversions separately in order to provide a consistency check. We find that the continuous curves drawn through the optimal averages in Figures 4 and 5 do not satisfy the perturbed hydrostatic equations, and therefore we conclude that our inversions are not consistent. This is perhaps not surprising, in view of the systematic errors, evident in Figures 1 and 2, in our inversions of the eigenfrequencies of the homogeneous model. We have not yet carried out a systematic investigation either to assess whether to put more trust in the inferred values of δc^2 or in the values of $\delta\rho$ (though one might naively suspect that it would be more prudent to trust δc^2, since the physics of acoustic wave propagation throughout all but the surface layers of the sun is more directly related to the sound speed than to the stratification of density), or to judge how sensitive our subsequent conclusions are to the errors whose existence the inconsistency reveals.

The inversion for the deviation δX of the hydrogen abundance is even less reliable. Not only is it susceptible to all the uncertainties in the inversions for $\delta\rho$ and δc^2, but also it requires a knowledge of the nuclear reaction rates and the opacity. Moreover, and perhaps most important of all, it depends on the assumption of thermal balance, expressed by equations (2.8) and (2.9). Judging from the inversion of the eigenfrequencies of the homogeneous model, illustrated in Figure 3, our procedure might appear to be no worse for the secondary variable δX than it is for the fundamental state variables. However, it is important to realize that this figure provides a test of only the procedure for inversion, and not of the results themselves, because, like the reference model, the homogeneous solar model is known to be in thermal balance, and, unlike the sun, was constructed with identical physics to the reference model. Of course δX could have been computed from $\delta\rho$ and δc^2, simply by integrating the perturbed thermal-balance equations. However, the outcome would depend not only on the errors in the inversions required to obtain $\delta\rho$ or δc^2, but also on the errors in drawing the smooth curve through the averages and extrapolating it into the regions where no optimal averages have been determined. It also depends on the validity of equating optimal averages with point values. Therefore we have preferred to obtain averages of δX by inverting the constraints (2.10) directly.

From the determinations of $\delta\rho$, δc^2 and δX we have estimated the variation of temperature in the solar core. This is compared with the temperature in the standard theoretical model in Figure 7. Although in the core of the sun electrons are partially degenerate and the perfect gas law should not be (and, indeed, was not) used in constructing the reference model, equation (4.2) is adequate, at the present level of reliability of the inversions, for estimating the partial derivatives of T with respect to c^2 and X required for determining the quite small relative temperature deviations $\delta T/T$. In the central region of the core the magnitude of the inferred solar temperature gradient is lower than in the theoretical model.

As we have already pointed out, the shape of the profile of $\delta X/X$ plotted in Figure 6 might be regarded as evidence for there having been some degree of material redistribution in the core during the main-sequence evolution. Diffusive mixing of the kind discussed by Schatzmann et al. (1981), however, is not consistent with the seismic data (e.g. Gough, 1983), and leads to a lowering of the central density (see also Kosovichev and Severny, 1985; Christensen-Dalsgaard, 1988) which is inconsistent with the inversion plotted in Figure 4. If energy transport in the core of the sun were enhanced by wimps the density would be increased, but the resultant sound speeds in the theoretical models that have been published are not in accord with our inferences in Figure 5 (Gough and Kosovichev, 1988).

The procedure we have outlined has yielded estimates of the thermodynamic state and the hydrogen abundance in the energy-generating core of the sun. From that we have been able to estimate the factor by which the solar neutrino flux should differ from the standard-model value: about 0.7 if the continuous curves in Figures 4-6 are adopted. Because we know that our independent estimates of $\delta\rho$

and δc^2 plotted in Figures 4 and 5 are not consistent with hydro-
statics, we have recomputed the flux from an estimate of $\delta \rho$ derived
from the sound-speed deviation plotted in Figure 5 and the hydrostatic
constraint (2.3) and (2.4). To accomplish that it was necessary to
know the sound-speed deviation throughout the model. In the radiative
envelope we used the results of an inversion of intermediate-degree p
modes [specifically, we used the dashed curve in Figure 15 of Gough
and Kosovichev (1988); this is consistent with other recent inversions
by Vorontsov (1988), Christensen-Dalsgaard et al. (1989) and
Shibahashi and Sekii (1988)], matched smoothly onto the optimal averages plotted in Figure 5. The resulting factor by which the standard-
model flux should be multiplied is 0.83, yielding $F_\nu \simeq 6.6$ snu if
Bahcall and Ulrich's (1988) standard value of 7.9 snu is adopted. The
difference between this value and the value (5.5 snu) obtained by
taking the independent inversions plotted in Figures 4 and 5 provides
some indication of the degree of uncertainty in our result. [Had we
adopted the value of the neutrino flux (5.8 snu) obtained from the
standard solar model of Turck-Chièze et al. (1988) to be representative of our reference model, we would have obtained 4.1 snu and 4.9
snu as our estimates of the neutrino flux from the sun. Of course,
neither the model of Bahcall and Ulrich (1988) nor that of Turck-
Chièze et al. (1988) is identical to Christensen-Dalsgaard's (1982)
standard model 1 which we have used as a reference, so our procedure
for estimating F_ν from the factors λ_i / λ_{im} is not strictly valid; what
we should do is either use the perturbation theory in the manner we
have described but in addition take due account of the structural
differences between the theoretical solar models, or alternatively
compute F_ν directly from the distribution of ρ, X and T obtained from
our inversions. We intend to carry out a more consistent analysis and
report on the results in the near future.] We must emphasize, however, that these estimates depend on our having assumed, as is the
case in standard stellar-evolution theory, that the sun is in thermal
balance and that the fast nuclear reactions in the p-p chain and the
CNO cycle are in equilibrium with the slowest reactions which control
the overall reaction rate. They also depend on the assumption that
the heavy-element abundance Z is not in serious error. In making
these estimates we have not linearized equation (4.3) and its rela-
tives determining the other contributions to the neutrino flux,
because the exponents of T in the formulae for ϕ_i are too large for
linearization to be valid.

One can also try to estimate how much hydrogen has been consumed
during the main-sequence evolution. From the variation of X in the
core alone, adopting the continuous curve drawn in Figure 6, it
appears that that exceeds the standard value by about 0.005M, suggesting that the sun has a somewhat greater evolutionary age than the
standard value. This result is consistent with previous discussions
(Gough, 1983; Christensen-Dalsgaard, 1988) based on the mean frequency
separation $\omega_{n,\ell} - \omega_{n-1,\ell+2}$ of low-degree modes. However, this result
does depend critically on the value of the asymptote of $\delta X/X$ as r/R
increases away from the core, which is ill determined by the data. It
is therefore quite uncertain. It also depends, as does our estimate
of the neutrino-flux factor, on the assumptions of thermal and nuclear

340

balance used in the determination of $\delta X/X$, and which, as we have already pointed out, are brought into question by the inferred profile of X.

We gratefully acknowledge support from the SERC and from NASA grant NSG-7511.

References
Backus, G. and Gilbert, F. (1968) Geophys J. R. astr. Soc., **16**, 169.
Bahcall, J. N. and Ulrich, R. K. (1988) Rev. Mod. Phys., **60**, 297.
Christensen-Dalsgaard, J. (1982) Mon. Not. R. astr. Soc., **199**, 735.
Christensen-Dalsgaard, J. (1988) Seismology of the sun and sun-like stars (ed. E. J. Rolfe, ESA SP-286, Noordwijk) 431.
Christensen-Dalsgaard, J. and Gough, D. O. (1980) Nature, **288**, 544.
Christensen-Dalsgaard, J. and Gough, D. O. (1981) Astron. Astrophys., **104**, 173.
Christensen-Dalsgaard, J., Gough, D. O. and Thompson, M. J. (1989) Mon. Not. R. astr. Soc., **238**, 481.
Däppen, W., Gough, D. O. and Thompson, M. J. (1988) Seismology of the sun and sun-like stars (ed. E. J. Rolfe, ESA SP-286, Noordwijk) 505.
Gough, D. O. (1983) Primordial helium (ed. P. A. Shaver, D. Kunth and K. Kjär, ESO, Garching) 117.
Gough, D. O. (1984) Mem. Soc. astr. Italiana, **55**, 13.
Gough, D. O. (1985) Solar Phys., **100**, 65.
Gough, D. O. (1988) Solar-terrestrial relationships and the Earth environment in the last millenia (ed. G. Cini-Castagnoli, Varenna Summer School, Soc. Italiana Fisica) **95**, 90.
Gough, D. O. (1989) Dynamiques des fluides astrophysiques (ed. J-P. Zahan and J. Zinn-Justin, Les Houches Session XLVII, 1987, Elsevier) in press.
Gough, D. O. and Kosovichev, A. G. (1988) Seismology of the sun and sun-like stars (ed. E. J. Rolfe, ESA Publ. SP-286, Noordwijk) 195.
Henning, H. M. and Scherrer, P. H. (1986) Seismology of the sun and the distant stars (ed. D. O. Gough, Reidel, Dordrecht) 55.
Jiménez, A., Pallé, P. L., Roca Cortes, T. and Domingo, V. (1988) Astron. Astrophys., **193**, 298.
Kosovichev, A. G. and Severny, A. B. (1985) Bull. Crimea. Astrophys. Obs., **72**, 188.
Kosovichev, A. G. (1986) Bull. Crimea Astrophys. Obs., **75**, 40.
Kosovichev, A. G. (1988) Bull. Crimea Astrophys. Obs., **80**, 175.
Marchuk, G. I. (1977) Methods in Computational Mathematics, Nauka, Moscow
Masters, G. (1979) Geophys. J. Roy astr. Soc., **57**, 507.
Schatzmann, E., Maeder, A., Angrand, F. and Glowinski, R. (1981) Astron. Astrophys., **96**, 1.
Shibahashi, H. and Sekii, T. (1988) Seismology of the sun and sun-like stars (ed. E. J. Rolfe, ESA ESP-286, Noordwijk) 471.
Turck-Chièze, S., Cahen, S. and Cassé, M. (1988) Astrophys. J., **335**, 415.
Vorontsov, S. V. (1988) Seismology of the sun and sun-like stars (ed. E. J. Rolfe, ESA ESP-286, Noordwijk) 475.

MAGNETIC FIELD IN THE SUN'S INTERIOR
FROM OSCILLATION DATA

W.A.DZIEMBOWSKI
Copernicus Astronomical Center
Warszawa, Poland

P.R.GOODE
Department of Physics, N.J.I.T.
Newark, NJ, USA

ABSTRACT. Solar oscillations provide a probe of the internal magnetic field of the Sun if the field has sufficient intensity. Using the oscillation data of Libbrecht, we find evidence for a 2 ± 1 megagauss quadrupole toroidal field centered at 0.7 of the solar radius which is barely beneath the base of the convection zone. This field, by its location and symmetry, may be associated with the dynamo that drives the Sun's 22-year activity cycle.

1. Introduction

Helioseismology is the science by which solar oscillations are used to determine internal properties of the Sun. In particular, helioseismological techniques have yielded detailed information about the interior rotation of the Sun [Duvall, et al (1984), Duvall, Harvey and Pomerantz (1986), Brown, et al (1989), Christensen-Dalsgaard and Shou (1988), and Dziembowski, Goode, and Libbrecht (1989)] and the radial dependence of the speed of sound [Christensen-Dalsgaard, et al (1985) and Brodsky and Vorontsov (1988)]. It is our purpose here to discuss the use of solar oscillations to extract information about the Sun's internal magnetic field. In the long run, the study of solar oscillations is our only real hope for determining the internal magnetism of the Sun.

Before the advent of more detailed solar oscillation data, the size of the magnetic field throughout the radiative interior was more open to question. At one extreme, Ulrich and Rhodes (1985) assumed a relic field of 300 MG, centered in the core, in an effort to describe some gross structures in oscillation data. At the other extreme, Spruit (1987) has argued that magnetic torques would preclude a steady-state field of more than a few Gauss in the interior.

Recent oscillation data like those due to Duvall, Harvey and Pomerantz (1986), Brown and Morrow (1987), and Libbrecht (1989) have shed some light on this situation. These data best sample the region near the base of the convection zone (0.73 of the radius). From these data, Brown, et al (1989), Christensen-Dalsgaard and Shou (1988) and Dziembowski, Goode, and Libbrecht (1989) calculated rotation laws which show no significant gradient above the base of the convection zone. Rather, near the base of the convection zone, there is a fairly sharp transition from surface-like differential rotation just above to solid body-like rotation just

G. Berthomieu and M. Cribier (eds.), Inside the Sun, 341–348.
© 1990 *Kluwer Academic Publishers. Printed in the Netherlands.*

beneath. Brown, et al (1989) used this result to suggest that the dynamo driving the 22-year activity cycle is seated just beneath the convection zone. Since differential rotation may persist into the outer radiative zone, there one may naively expect that a small poloidal field could be sheared into a sizeable quadrupole toroidal field. In fact, Dziembowski and Goode (1989) used the data of Libbrecht (1989) to calculate such a toroidal field of amplitude 2 ± 1 MG centered just beneath the convection zone. This field may be associated with the dynamo since it has the proper symmetry and proximity. However, the calculated field is two orders of magnitude more intense than generally expected for the dynamo field [Parker (1987)]. Since the 10^4G dynamo field is expected at the base of the convection zone, the calculated field could be a slightly deeper lying reservoir for the dynamo. We shall see that that oscillation data can be used, as well, to place limits on the toroidal field considerably deeper than the base of the convection zone.

A megagauss toroidal field fed by differential rotation is not in steady state and, therefore, is not precluded by Spruit's argument. Kuhn (1988) has argued that the symmetric part of the oscillation spectrum changes over the 22-year solar activity cycle. If he were correct, synchronous, large variations in the calculated field would be a source of the changes.

2. Solar Oscillations in the Presence of a Magnetic Field

Global solar oscillations are sound waves which sample the interior of the Sun. An individual oscillation is characterized by the product of a radial part, labelled by n the radial order, and a single $Y_\ell^m(\vartheta, \varphi)$, where ℓ is it angular degree and m is it azimuthal order. In the absence of a perturbing force the (n, ℓ)-multiplets are $(2\ell + 1)$-fold degenerate in m. A magnetic field lifts this degeneracy. For an axially symmetric field, the structure of each (n, ℓ)-multiplet is described by a polynomial in m^2 having its order equal to the multipole order of the field. The coefficients of this polynomial are a quadratic function in, separately, the poloidal and toroidal field components. Consistent with the oscillation data, it will be assumed here that the axis of symmetry of the field is aligned with the rotation axis. Then, contributions to the coefficients come from the quadratic and higher effects of rotation, as well.

3. Data

We used the data obtained by Libbrecht (1989). His frequency splittings, $\nu_{n\ell m} - \nu_{n\ell 0}$, are given in terms of the a_i-coefficients from

$$\nu_{n\ell m} - \nu_{n\ell 0} = \ell \sum_{i=1}^{5} a_i P_i \left(\frac{m}{\ell} \right) ,$$

where P_i is a Legendre polynomial. This Legendre expansion is a convenient way of representing the power series in m which includes the m^2 terms discussed in the previous section. The data set covers the ℓ-range $10 - 60$ and the frequency range $1.5 - 4.0$ mHz. These data best sample the region between $x = 0.5$ and 0.9, where $x = r/R$. Lower degree oscillations would sample more deeply and higher degree oscillations would provide the contrast

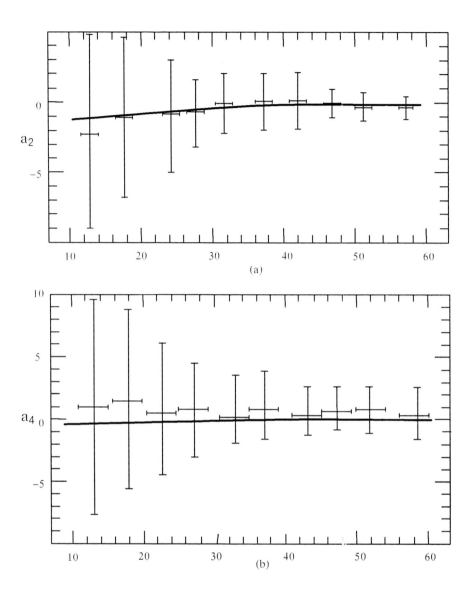

Figure 1. Weighted averages of Libbrecht's a_2 – and a_4 – coefficients (nHz) vs l. The solid lines represent the calculated effect of distortion.

to probe closer to the surface. Libbrecht's odd a-coefficients were used by Christensen-Dalsgaard and Shou (1988) and Dziembowski, Goode and Libbrecht (1989) to calculate the internal rotation of the Sun as a function of radius and latitude from $x = 0.4$ to 1.0. The even a-coefficients reflect centrifugal distortion and, for instance, an aligned axisymmetric magnetic field.

Figure 1 shows the measured values of the a_2- and a_4-coefficients averaged over n and grouped in bins 5 ℓ-values wide. The solid line represents the calculated effect of rotational distortion – this effect is subsequently subtracted from each multiplet. From the original data set covering 678 multiplets we removed 22 corresponding to the modes that could have been effected by an accidental degeneracy. We emphasize that if Libbrecht's data were averaged over n, like the other available sets, we could not have determined a magnetic field from it.

4. The Inverse Problem for a Toroidal Magnetic Field

The aligned, axisymmetric toroidal magnetic field we assumed is given by

$$B_\varphi^2 = 4\pi p(x) \sin^2 \vartheta \sum_{k=1} \beta_k(x) \cos^{2k-2} \vartheta,$$

where $p(x)$ is the gas pressure. We assumed that the β_k and their derivatives vanish at $x = 0$ and 1. The lowest order linear perturbation of the hydrostatic equation arising from the toroidal field yields our inverse problem

$$a_{2i} = \sum_{k \geq i} \int_0^1 \left(\beta_k \mathbf{E}_{2i,k} + x \frac{d\beta_k}{dx} \mathbf{D}_{2i,k} \right) dx.$$

The kernels $\mathbf{E}$ and $\mathbf{D}$ were determined from the mode eigenfunctions evaluated for a standard solar model. Examples of such kernels are shown in Figure 2 for $i = 1$ and 2 and $k = 2$. If we eliminate the $d\beta_k/dx$- term by integrating by parts, the resulting kernel,

$$\mathbf{F}_{2i,k} = \mathbf{E}_{2i,k} - \frac{d}{dx}(x\mathbf{D}_{2i,k}),$$

exhibits rapid sign changes – see the dashed lines shown in Figure 2. This latter property makes the kernel for each multiplet more orthogonal to those for the other multiplets. Since the success of the helioseismology depends critically on a differential sampling of the interior, the rapid sign changes greatly increase the prospects for the seismology. The inverse problem for the field closely resembles the one for differential rotation. Having two sets of even-a coefficients, we could have attempted to first determine β_2 from the a_4-coefficients and, subsequently, β_1 from β_2 and the a_2-coefficients. We found, however, that within the errors the β_2-term suffices. In the following, we consider this term alone and drop the subscript k from β and $\mathbf{F}$.

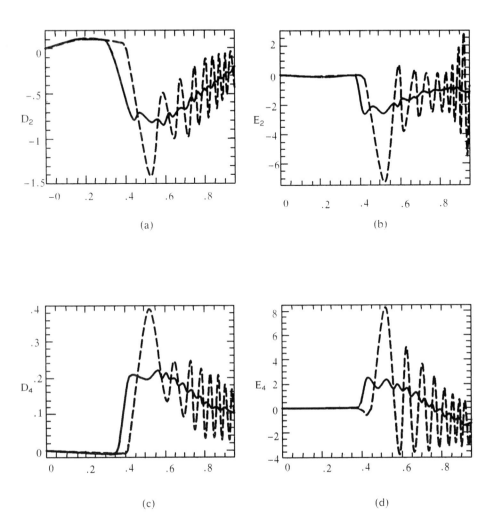

Figure 2. Sample a) $D_2 -$, b) $E_2 -$, c) $D_4 -$ and d) $E_4 -$ kernels vs. the fractional radius. The kernels are for $l = 20$ and are given in μHz. The dashed line (- - - -) is for $n = 14$ and the solid (————) is for the average over n in the five – minute period oscillation band.

5. Evidence for a Megagauss Magnetic Field Near the Base of the Solar Convective Envelope

The inverse problem for the toroidal field was solved using two methods. Our primary approach employed the method of Backus and Gilbert (1970) in which a unimodular kernel, $\mathbf{K}$, is constructed from a linear combination of the $\mathbf{F}$ kernels which are calculated for each mode present in the data set. The linear combination must be localized at a selected point, x_0. We thus write

$$\mathbf{K}_{2i}(x, x_0) = \sum_j c_{2i,j}(x_0)\mathbf{F}_{2i,j}(x),$$

where j is the mode counter. The two sets of c-coefficients for each x_0 - one from the a_2-data and the other from the a_4-data - were determined by a compromise between maximal sharpness and minimal amplification of the errors from the data. If $\mathbf{K}$ is indeed sharp, one may expect that

$$\beta(x_0) = \sum_j c_{2i,j}(x_0)a_{2i,j}.$$

However, if the field is more confined than the $\mathbf{K}$-kernel sampling it, then the field will always be underestimated. The results for inversions at three selected locations are shown in Figure 3. Only the results for $x_0 = 0.7$ are significant, where we have

$$\beta = (4.1 \pm 1.1) \times 10^{-3} \quad \text{and} \quad (4.9 \pm 1.7) \times 10^{-3}$$

following from the a_2- and a_4-data, respectively. These values correspond to a quadrupole toroidal field of amplitude 2 ± 1MG. At $x_0 = 0.55$ and 0.90, the $\mathbf{K}$-kernels are not so sharp and the average β-values are consistent with zero, however, the calculated errors in β may be used to place limits on the average quadrupole toroidal field. At $x = 0.55$, the limit is 4 MG and at $x = 0.9$ the limit is 0.2 MG.

In the second method of inversion, we employed a discretization of $\beta(x)$ in terms of cubic spline functions and solved the inverse problem by a least squares method. Since the functional form of $\beta(x)$ was assumed, we calculated $d\beta/dx$ explicitly rather than performing an integration by parts. The significant results of the inversions occur at $x_0 = 0.7$ and are $\beta = (1.7 \pm 0.6) \times 10^{-3}$ and $(3.2 \pm 1.6) \times 10^{-3}$, respectively. These numbers are in reasonable agreement with those from the Backus-Gilbert method.

6. Implications

If a megagauss toroidal field were centered just beneath the convection zone, it would cause no observable dynamical effect. The ratio of the magnetic pressure to the gas pressure would be about 10^{-3} – having no perceptible effect on the speed of sound near the base of the convection zone. The quadrupole moment would be comparable to that which would result from a rigidly rotating Sun. On the other hand, such a field would have far reaching consequences on our understanding of solar activity. In this regard, we emphasize that the magnetic energy density for $\beta = 10^{-3}$ at $x = 0.7$ is six times the corresponding rotational energy density. Such a field may have consequences for neutrino propagation – see the review by Smirnov (1989) in these proceedings.

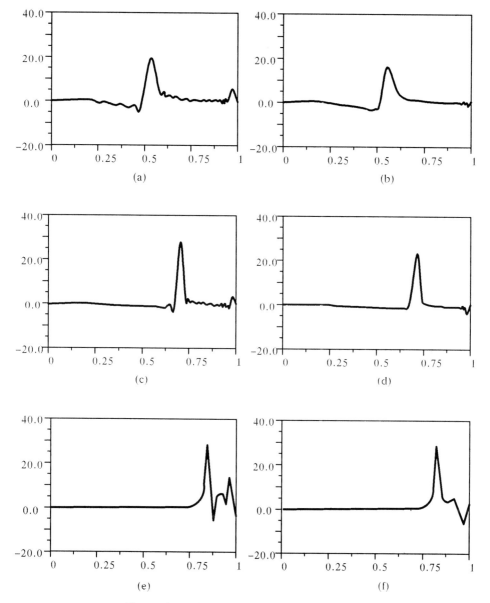

Figure 3. Unimodular Backus – Gilbert kernels,
a) $K_2(x,0.55)$, b) $K_4(x,0.55)$, c) $K_2(x,0.7)$,
d) $K_4(x,0.7)$, e) $K_2(x,0.9)$, f) $K_4(x,0.9)$
vs the fractional radius.

References:

Backus, G. and Gilbert, F. 1970, *Phil. Trans Roy. Soc.* London **A266**, 12.

Brodsky, M.A. and Vorontsov, S.V. 1988, *Advances in Helio- and Asteroseismology*, 137, J.Christensen-Dalsgaard and S. Fransden (Eds.), Reidel, Dordrecht, The Netherlands.

Brown, T.M., Christensen-Dalsgaard, J., Dziembowski, W.A., Goode, P.R., Gough, D.O., and Morrow, C.A.1989, *Ap.J.*, in press.

Brown, T.M. and Morrow, C.A. 1987, *Ap. J. Lett.* **314**, L21.

Christensen-Dalsgaard, J., Duvall, T.J., Gough, D.O., Harvey, J.W. and Rhodes, E.J. 1985, *Nature* **315**, 378.

Christensen-Dalsgaard, J. and Shou, J. 1988, *Seismology of the Sun and Sun-like Stars*, 149, E.J. Rolfe (Ed.), ESA, Noordwijk, The Netherlands.

Duvall, T.J., Dziembowski, W.A., Goode, P.R., Gough, D.O., Harvey, J.W. and Leibacher, J.W. 1984, *Nature* **310**, 22.

Duvall, T.J., Harvey, J.W. and Pomerantz, M.A. 1986, *Nature* **321**, 500.

Dziembowski, W.A. and Goode, P.R. 1989, *Ap. J*, in press.

Dziembowski, W.A.,Goode, P.R., Libbrecht, K.G. 1989, *Ap.J.*, **337**, L53.

Kuhn, J.R. 1988, *Ap. J. Lett.* **331**, L131.

Libbrecht, K.G. 1989, *Ap.J.*, **336**, 1092.

Parker, E.N. 1987, *The Internal Solar Angular Velocity*, 289, B. Durney and S. Sofia (Eds.), Reidel, Dordrecht, The Netherlands.

Smirnov, A.Y. 1989, *this volume.*

Spruit, H.C. 1987, *The Internal Solar Angular Velocity*, 185, B. Durney and S. Sofia (Eds.), Reidel, Dordrecht, The Netherlands.

Ulrich, R.K. and Rhodes, E.J. 1983, *Ap. J.* **265**, 551.

VARIATIONS OF THE LOW *l* SOLAR ACOUSTIC SPECTRUM CORRELATED WITH THE ACTIVITY CYCLE

P.L. PALLÉ, C. RÉGULO, T. ROCA CORTÉS
Instituto de Astrofísica de Canarias
38200 La Laguna. Tenerife
Spain

ABSTRACT. Solar cycle variation of the frequencies and of the power of solar acoustic oscillations are investigated. Integrated sunlight data from 1977 to 1988 obtained at the Observatorio del Teide (Izaña, Tenerife), using a resonant scattering spectrophotometer, is analyzed in 60 day time strings and their power spectra are calculated from 2 to 3.8 mHz. To study the frequency variation, each power spectrum is cross-correlated with the one corresponding to the 1981 series and the shifts of the centroids of the cross-correlation peaks are calculated. The results show a clear variation in frequency of the cross-correlation peaks of -0.37 ± 0.04 μHz peak to peak as solar activity cycle goes from maximum to minimum. Moreover, this effect is found to depend on the *l* value of the modes, being absent for $l = 0$ and of 0.42 ± 0.06 μHz for $l = 1$. These results can be interpreted as an amplitude modulation between modes of the same multiplet, probably as a consequence of the action of strong magnetic fields. As low *l* modes penetrate deeply into the Sun's interior, these observations suggest changes in its structure correlated with the solar activity cycle. When the power of the modes is calculated, using the same series as before, and its change along the solar cycle is studied, a variation of $\sim 40\%$ is found, the power being higher when solar activity is at its minimum. If this effect is independent of the *l* value of the p-modes, the results can be interpreted in terms of a change in the efficiency of the excitation mechanism of such modes. Indeed, if turbulent convection is such a mechanism, a change in the characteristic size of the granulation would account for the observed effect. Alternatively, another explanation could be a selective change in the efficiency of the excitation and/or damping mechanisms of the $l \leq 3$ modes in front of other *l* value modes.

1. Variations of the frequencies of low degree acoustic modes

Velocity data obtained from 1977 through 1988 is analyzed to look for variations in the frequencies of the acoustic p-modes (Pallé et al,1989). No useful disk-integrated light observations are available in 1979 and, in 1983, the observations were somewhat noisy, due to problems with the electronics. Data taken each day is individually analyzed and their residuals are joined together in sets of time strings of 60 continuous days (see Table 1). Power spectra are calculated using an interactive sine wave procedure; the amplitude squared and phase for each frequency, from 2 to 3.8 mHz, are calculated at intervals of 0.1 μHz. Then, the background noise of the spectra is calculated and subtracted. To compare the different power spectra, a cross-correlation technique is used.

349

G. Berthomieu and M. Cribier (eds.), Inside the Sun, 349–355.

Years	Period of Year		Useful-days	Duty cycle
1977	12- 7; 25- 8	*	38	30%
1978	31- 7; 9- 9	*	25	25%
1980	21- 7; 17- 8	*	28	40%
1981	25- 6; 22- 8	*	56	42%
1982	11- 5; 7- 7		54	39%
1982	30- 6; 28- 8	*	58	40%
1982	8- 7; 3- 9		56	39%
1983	4- 6; 2- 8	*	52	35%
1984	1- 5; 29- 6		55	37%
1984	12- 5; 8- 7		55	37%
1984	30- 6; 28- 8	*	59	42%
1984	10- 7; 7- 9		58	40%
1984	11-11; 9- 1		48	24%
1985	31- 3; 29- 5		46	30%
1985	28- 5; 23- 7		55	40%
1985	30- 6; 28- 8	*	53	35%
1985	26- 7; 18- 9		50	35%
1985	15- 9; 9-11		43	23%
1986	1-12(85);25- 1		40	21%
1986	8- 2; 31- 3		40	28%
1986	2- 4; 31- 5		50	33%
1986	16- 7; 4- 9	*	51	40%
1986	1- 9; 31-10		52	28%
1986	1-11; 31-12		45	23%
1987	3- 1; 28- 2		44	26%
1987	1- 3; 30- 4		37	20%
1987	1- 5; 30- 6		52	34%
1987	8- 6; 6- 8	*	54	40%
1987	1- 8; 30- 9		38	20%
1987	1-10; 30-11		35	13%
1988	12-12(87);31-1		38	24%
1988	1- 2; 31- 3		36	17%
1988	1- 4; 31- 5		52	35%
1988	1- 6; 31- 7	*	55	40%

TABLE 1.- *Time series (maximum 60 consecutive days long) analyzed in this work; an asterisk (*) denotes the summer series whose cross-correlation functions are shown in Fig. 1. The duty cycle is calculated as the percentage of hours of observations as opposed to total possible hours of observations (including nights).*

The power spectrum of 1981 data series is used as the reference against which all the others are cross-correlated.

The cross-correlation functions obtained using the summer months of each year are plotted in Fig. 1. A displacement of the cross - correlation peak of all functions to the left can be seen. To measure the position of the cross-correlation peak, the centroid is used. In Fig. 2 the centroids from all the series in Table 1 are shown. A variation from year to year is clearly evident. Taking the difference between the maximum (around 1981) and the minimum (1985/86), a peak to peak variation of $-0.37 \pm 0.04 \,\mu$Hz is found.

In order to find if this result changes with the degree of the modes, four spectra were obtained from each calculated power spectrum, each one containing only the information around the "isolated" first order sideband of modes $l = 0, 1, 2$ and 3. The same procedure as before has then been repeated. The resulting centroids are presented in Fig. 3. The peak to peak variations are : -0.04 ± 0.1, -0.42 ± 0.06, -0.38 ± 0.14 and $-0.47 \pm 0.1 \,\mu$Hz for $l = 0, 1, 2$ and 3. Thus, unlike $l > 0$ modes, $l = 0$ modes show very little, or no, trend at all; the mean values of all available points for l_0 is of $-0.18 \pm 0.03 \,\mu$Hz.

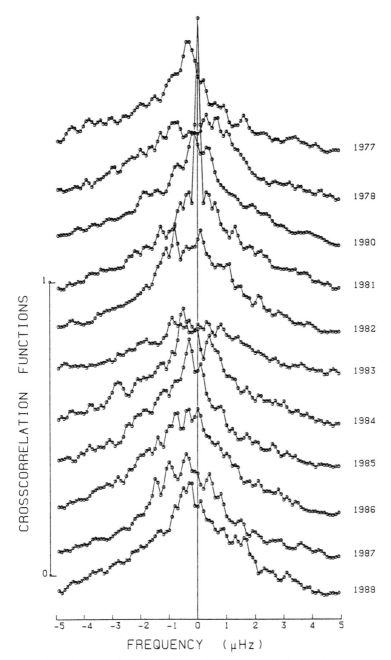

FIGURE 1.- *Cross-correlation functions of the power spectra of the summer months series for each year (see Table 1), with respect to the one for summer 1981.*

352

The dependence of this variation on the l value of the modes, the variation being absent for $l = 0$, suggest that other interpretations than merely a shift of frequencies in the acoustic mode spectrum across the solar cycle are plausible: an amplitude modulation between modes of the same multiplet and/or an asymmetric change of the splitting (probably due to magnetic fields) through the solar cycle. Therefore, the results found might come partially from an actual dynamic effect of an internal field (which would also shift the $l = 0$ mode), amplified due to the solar cycle amplitude modulation, as unresolved magnetically split lines weight differently to the rotationally split ones. A straightforward numerical simulation of such an explanation has been performed; it is found that a 25% amplitude variation of the modes within a multiplet across the solar cycle would yield the observed results.

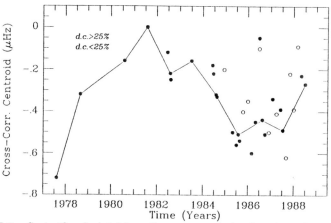

FIGURE 2.- Centroids calculated from the cross-correlation functions of power spectra of all available data (see Table 1). Unfilled circles stand for those series with less than 25% duty cycle.

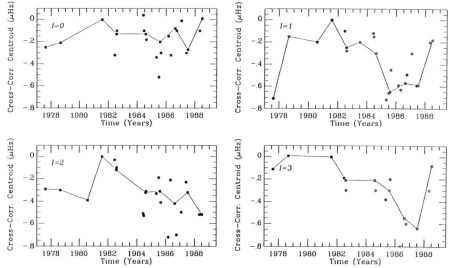

FIGURE 3.- Centroids calculated from the cross-correlation of power spectra for: a) $l = 0$, b) $l = 1$, c) $l = 2$, d) $l = 3$, with respect to 1981. Each spectrum used keeps only the information around the first "isolated" sideband of each mode.

2. Variation of the power of low degree acoustic modes

Using the same 60 day time series already studied (see Table 1) the power in each mode has been calculated and its possible variation along the solar cycle investigated. The power, rather amplitude squared, has been calculated as follows:

$$P = \sum_{l=0}^{2} \frac{P_l}{S_l} = \sum_{l=0}^{2} \frac{1}{S_l} \sum_{n=13}^{25} \sum_{i=-k}^{k} (A_{i,l,n}^2 - A_{noise}^2)$$

where:

$A_{i,l,n}$ is the amplitude at frequency ν_i , around a peak representing the mode of degree l and order n, as found by Jiménez et al (1988).

A_{noise} is the mean amplitude of the noise level for 136μ Hz around a given set of peaks.

k has been set at $2\delta\nu$, where $\delta\nu$ is line width as measured in Elsworth et al (1989).

S_l is the sensitivity of the integrated sunlight velocity measurements as defined in Pallé et al (1989).

n stands for the order of the modes.

The values found from P and P_l for $l = 0$ to 3, are shown in Fig. 4 and 5. It must be born in mind that, in these 60 day series non negligible first order side bands exist for every peak present in the spectra, due to the observing window function. Moreover, the worse the duty cycle, the more power goes into the sidebands. This explanation is important, because due to the separation between l_0 and l_2 ($\sim 9\,\mu$Hz), its relative power is, 1:1 and the rotational splitting for l_2 modes some of the sidebands power can contribute to the power per mode calculated. As far as $l = 1$ and 3 are concerned, this problem is somewhat less severe because their separation is $\sim 15\,\mu$Hz and their relative power is $\sim$ 10:1; therefore, the most reliable measurement is for $l = 1$ where this effect is very small. From the results shown in Fig. 4 and 5 a $\sim 40\%$ variation can be deduced, well correlated with solar activity cycle, the power being higher when the solar activity is at minimum, confirming a previous trend found in Jiménez et al(1988).

The interpretation of this effect could be attributed to the absorption of mode power by magnetic structures (sunspots, active regions, etc...) already found for higher l modes by Braun et al (1988), but it seems unlikely. The reason being that, if magnetic structures absorb the same amount of p-mode power for $l \leq 3$ as for higher l modes, which is $\sim 50\%$, then, since at the maximum of solar activity, the surface covered by active regions can be 10% at most hence the power absorbed by them would be less than 5%, which is clearly not enough to explain the observed effect.

Therefore, it is more likely that changes in the efficiency of the excitation mechanism of such modes could be the cause. Assuming turbulent convection to be the responsible for exciting these modes with energies $E \sim c^2 \rho L^2 H$(Librbrecht ,1988), where ρ is the density, c the sound speed, H the scale height and L the characteristic size of the granules, then a maximum variation of $\sim 20\%$ in L along the solar cycle would account for the observed effect. Observations of such variations have already been made by Muller and Roudier (1985) with similar numerical results (in fact their observations would imply a change of only 10% on L).

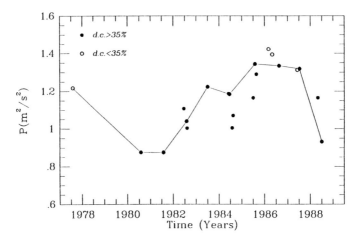

FIGURE 4.- *Power per mode summed over all modes, with $l = 0, 1, 2$ and within the interval (2.00, 3.84) mHz, multiplied by its sensitivity function (see text). A line joins the values corresponding to summer series (see Table 1).*

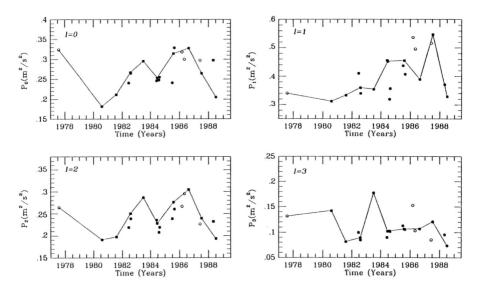

FIGURE 5.- *Power per mode summed ever all modes within the interval (2.00,3.84) mHz and different $l = 0, 1, 2$ and 3. The symbols are the same as in Fig. 4.*

References

• Braun D.C., Duvall Jr. T.L., La Bonte B.J., (1988),'Acoustic absorption by sunspots', Ap. J.,**335**, 1015.

• Elsworth Y., Isaak G.R., Jefferies S.M., McLeod C.P., New R., Pallé P.L., Régulo C., Roca Cortés T.,(1989), 'Linewidth of low degree acoustic modes of the Sun'. M.N.R.A.S., (in press).

• Jiménez J.A., Pallé P.L., Pérez J.C., Régulo C., Roca Cortés T., Isaak G.R., McLeod C.P., van der Raay H.B.,(1988) in J. Christensen-Dalsgaard and S. Frandsen (eds.) 'Advances in Helio and Asteroseismology', Reidel, 205.

• Libbrecht K., (1988). 'The excitation and damping of solar p-modes'.in E.J. Rolfe (ed.) 'Seismology of the sun and sun-like stars', ESA SP-286,3

• Muller R. and Roudier T., (1984).'Variability of the structure of the granulation over the solar activity cycle' in 'The Hydromagnetics of the Sun",ESA SP-220, 51.

• Pallé P.L., Régulo C. and Roca Cortés T.,(1989), 'Solar cycle induced variations of the low l solar acoustic spectrum', Ast. & Ap.(in press).

• Pallé P.L., Pérez Hernández F., Roca Cortés T., Isaak G.R., (1989), ' Observation of solar p-modes with $l \leq 5$ '. Astron. & Astrophys. , (in press)

STELLAR SEISMOLOGY

WERNER DÄPPEN
Space Science Department of ESA
ESTEC
2200 AG Noordwijk
The Netherlands

ABSTRACT. *Stellar* acoustic oscillation frequencies will likely be accurately observed in the near future, in analogy to the well-known *solar* five-minute oscillation frequencies. Of course, we will never expect the wealth of solar data, which is a result of spatial resolution. We will therefore not be able to solve the inverse problem, that is to probe physical quantities as functions of depth, and the low number of anticipated observed frequencies will make an unambiguous mode identification difficult. Despite this restriction to the forward problem, however, observed stellar oscillation frequencies will become valuable constraints for the determination of stellar parameters. One should not forget that the present knowledge of stellar ages and compositions relies on the calibration of theoretical models (matching effective temperature and luminosity). Additional observational constraints will improve these calibrations, even if the theoretical models themselves are not questioned. We hope, however, that the observation of stellar oscillation frequencies will also lead to improvements in the *physics* of stellar models, in analogy to the solar case. Again, of course, stellar seismologists will be less ambitious than helioseismologists, since there are more open parameters in stellar models. However, stellar observations will allow tests of models with different age and composition.

1. Introduction

Helioseismology has been extremely successful in probing the solar interior. The precisely observed p-mode oscillation frequencies act as a window, enabling us to peep into the Sun (see *e.g.* Christensen–Dalsgaard, these proceedings; for reviews see *e.g.* Christensen–Dalsgaard and Berthomieu 1990, Deubner and Gough 1984). The high spatial resolution of the solar disk has allowed to observe, and unambiguously identify, p modes in a range of angular degree l from zero (radial modes) to a few thousand. This in turn has lead to direct inversion of the oscillation frequencies to obtain physical quantitities (such as sound speed) as a function of depth – without recourse to models.

It is clear that little of this beautiful picture will remain in the stellar case. The lack of spatial resolution will restrict the observable modes to those of low l (not greater than 3 or 4), with higher-degree modes disappearing in the integrated light. It is worth mentioning that clever techniques exist, which are based on line profile variations (*e.g.* Smith 1985;

G. Berthomieu and M. Cribier (eds.), Inside the Sun, 357–370.

further references are found in Baade 1988a,b). They promise observation of some nonradial modes of somewhat higher angular degrees (l up to 8 or perhaps 16). But even these few extra modes can never bring the wealth of solar seismological data (but never say never!).

As a consequence there will be no direct inversions of stellar oscillations frequencies. Nevertheless, despite this lack of analogy, stellar seismology does have exciting prospects. The reason is that stars can offer something that the Sun can't: their large number. The Sun can be subjected to helioseismological tests as thoroughly as wanted, but as far as they are to test the theory of stellar evolution, they do it merely for the case of one mass, age, and chemical composition. In stars, however, we see these parameters differing from one object to another. Without seismology, the number of observable stellar quantities that can serve as tests of the theory of stellar evolution is not that large (one knows effective temperatures quite well, luminosities for nearby stars with known parallaxe, and mass in the case of close binary system). Though the theory of stellar evolution has helped enormously to decipher the phenomenology of these observations, there are still sizeable uncertainties in the physics of the stellar models used. These uncertainties matter: our cosmic distance-scale, as well as the age determination of globular clusters, relies on good models of variable stars (Cepheids and RR Lyrae; for a recent review see van den Berg 1989).

Despite the considerable precision that solar models have attained one should not forget the underlying standard assumptions of stellar evolution, which are, *e.g.* , an initially chemically homogeneous gas cloud, nuclear reaction rates that are determined or extrapolated from laboratory values, opacities from elaborate theoretical calculations, plus a crude mixing-length formalism for convective heat transport. But the fact that solar standard models work quite well is no guarantee to use the same assumptions for stars with different parameters. Helioseismology does not equally strongly test all these assumptions. Indeed, solar physicists are happy about those uncertainties in the physics of the models that are not so important; these uncertainties do not obstruct the way towards constructing a 'good' solar model. As an example I mention opacity in the bulk of the solar convection zone. Since there the temperature gradient is to a very good approximation given by the thermodynamic adiabatic gradient, opacity doesn't play a role. This is good for solar modellers, but bad for those who test stellar opacitites. For certain stars the same opacities that have little importance for the Sun do matter. So, for instance, in the case of stars that are slightly more massive than the Sun, where the convection zone is much shallower, and where thus opacity at temperatures and densitites corresponding to the solar convection zone significantly influences the structure of the star.

In the spirit of this interdisciplinary conference, I will stress the common mathematical language with other physical disciplines. Review articles that are more astrophysically slanted are found elsewhere (*e.g.* Christensen–Dalsgaard 1984, or Däppen *et al.* 1988). One of the goals of this article is to show that computing stellar oscillation frequencies is a relatively easy part of stellar seismology, while the hard work is being done in the computation of equilibrium models of evolved stars, where the nonlinear partial differential equations of stellar evolution must be solved. Furthermore, to interrogate the physics of stellar interiors by oscillation frequencies, one usually has to compare models that are the same up to the physical quantity to be tested (notice that we are restricted to the forward problem). Again, computing the series of similar evolutionary sequences is time-consuming. In contrast, the physical description of the pulsation itself remains unchanged, as long as one is content with the adiabatic approximation, which is sufficient for interpreting oscillation frequencies. Of course the situation becomes more complicated if other pulsational features are examined, such as exitation, life-time, or amplitudes.

2. Modelling Stellar Oscillations

In this section, I present three different approaches. Since I do not consider nonlinearities here, the problem of stellar oscillations is equivalent to finding all normal modes. The first approach will be an outline of the complete numerical solution. The similarity with other eigenvalue problems in mathematical physics is stressed (*e.g.* vibrating strings, bound states in quantum mechanics). The second approach will be based on a simplified wave equation and on propagation diagrams that reflect local conditions in a star. This approach is best suited for a qualitative discussion of the frequency spectrum of modes. Further characteristics of the modes, like their type (p mode or g mode, see below) or penetration depth are also directly visible in propagation diagrams. Finally, I will present yet an other approach to represent oscillation frequencies, namely asymptotic theory for high-order modes. Its advantage is a relatively easy, quantitative link between the observed frequency spectrum and the mass and age of the star.

Before discussing these three different approaches, I begin with some very basic facts from stellar evolution, because it is clear that one needs some ideas about the equilibrium solution before one can study the deviations from it.

2.1. SOME ELEMENTARY PREREQUISITES FROM STELLAR EVOLUTION

The basic framework to model stellar oscillations is given by the hydrodynamic equations of motion for the stellar fluid moving around its equilibrium. This equilibrium solution is assumed as given; it results from a calculation of stellar structure and evolution. A still excellent introduction to the basic principles of stellar evolution is the book by Schwarzschild (1958) (though the book can't of course cover all the fascinating progress that has been made possible since, especially thanks to a greatly increased computing power). For our purposes we need to know that stars in the equilibrium are self-gravitating gas balls, with a pressure gradient balancing the local gravitational force, and with a temperature gradient associated to the heat flux going from the nuclear-burning central regions to the (much cooler) outside. These two gradients are at the base of the nonlinear stellar structure equations.

As long as the star still lives from its hydrogen supply in the center, things happen very slowly (the Sun, *e.g.* , has been around for about $5 \, 10^9$ years, with now about half of its hydrogen fuel used). More massive stars have a much shorter life, because their energy output (*i.e.* their luminosity) increases with at least the third power of the mass of the star; therefore the hydrogen-burning phase of a 100 solar-mass star is only of the order of

10^6 years. Less massive stars make much more economical use of their resources; their life time easily exceeds the currently assumed age of the universe.

During this hydrogen-burning phase stars are on the so-called main sequence, which is a phenomenological name, originating in the fact that if stellar luminosity is plotted against surface (effective) temperature (in the so-called Hertzsprung–Russel diagram), then most stars lie on a line, *i.e.* the main sequence. It was an important success of the theory of stellar evolution to identify this observational fact with hydrogen burning. Another important result of the theory of stellar evolution is that during the hydrogen-burning phase a star moves the main sequence slightly upwards (the Sun is generally thought to have been about 25% less luminous at the begin of its main-sequence life). Stars thus climbing up the main sequence look practically the same as more massive but younger ones, if the only discriminating observables are temperature and luminosity. It is precisely one of the goals of stellar seismology to lift this degeneracy by providing additional observables (see section 3).

2.2. OUTLINE OF NUMERICAL COMPUTATIONS OF STELLAR OSCILLATIONS

The purpose of this subsection is to convince the reader that calculations of stellar oscillations are a relatively simple affair, if reasonable assumptions and simplifications are made. As briefly mentioned before, it is the problem of finding the stellar equilibrium solution that is the difficult part.

Assuming the existence of a stable and constant equilibrium configuration, to discuss the pulsational motion of a star we start out from the hydrodynamic equations for compressible fluids

$$\frac{\partial \mathbf{v}}{\partial t} + \mathbf{v}\nabla\mathbf{v} = -\frac{1}{\rho}\nabla p + \nabla\psi \ . \tag{1}$$

Here, $\mathbf{v}$ is the (Eulerian) velocity field, p and ρ pressure and density, respectively, and ψ is the (self-) gravitational potential. For simplicity, we have disregarded viscosity. To this equation, one must add the usual equation of continuity and also an energy equation, which - in the simplest case - is replaced by a condition of adiabaticity, normally expressed in the form of constant co-moving specific entropy (per mass). Under the assumption of adiabaticity, stellar pulsation is frictionless and energy conserving. It is clear that this assumption precludes a discussion of mode excitation (or damping), necessary to understand why we see this pulsation mode but not the other. Nevertheless, this hypothesis of adiabaticity still leads to many important results by telling, *e.g.* , what the linear eigenfrequencies are. Virtually the whole success of helioseismology has been so far in the framework of adiabatic pulsations; only very recently have serious attempts been made to go beyond and to address questions like mode excitation, damping, and amplitudes (for a review see Cox *et al.* 1990).

In the following I discuss a few of the key steps used in manipulating Equation (1). I will be very brief, but details can be found, *e.g.* , in the book by Unno *et al.* (1979). We restrict ourselves to a simple, but still relevant case, in which the equilibrium configuration is assumed to be at rest (*i.e.* no rotation) and spherically symmetric. The gravitational potential is assumed to be static, *i.e.* its distortions due to the stellar pulsation itself is neglected (this is the so-called Cowling approximation, see Unno *et al.* 1979). Restricting even more to *small* perturbations, we introduce *linearization* of Eq. (1).

The static equilibrium and the linearized equations allow the separation of the time dependence in the form of $\exp(i\omega t)$. The spherically symmetric equilibrium configuration allows expressing the field variables (displacement vector, pressure, and density) as a series of spherical harmonics Y_l^m of angular degree l and azimuthal order m, so that each term in the series is itself a solution of the equations. The adiabatic assumption is used to link pressure and density fluctuation with the help of a thermodynamical quantity (the adiabatic exponent), which is a given quantity of the equilibrium model. One arrives therefore at a fourth-order system for the (independent) three displacement-vector components and the pressure fluctuation. The Cowling approximation allows yet another simplification; with it, the tangential part of the displacement field becomes proportional to the tangential component of the gradient of the (Eulerian) pressure fluctuation, and thus the only independent fields remaining are the radial component of the displacement vector and the pressure fluctuation. Their amplitudes are governed by the following (schematic) system of ordinary differential equations

$$\frac{dy_1}{dr} = f_{11}(r)y_1 + f_{12}(r;l;\omega^2)y_2$$

$$\frac{dy_2}{dr} = f_{21}(r;\omega^2)y_1 + f_{22}(r)y_2 \qquad (2)$$

$$y_1 \equiv \frac{\xi_r}{r}$$

$$y_2 \equiv \frac{p'}{gr\rho} \ .$$

Here, the functions $f_{ij}(r;\omega^2)$ are expressions involving quantities of the equilibrium model (like pressure, density, local gravity, sound-speed, etc.). As is standard practice, the labels l and ω are dropped in the y_i's. The prime $'$ denotes the Eulerian (first-order) displacement from equilibrium, (and not a derivative as everywhere else in the paper).

Adding boundary conditions to equation (2) leads to an eigenvalue problem. The one at the *centre* is the usual regularity condition due to the singular nature of Eq. (2). From the specific nature of the coefficients one knows (see Unno 1979) that Eq. (2) has a regular-singular point at $r = 0$, and so there is a regular and a singular solution. Picking the regular one gives the boundary condition (this is explicitly done by a standard power-series development). The *outer* boundary condition is not so simple. In principle, one would have to put a good stellar atmosphere (for which there are elaborate models) at the outer end, and impose smooth matching as the outer boundary condition. Until now, nobody has succeeded in doing that, and simpler approaches must be chosen. Often an isothermal atmosphere is assumed, and the outer boundary condition is determined by a discussion of propagation and reflection of sound waves in a stratified atmosphere analogous to Lamb (1932) (we come to that in 2.3.). Here, we can afford something even simpler, namely a mechanical boundary condition of the form $\delta p = 0$ where δ stands for the Lagrangian displacement. Such a boundary condition is the three-dimensional analogon of a frictionless lid. In terms of our variables y_1 and y_2, the condition $\delta p = 0$ is given by $y_1 + y_2 = 0$.

I hope that I have convinced everyone that computing stellar oscillation frequencies is the easy part of seismology, if one can borrow a good equilibrium model from a friend. Finding the eigenvalues of Eq. (2) is really not much harder than those of a vibrating string, whose equation is $d^2y/dx^2 + \omega^2 y = 0$ with boundary conditions $y = 0$ at two different x values. The only complication in Eq. (2) comes from the nonconstant coefficents, but numerically it is still an easy task. The hard part is therefore finding the equilibrium solution which delivers the coefficients for Eq. (2). In studies of the parameter dependence of oscillation frequencies one generates series of similar evolutionary models, with slightly different stellar parameters (age, mass, composition, or quantities from basic physics like the opacity). This can be time consuming, especially if rather evolved stars are considered, where many time steps are required for the solution.

2.3. QUALITATIVE DISCUSSION USING PROPAGATION PROPERTIES

We adopt the asymptotic discussion of Deubner and Gough (1984), which itself is similar to the treatment of acoustic waves by Lamb (1932) [see also Christensen-Dalsgaard (1986)]. For wavelengths much shorter than the solar radius, normal oscillation modes can be quite accurately discussed using the simplified wave equation

$$\Psi'' + K^2(r)\Psi = 0 \ . \tag{3}$$

Here, $\Psi = \sqrt{\rho}c^2 \text{div}(\delta\mathbf{R})$, where ρ and c are density and sound speed of the equilibrium configuration, and $\delta\mathbf{R}$ is the fluid displacement vector. The local wave number is given by

$$K^2(r) = \frac{\omega^2 - \omega_c^2}{c^2} + \frac{l(l+1)}{r^2}\left(\frac{N^2}{\omega^2} - 1\right) , \tag{4}$$

with the acoustic cut-off frequency defined by

$$\omega_c^2 = \frac{c^2}{4H^2}(1 - 2\frac{dH}{dr}) , \tag{5}$$

and the Brunt-Väisälä frequency N by

$$N^2 = g(\frac{1}{H} - \frac{g}{c^2}) , \tag{6}$$

where H is the density scale height and g the local gravity. From the form of Eq. (3) (to which upper and lower boundary conditions must be added), one immediately realizes that in propagation zones necessarily $K^2 > 0$.

Our present qualitative discussion of the influence of mass and evolution on oscillation frequencies aims at showing the maximum of effects with a minimum of curves. Here, we restrict ourselves to the role of N^2 and ω_c^2 in K. For finer details we refer the reader to Deubner and Gough (1984), Christensen-Dalsgaard (1984) or Gough (1985).

With the convenient definition of the Lamb frequency

$$S_l^2 = \frac{l(l+1)}{r^2}c^2 , \tag{7}$$

we obtain the simplified necessary conditions for propagation of an acoustic wave, $\omega > \omega_c$ and $\omega > S_l$. Additionally, in order to have a trapped standing wave, it is also necessary that in some surface layer ω_c becomes greater than ω. This happens indeed; the height of this (outer) ω_c mountain is the greater, the cooler the local temperature at the edge of the star is [this is seen from Eq. (5)]. 'Mathematical' stars with zero temperature at the outer boundary have an infinitely high ω_c mountain; they can therefore trap modes of arbitrary high frequency. Real stars have an 'inversion' temperature (just above the photosphere); further up temperature begins to rise again. The maximum p-mode frequency 'measures' this inversion temperature.

In propagation zones (if a constant adiabatic exponent of 5/3 is assumed), a further simplification follows from the fact that $\omega > g/c$ implies $\omega > \omega_c$. And finally, we choose the approximation of identifying (the absolute value of) g/c with N, which certainly gives the correct order of magnitude in radiative zones (but would be entirely wrong in convection zones, where $N \approx 0$). The advantage of this choice is that the same curves will give information about g modes, too.

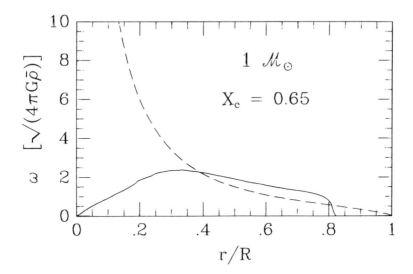

Figure 1. Critical frequencies as functions of the fractional radius r/R. The solid line denotes the Brunt-Väisälä frequency N, the dashed line the Lamb frequency S_l for $l = 1$. The model parameters are: hydrogen abundance $X = 0.70$, heavy-element abundance

$Z = 0.01$, and the mixing-length parameter $l/H_p = 1.5$. Stellar age is indicated by the central hydrogen abundance X_c.

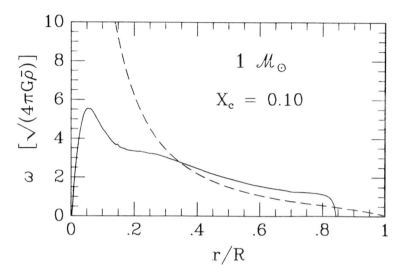

Figure 2. Same as Figure 1, but for a more evolved model.

Let us now consider S_l (we choose the representative case $l = 1$) and N in a model of a 1 $M_\odot$ star (Figure 1). Due to the rather deep convection zone, N cannot represent the increase of ω_c close to the surface, and so the diagram does not show the upper turning point that is caused by the large spatial inhomogeneity near the surface. In Figure 1, S_1 defines the penetration depth of the $l = 1$ modes; for $l > 1$ the corresponding curves would be shifted to the right.

Figure 1 also shows how to distinguish p modes from g modes. The distinction is only asymptotic. Modes with $\omega \to \infty$ would become true p modes (but they cease to be trapped above a certain frequency, see above). Modes with $\omega \to 0$ become true g modes [they have a large K^2 due to N/ω, see Eq. (4)]. Outside the asymptotic regime one still speaks of p modes and g modes, but they are not 'pure', though most of them are of a dominant type. Only in highly evolved stars do genuine dual-status modes appear (see section 3.).

2.4. ASYMPTOTIC THEORY OF OSCILLATION FREQUENCIES

The simplest theoretical analysis of oscillation frequencies is asymptotic theory (Tassoul 1980), which - to second order - yields the following expression for the frequencies $\nu_{n,l}$

$$\nu_{n,l} = (n + l/2 + \sigma)\nu_0 + \epsilon_{n,l} \ . \tag{8}$$

Here, σ is a constant of order unity, ν_0 will be defined below (Eq.10), and $\epsilon_{n,l}$ is small compared with $\nu_{n,l}$. (This expression is found by manipulating Eq. (2), exploiting the simplifications due to the high-order modes with short wave-length). Given a set of observed oscillation frequencies, the constants appearing in Equation (8) can be determined. As shown below, they are related to integral quantities of the star.

At this point it is useful to introduce two definitions pertaining to the structure in the periodogram of high order pulsators.

(i) Large frequency-separation:

$$D_{n,l} \equiv \nu_{n+1,l} - \nu_{n,l} \ . \tag{9}$$

To first order asymptotic theory it is well known that

$$D_{n,l}{}^{-1} \approx \nu_0{}^{-1} \equiv 2 \int_0^R \frac{1}{c} dr \ , \tag{10}$$

with c denoting local sound speed and R the radius of the star. In simplified stellar models (polytropes) it is easy to show that

$$\nu_0 \propto \sqrt{\frac{g}{R}} = \sqrt{\frac{GM}{R^3}} \ . \tag{11}$$

(ii) Small frequency-separation:

$$d_{n,l} \equiv (\nu_{n,l+1} - \nu_{n,l}) - \frac{1}{2}(\nu_{n+1,l} - \nu_{n,l}) \ . \tag{12}$$

The small separation serves to cancel the first-order term of Eq. (8), and thus reveals second-order details. The ratio between small and large separation pertains to the central regions of the star, and is, to second-order asymptotic theory, given by (Tassoul 1980)

$$\frac{d_{nl}}{D_{nl}} \approx \left(\frac{l+1}{2\pi^2 \nu_{nl}}\right) \int_0^R \frac{dc}{dr} \frac{1}{r} dr \ . \tag{13}$$

This equation is somewhat simplified; for a more thorough discussion see Gabriel (1990). Since sound speed increases steeply from the surface to the centre of a star, $D_{n,l}$ probes more the surface regions and $d_{n,l}$ more the central regions.

3. Determination of Stellar Age and Mass from Seismology

3.1. QUALITATIVE DISCUSSION

The two Figures 1-2 show the principal effects of evolution and mass on oscillation frequencies, thus demonstrating the power of the qualitative tools of section 2.3. The first effect of evolution is an increase of the inner 'N-mountain', caused by the growing spatial chemical inhomogeneity in the central regions. The inner mountain is capable eventually to prevent radial modes from penetrating to the center. Note that this might sound paradoxical, because Eq. (4) tells us that N does not influence modes with $l = 0$. But recall that in our simplified figures we also use N to represent the (inner) ω_c mountain, which is of the same order as N.

In evolved stars, modes with $l \geq 1$ can acquire 'dual status', i.e. they become gravity modes in the core and remain acoustic modes in the envelope. Thus they penetrate deeper into the interior, while simultaneously their frequency spectrum becomes denser. The main effect of mass (in the range around and slightly above 1 $M_\odot$) on the frequency spectrum is related to the disappearance of the convective envelope and the forming of a convective core. This has a major influence on g-mode propagation.

3.2. QUANTITATIVE DISCUSSION

The small separation (Eq. 12) carries an important signature of stellar age, because as hydrogen is converted into helium in the stellar core, the mean molecular weight increases, which causes a decrease of sound speed and its gradient, thus reducing the small separation. An excellent diagnostic diagram that extracts the information contained in the small and large separation has been invented by Christensen-Dalsgaard (1984) (for a more detailed calculation see Christensen-Dalsgaard 1988). In this diagram [hereinafter JCD diagram (for Jørgen Christensen–Dalsgaard)], contours of constant stellar mass and age are plotted against the theoretically computed large and small separations. Since the mass and age contours are rather perpendicular than parallel to each other, they reveal the considerable diagnostic potential of these diagrams.

Going a step further, Gough (1987) discussed the accuracy of seismological mass and age determination, using JCD diagrams and stellar models computed by Ulrich (1986, 1988). Gough's discussion was purely formal: taking the theoretical model for granted, he computed the uncertainty in the mass and age determination, assuming given errors for the observed stellar parameters (effective temperature, luminosity, heavy-element abundance, large and small frequency separation). Gough's (1987) result is that mass and age determination are so sensitive to the heavy-element abundance that they cannot be carried out in this way, unless other stellar parameters are known by independent means. If, for instance, in the case of a binary system we can determine mass, or if we can estimate it from surface gravity, then the large separation can reveal the evolutionary information contained in the deviation from the simple relation (11) (otherwise the large separation mainly fixes M/R^3). Thus a more accurate age determination could become possible (see Gough 1987).

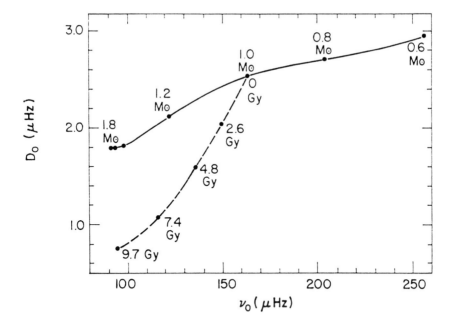

Figure 3. The location of a ZAMS (zero-age main sequence) (*solid line*) and of a 1 $M_\odot$ evolutionary sequence (*dashed line*) in a (D_0, ν_0) JCD diagram. Here, ν_0 is as in Eq. (8), and D_0 is a suitably defined average over small separations $d_{n,l}$ (for details see Christensen–Dalsgaard 1984, from which this figure is taken).

4. The Observational Situation: Procyon, α Centauri, and ϵ Eridani

Gelly et al. (1986) have reported p modes in Procyon and α Centauri. Only the large separation has been observed. Noyes et al. (1984) have reported three individual p-mode frequencies and the large separation for ϵ Eridani. To illustrate the potential, and the difficulties, of such observations for testing stellar structure and evolution, consider the recent controversial theoretical articles dealing with Noyes *et al.* 's (1984) observations. While Guenther and Demarque (1986) have concluded that a model of a very old star (12 Gyr) fits the data best (though they are aware of indications of stellar activity that speak against such a high age), Soderblom and Däppen (1989) conclude that a model of a very young star (1Gyr) is equally well suited, provided that one accepts the unusually small value of the mixing-length parameter of 0.45. The discrepancy of the two interpretations is well in line with the aforementioned error analysis by Gough (1987). However, an erroneous frequency determination could also have been the source of these difficulties. More and better resolved frequencies will be needed.

368

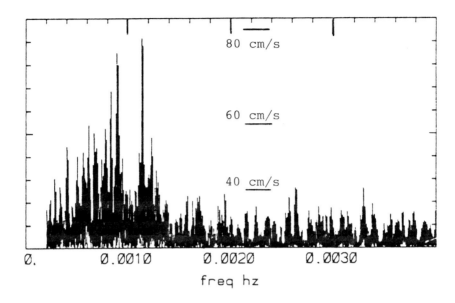

Figure 4. Power spectrum of Procyon, obtained during 6 nights of observation. Power is in arbitrary units, the corresponding velocities are indicated in the figure.

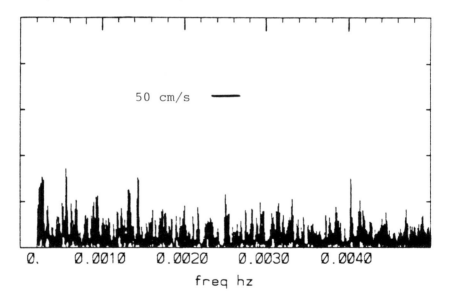

Figure 5. Same as Figure 4, but for the Sun (5 days of observation).

We end this section with a recent very promising result. Brown, Gilliland, Noyes, and Ramsey (private communication) have carried out Doppler velocity measurements of Procyon and obtained a clear signal around and below 1 mHz (Figures 4 and 5).

5. Prospects

Despite little analogy with the solar case, and despite the fact that not much has been achieved yet, stellar seismology has excellent prospects. Together with new observational techniques (like high S/N spectroscopy, astrometrical space missions), which will allow better determinations of stellar parameters (age, mass, chemical composition), the expected seismological information will put additional constraints on theoretical models of stellar evolution. The observational data will enable us to develop better physical models for the theory of stellar evolution (equation of state, convection, opacity, nuclear reactions, etc.). Since many of the diagnostically powerful small frequency-separations lie around 10 μHz, which is close to the diurnal frequency (11.6 μHz), it will greatly help to go to space or to coordinate several ground-based telescopes.

ACKNOWLEDGMENTS: I am grateful to Tim Brown, Ron Gilliland, Robert Noyes, and Lawrence Ramsey for their permission to show the results on Procyon.

References

Baade, D. 1988a, in *The Impact of Very High S/N Spectroscopy on Stellar Physics*, eds. G. Cayrel de Strobel & M. Spite (IAU Symp. 132, Kluwer, Dordrecht), 193-198.

Baade, D. 1988b, in *The Impact of Very High S/N Spectroscopy on Stellar Physics*, eds. G. Cayrel de Strobel & M. Spite (IAU Symp. 132, Kluwer, Dordrecht), 217-222.

Christensen-Dalsgaard, J. 1984, in *Proc. workshop on space research prospects in stellar activity and variability*, eds. A. Mangeney & F. Praderie (Paris Observatory Press), 11-45.

Christensen-Dalsgaard, J. 1986, in *Seismology of the Sun and the Distant Stars*, ed. D.O. Gough (NATO ASI Series, Reidel, Dordrecht), 3-18.

Christensen-Dalsgaard, J. 1988, in *Advances in Helio- and Asteroseismology*, eds. J. Christensen-Dalsgaard and S. Frandsen (IAU Symp. 123, Reidel, Dordrecht), 3-18.

Christensen-Dalsgaard, J., Berthomieu, G. 1990, in *Solar Interior and Atmosphere*, eds. A.N. Cox, W.C. Livingston, & M. Matthews (Space Science Series, University of Arizona Press), in press.

Cox, A.N., Chitre, S.M., Frandsen, S., Kumar, P. 1990, in *Solar Interior and Atmosphere*, eds. A.N. Cox, W.C. Livingston, & M. Matthews (Space Science Series, Unversity of Arizona Press), in press.

Däppen, W., Dziembowski, W., Sienkiewicz, R. 1988, in *Advances in Helio- and Asteroseismology*, eds. J. Christensen–Dalsgaard and S. Frandsen (IAU Symp. 123, Reidel, Dordrecht), 233-247.

Deubner, F.-L., Gough, D.O. 1984, *Ann. Rev. Astron. Astrophys.*, **22**, 593-619.

Gabriel, M. 1990, The l-dependent part of D_{nl} and the structure of the solar core, *Solar Physics*, (submitted).

Gelly, B., Grec, G., Fossat, E. 1986, *Astron. Astrophys.*, **164**, 383-394.

Gough, D.O. 1985, in *Future Missions in Solar, Heliospheric, and Space Plasma Physics*, ed. E. Rolfe, ESA SP-235, European Space Agency, Noordwijk), 183-197.

Gough, D.O. 1987, *Nature*, **326**, 257-259.

Guenther, D.B., Demarque, P. 1985, *Astrophys. J.*, **301**, 207.

Lamb, H. 1932, *Hydrodynamic*, (Cambridge Univ. Press, Cambridge).

Noyes R.W., Baliunas, S.L., Belserene, E., Duncan, D.K., Horne, J., Widrow, L. 1984, *Astrophys. J. Letters*, **285**, L23.

Schwarzschild, M. 1958, *Structure and Evolution of the Stars*, (Princeton Univ. Press, Princeton).

Smith, M. 1985, *Astrophys. J.*, **297**, 206-223.

Soderblom, D., Däppen, W. 1989, *Astrophys. J.*, (in press).

Tassoul, M. 1980, *Astrophys. J. Suppl. Ser.*, **43**, 469-490.

Ulrich, R.K. 1988, in *Advances in Helio- and Asteroseismology*, eds. J. Christensen–Dalsgaard and S. Frandsen (IAU Symp. 123, Reidel, Dordrecht), 299-302.

Ulrich, R.K. 1986, *Astrophys. J.*, **306**, L37-L40.

Unno, W., Osaki, Y., Ando, H., Shibahashi, H. 1979, *Nonradial Oscillations of Stars*, (University of Tokyo Press).

van den Bergh, S. 1989, *Astron. Astrophys. Rev.*, **1**, 111-139.

SOLAR AND STELLAR CYCLES

GAETANO BELVEDERE
Istituto di Astronomia
Universita di Catania
Italy

ABSTRACT. The increasing observational evidence offered by photometric and spectroscopic data of magnetic cycles in lower main sequence stars, has confirmed the general expectation that the same basic dynamo mechanism operates in the Sun and main sequence in stars with outer convective envelopes.

Unfortunately, no clear correlation has been found, up to date, with stellar parameters as mass, rotation rate and age, even if irregular activity and shorter cycle periods seem to be characteristic of stars more massive than the Sun, while hyperactive fast rotating components of binary systems like RS CVn's and BY Dra's show a tendency for cycles as long as several decades.

Although dynamo theory has probably captured the essential physics of the convection-rotation interaction giving rise to stellar magnetic activity, as evidenced, for instance, by the correlation between proxy activity, indicators and the Rossby parameter related to the dynamo number, the reliability of the present theoretical background should be measured by its capacity of interpreting and predicting characteristics and periodicities (or aperiodicities) of stellar cycles. This should be done in the framework of the nonlinear approach, which, in principle can describe multimodal dynamo behaviour with a variety of time scales.

The fundamentals of the theory must be tested, however, in the closest astrophysical laboratory, our Sun. Serious problems to a dynamo mechanism operating in the convection zone have been posed by most recent helioseismological results, which, on the other hand, do not rule out the possibility of dynamo action in the transition layer between the convective and the radiative zones, which is suggested independently by the global solar cycle features. Indeed, assuming the correct sign of helicity in the transition layer, the helioseismological data on the radial gradient of angular velocity support both equatorward propagation of dynamo waves at lower latitudes and poleward propagation at higher latitudes, which is evidenced by different tracers of the solar cycle.

1. The Principle of Solar-Stellar Connection

This is an unification principle. It states: (i) that the phenomenology of solar and stellar magnetic activity is essentially the same despite of different strength, topology and time scale of active phenomena occurring on the Sun and distant stars; and (ii) that this phenomenology can be interpreted in terms of the same fundamental mechanism, the dynamo mechanism, operating in stellar convection zones, although differences in the dynamo operation mode are expected,depending on stellar mass, rotation rate and age, as well as on the dynamo number, a parameter which characterizes the

G. Berthomieu and M. Cribier (eds.), Inside the Sun, 371–382.
© 1990 *Kluwer Academic Publishers. Printed in the Netherlands.*

strength of dynamo action - the last point being clearly evidenced by the non-linear analysis of dynamo equations.

Hence, the complementarity of solar and stellar observations to highlight the basic phenomena and mechanisms of stellar activity:

- in a way, solar data give the possibility of detailed study of activity features and phenomena in the closest astrophysical laboratory (the Rosetta Stone of Astrophysics), provide a guide-line to explore stellar activity on the basis of what we learn from the Sun, and allow to test models of stellar activity "at home"

- on the other side, stellar observations do offer a large sample of phenomenology on a multiplicity of astrophysical situations and time scales, suggesting a more general scenario and widening out our ideas about the operation modes of stellar activity.

Unity in multiplicity is therefore the basic concept of the solar - stellar connection: unity as for the mechanism, multiplicity as for its operation modes depending on stellar parameters.

But, warning! Some caution is necessary when making use of the analogy to the Sun to understand phenomena in stars of different activity level. For instance :

- extrapolating from slow rotators (hypoactive stars) to fast rotators (hyperactive stars) may be seriously dangerous

- the dynamo operation mode may be different in different activity level stars, even as to location (the convection zone(c.z.) or the boundary layer at the base of the c.z. or even the stellar core) and driving mechanism itself (radial or latitudinal angular velocity gradient in the stellar interior)

- significant differences in the activity signatures of very similar stars are expected when the analysis is made in the non - linear magnetohydrodynamic regime (Weiss et al. 1984, Belvedere et al. 1989, Belvedere and Proctor 1989). This is confirmed on the observational side (Rodonò 1987).

Therefore, dynamo theory must be tested both on the Sun and distant stars and is reliability measured by its capacity of interpreting and predicting different activity signatures in different stars. In this context, it appears to be of outstanding relevance to confront theory and observations as to periodicities or aperiodicites in the time evolution of activity, with particular reference to stellar cycles.

2. Some Relevant Features of the Solar Cycle

As is known, proxy records of solar activity, as C's fluctuations in tree rings or Be's in polar ice caps (Stuiver and Quay 1980) allow us to explore the activity of the Sun back to about nine millenia ago. Note that the layered Precambrian sediments found in South Australia, which were previously interpreted as an evidence for solar cyclic activity some 700 million years ago, were due to tidal processes (Williams 1989; Weiss 1989).

Human sporadic sunspot observations date back to two millenia ago, but systematic spot counts only to mid - seventeenth century.

It has clearly been stated on the basis of spot counts (Wolf number) that the mean periodicity of the solar cycle is 11.2 years, with range from 8 to 15 years. More correctly, one has to take the double (22y) as the real solar cycle period, this being the time interval between two successive appearances of the same magnetic field polarity at the solar poles. Also long - term periodicities in the sunspot counts have been evidenced: the 80-90 y Gleissberg cycle and the 190 y Grand Cycle. For details on the solar cycle features see e.g. Stix (1981).

Curiously, it is not the modulation of activity which is more relevant to understand the solar cycle (and the underlying dynamo mechanism) but the nearly absence of activity at certain epochs, the so called Grand Minima (Eddy 1976, 1983): Oort's Grand Minimum (1010-1050 A.D.), Wolf's (1280-1340), Spörer's (1420-1530) and Maunder's (1645-1715).

Note also that in the dendrochronologic record episodic cycle suppression occurs roughly at 250 year intervals and lasts about one third of this time interval (Damon 1977).This indicates that the solar cycle is really intermittent or, in other words, it exhibits chaotic behaviour with different time scales and episodes of reduced or zero activity, namely a real aperiodicity rather than a quasi - periodicity (Williams 1981, Wallenhorst 1982). To this regard one may question how significant using the term "solar cycle" is, at least referring to suitably extended time intervals.

However, interpreting the solar cycle anomalies in terms of stochastic behaviour instead of deterministic chaos cannot be excluded. The matter is still controversial, but, to this regard, the relation between mean magnetic field and stochastic turbulent convection may be relevant (Hoyng 1987, 1988).

Now, what about stellar cycles? Before entering this problem, let us examine the indicators of magnetic activity in stars and their relations to the basic stellar parameters as mass, rotation rate and age.

3. Stellar Activity Indicators

3.1 MAGNETIC FIELD MEASUREMENTS

Robinson's (1980) method allows to measure non-averaged magnetic field strengths and filling factors by means of comparing magnetic sensitive and insensitive lines. This avoids cancelling out of opposite polarities, which occurs in conventional Zeeman polarization measurements. Results for 29 G-K stars (with 19 detections) have been reviewed by Marcy (1984), indicating magnetic field intensities B as high as $1500 \div 2000$ Gauss, filling factors f in the range $0.3 \div 0.8$ (>> than the solar value ~ 0.01) and a correlation between magnetic flux density, effective temperature and rotational velocity of the type $\Phi \approx T_{eff}^{-2.8} \, V_{rot}^{0.55}$. Note that the Sun does not fit in with this relationship! Gray (1984) found a similar trend. Saar and Linsky (1986), after measuring 20 main sequence stars, found that the magnetic field intensity B correlates with the surface pressure (related to T_{eff}), as suggested by theoretical arguments on flux tube equilibrium, but does not with the rotational velocity: it is indeed the magnetic flux $\Phi \approx fB$ which correlates with V_{rot}. This is an important observational result, as it shows: (i) that the relevant quantity to be related to the rate of rotation in dynamo theories is the magnetic flux; and (ii) that for similar stars B and f may be inversely proportional. More recently (Saar 1989), it has been shown that Φ is proportional to the product $\tau_c \omega$, where τ_c s the convective turnover time (see later) and ω the angular velocity of rotation, and declines exponentially with stellar age.

Future improvement of the observational techniques will check the previous results about magnetic field strengths and filling factors, which are essential data for testing dynamo theory. To this regard, we look with interest at the recent Zeeman - Doppler Imaging Method (Donati, Semel and Praderie,1988; Semel 1989) which can give both intensity and cartography of magnetic fields on stellar surfaces, by measuring both I and V Stokes parameters. In principle, this is an extension of the Doppler Imaging Method (Vogt 1983; Vogt et al. 1987), which allows to reveal spotted regions on stars.

3.2. PHOTOMETRIC VARIABILITY

This evidences the presence of dark spots on stellar surfaces and can in principle give us information on periodicities or aperiodicities in stellar activity as well as on differential rotation of stellar envelopes (large and short scale variation in time of stellar luminosity). We do not go into details here, referring the reader to Byrne and Rodonò (1983), Baliunas and Vaughan (1985) and Rodonò (1987). We just outline that three levels of activity have been evidenced

- I level (Hyperactivity)

This is characteristic of the fast rotating RS Canum Venaticorum and BY Draconis type stars, with luminosity variation amplitude A up to 0.6÷0.7 mag, filling factor f up to several tenths (>50%) and time scale (cycle period?) of the order of 50÷60 y

- II level (Mid - activity)

This refers essentially to young and fast rotating stars in Pleiades and Hyades, with A~0.1 mag, f~0.1

- III level (Hypoactivity)

This is the class to which most stars, generally slow rotators, belong, including our Sun, with A~0.01÷0.05, f~0.01 and time scale (cycle period) ~10y.

Before entering the next subsection, we point out the important fact that both photospheric flux modulation amplitude and chromospheric emission flux do increase from level III to level I, this giving consistency to our global understanding of stellar activity.

3.3. SPECTROSCOPIC VARIABILITY

We refer to $Ca_{II}H,K$ and $Mg_{II}h,k$ line emission cores, which are a powerful tool to investigate stellar activity, as clearly shown since the pioneeristic work by Wilson (1978). For all details about the method and observational results we address the reader to Catalano (1984), Baliunas and Vaughan (1985), Hartmann and Noyes (1987), Baliunas et al. (1989), and references therein. Chromospheric fluxes (emission cores), originating from active regions are very good proxy indicators of stellar activity, being correlated to the magnetic field strength and filling factor (Skumanich et al. 1975).

The flux modulation in time allows to measure both rotation rates (of course, independent of V sin i) and activity time-scales, including cycle periods. The reliance on proxy indicators of stellar activity derived from the difficulty of directly measuring the magnetic field topology and intensity.

However, due to the progress in the techniques of investigation, calibrating the emission flux in terms of the magnetic flux density $\Phi \approx fB$ is now possible. For the Sun we have (Schrijver et al. 1989) :

$$log\ \Delta F_{H,K} = 0.6\ log\ (fB) + 4.8$$

and, for a sample of active stars (Saar and Schrijver, 1987; Schrijver, 1987) :

$$\Delta F_{H,K} \sim (fB)^{0.62\pm0.14}\ for\ (fB) < 300\ Gauss$$

where $\Delta F_{H,K}$ is the "excess" emission flux, defined in the next subsection.

The substantial agreement between the relationships found for the Sun and the sample of stars is a further evidence of the correctness of the solar-stellar connection principle.

3.4. THE DEPENDENCE OF CHROMOSPHERIC ACTIVITY ON STELLAR PARAMETERS

Different proxy indicators of chromospheric activity have been used in recent years :
(i) the emission flux $F_{H,K}$
(ii) the ratio $R_{H,K}$ of the emission flux to the bolometric flux

(iii) the "excess" emission flux $\Delta F_{H,K}$, namely the difference between the total flux and the "basal" flux, the latter representing the contribution of the quiet chromosphere

(iv) the emission luminosity $L_{H,K}$

Noyes et al. (1984a) found a correlation between $\log R_{H,K}$ and $\log P_{rot}$, where P_{rot} is the rotational period.

However, they noted that the scatter is minimized when correlating $\log R_{H,K}$ and $\log P_{rot}/\tau_c$, where $R_0 = P_{rot}/\tau_c$ is the Rossby parameter and τ_c is the convective turnover time, namely the ratio of the scale height at the base of the convection zone (c.z.) to the characteristic convective velocity in the deep c.z. This is the order of magnitude of the time necessary for an element of fluid to describe a closed path in a rotating convection zone. Note that τ_c is stellar mass dependent, as the structure and dynamics of the convective zone of a star depend on its mass (or colour index, (B-V)).

Therefore the Rossby parameter is a function of both mass and rotation rate. Note that this parameter is a "hybrid", as P_{rot} is an observed quantity and τ_c a theoretical one, computed by Gilman (1980).

But the important point is the following. Increasing R_0 means increasing the rotational period and/or decreasing τ_c i.e. the thickness of the stellar c.z. In other words, R_0 increases for slower rotators and/or earlier spectral types, and the observed correlation between $R_{H,K}$ and R_0 shows that activity declines with increasing R_0. This is consistent with dynamo theory : in fact $R_0 \sim 1/\sqrt{D}$, (see e.g. Belvedere 1985), where D is the dynamo number which parametrizes the effect of the interaction of convection and rotation giving rise to dynamo action, the strength of which increases with increasing D. Therefore, the larger is R_0 , the weaker is dynamo action, in agreement with the observational evidence. Note that this argument was already in Durney and Latour (1978) and Belvedere et al. (1980a).

Inverting the argument one can conclude that stellar activity gets stronger in faster rotators and later spectral types.

Zwaan (1986), Rutten and Schrijver (1986), using basically the same data, but $\Delta F_{H,K}$ as the proxy indicator, did not find a unique correlation between chromospheric activity, rotation rate and mass. They indeed obtained a family of curves in the plane ($\log \Delta_{H,K}$, $\log P_{rot}$), with slopes depending on (B-V). A detailed discussion on this discrepancy can be found in Hartmann and Noyes (1987).

Moreover, Marilli et al. (1986), using another proxy indicator, the chromospheric emission luminosity, found a correlation such as :

$$L_{H,K} \sim e^{-\alpha P_{rot}}$$

where the interesting fact is that a, derived by observations nearly coincides with $1 / \tau_c$, computed by Gilman (1980).

At the present, all the authors here mentioned claim to have used the best chromospheric activity physical indicator. We do not enter upon this subject here, but stress the fact that, independently of the particular indexes selected as tracers (spots, active chromospheric regions) or proxy indicators ($R_{H,K}$, $\Delta F_{H,K}$, $L_{H,K}$) the observational evidence shows that activity increases with increasing rotation rate and c.z. thickness. This is also confirmed by direct magnetic field strength and filling factor measurements, and is in agreement with the predictions of dynamo theory.

4. Stellar Activity Cycles : Observations and Theory

In the last twenty years, a great amount of data on chromospheric variability of active stars have been collected on a sample of about 100 objects (see Baliunas and Vaughan 1985, Baliunas 1986, Baliunas et al. 1989).

The results indicate that 15% of the observed stars show no variability, 25% show chaotic or erratic time behaviour and the remaining 60% show cyclic time behaviour, with periods of the order of 10y, mostly ranging from 5 to 15 years. Also most cycle characteristics are solar - like, as to smoothness and shape. Further a Maunder minimum - like behaviour has been detected on the star HD 10700.

Cycle periods <5 y have been observed only on stars with mass $M > M_o$, this having some relevance as for the comparison between theory and observations, as pointed out later. However, and this is a very important point, no clear correlation has been found between the cycle period and stellar parameters as rotational period, Rossby parameter, mass (or spectral type), chromospheric activity strength (i.e. magnetic flux density) and age.

Perhaps the observational time span is too short (?) and the sample too limited (?), so that we have to wait for new observations prior to drawing any conclusions. If the present lack of correlation is confirmed, dynamo theory will confront a new challenge, since up-to-date predictions are few and controversial, and disagree with the observational evidence.

Let us have a look at the theoretical investigation on stellar cycles and their relation to stellar parameters.

According to Belvedere et al. (1980c), the cycle period should increase with (B-V), this result being obtained in the case of marginal dynamo instability for which the dynamo wave period is of the order of the diffusion time in the stellar convection zone.

This result is just the opposite of what has been found by Durney and Robinson (1982), according to whom the period should decrease with (B-V), assuming the dynamo wave period to be of the order of the magnetic field amplification time in the convective layers.

As to the P_{rot}-dependence, both models predict $P_{cycle} \sim P_{rot}$ (see e.g. Belvedere 1985). More recently Noyes et al.(1984b) found the empirical relationship $P_{cycle} \sim (P_{rot} / \tau_c)^{1.25}$ and discussed it in the framework of a simple nonlinear dynamo model. However, the number of stars was too limited (12 objects), so that their result was not confirmed once the sample was extended to all the measured stars (Baliunas and Vaughan 1985, Baliunas et al. 1989).

Therefore these models not only disagree with each other, but, further, have scarce or no observational support, since no cycle dependence on the spectral type or the rotational period has been found up to now.

Perhaps, however, the limited evidence of short cycle periods in late F-type stars and long cycle periods in RS CVn and BY Dra gives some slight support to Belvedere et al. (1980c), as well as the P_{rot} independence of P_{cycle} is consistent with the theoretical result of Kleeorin et al. (1983) for the case of rapid rotators.

In the light of these non-encouraging theoretical predictions on stellar cycles, a question naturally arises: is dynamo theory well founded and at least partially reliable, even if same caution is undoubtedly necessary? To explore the implications of this question and try to answer it, let us come back home, to the solar dynamo.

5. The Present Status of Solar Dynamo Theory

As is known, dynamo theory attempts to explain the generation and evolution of cosmical magnetic fields in terms of induction effects in rotating conducting fluid masses. For the development of the main concepts and exhaustive detailed information, we address the reader to the books of Moffatt (1978), Parker (1979), Krause and Rädler (1980), and to the reviews of Stix (1981), Schüssler (1983), Belvedere (1985).

Rather a lot of solar dynamo models have been dished up in recent and less recent years, most based on the original Parker's (1955) formulation in terms of the so called α–ω mechanism: toroïdal magnetic fields are generated from poloïdal ones by differential rotation (ω-effect), while cyclonic turbulence (α-effect) regenerates poloïdal fields from toroïdal ones (e.g. Steenbeck and Krause 1969 a,b; Yoshimura 1975, 1983; Belvedere et al. 1980b, 1980c, 1987, 1989; Schüssler 1984; Gilman 1983, 1986; Gilman and Miller 1981, Gilman et al. 1989; Durney and Robinson 1982, Ruzmaikin 1984; Durney 1988, Brandenburg et al. 1988, 1989, Schmitt and Schüssler 1989).

Dynamo models may be separated into two classes: cinematic, linear dynamos and hydromagnetic, non-linear dynamos. In the latter the back-reaction of the magnetic field on the velocity field is taken into account, and the whole system of the magnetohydrodynamic equations is solved simultaneously, assuring internal consistency.

Although linear models seem to have captured the essential physics of the convection-rotation interaction giving rise to dynamo action, as is shown by their capacity of reproducing the solar cycle, there is no doubt, however, that present and future research has to be carried out in the framework of the nonlinear approach, which in principle allows to describe a multiplicity of dynamo operation modes and to predict magnetic field strengths.

Here we do not enter upon the subject of solar activity models, but outline some recent basic problems and developments of solar dynamo.

5.1. THE LOCATION AND WORKING MECHANISM OF DYNAMO IN THE LIGHT OF HELIOSEISMOLOGY

The location of dynamo action has vigorously been debated in recent years. Three possible locations have been suggested in modelling solar dynamo:
(i) the whole convection zone
(ii) the base of the convection zone (1st scale height)
(iii) the boundary (overshoot) layer between the convective and the radiative zones.

The last possibility seems to be the most realistic on the basis of the following argument.Spatial separation of the ω-effect and the α-effect is not plausible in so far as it would imply problems of upward and downward field transport in a turbulent medium. On the other hand, since stability of magnetic flux tubes against magnetic buoyancy suggests that the ω-effect is located deep in the convection zone or better in the boundary layer (Parker 1979, Spiegel and Weiss 1980, Spruit and Van Ballegooijen 1982), also the α-effect is expected to operate mainly at deep levels. This is also supported by the argument that the α-effect on flux tubes rapidly rising to the surface would be ineffective in the top half of the convection zone. (Golub et al. 1981).

The location of dynamo in the boundary layer is indirectly supported by the helioseismic results. The most recent helioseismological data (Duvall et al. 1986, Brown and Morrow 1987; Christensen - Dalsgaard and Shou 1988; Harvey 1988; Libbrecht 1988, 1989; Dziembowski et al. 1989; Brown et al. 1989) seem to agree as to the following (provisional) scenario:

(i) the surface angular velocity ω (R_O, θ) persists throughout the convection zone (1 R_O →0.7 R_O): i.e. $(d\omega/dr)_{c.z.}$ ~ 0 or slightly >0

(ii) beneath the convection zone (0.7 R_O →0.65 R_O), rigid rotation dominates with $\omega_0 = 2.7 \ 10^{-6}$rad/s which is the surface value at latitude ~ 37°. All this implies that the equatorial and polar rotation rates do converge to the intermediate value ω_0 below the convection zone.

Let us see what the implications to solar dynamo are :
(i) radial shear driven dynamo, operating in the convection zone might be unrealistic

(ii) radial shear driven dynamo operating in the boundary layer is still supported by helioseismological data. The argument is the following (with reference to the northern hemisphere).

Since α<0 in the boundary layer (Yoshimura 1975, Gilman 1983, Glatzmaier 1984, 1985a,b, Gilman et al. 1989, Parker 1989), and, interpolating the helioseismological results in the boundary layer :

$$\frac{d\omega}{dr} < 0 \text{ at high latitudes } (\gtrsim 37°)$$

$$\frac{d\omega}{dr} > 0 \text{ at low latitudes } (\lesssim 37°)$$

we get poleward migration of the toroïdal field at high latitudes ($\alpha\frac{d\omega}{dr} > 0$) and equatorward migration at low latitudes ($\alpha\frac{d\omega}{dr} < 0$), in agreement with the observational evidence shown by different tracers of the solar cycle (polar faculae and prominences on the one side, spots and faculae on the other side).

This means that dynamo can still work and reproduce the observations if is located in the boundary layer.

At present, however, all must be taken with great caution! Indeed :
- the helioseismic inversion data are not conclusive. For instance, noise may mask the effect of rotation (Brown et al. 1989).
- discrepancies do exist between different sets of data (see e.g. Rhodes et al. 1987, 1988)
- convection zone dynamo models which may work, have still been proposed: magnetostrophic wave dynamo (Schmitt 1987) cylindrical isorotation dynamo (Durney 1989a,b) latitudinal shear driven dynamo (Parker 1989).

6. Concluding Remark

Dynamo theory has some obscure and bright aspects. The next decade will probably decide on its validity.

We refer above all to :
- the impact of helio- and asteroseismology
- the impact of the recently observed resonant structure of the magnetic pattern of the solar cycle (Stenflo and Vogel 1986, Stenflo 1988),characterized by different stability of odd and even parity modes with respect to the equator and different l dependences of the mode frequency and then, of the associated sinusoidal period, for odd and even modes (l is the spherical harmonic degree)
- the impact of the present no - correlation between stellar cycle periods and stellar parameters, or, in the future, of any (expected?) correlation found when a larger sample of stellar data is available.

The one we are living is probably a period of transition, in which a pause of meditation and reflection is perhaps necessary.

Acknowledgements

I want to express my friendly thanks to many colleagues for kindly sending me their recent work: S.L.Baliunas, A.Brandenburg, T.M.Brown, B.R.Durney, P.Foukal, P.A.Gilman, B.M.Haisch, P.Hoyng, R.W.Noyes, E.N.Parker, L.Paternõ, F.Praderie, R.Rosner, R.J.Rutten, M.Schüssler, D.R.Soderblom, H.C.Spruit, J.O.Stenflo, M.Stix, I.Tuominen, G.S.Vaiana, A.A.Van Ballegooijen, H.B.Van der Raay, S.S.Vogt, N.O.Weiss, M.F.Woodard, C.Zwaan.

I sincerely apologize to those ones whose papers I didn't quote, due to the limited extent of this review.

I also enjoyed helpful e-mail and letter discussions with Bernard Durney and Peter Hoyng.

References

Baliunas, S.L. and Vaughan, A.H.: 1985, Ann. Rev. Astron Astrophys. 23, 379
Baliunas, S.L.: 1986, in *Cool Stars, Stellar Systems and the Sun*, M.Zeilik and D.Gibson eds., Springer, Berlin, p.3
Baliunas, S.L. Donahue, R.A., Horme, J.H, Noyes,R.W., Porter, A., Gilliland, R.L. Duncan, D.K., Frazer, J. Lanning, H., Misch, A., Mueller, J., Soyumer, D., Vaughan, A.H., Wilson, O.C. and Woodard, L.A.: 1989, Astrophys. J., in press
Belvedere, G.: 1985, Solar Phys. 100 , 363
Belvedere, G. and Proctor, M.R.E.: 1989, in *Solar Photosphere : Structure, Convection and Magnetic Fields*, J.O. Stenflo ed., Kluwer, Dordrecht, in press
Belvedere, G., Paternõ, L. and Stix, M.: 1980a, Astron. Astrophys. 86 , 40
Belvedere, G., Paternõ, L. and Stix, M.: 1980b, Astrophys. 88, 240
Belvedere, G., Paternõ, L. and Stix, M.: 1980c Astron. Astrophys. 91 , 328
Belvedere, G., Pidatella, R.M. and Proctor, M.R.E.: 1989, Geophys Astrophys. Fluid Dyn, in press
Belvedere, G., Pidatella R.M. and Stix, M.: 1987, Astron. Astrophys. 177, 183

380

Brandenburg, A. and Tuominen, I.: 1988, Adv. Space Res. 8,

Brandenburg, A. Krause, F., Meinel, R.,Moss, D. and Tuominen, I.: 1989
  Astron. Astrophys 213, 411

Brown, T.M. and Morrow, C.A.: 1987, Astrophys. J. 314, L21

Brown, T.M. Christensen-Dalsgaard, J.,Dziembowski, W., Goode, P.R., Gough, D.O. and
  Morrow C.A.: 1989, Astrophys. J. 343, 526

Byrne P.B. and Rodonõ, M. (eds.): 1983, *Activity in Red-Dwarf Stars*, Reidel, Dordrecht.

Catalano, S.: 1984, in *Space Research Prospects in Stellar Activity and Variability*, A. Mangeney
  and F. Praderie eds., Observatoire Meudon, p. 243

Christensen - Dalsgaard, J. and Shou, J.: 1988, in *Seismology of the Sun and Sun-like Stars*,
  E.J. Rolfe ed., ESA-SP 268, p. 55

Damon, P.E.: 1977, in *The Solar Output and Its Variation*,O.R. White ed., Colorado. Ass.
  Univ. Press., Boulder, p. 429

Donati, J.F., Semel, M. and Praderie, F.: 1989, Astron. Astrophys in press

Durney, B.R.: 1988, Astron. Astrophys., 191, 374

Durney, B.R.: 1989a, Astrophys. J, 338, 509

Durney, B.R.: 1989b, Solar Phys., in press

Durney, B.R. and Latour. J: 1978, Geophys. Astrophys. Fluid Dun. 9, 241

Durney, B.R. and Robinson, R.D.: 1982, Astrophys. J. 253, 290

Duvall, T.L., Harvey, J.W. and Pomerantz, M.A.: 1986, Nature 321, 500

Dziembowski, W.A., Goode, P.R. and Libbrecht, K.G. 1989, Astrophys. J. 337, L53

Eddy, J.A.: 1976, Science 286, 1189

Eddy, J.A.: 1983, Solar Phys. 89, 195

Gilman, P.A.: 1980, in *Stellar Turbulence*, D.F. Gray and J.L.Linkly eds.,
  Springer, Berlin, p. 19

Gilman, P.A.: 1983, Astrophys. J.Suppl. 53, 243

Gilman, P.A.: 1986, Geophys. Astrophys. Fluid Dyn. 31, 137

Gilman, P.A. and Miller, J.: 1981, Astrophy. J. Suppl. 46,

Gilman, P.A., Morrow, C.A. and De Luca, E.E.: 1989, Astrophys. J. 338, 528

Glatzmaier, G.A.: 1984, J.Comp. Phys. 55, 461

Glatzmaier, G.A.: 1985a, Geophys. Astrophys. Fluid Dyn. 31, 137

Glatzmaier, G.A.: 1985b, Astrophys. J. 291, 300

Golub, L., Rosner, R., Vaiana, G.S., and Weiss, N.O.: 1981, Astrophys. J. 243, 309

Gray, D.F.: 1984, Astrophys. J. 277, 640

Hartmann, L.W. and Noyes, R.W.: 1987, Ann. Rev. Astron. Astrophys. 25, 271

Harvey, J.W.: 1988, in *Seismology of the Sun and Sun-like Stars*, E.J. Rolfe ed.,
  ESA-SP 286, p.55

Hoyng, P.: 1987a, Astron. Astrophys. 171, 348

Hoyng, P.: 1987b, Astron. Astrophys. 171, 357

Kleeorin, N.I., Ruzmaikin, A.A. and Sokoloff, D.D.: 1983, Astrophys, Space Sci. 95, 131

Krause, F. and Rä dler, K.H.: 1980, *Mean Field Magneto hydrodynamics and Dynamo
  Theory*, Pergamon Press, Oxford

Libbrecht, K.G.: 1988, in *Seismology of the Sun and Sun-like Stars*, E.J. Rolf ed.
  ESA-SP 286, p.131

Libbrecht, K.G.: 1989, Astrophys. J. 336, 1092

Marcy, G.W.: 1984, Astrophys. J. 276, 786

Marilli, E., Catalano, S. and Trigilio, C.: 1986, Astron. Astrophys. 167, 297

Moffatt, H.K.: 1978, *Magnetic Field Generation in Electrically Conducting Fluids*, Cambridge Univ.Press

Noyes, R.W., Hartmann, L.W., Baliunas, S.L., Duncan, D.K. and Vaughan, A.H.: 1984a, Astrophys. J. 279, 763

Noyes, R.W., Weiss N.O. and Vaughan, A.H.: 1984b, Astrophys. J. 287, 769

Parker, E.N.: 1979, *Cosmical Magnetic Fields*, Clarendon, Oxford

Parker, E.N.: 1987, Solar, Phys. 110, 11

Parker, E.N.: 1989, in *Solar and Stellar Flares*, B.M.Haisch and M.Rodonõeds., Solar Phys., in press

Rhodes, E.J., Cacciani, A., Woodard, M.F. Tomczyk, S., Korzennik, S. and Ulrich, R.K.: 1987, in *The Internal Solar Angular Velocity*, B.R. Durney and S. Sofia, eds. Kluwer, Dordrecht, p. 75

Rhodes, E.J. Cacciani A., Tomczyk, S., Ulrich, R.K. and Woodard, M.F.: 1988, in *Seismology of the Sun and Sun-like Stars*, E.J. Rolfe ed., ESA-SP 286 p.73

Robinson, R.D.: 1980, Astrophys. J. 239, 261

Rodonõ, M.: 1987, in *Solar and Stellar Physics*, E.H. Schröter and M. Schüssler eds., Springer, Berlin, p.39

Rutten, R.G.M. and Schrijver, C.J.: 1986, in *Cool Stars, Stellar Systems and the Sun*, M.Zeilik and D.Gibson eds., Springer, Berlin, p.120

Ruzmaikin, A.A.: 1984, in Proceed. 4th European Meeting on Solar Phys., T.D. Guyenne and J.J. Hunt eds., ESA SP-220, p.85

Saar, S.H.: 1989, in *Solar Photosphere : Structure, Convection and Magnetic Fields*, J.O. Stenflo ed., Kluwer, Dordrecht, in press

Saar, S.H. and Linsky, J.L.: 1986, Adv. Space Res. 6, 235

Saar, S.W. and Schrijver, C.J.: 1987, in *Cool Stars, Stellar Systems and the Sun*, J.L. Linsky and R.E. Stencel eds., Springer, Berlin, p. 38

Schmitt, D.: 1987, Astron. Astrophys. 174, 281

Schmitt, D. and Schüssler, M.: 1989, Astron. Astrophys. submitted

Schrijver, C.J.: 1987, Astron. Astrophys. 172, 111

Schrijver, C.J., Coté, J., Zwaan C. and Saar S.H.: 1989, Astrophys. J. 337, 964

Schüssler, M.: 1983, in *Solar and Stellar Magnetic Fields : Origin and Coronal Effects*, J.O.Stenflo ed., Reidel, Dordrecht, p. 213

Schüssler, M.: 1984, in Proceed. 4th European Meeting on Solar Physics, T.D. Guyenne and J.J. Hunt eds., ESA SP-220, p. 67

Semel, M.: 1989, Astron. Astrophys., submitted

Skumanich, A., Smythe, C., Frazier, E.N.: 1975, Astrophys., J. 200, 747

Spiegel, E.A. and Weiss, N.O.: 1980, Nature 287, 616

Spruit, H.C. and Van Ballegooijen, A.A.: 1982, Astron. Astrophys. 106, 58

Steenbeck, M. and Krause, F.: 1969, Astron. Nachr. 291, 49

Steenbeck, M. and Krause, F.: 1969, Astron. Nachr. 291, 271

Stenflo J.O.: 1988, Astrophys. Space Sci. 144, 321

Stenflo J.O. and Vogel, M.: 1986, Nature 319, 285

Stix, M.: 1981, Solar Phys. 74, 79

Stuiver, M. and Quay, P.D.: 1980, Science 207, 11

Vogt, S.S.: 1983, in *Activity in Red-Dwart Stars*, P.B. Byrne and M. Rodonõ eds., Reidel, Dordrecht, p. 137

Vogt, S.S., Penrod, G.D. and Hatzes, A.P.: 1987, Astrophys J. 321, 496

Wallenhorst, S.G.: 1982, Solar Phys. 80, 379

382

Weiss, N.O.: 1989, in *Accretion Disks and Magnetic Fields in Astrophysics*, G.Belvedere ed., Kluwer, Dordrecht, p.11

Weiss, N.O., Cattaneo, F. and Jones, C.A.: 1984, Geophys. Astrophys. Fluid Dyn. 30, 305

Williams, G.E.: 1981, Nature 291, 624

Williams, G.E.: 1989, J.Geol. Soc. 146, 97

Wilson, O.C.: 1978, Astrophysical J. 226, 379

Yoshimura, H.: 1975, Astrophys. J. Suppl. 29, 467

Yoshimura, H.: 1983, Astrophys. J. Suppl. 52, 363

Zwaan, C.: 1986, in *Cool Stars, Stellar Systems and the Sun,* M.Zeilik and D.Gibson eds., Springer, Berlin, p.19

PART 4

SOLAR CYCLE, DYNAMO AND TRANSPORT PROCESSES

THE SOLAR DYNAMO

K.-H. RÄDLER
Sternwarte Babelsberg
DDR-1591 Potsdam
GDR

ABSTRACT. The phenomena of solar activity are connected with a general magnetic field of-the Sun which is due to a dynamo process essentially determined by the α-effect and the differential rotation in the convection zone. A few observational facts are summarized which are important for modelling this process. The basic ideas of the solar dynamo theory, with emphasis on the mean-field approach, are explained, and a critical review of the dynamo models investigated so far is given. Although several models reflect a number of essential features of the solar magnetic cycle there are many open questions. Part of them result from lack of knowledge of the structure of the convective motions and the differential rotation. Other questions concern, for example, details of the connection of the α-effect and related effects with the convective motions, or the way in which the behaviour of the dynamo is influenced by the back-reaction of the magnetic field on the motions.

1. Introduction

It is widely believed that the phenomena of solar activity are closely coupled with a general magnetic field of the Sun, and that the 11 years activity cycle is connected with a magnetic cycle. Let us first give a crude idea of the general magnetic field of the Sun. Its dominant constituent is a toroidal, that is, belt-shaped field confined to the convection zone. It is in the main symmetric about the rotation axis and possesses opposite orientations in the two hemispheres or, in other words, shows antisymmetry about the equatorial plane. In addition to the toroidal field there is a much weaker poloidal field, which penetrates not only the convection zone but continues in the atmosphere. To have a very simple picture it may be assumed to be again symmetric about the rotation axis and to show antisymmetry about the equatorail plane as, for example, dipole or octupole fields do; this assumption ignores, of course, any sectorial structure of the general magnetic field. The magnetic cycle occurs as an interplay of the toroidal and the poloidal field, which change their magnitudes and their orientations with a period of 2×11 years, that is, 22 years.

The general magnetic field of the Sun is usually attributed to dynamo processes, that is, to the interaction of fluid motions and magnetic fields in the convection zone. This idea was already proposed by Larmor (1919). Its elaboration, however, proved to very complicated. A part of the difficulties, as is attested by Cowling's theorem (1934), is due to the fact that a dynamo cannot work with completely axisymmetric magnetic fields.

G. Berthomieu and M. Cribier (eds.), Inside the Sun, 385–402.

It is easy to understand that, as a consequence of the differential rotation of the convection zone, a toroidal magnetic field is generated from a given poloidal field. The crucial question is the regeneration of the poloidal from the toroidal field. Parker (1955) elaborated the idea that the regeneration is achieved by the cyclonic motions in the convection zone. Within the framework of mean-field electrodynamics established by Steenbeck, Krause and Rädler (1966) the influence of small-scale motions on large-scale magnetic fields has been systematically investigated. Within this scope the effect of cyclonic motions occurs as a contribution to the α-effect, that is, to a large-scale electromotive force caused by small-scale helical motions, which possesses components parallel or antiparallel to the large scale magnetic field. The α-effect is not only able to generate a poloidal from a toroidal field but also a toroidal from a poloidal field.

The α-effect and the differential rotation are considered as the basic elements of the solar dynamo. The resulting scheme of couplings between the toroidal and poloidal magnetic fields includes the possibilities of the $\alpha\omega$ and the α^2-dynamos, which correspond to the cases where the α-effect or the differential rotation is negligible in the generation of the toroidal field, respectively. There are good reasons to assume that the solar dynamo is of $\alpha\omega$-type rather than α^2-type.

In the following we first summarize a few characteristics of the solar convection zone and some observational results on solar magnetic fields. We then turn to kinematic solar dynamo models, in which the back-reaction of the magnetic field on the fluid motion is ignored, explain the basic ideas of mean-field electrodynamics and discuss some general aspects of solar mean-field dynamo models as well as some problems that occur in this context. Finally we consider dynamo models which include the back-reaction of the magnetic field, discuss the extension of the mean-field concept to such models and mention a few attempts to gain understanding of the complex nonlinear behaviour of dynamos.

Being aware that our paper covers only a few aspects of the solar dynamo problem we want to point out other review articles, e.g., by Parker (1979), Stix (1981), Belvedere (1983), Ruzmaikin (1985) or Gilman (1986).

2. Convection zone characteristics

2.1. GENERAL

Unfortunately only the upper boundary of the solar convection zone is more less accessible to direct observations. Most of our knowledge on the deeper layers originates from stellar structure calculations, mainly based on mixing length theory. According to such calculations it seems reasonable to assume that the convection zone extends over the outer 30% of the solar radius.

For the solar dynamo process the convective motions and the differential rotation of the convection zone are of particular interest. We summarize here a few relevant findings. Recent reviews of these subjects are, e.g., by Gilman (1986) and Stix (1989).

2.2. THE CONVECTIVE MOTIONS

There are two clearly defined patterns of convective motion visible at the solar surface. They correspond to granulation and supergranulation. For the granulation the typical length scales are about 1 to 2.10^3 km, the time scales about 5 min, and the velocities about 1 km/s. For the supergranulation the corresponding values are 3.10^4 km, 20 h, and 0.3 to 0.5 km/s.

Moreover, there is some evidence of another pattern, called mesogranulation, with a typical length scale between those of granulation and supergranulation, namely about 5 to 10.10^3 km, a time scale of about 2 h, and rather low velocities of about 60 m/s.

We note that there have been many attempts to identify much larger structures like "giant cells" but their existence is still questionable. Recently an interpretation of observational results by "toroidal convection rolls" was proposed (Ribes et al. 1985, Ribes and Laclare 1988).

The convective motions in deeper layers are a matter of theoretical investigation only, and there are still considerable uncertainties.

2.3. THE DIFFERENTIAL ROTATION

As can be seen from the motion of several tracers, sunspots for example, the equatorial regions of the solar surface rotate faster than the polar regions. The angular velocity Ω of the surface, as inferred from Doppler shift measurements, can be represented by

$$\Omega = A - B \cos^2\theta - C \cos^4\theta \qquad (2.1)$$

where A, B, C are positive coefficients, which slightly vary with the phase of the cycle, and θ is the colatitude.

Roughly, we have B/A ~ 0.1 and C/A ~ 0.2, which implies that the angular velocity near the poles is by about 30% lower than at the equator. The dependence of A, B and C on the phase corresponds to torsional oscillations.

The angular velocity inside the convection zone has so far been only a matter of theoretical investigations. Although at least a part of the theories is well elaborated they contain assumptions that remain to be checked. The results are controversal even with respect to the sign of the radial derivative of Ω. Recently some conclusions concerning the angular velocity inside the convection zone and below could be drawn from helioseismological observations. These observations seem to rule out the concept in which Ω is constant on cylindrical surfaces. They are, however, compatible with the assumption that Ω shows no radial variation inside the convection zone but, within a small layer at its bottom or below, turns into a value corresponding to a rigid body rotation. In this layer there inevitably exist strong radial derivatives of Ω which change sign in dependence on the latitude (see Gilman et al. 1989).

3. Observational results on magnetic fields

3.1. GENERAL

Let us proceed now to observational results which provide us with some information about the toroidal and poloidal constituents of the general magnetic field of the Sun. We shall explain a few important aspects and refer again to the comprehensive presentations by Gilman (1986) and Stix (1989).

3.2. MAGNETIC FLUX TUBES

We first mention the remarkable discovery that the magnetic flux at the solar surface, and presumably also in deeper layers, is not more or less homogeneously distributed over the solar plasma but concentrated in very thin flux tubes (Stenflo 1973). These flux tubes, whose diameters are small compared with the length scales of granulation, can generally not be resolved by observing the Zeeman splitting of a single line. Their properties have been concluded by measuring ratios of Zeeman splittings of neighbouring lines.

Typical diameters of these flux tubes are 100 to 300 km, typical life times 5 min to 2 h, and typical flux densities 2000 gauss. There are estimates according to which 90% or more of the total flux emerging from the Sun is contained in these flux tubes.

3.3. THE TOROIDAL MAGNETIC FIELD

Essential conclusions concerning the toroidal magnetic field can be drawn from the distribution of sunspots on the solar surface. Sunspots are assumed to occur if the magnetic flux concentration immediately under the solar surface, which is mainly determined by the toroidal field, exceeds a certain bound. They are formed with flux tubes which, as a result of magnetic bouyancy, emerge at the surface. The diameters of sunspots are comparable to the length scale of the supergranulation, that is, 3.10^4 km, the life times are about 1 or 2 months, and the magnetic flux densities about 2000 gauss.

The distribution of the sunspots with respect to latitude and time is represented by the Maunder butterfly diagram. It shows that the occurrence of sunspots is restricted to lower latitudes up to about ± 40 degrees, with a maximum at ± (15...20) degrees. In the course of an 11 years activity cycle sunspots occur at first mainly at the highest of these latitudes but later at lower latitudes. This indicates that the belts of maximum toroidal field are in latitudes lower than ± 40 degrees and show an equatorward migration.

The magnetic polarities of sunspots in bipolar groups follow Hale's rules. The leading and the following spots show opposite magnetic polarities. At a given time, the polarities of, for example, the leading spots coincide in one hemisphere but differ from those in the other hemisphere. In a given hemisphere, these polarities change from one 11 years activity cycle to the next. This indicates that the toroidal field belts in the two hemispheres have opposite orientation and that the orientations change with a 22 years period, which defines the magnetic cycle.

The sunspot activity can be measured, for example, by the Wolf relative sunspot number. This number shows no completely periodic time dependence. There are significant differences between the cycles and typically stochastic features. In particular, there are long-term variations clearly indicated by the grand minima as, for example, the Maunder and Spörer minima in the 17th and 15 century. These variations of the sunspot activity presumably indicate variations of the underlying magnetic field.

The sunspots are not the only indicators of the toroidal magnetic field. There are, for instance, the ephemeral active regions, which are small bipolar regions with life times and magnetic fluxes much smaller than those of sunspots, and which occur also in higher latitudes, namely up to ± (50...60) degrees.

3.4. THE POLOIDAL MAGNETIC FIELD

The poloidal magnetic field, by contrast with the toroidal field, is very hard to observe, for its flux density is 1 to 2 gauss only.

One interesting feature of the poloidal field known from observation is the polarity reversals at the poles. They have been observed directly since 1957/58. There is, in addition, some indirect evidence of such reversals. They can be concluded from the motion of lines of zero radial field towards the poles. These lines are the places were prominences occur, and systematic observations of prominences are available since 1870.

A simple relation has been found between the polarities of the poloidal field at the poles and the orientations of the toroidal field in the latitudes of maximum sunspot activity during the reversal. If, at a pole, negative polarity changes into positive polarity, the toroidal field in the same hemisphere points eastward, and the same applies with opposite field directions. Let us denote the magnetic flux density of the general field by $\overline{B}$ and refer to spherical coordinates r, θ, ø, corresponding to radius, colatitude and longitude. Then we have during a reversal

$$(\partial B_r/\partial t)_{pole} \cdot (B_\phi)_{maximum} > 0 \qquad (3.1)$$

Direct measurements of the poloidal magnetic field at the whole solar surface have been carried out at the Mount Wilson and Kitt Peak observatories since 1959. The results revealed a phase relation between the poloidal and the toroidal magnetic field in the latitudes of maximum sunspot activity. The maximum of the radial field component precedes that of the toroidal field by a phase angle of (0.85...1) π. We therefore have almost always

$$(B_r \cdot B_\phi)_{maximum} < 0 \qquad (3.2)$$

The observational material on the poloidal field gained since 1959 has been used for an analysis of the space and time structure of this field (Stenflo and Vogel 1986, Stenflo and Güdel 1987). In this analysis the radial field component was represented by

$$B_r(\theta,\varphi,t) = \sum_{l \geq 0} \sum_{|m|\leq 1} \int_{-\infty}^{\infty} \frac{d\omega}{2\Pi} c_l^m(\omega) \, Y_l^m(\theta,\varphi) \, exp(im\varphi) \qquad (3.3)$$

where the $Y_l^m(\theta,\varphi)$ are spherical harmonics, and ω the frequency that occurs in the Fourier analysis of the time dependance. Conclusions were mainly drawn from the power spectrum of the coefficients $c_1^m(\omega)$.

Let us first consider the axisymmetric part of B_r, given by the contributions with m=0. For this part there is a striking difference between the contributions with odd or even 1, which correspond to antisymmetry or symmetry about the equatorial plane, respectively. The contributions with odd 1 are generally larger compared with those with even 1. In particular, the contributions with 1=5, 7, and 9 are remarkably large. For odd 1 there is a clearly dominant period near 22 years. For even 1 this period also appears but is by no means dominant compared with some much shorter periods.

Since the differential rotation renders the definition of the longitude coordinate system difficult the situation with the non-axisymmetric part of B_r, that is, with the contributions with m≠0, is rather complex. These contributions are related to the sectorial structure of the solar magnetic field. No noticeable difference between odd and even 1 has been found. The same applies to odd or even 1-m, which correspond to antisymmetry or symmetry about the equatorial plane, respectively.

4. Cinematic dynamo models

4.1. BASIC EQUATION

When studying the dynamo process in the Sun theoretically, we consider the solar plasma simply as an electrically conducting fluid and use Maxwell's equations in the magnetohydrodynamic approximation. The magnetic flux density B is then governed by the equations

$$curl\ (\eta curl\ B - u \times B) + \frac{\partial B}{\partial t} = 0 \tag{4.1a}$$

$$div\ B = 0 \tag{4.1b}$$

where ε is the magnetic diffusivity and u the velocity of the fluid motion. At first we consider the motion to be given, that is, restrict ourselves to the kinematic aspect of the dynamo problem.

4.2. MEAN-FIELD ELECTRODYNAMICS

As explained above, the solar magnetic field as well as the motions of the convective zone show complex structures in space and time, which can hardly be taken into account in detail. It proved useful to discuss the dynamo process in terms of mean-field electrodynamics. We briefly explain some basic ideas; for more details see, e.g., Krause and Rädler (1980) or Rädler (1980).

Within this framework each field quantity F, like $\overline{\mathbf{B}}$ or $\overline{\mathbf{u}}$, is understood as a superposition of a mean part, $\overline{F}$, showing more or less smooth dependence on the space and time coordinates, and a fluctuating part, F'. The mean part $\overline{F}$ is defined as a proper average of F, which may be over space or time, or may also be an ensemble average. It has to be required that certain averaging rules, the Reynolds rules, apply. This, in turn, generally implies requirements concerning the length or time scales of $\overline{F}$.

From equations (4.1) we obtain equations for the mean magnetic flux density $\overline{\mathbf{B}}$,

$$curl\ (\eta curl\ B - \overline{u} \times \overline{B} - \mathscr{E}) + \frac{\partial \overline{B}}{\partial t} = 0 \tag{4.2a}$$

$$div\ \overline{\mathbf{B}} = 0 \tag{4.2b}$$

where $\overline{\mathbf{u}}$ is the velocity of the mean motion, and $\mathscr{E}$ is a mean electromotive force caused by the fluctuating parts of motion and magnetic field,

$$\mathscr{E} = \overline{u' \times B'} \tag{4.3}$$

This electromotive force $\mathscr{E}$ at a given point in space and time is determined by $\overline{u}$, u' and $\overline{B}$ in a certain neighborhood of this point, and it depends linearly on $\overline{B}$. Provided the variation of $\overline{B}$ in this neighbourhood is sufficiently weak, that is, the length and time scales of $\overline{B}$ are sufficiently large, $\mathscr{E}$ can, relative to Cartesian coordinates, be represented in the form

$$\mathscr{E} = a_{ij}\overline{B_j} + b_{ijk}\frac{\partial \overline{B}_j}{\partial x_k} \tag{4.4}$$

The tensorial coefficients a_{ij} and b_{ijk} are determined by $\overline{u}$ and u' but do no longer depend on $\overline{B}$.

In the very simple case in which u is equal to zero and u' corresponds to homogeneous isotropic turbulence, ε takes the form

$$\mathscr{E} = \alpha\overline{B} - \beta curl\,\overline{B} \tag{4.5}$$

with two constants α and β which are determined by u'. The term $\alpha\overline{B}$ describes the α-effect, that is, an electromotive force parallel or antiparallel to the mean magnetic field. The coefficient α and thus the α-effect are, however, only non-zero if the turbulence lacks mirror-symmetry. For turbulences on rotating bodies the mirror-symmetry is generally violated because of the action of Coriolis forces. The deviation from mirror-symmetry can, for example, be indicated by a non-zero helicity of the motions. The term $-\beta curl\overline{B}$ gives rise to introducing a "turbulent" electric conductivity or a corresponding"turbulent" magnetic diffusivity η_T defined by $\eta_T = \eta + \beta$.

In (4.2a), this contribution to η can be disregarded if η is replaced by $\eta_{.T}$

Returning to the general case we note that (4.4) can be written in the form

$$\mathscr{E} = -\alpha\overline{B} - \beta curl\,\overline{B} - \delta \times curl\,\overline{B} - ... \tag{4.6}$$

Now α and β are symmetric tensors, γ and δ vectors, all depending on $\overline{u}$ and u'. The term $-\alpha\overline{B}$ describes an anisotropic α–effect, and $-\gamma\times\overline{B}$ a transport of mean magnetic flux different from that by a mean motion, covering also effects discussed as "turbulent diamagnetism" or "pumping" of magnetic flux. The term $-\beta.curl\overline{B}$ gives rise to introduce an anisotropic "turbulent" electric conductivity or corresponding magnetic diffusivity, and $-\delta \times curl\overline{B}$ covers in particular the "ωxj-effect", which, like the α-effect, in combination with differential rotation allows dynamo action. We note that $curl\overline{B}$ is connected with the antisymmetric part of the tensor $\partial\overline{B}_i/\partial x_j$. For the sake of simplicity the terms connected with the symmetric part of $\partial\overline{B}_i/\partial x_j$ are not given explicitly in (4.6).

An important problem in the elaboration of mean-field electrodynamics consists in the calculation of the electromotive force $\mathscr{E}$, or of the quantities α, β,... for given $\overline{u}$ and u'. Most of the results available have been derived in the second-order correlation approximation, often called "first-order smoothing", which neglects all higher-order correlations of u'.

4.3. TRADITIONAL SPHERICAL KINEMATIC MEAN-FIELD DYNAMO MODELS

On the basis explained here a number of spherical mean-field dynamo models have been elaborated. In all cases it is assumed that the mean magnetic flux density $\overline{B}$ in a rotating spherical fluid body is governed by equations (4.2) and (4.6) with more or less simple assumptions on $\overline{u}$, α, β,...As a rule, the surroundings of the body are considered to be free space, and it is required that $\overline{B}$ inside the body fits continuously to an irrotational solenoidal field in this external region.

In almost all cases it is assumed that $\overline{u}$, u' and η are symmetric about the equatorial plane and the rotation axis, and steady. For $\overline{u}$ and η this should be understood in the usual sense, for u' in the sense of the invariance of all averaged quantities under reflexions of the u'-field about the equatorial plane, under rotations about the rotation axis, and under time shift. If $\mathcal{E}$ contains, for example, contributions like $\alpha\overline{\mathbf{B}}$ or βcur1$\overline{\mathbf{B}}$, then, owing to these assumptions, α is antisymmetric and β symmetric about the equatorial plane, and both are symmetric about the rotation axis and steady.

These assumptions further imply that the solutions of the equations governing B have the form of the real part of

$$\widehat{B}exp(im\varphi + (\lambda+i\omega)t) \tag{4.7}$$

or are superpositions of such solutions. $\widehat{B}$ is a complex field which is either antisymmetric or symmetric about the equatorial plane, symmetric about the rotation axis and steady, m is a non-negative integer, φ again the longitudinal coordinate, and λ and ω are real constants. As usual the symmetry of such solutions is characterized by Am or Sm, where A or S stand for antisymmetry or symmetry about the equatorial plane and m is the integer introduced above describing the longitudinal variation. Simple examples of AO or S1-symmetry are the fields of central dipoles parallel or perpendicular to the rotation axis, and an example of SO-symmetry is the field of a central quadrupole symmetric about that axis. Depending on kind and intensity of the induction effects the solutions in general exponentially grow or decay; λ is the growth rate. Of course, the marginal case deserves special interest, in which the solutions neither grow nor decay, that is, $\lambda = 0$. In this case they may be steady or oscillatory, where ω defines the frequency. For axisymmetric solutions, m=0, oscillatory behaviour implies real changes of the field configuration in the course of a period, for non-axisymmetric solutions, m≠0, simply a rotation of the field configuration.

In the simplest models α-effect and differential rotation are taken into account in the form

$$\mathcal{E}= \alpha \overline{B}, \quad \alpha = -\widetilde{\alpha}(r)\ cos\theta \tag{4.8a,b}$$

$$\overline{u} = \Omega \times r, \quad \Omega = \Omega(r)\ \widehat{z} \tag{4.8c,d}$$

where θ is again the colatitude, r the radius vector and z the unit vector parallel to the rotation axis. The ratio of the two induction effects can be characterized by

$$q = \left|\frac{\alpha_0}{\Delta\Omega\ l}\right| \tag{4.9}$$

where α_0 and $\Delta\Omega$ are typical values of α and of the variation of Ω, and l is the thickness of the layer where the dynamo process takes place. Roughly speaking, the limits q << 1 and θ >> 1 correspond to the $\alpha\omega$ and α^2-regimes of a dynamo, in which α-effect or differential rotation, respectively, no longer contribute essentially to the generation of the toroidal field from the poloidal field.

Using those simple assumptions on $\mathcal{E}$ and $\overline{u}$ Steenbeck and Krause (1969a,b) elaborated $\mathcal{E}$ model of an $\alpha\omega$-dynamo, which reflects already some essential features of the solar magnetic fields, and also a model of an α^2-dynamo, which was discussed in view of planetary magnetic fields. Later on a number of other models of the same kind, or with somewhat more general assumptions concerning $\mathcal{E}$ and $\overline{u}$ have been investigated, e.g. by Deinzer and Stix (1971), Krause (1971), Stix (1971, 1973, 1976a,b), Levy (1972), Roberts (1972), Roberts and Stix (1972), Köhler (1973), Deinzer, von Kusserow and Stix (1974), Jepps (1975), Rädler (1975, 1986a), Yoshimura (1975b), Ivanova and Ruzmaikin (1977), Busse (1979), Busse and Miin (1979), Belvedere et al. (1980a,b) Rüdiger (1980), Weisshaar (1982),

Yoshimura et al. (1984a,b,c), Schmitt (1987), Brandenburg (1988), Brandenburg and Tuominen (1988) and Brandenburg et al. (1989e).

Let us briefly summarize a few features of $\alpha\omega$ and α^2-dynamos revealed by those investigations. In $\alpha\omega$-dynamos the generation of axisymmetric magnetic fields is in general favoured over that of non-axisymmetric fields (see also section 4.5). The toroidal part of an axisymmetric field is much stronger than its poloidal part. In the marginal case both steady and oscillatory axisymmetric fields proved to be possible. In α^2-dynamos, however, there is no such discrimination between the two field types. Depending on, e.g., the anisotropy of the α-effect axisymmetric or non-axisymmetric fields can be favoured. The toroidal and poloidal parts of a field have the same order of magnitude. Only in very exceptional cases oscillatory axisymmetric fields have been found (Rädler and Bräuer 1987).

We note that, in addition to the $\alpha\omega$ and α^2-mechanisms other dynamo mechanisms have been investigated, which are based on a combination of induction effects covered by the terms $-\beta.\mathrm{curl}\overline{B}$ or $-\delta\mathrm{xcurl}\overline{B}$ in (4.6), e.g., the "$\omega\mathrm{xj}$-effect", and differential rotation (Rädler 1969, 1976, 1980, 1986a, Stix 1976b). They are presumably less effective than the $\alpha\omega$ and α^2-mechanisms and therefore ignored in the following.

4.4. CONSTRAINTS ON SOLAR DYNAMO MODELS AND SOME CONCLUSIONS

Let us now list a few observational constraints which should be met by kinematic mean-field models of the solar dynamo.

(i) The magnetic field possesses mainly A0-symmetry. Its toroidal part is much stronger than its poloidal part.

(ii) The magnetic field is oscillatory, with a period of about 22 years.

(iii) The dependence of the toroidal magnetic field on latitude and time allow us to reproduce the butterfly diagram; in particular, the maxima of this field occur in   latitudes lower than ±40 degrees and migrate equatorward.

(iv) The toroidal field in the latitudes of maximum activity and the poloidal field near the poles during its reversal satisfy the relation (3.1).

(v) The toroidal and the poloidal field in the latitudes of maximum activity satisfy the phase relation (3.2).

(vi) If the radial field component is represented in the form (3.3) the coefficient $c_1{}^m$ are close to those derived from observation.

Of course, phenomena like the long-term variation of the sunspot activity cannot be comprehended by kinematic models. In a more general context observational constraints have recently been compiled by Tuominen et al. (1988).

The constraints (i) and (ii) are met by a dynamo working in the ω-regime rather than the α^2-regime. That is the generation of the toroidal field is mainly due to differential rotation.

In our further discussion of the constraints listed above we assume an $\alpha\omega$-dynamo for which $\mathscr{E}$ and $\overline{u}$ are defined by (4.8) and α and $d\Omega/dr$ do not change their signs all through the fluid body. In this case some remarkable conclusions can be drawn. Of course, these simple assumptions are presumably rather unrealistic but the conclusions can help us to understand more realistic models. We briefly return to the constraints (i) and (ii) and note that under the assumptions mentioned a preferrence of fields with A0-symetry and oscillatory behaviour has been observed so far only if the signs of a in the northern hemisphere and of dΩ/dr are opposite. As far as constraint (iii) is concerned we consider only the migration of the maxima of the toroidal field. According to general results on dynamo waves (Yoshimura 1975a,1976) an equatorward migration occurs just under the same condition concerning the signs of α and dΩ/dr. Whereas constraint (iv) does not lead us to a simple conclusion, constraint (v) can only be fulfilled if α is positive in the northern hemisphere (Stix 1976b). Since, as inferred from (i) and (ii), and also from (iii), α and dΩ/dr have there opposite signs, dΩ/dr should be negative, that is, the angular valocity increases inward. The constraint (vi) has been considered so far only for one model (Brandenburg 1988), and we see no simple conclusions of general nature.

4.5. SOME REMARKS CONCERNING THE FURTHER ELABORATION OF KINEMATIC MEAN-FIELD MODELS

We have seen that the mean-field concept provides us with some basis for the understanding of the solar dynamo process, and very simple models yielded some promising results. Further elaboration of mean-field models of the solar dynamo is however hampered by several difficulties. Part of them result from unsolved problems of the mean-field theory, others from lack of knowledge on convective motions, differential rotation, etc... There is at the moment no really comprehensive mean-field model of the solar dynamo. In the following we shall explain some of the problems that we are faced with in this context and some efforts to improve models.

Let us first recall that the mean fields have to be defined so that the Reynolds rules apply. For many purposes it seems reasonable to define them by a time average over 1 or 2 years (Stix 1976a, 1981). One should bear in mind that then the Reynolds rules hold only in some approximation. With averages over the longitude or stastitical averages these rules apply exactly but some other difficulties occur (e.g., Rädler 1981a). The finding that the magnetic flux is concentrated in thin tubes implies no objection to the applicability of the mean-fields more difficult.

While the dependance of $\mathscr{E}$ on $\overline{B}$ given by (4.4) or (4.6) seems to be justified to a certain extend (Rädler 1981a), there is considerable uncertainty as to how quantities like α,β... depend on the motions. As mentioned above, most of the results available were derived in the second-order correlation approximation, which is not well justified for the solar convection zone. A recent investigation (Nicklaus and Stix 1988) demonstrates that corrections by fourth order correlations may change the results considerably; even the sign of the scalar α occuring with isotropic turbulence can change if higher approximations are used. Despite several results on the α-effect and related effects under more realistic conditions (e.g., Walder et al. 1980, Stix 1983, Schmitt 1985, Durney 1988) are further efforts necessary.

Some discussions on the solar dynamo are based on the idea that the α-effect can simply be described by a scalar α which is essentially equal to $-(1/3)\overline{u' \, \mathbf{curl}(u')} \, \tau$, where $\overline{u' \, \mathbf{curl}(u')}$ is the helicity of the fluctuating motions and τ some turnover time. This idea implies, apart from the second-order correlation approximation and the restriction to the high-conductivity limit, the assumption of the isotropy of fluctuating motions and is thus rather questionable. In spherical dynamo models the generation of an axisymmetric toroidal field depends on the component a_{00} of the tensor α (Rädler 1981b). Even if we accept the second-order correlation approximation and the high conductivity limit it is not a_{00} but $(1/3)$ trace (α) which is equal to $(1/3)\overline{\mathbf{u}' \; \mathbf{curl}(\mathbf{u}')} \; \tau$, and these quantities coincide only in the isotropic case. In general, a_{00} may even vanish for non-zero trace (α).

The magnetic flux which penetrates the plasma of the convection zone contributes to its buoyancy. The motions due to this magnetic buoyancy reduce the mean flux in the convection zone (e.g., Krivodubski 1984). With this mind, in several cases on the right-hand side of the induction equation (4.2a) an additional term $-\kappa\overline{\mathbf{B}}$, with some positive coefficient κ, has been introduced (e.g., De Luca and Gilman 1986). We should like to emphasize that this is hardly the correct way to comprehend the effect of magnetic buyancy. According to the derivation of (4.2a) from Maxwwell's equations any effect of motions must be described by $\overline{\mathbf{u}}$ or $\mathscr{E}$.

Apart from the reasons explained above, the lack of knowledge on the convective motions and the differential rotation in deeper layers poses several problems for the construction of realistic solar dynamo models. There have been many attempts to meet observational constraints such as those formulated above, by more complex assumptions on the radial and latitudinal dependence of the α– effect and the differential rotation, and by taking into account anisotropies of the α-effect and related effects (e.g., Krause and Steenbeck 1969a, Stix 1973, 1976, Köhler 1973, Yoshimura 1975, Ivanova and Ruzmaikin 1977, Belvedere et al. 1980a, Yoshimura et al. 1984a,b,c, Schmidt 1987, Brandenburg 1988). We mention in particular a recent model in which in addition to an anisotropic α-effect several related effects are included and an angular velocity inferred from helioseismological measurements is used (Brandenburg and Tuominen 1988). This model reflects rather well the latitudinal dependence of the toroidal magnetic field as derived from observations.

It has been argued that the flux losses in the convection zone due to magnetic buoyancy do not allow a generation process of magnetic fields in this zone. For this and other reasons, models have been proposed in which this process is mainly localized at the bottom of the convective zone or in the overshoot layer below it (e.g., Glatzmaier 1985. De Luca and Gilman 1986, Gilman et al. 1989).

In several models of aω-dynamos in addition to axisymmetric also non-axisymmetric magnetic fields have been considered (Krause 1971, Stix 1971, Roberts and Stix 1972, Ivanova and Ruzmaikin 1986a, Ruzmaikin et al. 1988), which deserve some interest in view of the sectorial structure of the solar magnetic field. Under certain conditions comparable excitation conditions for axisymmetric and non-axisymmetric fields have been found. For general reasons this is only to be expected if the rotational shear is sufficiently small (Rädler 1986a,b). It turned out that some of the models mentioned (Krause 1971, Stix 1971, Ivanova and Ruzmaikin 1985, Ruzmaikin et al. 1988), which are in conflict to that, have to be revised (Rädler et al. 1989).

5. Dynamo models including the back-reaction of the magnetic field on the motion

5.1. BASIC EQUATIONS

So far we discussed the dynamo process assuming that the fluid motion is given. In particular, we ignored any back-reaction of the magnetic field on the motion. However, it is the back-reaction that determines the magnitude to which the magnetic field grows and it can influence the geometrical structure and time behaviour of the field considerably. Changing our view we now consider the motion no longer given and assume that the velocity $\overline{u}$ obeys the Navier-Stokes equation and the mass balance,

$$\rho(\frac{\partial u}{\partial t} + (u\bullet\nabla)u = -\nabla p + F + \frac{1}{\mu} curl(B\times B)$$

(5.1a)

$$\frac{\partial\rho}{\partial t} + div(\rho u) = 0$$

(5.1b)

Here ρ is the mass density, p the pressure, and F stands for viscous or any other but electromagnetic forces; if F only means viscous forces we have, in Cartesian coordinates,

$$F_i = \frac{\partial \tau_{ij}}{\partial x_j}$$

(5.2)

with some stress tensor τ_{ij}. The term $(1/\mu)$ $curl\overline{B}\times\overline{B}$ describes the Lorentz force, where μ is the permeability of free space. For compressible fluids, equations (5.1) have to be supplemented by the equation of state and the energy balance, which we do not write down here.

Compared to the kinematic dynamo problem posed by equations (4.1), the full dynamo problem defined by equations (4.1), (5.1) and some supplementary relations is much more complex. It is no longer a linear but an involved nonlinear problem.

5.2. MEAN-FIELD MAGNETOHYDRODYNAMICS

The mean-field concept introduced above with the kinematic dynamo problem can readily be extended to the full dynamo problem considered now; see, e.g., Rädler (1976). The equations (4.2) and (4.3) have then to be supplemented by corresponding equations derived from (5.1). We shall demonstrate here only the basic idea. To this end we restrict ourselves to incompressible fluids. From (5.1) we then obtain

$$\rho(\frac{\partial \overline{u}}{\partial t} + (\overline{u}\bullet\nabla)\overline{u} = -\nabla\overline{p} + \overline{F} + \frac{1}{\mu} curl(\overline{B}\times\overline{B}) + \mathscr{F}$$

(5.3a)

$$div(\rho\overline{u}) = 0$$

(5.3b)

where $\mathscr{F}$ is a mean force resulting from fluctuating motions and magnetic fields,

$$\mathscr{F} = -\rho \overline{(u'\bullet\nabla)u'} + \frac{1}{\mu} curl(\overline{B'\times B'})$$

(5.4)

In the case of vanishing magnetic field these equations coincide with the Reynolds equations of hydrodynamic turbulence theory. We note that F and f appear only as their sum, and that this can be represented in the form

$$\overline{F}_i + \overline{\mathscr{F}}_i = \frac{\partial\left(\overline{\tau}_{ij} + \tau_{ij}^R + \tau_{ij}^M\right)}{\partial x_j}$$

(5.5a)

$$\tau_{ij}^R = -\rho\overline{u'_i u'_j}$$

(5.5b)

$$\tau_{ij}^M = (\frac{1}{\mu})\overline{(B'_i B'_j)} - (\frac{1}{2}) \overline{B'^2}\delta_{ij}$$

(5.5c)

where τ_{ij}^{R} is the Reynolds stress tensor and τ_{ij}^{M} the Maxwell stress tensor resulting from the fluctuating magnetic field.

In the same sense in which $\mathscr{E}$ was considered a quantity depending on $\overline{B}$ and its derivatives, $\mathscr{E}$ and $\mathscr{F}$ are now to be understood as quantities depending on $\overline{B}$, $\overline{u}$ and their derivatives. We do not want to discuss the general aspects of such relations. We only mention that, for example, the dependence of ε on B is now no longer linear. To explain this in some more detail we return to the simple case considered above in which $\overline{u}$ is equal to zero, $\overline{u}$ ' corresponds to a homogeneous isotropic turbulence, and $\mathscr{E}$ then is given by (4.5). Consider this turbulence to be due to any forces independent of the magnetic field but admit, in addition, Lorentz forces. The latter will influence the intensity of the turbulence and disturb its isotropy. Under these circumstances $\mathscr{E}$ again contains a term $\alpha\overline{B}$ but α now depends on $\overline{B}$. It was shown that, under certain assumptions, α decreases with growing $|\overline{B}|$ (Rüdiger 1974a).

As already mentioned, in the case of vanishing magnetic field our equations reduce to Reynolds equations. we note that on this basis a comprehensive theory of the differential rotation has been developed (Rüdiger 1989).

5.3. NONLINEAR MEAN-FIELD DYNAMO MODELS

When starting from kinematic mean-field dynamo models and introducing the back-reaction of the magnetic field on the motion we have to consider a variety of new effects. All the coefficients defining the mean electromotive force $\mathscr{E}$, for example α,β,...occuring in (4.6), depend now on $\overline{B}$. This concerns not only their magnitudes but also their tensorial structures. In addition, the velocity $\overline{u}$ of the mean motion is influenced by $\overline{B}$. This may happen in a direct way via the Lorentz force $(1/\mu)$ curl$\overline{B}$ x$\overline{B}$ or in an indirect way via the force $\mathscr{F}$, that is, via the Reynolds stresses and via the Maxwell stresses of the fluctuating magnetic field. The mathematical problem posed by a kinematic dynamo model is linear in $\overline{B}$. All the effects mentioned here disturb this linearity.

Several mean-field dynamo models have been investigated which deviate from kinematic models with isotropic α-effect only in so far as α now depends on $\overline{B}$ and decays with growing $|\overline{B}|$, sometimes referrred to as "α-quenching", e.g., by Stix (1972), Rüdiger (1974b), Jepps (1975), Ivanova and Ruzmaikein (1977), Kleeorin and Ruzmaikin (1984), Krause and Meinel (1988), Brandenburg et al. (1988a,b,d) and Rädler and Wiedemann (1989). In αω-models of that kind which, in its zero-field limit, show the symmetries discussed above about equatorial plane and rotation axis, axisymmetric oscillatory magnetic fields have been found which show no longer pure A0 or S0-symmetry but contain both A0 and S0-parts (Brandenburg et al. 1989a,b,d). The magnitude of these fields and its composition of A0 and S0-parts vary, apart from the fundamental oscillation, also with a longer period. This is, of course, of interest in view of the long-term variation of the solar activity. However, in these models the possibility of non-axisymmetric fields has not been considered at all. As we know from a^2-models the stability of an axisymmetric field may well be disturbed by non-axisymmetric fields (Rädler and Wiedeman 1989).

In some other models the back-reaction of the magnetic field on the mean motions via Lorentz-forces of the mean field was studied, e.g., by Malkus and Proctor (1975), De Luca and Gilman (1986), Brandenburg, et al. (1989a,b,d) and Belvedere and Proctor (1989). A reduction of both the α-effect and the differential rotation by the back-reaction of the magnetic field was taken into account by Yoshimura (1978a,b).

We further refer to a model by Brandenburg et al. (1989c) in which, starting from assumptionss concerning the turbulent motions in the solar convection zone, both the magnetic field and the differential rotation are computed, taking into account the two types of back-reaction considered above of the magnetic field on the motions.

Incidentally, for some mean-field dynamo models the Lorentz-forces of the mean fields have been computed and discussed in view of the torsional oscillations (Yoshimura 1981, Rüdiger et al. 1986).

5.4. OTHER NONLINEAR DYNAMO MODELS

Interesting studies of the solar dynamo problem beyond the mean-field approach have been done by Gilman and Miller (1981) and by Gilman (1983). For a spherical shell modelling the solar convection zone the full set of the magnetohydrodynamic equations governing $\overline{B}$ and $\overline{u}$ in the Bussinesq approximation has been integrated numerically. The model reflects several features of solar magnetic phenomena. In particular, it shows a magnetic cycle. However , the period is too short for describing the solar cycle, and the equartorward migration of the toroidal field belts as indicated by sunspots is not reproduced. It is not clear whether this is to be ascribed to the Boussinesq approximation, to the unsufficient consideration of the interaction between magnetic field and motion in small scales, or to other reasons. Similar results have been obtained with somewhat different assumptions by Glatzmaier (1985).

There are also a few attempts to mimic features of the solar dynamo by a low-order set of non-linear ordinary differential equations. Ruzmaikin (1985) proposed a set of three first-order equations which form a Lorentz system and tried to find so some explanation of the stochastic nature of the solar activity. Preliminary results of an attractor analysis of the Wolf relative sunspot numbers obtained by Kurths (1987) indeed suggest that the underlying process can be described by a small number of equations, the number being estimated between 3 and 7. In this context an investigation by Weiss (1985) on the chaotic behavior of dynamos, in which a system of seven first-order ordinary differential equations is considered, deserves some interest.

I would like to thank my colleagues Drs. G. Rüdiger and J. Staude (Potsdam) as well as Dr. A. Brandenburg (Helsinki) for many important suggestions.

References

Belvedere, G. (1983) 'Dynamo theory in the Sun and stars', in P.B. Byrne and M. Rodonõ (eds.), Activity in Red-Dwarf Stars, D. Reidel Publishing Co., Dordrecht, pp. 579-599.

Belvedere, G., Paternõ, L. and Stix, M. (1980a) 'Dynamo action of a mean flow caused by latitude-dependent heat transport', Astron. Astrophys. 86, 40-45.

Belvedere, G., Paternõ, L. and Stix M. (1980b) 'Magnetic cycles of lower main sequence stars', Astron. Astrophys. 91, 328-330.

Belvedere, G. and Proctor, M.R.E. (1989) 'Nonlinear dynamo modes and timescales of stellar activity', submitted to Proceedings IAU-Symp. 138.

Brandenburg, A. (1988) 'kinematic dynamo theory and the solar activity cycle', Licenciate thesis, University of Helsinki.

Brandenburg, A., Krause, F., Meinel, R., Moss, D. and Tuominen, I. (1989a) 'The stability of nonlinear dynamos and the limited role of kinematic growth rates', Astron. Astrophys. 213, 411-422.

Brandenburg, A., Krause, F., and Tuominen, I. (1989b) 'Parity selection in nonlinear dynamos', in M. Meneguzzi et al. (eds.), Turbulence and Nonlinear Dynamics in MHD Flows, Elsevier Science Publishers, North Holland.

Brandenburg, A., Moss, D., Rüdiger, G. and Tuominen. I. (1989c) 'The nonlinear solar dynamo and differential rotation: A Taylor number puzzle?', submitted to Solar Physics.

Brandenburg, A., Moss, D. and Tuominen, I. (1989d) 'On the nonlinear stability of dynamo models', Geophys. Astrophys. Fluid Dyn., in press.

Brandenburg, A. and Tuominen, I. (1988) 'Variation of magnetic fields and flows during the solar cycle', Adv. Space Res. 8, No 7, (7)185 - (7)189.

Brandenburg, A., Tuominen, I. and Rädler, K.-H. (1989e) 'On the generation of non-axisymmetric magnetic fields in mean-field dynamos', Geophys. Astrophys. Fluid Dyn., in press.

Busse, F.H. (1979) 'Some new results on spherical dynamos', Physics Earth Planet. Inter. 20, 152-157.

Busse, F.H. and Miin, S.W. (1979) 'Spherical dynamos with anisotropic α-effect', Geophys. Astrophys. Fluid Dyn. 14, 167-181.

Cowling, T.G. (1934) 'The magnetic fields of sunspots', Mon. Not. Roy. Astr. Soc. 94, 39-48.

Deinzer, W. and Stix, M. (1971) 'On the eigenvalues of Krause-Steenbeck's solar dynamo', Astron. Astrophys. 12, 111-119.

Deinzer, W., von Kusserow, H.U. and Stix, M. (1974) 'Steady and oscillatory aω-dynamos', Astron. Astrophys. 36, 69-78.

Deluca, E.E. and Gilman, P.A. (1986) 'Dynamo theory for the interface between convection zone and the radiative interior of a star. Part I. Model equations and exact solutions', Geophys. Astrophys. Fluid Dyn. 37, 85-127.

Durney, B.R. (1988) 'On a simple dynamo model and the anisotropic α–effect', Astron. Astrophys. 191, 374.

Gilman, P.A. (1983) 'Dynamically consistent nonlinear dynamos driven by convection in a rotating spherical shell. II. Dynamos with cycles and strong feedbacks', Astrophys. J. Suppl. 53, 243-268.

Gilman, P.A. (1986) 'The solar dynamo: observations and theories of solar convection, global circulation, and magnetic fields', in P.A. Sturrock et al. (eds.), Physics of the Sun, D. Reidel Publishing Co., Dordrecht, pp. 95-160.

Gilman, P.A. and Miller, J. (1981) 'Dynamically consistent nonlinear dynamos driven by convection in a rotating spherical shell', Astrophys. J. Suppl. 46, 211-238.

Gilman, P.A., Morrow, C.A. and Deluca, E.E. (1989) 'Angular momentum transport and dynamo action in the Sun: Implications of recent oscillation measurements', Astrophys. J. 338, 528-537.

Glatzmaier, G.A. (1985) 'Numerical simulations of stellar convective dynamos. II. Field propagation in the convection zone', Astrophys. J. 291, 300-307.

Ivanova, T.S. and Ruzmaikin, A.A. (1975) 'A magnetohydrodynamic dynamo model of the solar cycle', Sov. Astron. 20, 227-234.

Ivanova, T.S. and Ruzmaikin, A.A. (1977) 'A nonlinear MHD-model of the dynamo of the Sun', Astron. Zh. (USSR) 54, 846-858 (in Russian).

Ivanova, T.S. and Ruzmaikin, A.A. (1985) 'Three-dimensional model for the generation of the mean solar magnetic field', Astron. Nachr. 306, 177-186.

400

Jepps, S.A. (1975) 'Numerical models of hydromagnetic dynamos', J. Fluid Mech. 67, 629-646.

Kleeorin, N.I. and Ruzmaikin, A.A. (1984) 'Mean-field dynamo with cubic non-linearity', Astron. Nachr. 305, 265-275.

Köhler, H. (1973) 'The solar dynamo and estimates of the magnetic diffusivity and the α-effect', Astron. Astrophys. 25, 467-476.

Krause, F. (1971) 'Zur Dynamotheorie magnetischer Sterne: Der   'symmetrische Rotator' als Alternative zum 'schiefen Rotator", Astron. Nachr. 293, 187-193.

Krause, F. and Meinel, R. (1988) 'Stability of simple nonlinear α^2-dynamos', Geophys. Astrophys. Fluid dyn. 43, 95-117.

Krause, F. and Rädler, K.-H. (1980) 'Mean-Field Magnetohydrodynamics and Dynamo Theory', Akademie-Verlag, Berlin and Pergamon Press, Oxford.

Krivodubski, V.N. (1984) 'Magnetic field transfer in the turbulent solar envelope', Sov. Astron. 28, 205-211.

Kurths, J. (1987) 'An attractor analysis of the sunspot relative number', Preprint PRE-ZIAP (Potsdam) 87-02.

Larmor, J. (1919) 'How could a rotating body such as the Sun become a magnet?' Rep. Brit. Assoc. adv. Sc. 1919, 159-160.

Levy, E.H. (1972) 'Effectiveness of cyclonic convertion for producing the geomagnetic field', Astrophys. J. 171, 621-633.

Malkus, W.V.R. and Proctor, M.R.E. (1975) 'The macrodynamics of α-effect dynamos in rotating fluids', J. Fluid Mech. 67, 417-444.

Nicklaus, B. and Stix, M. (1988) 'Corrections to first order smoothing in mean-field electrodynamics', Geophys. Astrophys. Fluid Dyn. 43, 149-166.

Parker, E.N. (1955) 'Hydromagnetic dynamo models', Astrophys. J. 122, 293-314.

Parker, E.N. (1979) 'Cosmical Magnetic fields', Clarendon Press, Oxford.

Rädler, K.-H. (1969) 'über eine neue Möglichkeit eines Dynamomechanismus in turbulenten leitenden Medien', Mber. Dtsch. Akad. Wiss. Berlin 11, 194-201.

Rädler, K.-H. (1975) 'Some new results on the generation of magnetic fields by dynamo action', Mem. Soc. Roy. Sc. Liege VIII, 109-116.

Rädler, K.-H. (1976) 'Mean-field magnetohydrodynamics as a basis of solar dynamo theory', in B. Bumba and J. Kleczek (eds.), Basic Mechanisms of Solar Activity, D. Reidel Publishing Co., Dordrecht, pp. 323-344.

Rädler, K.-H. (1980) 'Mean-field approach to spherical dinamo models', Astron. Nachr. 301, 101-129.

Rädler, K.-H. (1981a) 'On the mean-field approach to spherical dynamo models', in A.M. Soward (ed.), Stellar and Planetary Magnetism, Gordon and Breach Publishers, New York, pp. 17-36.

Rädler, K.-H. (1981b) 'Remarks on the α-effect and dynamo action in spherical models', in A.M Soward (ed.), Stellar and Planetary Magnetism, Gordon and Breach Publishers, New York, pp. 37-48.

Rädler, K.-H. (1986a) 'Investigations of spherical kinematic mean-field dynamo models', Astron. Nachr. 307, 89-113.

Rädler, K.-H. (1986b) 'On the effect of differential rotation on axisymmetric and non-axisymmetric magnetic fields of cosmical bodies', Plasma-Astrophysics, ESA SP-251, 569-574.

Rädler, K. -H. and Bräuer, H.-J. (1987) 'On the oscillatory behaviour of kinematic mean-field dynamos', Astron. Nachr. 308, 101-109.

Rädler, K.-H., Brandenburg, A. and Tuominen, I. (1989) 'On the non-axisymmetric magnetic-field modes of the solar dynamo', Poster IAU-Colloquium No 121, to be submitted to Solar Physics.

Rädler, K.-H. and Wiedemann, E. (1989) 'Numerical experiments with a simple nonlinear mean-field dynamo model',Geophys. Astrophys. fluid Dyn., in press.

Ribes, E., Mein, P. and Manganey, A. (1985) 'A large scale meridional circulation in the convective zone', Nature 318, 170-171.

Ribes, E. and Laclare, F. (1988) 'Toroidal convection rolls in the Sun', Geophys. Astrophys. Fluid Dyn. 41, 171-180.

Roberts, P.H. (1972) 'Kinematic dynamo models', Phil. Trans. Roy. Soc. A 272, 663-703.

Roberts, P.H. and Stix, M. (1972) 'α-effect dynamos, by the Bullard-Gellman formalism', Astron. Astrophys. 18, 453-466.

Rüdiger, G. (1974a) 'The influence of a uniform magnetic field of arbitrary strength on turbulence', Astron. Nachr. 295, 275-283.

Rüdiger, G. (1974b) 'Behandlung eines einfachen hydromagnetischen Dynamos mit Hilfe der Gitterpunktmethode', Pub. Astrophys. Obs. Potsdam 32, 25-29.

Rüdiger,G. (1980) 'Rapidly rotating α^2-dynamo models', Astron. Nachr. 301, 181-187.

Rüdiger, G. (1989) 'Differential Rotation and Stellar Convection', Akademie-Verlag, Berlin and Gordon and Breach Science Publishers, New York.

Rüdiger, G. Tuominen, I., Krause, F. and Virtanen, H. (1986) 'Dynamo generated flows in the Sun', Astron. Astrophys. 166, 306-318.

Ruzmaikin, A.A. (1985) 'The solar dynamo', Solar Physics 100, 125-140.

Ruzmaikin, A.A., Sokoloff, D.D. and Starchenko, S.V. (1988) 'Excitation of non-axially symmetric modes of the Sun's magnetic field', Solar Phys. 115, 5-15.

Schmitt, D. (1985) 'Dynamowirkung magnetischer Wellen', Thesis, Univ. Göttingen.

Schmitt, D. (1987) 'An α-dynamo with an α-effect due to magnetostrophic waves', Astron. Astrophys. 174, 281-287.

Steenbeck, M. and Krause, F. (1969a) 'Zur Dynamotheorie stellarer und planetarer Magnetfelder. I. Berechnung sonnenähnlicher Wechselfeldgeneratoren', Astron. Nachr. 291, 49-84.

Steenbeck, M. and Krause, F. (1969b) 'Zur Dynamotheorie stellarer und planetarer Magnetfelder. II. Berechnung planetenähnlicher Gleichfeldgeneratoren', Astron. Nachr. 291, 271-286.

Steenbeck, M., Krause, F. and Rädler, K.-H. (1966) 'Berechnung der mittleren Lorentz-Feldstärken vxB für ein elektrisch leitendes Medium in turbulenter, durch Coriolis-Kräfte beeinflubter Bewegung', Z. Naturforsch. 21a, 369-376.

Stenflo, J.O. (1973) 'Magnetic-field structure of the photospheric network', Solar Physics 32, 41-63.

Stenflo, J.O. and Vogel, M. (1986) 'Global resonances in the evolution of solar magnetic fields', Nature 319, 285.

Stenflo, J.O. and Güdel, M. (1987) 'Evolution of solar magnetic fields: Modal stucture', Astron. Astrophys. 191, 137.

Stix, M. (1971) 'A non-axisymmetric α-effect dynamo', Astron. Astrophys. 13, 203-208.

Stix, M. (1972) 'non-linear dynamo waves', Astron. Astrophys. 20, 9-12.

Stix, M. (1973) 'Spherical α-dynamos, by a variational method', Astron. Astrophys. 24, 275-281.

Stix, M. (1976a) 'Dynamo theory and the solar cycle', in V. Bumba and J. Kleczek (eds.), Basic Mechanisms of Solar Activity, D. Reidel Publishing Co., Dordrecht, pp. 367-388.

Stix, M. (1976b) 'Differential rotation and the solar dynamo', Astron. Astrophys. 47, 243-254.

Stix, M. (1981) 'Theory of the solar cycle', Solar Physics 74, 79-101.

Stix, M. (1983) 'Helicity and a-effect of simple convection cells', Astron. Astrophys. 118, 363-364.

Stix, M. (1989) 'The Sun', Springer-Verlag Berlin.

Tuominen, I., Rüdiger, G. and Brandenburg, A. (1988) Observational constraints for solar-type dynamos', in O. Havens et al. (eds.), Activity in Cool Star Envelopes, Kluwer Academic Publishers, London, pp. 13-20.

Walder, M., Deinzer, W. and Stix, M. (1980) 'Dynamo action associated with random waves in a rotating stratified fluid', J. Fluid Mech. 96, 207-222.

Weiss, N.O. (1985) 'Chaotic behaviour in stellar dynamos', Journal of Statistical Physics 39, 477-491.

Weisshaar, E. (1982) 'A numerical study of α^2-dynamos with anisotropic α-effect', Geophys. Astrophys. Fluid dyn. 21, 285.

Yoshimura, H. (1975a) 'Solar-cycle dynamo wave propagation', Astrophys. J. 201, 740-748.

Yoshimura, H. (1975b) 'A model of the solar cycle driven by the dynamo action of the global convection in the solar convection zone', Astrophys. J. Suppl. 29, 467-494.

Yoshimura, H. (1976) 'Phase relation between the poloidal and toroidal solar-cycle general magnetic fields and location of the origin of the surface magnetic fields', Solar Physics 50, 3-23.

Yoshimura, H. (1978a) 'Nonlinear astrophysical dynamos: The solar cycle as the non-linear oscillation of the general magnetic field driven by the non-linear dynamo and the associated modulation of the differential-rotation-global-convection system', Astrophys. J. 220, 692-711.

Yoshimura, H. (1978b) 'Nonlinear astrophysical dynamos: multiple-period dynamo wave oscillations and long-term modulations of the 22 years solar cycle', Astrophys. J. 226, 706-719.

Yoshimura, H. (1981) 'Solar cycle Lorentz force waves and the torsional oscillations of the Sun', Astrophys. J. 247, 1102-1112.

Yoshimura, H., Wang, Z. and Wu, F. (1984a) 'Linear astrophysical dynamos in rotating spheres: Differential rotation, anisotropic turbulent magnetic diffusivity, and solar-stellar cycle magnetic parity', Astrophys. J. 280, 865-872.

Yoshimura, H., Wang, Z. and Wu, F. (1984b) 'Linear astrophysical dynamos in rotating spheres: Mode transition between steady and oscillatory dynamos as a function of the dynamo strength and anisotropic turbulent magnetic diffusivity', Astrophys. J. 283, 870-878.

Yoshimura, H., Wu, F. and Wang, Z. (1984c) 'Linear astrophysical dynamos in rotating spheres: Solar and stellar cycle north-south hemisphere parity selection mechanism and turbulent magnetic diffusivity', Astrophys. J. 285, 325-338.

THE EVOLUTION OF THE SUN'S ANGULAR MOMENTUM

DAVID R. SODERBLOM
Space Telescope Science Institute
3700 San Martin Drive
Baltimore MD 21218 USA

JOHN R. STAUFFER
NASA–Ames Research Center
Moffett Field CA 94035 USA

ABSTRACT. We discuss the observational data that are available to illustrate the rotational history of the Sun, and those data that are particularly pertinent to the interiors of the Sun and stars.

1. Introduction

Observationally, rotation is a purely surface phenomenon, yet there are aspects of its study in solar-type stars that have implications for the study of stellar interiors. This review will emphasize those aspects. For other recent reviews, see Hartmann and Noyes (1987) and Stauffer and Soderblom (1989). The reader is also referred to the forthcoming text by Giampapa and Sonett (1989) for discussions pertinent to solar rotation. Among the questions that are relevant to this meeting that rotation studies can address are:
Is the Sun typical for its mass and age?
Do planets play a role in the angular momentum history of the Sun?
What is the rotational history of a star like the Sun?
What surface phenomena are useful diagnostics of the Sun's interior?
Differential rotation.
Rotation and activity cycles.
Rotation and lithium depletion.
Is there stellar evidence for a rapidly rotating core in the Sun?

2. The Sun Among the Stars

Classical studies of stellar rotation using photographic plates reached their limit with the work of Kraft (1967), who demonstrated the connections between rotation, age, and chromospheric activity for stars like the Sun. Skumanich (1972) quantified Kraft's results in his $t^{-1/2}$ laws between rotation or activity and age. (Note that Skumanich based this on three data points.) Soderblom (1983) showed that the Sun has an average $v \sin i$ for an old solar-type star.

But the most important data for studying rotation in low mass stars has resulted from the

403

G. Berthomieu and M. Cribier (eds.), Inside the Sun, 403–413.

intensive work on modulation of star light by stellar surface inhomogeneities, especially that done by the Mount Wilson group. They have now determined rotation periods for about 60 stars (Baliunas *et al.* 1983). These stars cover a broad range of age and mass, although there tend to be few old stars because the activity is so low in such stars, reducing chromospheric contrast to an unobservably low level. Thus there are not enough old solar-type stars to form a meaningful comparison sample. However, there are enough stars to calibrate a relationship between the level of chromospheric activity (quantified by R'_{HK}, the ratio of the star's Ca II H and K flux to its bolometric luminosity) and the Rossby number (the ratio of the rotation period to the convective turnover time). For details of this, see Noyes *et al.* (1984) and Gilliland (1987). Using such a relation, one or two observations of the HK flux are sufficient to determine the rotation period to an accuracy of 20% (Soderblom 1985).

If this relation is applied to the Vaughan and Preston (1980) survey of HK emission among solar-type stars, one derives the distribution of rotation rates for stars near 1 $M_\odot$ (Fig. 1). For a discussion of this, see Soderblom (1985). The essential points are: (1) the Sun is quite typical for an old star of its color; (2) the range in rotation is about one dex at any one color (roughly the same mass), but most of the stars fall within a range of only a factor of 3 to 4; (3) the mass range represented is about 0.8 to 1.2 $M_\odot$, and the distribution of stars here suggests that all that mass range has about the same relation between rotation and age; (4) for the redder stars, the minimum is $\sim 0.5\Omega_\odot$, probably because at these low masses the evolutionary time scales exceed the age of the Galactic disk and these stars have not yet been able to evolve to very low velocities; and (5) the sharp bottom among the bluest stars is a selection effect because stars of those masses evolve away from the main sequence quickly, and are no longer included in a survey such as this (they are classified as

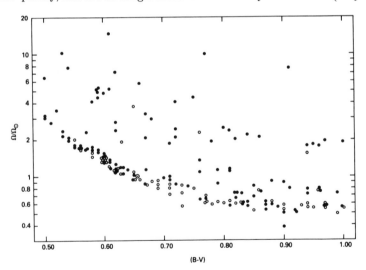

Figure 1. Predicted angular velocity (in solar units) versus $(B - V)$ color for solar-type stars of the Vaughan-Preston survey (from Soderblom 1985). The Sun is shown at $(B - V)$ = 0.656 near the lower bound of the stars. Note that rotation appears to flatten out at about $0.5\Omega_\odot$ for the lowest masses (reddest stars), and that the range at any one color is about one dex. The open circles represent high-velocity stars of the old disk population.

subgiants instead of main sequence stars).

All of the above is consistent with the Sun having arrived on the main sequence rotating at about 4 times its present rate, and then having lost angular momentum by the $t^{-1/2}$ law, but the details are highly uncertain due to a lack of stars we can observe whose ages are known with confidence.

3. A Brief History of Solar Rotation

The field star studies outlined in the preceding section suggest that the rotational evolution of the Sun and other stars of similar mass stars is fairly sedate and well defined over most of their main sequence lifetimes. This suggests either that the star formation process enforces uniformity or that angular momentum evolution during the pre-main sequence and early main sequence period significantly decreases variations in initial angular momentum among low mass stars of a given mass. Recent observational studies indicate that the latter explanation is probably the correct one. In particular, stars in the Pleiades (age 70 Myr) exhibit an appreciable scatter in rotation rate at any one mass, a scatter that disappears by the age of the Hyades (age 0.6 Gyr) – see below.

Low mass stars first become visible to us at an age of about 1 Myr, and we call them T Tauri stars. Their characteristics include veiling of their spectra, excess infrared radiation, very high levels of stellar activity, etc. (see Bertout 1989 for a recent review). The veiling gradually disappears and the activity subsides, even while remaining much higher than is seen in main sequence stars. The veiling and IR excesses of T Tauri stars are plausibly ascribed to the presence of disks around the T Tauri stars. These disks may also play a role in the angular momentum evolution of these stars – either by accretion from the disk adding angular momentum (and perhaps serving as a source of fresh lithium for the outer

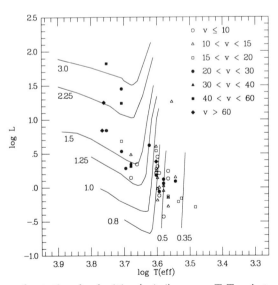

Figure 2. Observed rotational velocities ($v\sin i$) among T Tauri stars (from Hartmann *et al.* 1986). Note that at and below 1 $M_\odot$ no $v\sin i > 30$ km s^{-1} is seen, except for one star.

envelope of the star), or by winds from a boundary layer helping extract angular momentum from the star. The disk also is the eventual source for planets (which, if our solar system is an example, can hold significantly more angular momentum than the stars which they orbit).

Among the T Tauris, the observed rotation rates at 1 $M_\odot$ are about 6 to 30 km s^{-1}, or 3 to 10 $\Omega_\odot$ (Fig. 2). A number of rotation periods have been determined from variability of broad-band light, but there is a selection effect against long periods (it is difficult to obtain observational data over a period much exceeding a week or two at a time). The T Tauris are typically 3 times the size of the Sun, and much less centrally condensed, so there is much more angular momentum in one than in a main sequence star rotating at 3 times the solar rate. Until fairly recently it was thought that this angular momentum was lost gradually during the several tens of Myr that the star took in approaching the main sequence.

It then came as a major surprise a few years ago when some stars were discovered in the Pleiades that are rotating at as much as 100 times the solar rate yet are barely on the Zero-Age Main Sequence. These ultra-fast rotators constitute a significant fraction of the stars in the mass range in which they're seen, and are not just a few freaks (Fig. 3).

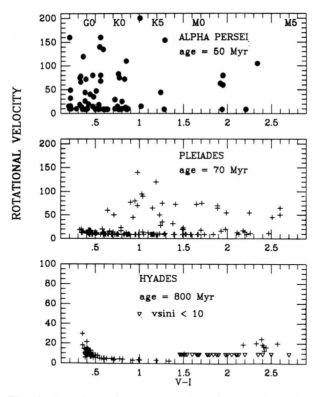

Figure 3. Distribution of rotational velocity ($v\sin i$) versus $(V - I)$ color for solar-type stars in the α Persei and Pleiades clusters (adapted from Stauffer and Hartmann 1987). The older stars of the Hyades are also shown. A $(V - I)$ color of 0.5 corresponds to about 1 $M_\odot$, while 1.0 corresponds to about 0.8 $M_\odot$.

Subsequent observations of a somewhat younger cluster (α Persei) showed that the presence of these unusual objects in the Pleiades was not just a fluke.

There is good evidence that the duration of this main sequence rapid rotation phase is a strong function of mass. For the youngest, well-studied cluster (α Per, age 50 Myr), rapid rotators are found among all spectral types for low mass stars. For the slightly older Pleiades cluster (70 Myr), the G dwarfs are all relatively slow rotators, with rapid rotators only among the K and M dwarfs. The oldest, well-studied cluster (the Hyades, age 0.6 Gyr) shows rapid rotators only among the very-low-mass M dwarfs. Thus the spin-down time scales must increase with decreasing stellar mass. The open cluster data allow a quantitative estimate of the spin-down time scales for stars of various masses. 1 $M_\odot$ rapid rotators must lose a significant fraction of their angular momentum in only ~ 20 Myr; 0.7 $M_\odot$ stars must have a characteristic spin-down time of order 100 Myr; and 0.5 $M_\odot$ stars apparently must have a spin-down time scale of several hundred million years.

Just recently, Stauffer *et al.* (1989) have observed the very young (age 30 Myr) cluster IC 2391, which shows a spread in rotation rates that is similar to that of the Pleiades, yet IC 2391 has a very small age spread (no more than 20 Myr). Stauffer *et al.* thus argue that an age spread cannot account for the distribution of Pleiades rotation rates at any one mass, despite the arguments of Butler *et al.* (1987), who saw commensurate differences in lithium abundances. Perhaps these differing rotation rates reflect the initial distribution of angular momentum among stars, as suggested by Pinsonneault, Kawaler, and Demarque (1989).

Note that it is just possible to account for the fastest rotators seen in the Pleiades and α Persei clusters if no substantial angular momentum loss is tolerated between the T Tauri phase and the ZAMS (Stauffer and Hartmann 1987). This is itself a problem. That is, if the spindown of these stars is so fast and efficient on the main sequence, how do they achieve such high rotation rates to begin with? Consideration of this problem has led some theorists to suggest the possibility that these stars rotate as solid bodies while approaching the ZAMS (such is necessary to transfer angular momentum from the core to the surface to explain the highest velocities), but upon reaching the ZAMS the convective envelope decouples from the core (Endal and Sofia 1981). Thus much less mass is spun down, leaving the core spinning at a high rate. Presumably the core is at least weakly coupled to the outer convective envelope so that over some longer time scale the core comes into corotation with the envelope.

Another interesting result of the study of open clusters is the evidence that the rotational evolution of low mass stars is a convergent process. That is, whereas the young clusters (viz., the Pleiades and α Per) show a wide range of rotational velocities at a given spectral type, this range has essentially disappeared for G and K dwarfs by the age of the Hyades (Radick *et al.* 1987). This is consistent with the angular momentum loss scenario usually envisioned in which the interaction of rotation and convection drive a magnetic dynamo. The resulting magnetic field can force corotation of an ionized wind beyond the stellar surface, leading to angular momentum loss. This mechanism naturally incorporates feedback.

4. Rotation and Activity Cycles

The Sun's 11-year activity cycle has been going on now for 300 years or more. Other old solar-type stars almost always show some kind of similar cyclic behavior (Wilson 1978), indicating that the solar cycle is a long-term phenomenon. As such, it would then seem natural that its 11-year period is an inevitable one for a star of the Sun's mass, age, and composition.

But it doesn't work that way. These stellar activity cycles have now been observed at Mount Wilson for more than 20 years, and cycles of various periods have been seen in most of the stars observed – about 100 of them – yet no clear trends emerge. There is no correlation between, for example, the cycle period and the rotation rate, nor for the Rossby number (the ratio of the rotation period to the convective turnover time). Baliunas has pointed out in her Pierce Prize lecture that if the Sun itself were observed for only some arbitrary 20 years out of the last 300, the range of periods one would see would be quite similar to that seen among the Mount Wilson sample. In other words, we can reasonably hope that at present the "signal" has not yet appeared from the "noise," but that it will do so in another 20 years or so. Soderblom and Baliunas (1987) give a fuller discussion of activity cycles. Figure 4 illustrates the data relating cycle period and Rossby number.

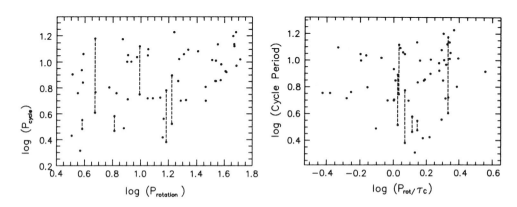

Figure 4. Observed long-term activity cycle periods (in years) versus rotation period and Rossby number (the ratio of the rotation period to the convective turnover time) for the solar-type stars observed at Mount Wilson (adapted from Soderblom and Baliunas 1987).

5. Differential Rotation

Relatively little is known of differential rotation in solar-type stars. For hyperactive stars like those of the RS Canum Venaticorum and BY Draconis types, there is some evidence for long-term variability that is accompanied by poleward migration of spots, in the opposite sense as is seen on the Sun (Vogt 1988). Such stars seem to commonly have enormous spots in their polar regions, which is also quite unlike the Sun.

The careful chromospheric monitoring of solar-type stars done at Mount Wilson has resulted not only in the delineation of the activity cycles just mentioned, but also for some evidence for differential rotation. That evidence (Baliunas et al. 1985) consists of changing or double peaks in the power spectrum of the spectrophotometry from season to season. Some of these cases are illustrated in Figure 5. The period changes seem unambiguous, but there is one curious feature of them: increasing activity in the star is accompanied by longer rotation periods. This is the opposite of the behavior of the Sun.

Ultimately, we must know more about differential rotation in stars like the Sun in order to better understand the dynamo mechanism and ultimately understand magnetic activity on the Sun itself. It is, after all, *differential* rotation which is instrumental in driving the dynamo, not just rotation by itself.

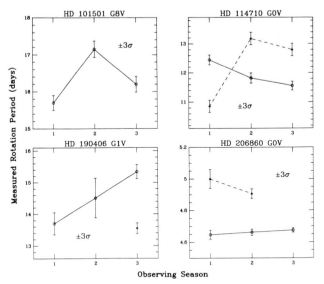

Figure 5. Rotation period changes in four solar-type stars (from Baliunas *et al.* 1985). In two cases the power spectrum is double-lined, indicating two periods.

6. Does the Sun Have a Rapidly Spinning Core?

Helioseismology is neutral on this subject, although it appears unlikely that the solar core could be rotating at a very high rate. Nevertheless, observations of 1 $M_\odot$ that are leaving the main sequence could help one understand this problem because the convective envelopes of such stars get very deep. Thus one might see higher-than-expected rotation rates in such stars as the angular momentum in the core found its way to the surface.

Consider as an example Arcturus, which is a K giant star. It is single, and therefore its mass is not known with precision, but many attempts have been made to estimate its gravity spectroscopically. Those results are very consistent in indicating that Arcturus is no more massive than the Sun (the star is very close so that the radius is accurately known; it is about 20 solar radii). The problem is this: Arcturus has a measurable $v\sin i$ of about 2.5 km s^{-1}. If one collapses Arcturus back to one solar radius, the surface velocity is going to depend on how angular momentum is redistributed during post-main sequence evolution, but the star would clearly leave the main sequence with a rotation rate that is high, ~ 50 km s^{-1}. But the Sun, which is typical of an old solar-type star, rotates at only 2 km s^{-1}. Where did Arcturus get this angular momentum?

There are several possible solutions to this dilemma. First, perhaps all those spectroscopic measurements are wrong and the mass is really closer to 2 $M_\odot$. This would account for the rotation, but it seems unlikely that so many independent results could be so far off. Second, perhaps the $v\sin i$ determinations are wrong, and the real $v\sin i$ is really close to zero. The determinations that were done were done carefully, but it is conceivable that plausible alterations in the model atmosphere for the star could lead to a zero $v\sin i$. This is currently being investigated. Third, perhaps the observations should be taken at face value, and Arcturus has somehow managed to dip into a rapidly rotating core to achieve its present rotation rate. Fourth, the observations are OK, but Arcturus left the main sequence with twice its current mass. Some mass loss after leaving the main sequence is not unheard of, although not quite at that level. None of these solutions is entirely satisfactory. Evolved stars with accurately known masses and in which the components do not interact are rare, but there is an interesting problem here.

7. Rotation and Planets

Stars more massive than about 1.5 $M_\odot$ obey a power-law relation between their mean specific angular momenta and mass: $J/M \propto M^{-2/3}$. Below 1.5 $M_\odot$, observed angular momenta fall about two orders-of-magnitude below this relation, but many have noted that if, for example, the angular momentum in the orbits of the planets is added to that of the Sun, the total falls very near the mean power-law relation. Is it possible that the primary reason for the Sun's slow rotation is that it has deposited its original angular momentum into planets?

One should first note that it is not necessary to invoke such a process to account for the slow rotation of the Sun and similar stars, for we observe the Sun to be losing angular momentum through its wind, and we observe a steady decline of rotation with age among solar-type stars (Soderblom 1983). Furthermore, no satisfactory mechanism has been put forth to transfer the angular momentum from the star to the circumstellar material. This difficulty is compounded because it appears that solar-type stars lose their disks well before reaching the ZAMS (Strom *et al.* 1989). It is at the ZAMS that the very high rotation rates are seen that make one consider unusual angular momentum loss mechanisms, yet deposition of angular momentum from a star into cold, solid matter would be inexplicable.

Our knowledge of the material surrounding young stars is still sufficiently poor that a mechanism involving the transfer of angular momentum to planets or a proto-planetary disk cannot be completely ruled out, but that appears a most unlikely possibility. High-spatial-resolution observations of many stars of different ages by, e.g., the Hubble Space Telescope, may help.

8. Rotation and Lithium Depletion

The gradual depletion of lithium exhibited by stars near 1 $M_\odot$ is surely providing some clues to processes taking place of the convective envelopes of stars. Li is destroyed at a temperature of about 2.4 MK, which is reached somewhat below the base of the solar convective zone. Thus some process such as convective overshooting may reasonably account for the very slow depletion of Li, on a time scale of roughly 1 Gyr.

Observations of Li in solar-type stars are confusing, but the following are particularly pertinent here: First, the Pleiades cluster (age 70 Myr) shows a pronounced spread in rotation rates at any one mass that we have already remarked on, but there is a similar spread in the Li. This spread grows with decreasing mass (Soderblom, Jones, and Stauffer 1990). In general, the ultra-fast rotators have about 1 dex more Li than their slowly rotating siblings, an observation that led Butler *et al.* (1987) to suggest that the spread in rotation and Li were both indicative of an age spread in the Pleiades. It is also relevant that the most

Li-rich Pleiades stars appear to have the primordial Li abundance.

In the Hyades, at an age of 0.6 Gyr, things are quite different. Not only have rotation rates converged at a given mass to a uniform value, but so have the Li abundances (Soderblom *et al.* 1990). The convergence in rotation is expected because of feedback in the angular momentum loss process: stars that rotate faster should have stronger dynamos, hence larger angular momentum loss rates.

It is difficult to see how a star "knows" how to make its Li abundance converge to the "correct" value by the age of the Hyades unless there is some fairly direct connection between angular momentum loss and Li depletion. Such a connection is plausible if rotationally-induced turbulence at the base of the convective envelope is the mechanism responsible for Li depletion. Pinsonneault *et al.* (1989) and Pinsonneault, Kawaler, and Demarque (1989) have incorporated such a mechanism in their models for the evolution of the Sun and solar-type stars. They suggest that the spread in rotation rates seen in young clusters is indicative of a spread in the angular momenta with which the stars formed, rather than representing an age spread. In other words, all the stars of the Pleiades would then be the same age but have been formed with different initial conditions. The recent observations of rotation in IC 2391 by Stauffer *et al.* (1989) would seem to support that hypothesis (see remarks above).

Some other conclusions about Li depletion can be drawn from the available data: Li depletion probably starts at the Zero-Age Main Sequence, not before, at least for $M \gtrsim 0.8 M_\odot$, because such stars in the Pleiades appear to have the primordial abundance. The situation for lower masses is not clear. Second, the depletion rate slows after the age of the Hyades. In other words, the Sun lost the greater part of its initial Li by an age of 0.6 Gyr, and then lost more later, but at a very slow rate.

Work now in progress should help significantly to understand how stars like the Sun have the lithium abundances that they do. Many questions arise: 1) Are all stars in the current epoch formed with the same Li abundance, or is there an intrinsic spread? If there is a spread, is it just between clusters, or within individual clusters? 2) Do stars alter their Li before reaching the ZAMS? As mentioned above, the quantities of Li seen in Pleiades G dwarfs suggests that the answer to this question is "no," at least for stars near $M_\odot$, but other observations (Strom *et al.* 1990) and models (Proffitt and Michaud 1989) suggest otherwise. It is possible, for example, that the high activity levels of pre-main sequence stars lead to autogenesis of Li, which could replace some burned and lead to star-to-star differences. 3) Can a star exhibit Li variability? Very high quality observations of an active young star (Boesgaard 1988) show no evidence for variability, but this cannot be ruled out for the T Tauris. 4) What factors influence Li depletion and the rate at which it occurs? Clearly the star's mass is critical, but how does metallicity enter? How about rotation? 5) If these factors are specified, is there a unique Li abundance implied for a star, or does there remain significant uncertainty because of stochastic effects?

One of us (DS) would like to thank Cerro Tololo Interamerican Observatory for its hospitality during the completion of this review and for help with the illustrations. S. Baliunas also provided helpful information.

9. References

Baliunas, S.L. *et al.* (1983) 'Stellar Rotation in Lower Main-Sequence Stars Measured from Time Variations in H and K Emission-Line Fluxes. II. Detailed Analysis of the 1980 Observing Season Data,' *Astrophys. J.*, **275**, 752-772.

Baliunas, S.L., *et al.* (1985) 'Time-Series Measurements of Chromospheric Ca II H and K Emission in Cool Stars and the Search for Differential Rotation,' *Astrophys. J.*, **294**, 310-325.

Bertout, C. (1989) 'T Tauri Stars: Wild as Dust,' *Ann. Rev. Astron. Astrophys.*, **27**,

Boesgaard, A.M. (1988) in G. Cayrel de Strobel and M. Spite (eds.) *The Impact of Very High Signal-to-Noise Spectroscopy on Stellar Physics,* IAU Symp. 132, Dordrecht, Kluwer, p. 386.

Butler, R.P., Cohen, R.D., Duncan, D.K., and Marcy, G.W. (1987) 'The Pleiades Rapid Rotators: Evidence for an Evolutionary Sequence,' *Astrophys. J. Lett.*, **319**, L19-L22.

Endal, A.S., and Sofia, S. (1981) 'Rotation in Solar-Type Stars. I. Evolutionary Models for the Spin-down of the Sun,' *Astrophys. J.*, **243**, 625-640.

Giampapa, M.S., and Sonett, C.P. (1989) *'The Sun in Time,'* Tucson, Univ. of Arizona Press, in press.

Gilliland, R.L. (1987) 'The Relation of Chromospheric Activity to Convection, Rotation, and Evolution Off the Main Sequence,' *Astrophys. J.*, **299**, 286-294.

Hartmann, L.W., Hewett, R., Stahler, S., and Mathieu, R. 'Rotational and Radial Velocities of T Tauri Stars,' *Astrophys. J.*, **309**, 275-293.

Hartmann, L.W., and Noyes, R.W. (1987) 'Rotation and Magnetic Activity in Main and Sequence Stars,' *Ann. Rev. Astron. Astrophys.*, **25**, 271-301.

Kraft, R.P. (1967) 'Studies of Stellar Rotation. V. The Dependence of Rotation on Age Among Solar-Type Stars,' *Astrophys. J.*, **150**, 551-570.

Noyes, R.W., Hartmann, L.W., Baliunas, S.L., Duncan, D.K., and Vaughan, A.H. (1984) 'Rotation, Convection, and Magnetic Activity in Lower Main-Sequence Stars,' *Astrophys. J.*, **279**, 763-777.

Pinsonneault, M.H., Kawaler, S.D., and Demarque, P. (1989) 'Rotation of Low-Mass Stars: A New Probe of Stellar Evolution,' preprint.

Pinsonneault, M.H., Kalwaler, S.D., Sofia, S., and Demarque, P. (1989) 'Evolutionary Models of the Rotating Sun,' *Astrophys. J.*, **338**, 424-452.

Proffitt, C.R., and Michaud, G. (1989) 'Pre-Main Sequence Depletion of ^{6}Li and ^{7}Li,' *Astrophys. J.*, in press.

Radick, R.R., Thompson, D.T., Lockwood, G.W., Duncan, D.K., and Baggett, W.E. (1987) 'The Activity, Variability, and Rotation of Lower Main-Sequence Hyades Stars,' *Astrophys. J.*, **321**, 459-472.

Skumanich, A. (1972) 'Time Scales for Ca II Emission Decay, Rotational Braking, and Lithium Depletion,' *Astrophys. J.*, **171**, 565-567.

Soderblom, D.R. (1983) 'Rotational Studies of Late-Type Stars. II. Ages of Solar-Type Stars and the Rotational History of the Sun,' *Astrophys. J. Suppl.*, **53**, 1-15.

Soderblom, D.R. (1985) 'A Survey of Chromospheric Emission and Rotation Among Solar-Type Stars in the Solar Neighborhood,' *Astron. J.*, **90**, 2103-2115.

Soderblom, D.R., and Baliunas, S.L. (1987) 'The Sun Among the Stars: What the Stars Indicate About Solar Variability,' in F.R. Stephenson and A.W. Wolfendale (eds.), *Secular Solar and Geomagnetic Variations in the Last 10,000 Years*, NATO

Advanced Research Workshops, Ser. C., v. 236, 25-45.

Soderblom, D.R. Jones, B.F., and Stauffer, J.R. (1990) 'The Evolution of the Lithium Abundances of Solar-Type Stars. II. The Pleiades,' *Astron. J.*, in preparation.

Soderblom, D.R., Oey, M.S., Johnson, D.R.H., and Stone, R.P.S. (1990) 'The Evolution of the Lithium Abundances of Solar-Type Stars. I. The Hyades and Coma Berenices Clusters,' *Astron. J.*, in press.

Stauffer, J.R., and Hartmann, L.W. (1987) 'The Distribution of Rotational Velocities for Low-Mass Stars in the Pleiades,' *Astrophys. J.*, **318**, 337-355.

Stauffer, J.R., Hartmann, L.W., Jones, B.F., and McNamara, B.R. (1989) 'Pre-Main-Sequence Stars in the Young Cluster IC 2391,' *Astrophys. J.*, **342**, 285-292.

Stauffer, J.R., and Soderblom, D.R. (1989) 'The Angular Momentum History of the Sun,' in M. Giampapa and C. Sonett (eds.), *The Sun in Time*, Univ. of Arizona Press, in press.

Strom, K.M., Strom, S.E., Edwards, S., Cabrit, S., and Skrutskie, M.F. (1989) 'Circumstellar Material Associated with Solar-Type Pre-Main Sequence Stars: A Possible Constraint on the Timescale for Planet Building,' *Astron. J.*, **97**, 1451-1470.

Strom, K.M., Wilkin, F.P., Strom, S.E., and Seaman, R.L. (1990) 'Lithium Abundances Among Solar-Type Pre-Main Sequence Stars,' preprint.

Vogt, S.S. (1988) 'Doppler Images of Spotted Late-Type Stars,' in G. Cayrel de Strobel and M. Spite (eds.) *The Impact of Very High Signal-to-Noise Spectroscopy on Stellar Physics*, IAU Symp. 132, Dordrecht, Kluwer, p. 253-272.

Wilson, O.C. (1978) 'Chromospheric Variations in Main-Sequence Stars,' *Astrophys. J.*, **226**, 379-396.

ANGULAR MOMENTUM TRANSPORT AND MAGNETIC FIELDS IN THE SOLAR INTERIOR

H.C. SPRUIT
Max Planck Institut für Physik und Astrophysik
Karl-Schwarzschildstr. 1, 8046 Garching West Germany

ABSTRACT. The possible mechanisms of angular momentum transport in convectively stable regions of a star are reviewed, with emphasis on transport by magnetic torques. The strength and configuration of the field in such layers is quite uncertain, because it is not known if the field can reach a dynamically stable configuration. A lower limit to the field strength is obtained by assuming that the field is always dynamically unstable, and decaying at the (rotation modified) dynamical time scale. The present field in the sun would then be of the order 1G, with poloidal and toroidal components of similar strength. The differential rotation in the core, if due only to the solar wind torque, would be very small for this field strength, and instead would more likely be governed by magnetic coupling to the differential rotation of the convection zone. If small scale hydrodynamic transport mechanisms are present, their properties would also be influenced by a field of this strength.

1. Introduction

The internal rotation of the sun in the convectively stable core is remarkably uniform (Duvall and Harvey, 1984; Duvall, this volume; Brown et al. 1989). Though the lowest degree p-modes (l=1,2,3) indicate a possible rise in rotation speed inside $r = 0.3\ R_\odot$ (Duvall and Harvey, 1984), the statistical significance of this rise is not large. In any case, the present limits on the differential rotation already put rather stringent constraints on proposed mechanisms for angular momentum transport in the core. These mechanisms come in roughly four flavors, (i) turbulent stress associated with *shear instabilities* generated by the differential rotation itself, (ii) *meridional circulations* due to the rotational deformation of the star or the external torque, (iii) angular momentum transport by *waves* generated by the convection zone, and (iv) *magnetic stresses*. These mechanisms are each discussed briefly below. The upshot will be that magnetic fields are arguably the most effective in enforcing uniform rotation, but that they are also the least amenable to quantitative theory at this time.

1.1 A CLUE: THE LITHIUM DEPLETION

The solar Lithium abundance is a factor of 100 below its primordial value. Stars like the sun

G. Berthomieu and M. Cribier (eds.), Inside the Sun, 415–423.
© 1990 *Kluwer Academic Publishers. Printed in the Netherlands.*

show a gradual decline of Lithium abundance with age (Soderblom 1988). This is believed to be due to a slow mixing process in outer parts of the radiative core, which transports the Lithium from the convection zone down until it reaches its burning temperature ($\approx$ 2.8 10^6 K) (Baglin, Morel, Schatzman, 1985; Vauclair, 1988). Such a mixing process is also likely to transport angular momentum and the relative efficiency of mixing and angular momentum transport depends on the mechanism. Hence if the low internal rotation and the Lithium depletion are the consequence of a single process, combining these observations would give a clue on its nature. Small scale turbulence for example is likely to have a turbulent diffusivity and turbulent viscosity of the same order of magnitude. Magnetic fields on the other hand can in principle exert torques without causing any mixing.

2. Hydrodynamics

The solar wind exerts, through its coupling with the atmospheric magnetic fields generated by the convection zone, a torque (Schatzman, 1962) that would spin the sun down on a time scale (Pizzo *et al.* 1984) similar to its present age. The radiative layers below the convection zone will not take part in this spindown unless there is a mechanism that couples them to the convection zone.

2.1 SHEAR INSTABILITIES

The radial gradient of rotation rate that builds up during spindown may become unstable to various shear instabilities. The thermal gradient is strongly stabilizing (the buoyancy frequency being much larger than the rotation rate). On sufficiently small scales the high thermal conductivity due to radiation (low Prandtl number) cancels this stabilizing effect. Another stabilizing effect is the inward increase of the mean weight per particle (Helium inside, Hydrogen outside). This stabilization remains effective also on small scales. An approximate condition for the onset of shear instability taking these factors into account has been derived by Zahn (1974, 1983). In addition, a number of more subtle instabilities exists (for reviews see Zahn, 1983; Spruit, 1987a; Zahn, this volume). A minimum rotation curve for the present sun, assuming the minimum gradient needed to keep at least one of the instabilities acting, was derived by Spruit et al. (1983). It is not compatible with the observed p-mode splittings (Duvall and Harvey 1984), which require significantly lower rotation rates. Thus, the known shear instabilities can not account for the inferred internal torques in the core. Tassoul (1989), referring to conditions in the earth oceans, has argued that the turbulent motions could be highly anisotropic (due to the strong influence of the stabilizing buoyancy force) and that the measured diffusivity for momentum in these cases is significantly higher than that for composition. In the cases referred to, wave motions (internal gravity waves) make up a sigificant fraction of the 'turbulence', and the transport properties of these waves are indeed quite different from isotropic turbulence. To make a valid comparison with the solar interior one has to establish what sources of wave motion exist in the sun, and how their transport properties differ in the sun's low Prandtl number climate.

2.2 Circulations

The deformation of the sun due to its rotation induces a slow meridional circulation, the Eddington-Sweet-Vogt circulation. It is driven by a small difference in heat flux between the pole and the equator (*e.g.* Zahn, 1983). It has been too weak over most of the sun's history to transport much angular momentum. It may perhaps produce a low level of turbulence through Zahn's scenario (see Zahn, this volume), enough to be relevant for the Lithium depletion. Another circulation is an Ekman flow driven by a difference in rotation rate between the convection zone and the interior. It has been studied in detail (Benton and Loper, 1970; Benton and Clark, 1974; Sakurai, 1975; Osaki, 1982). Its effect is negligible except in a very thin layer below the convection zone.

2.3 Waves

The sun is traversed by sound waves excited in the top of the convection zone. In addition, the unsteady nature of the convective flow at the base of the convection zone is expected to generate internal gravity waves in the core, though such waves have not been measured yet. Waves can exert torques by being refracted or absorbed at some distance from the place they were excited. If the wave field is excited asymmetrically with respect to the direction of rotation (eastward moving waves having slightly different amplitudes or propagation speeds than westward waves for example), they can exert a selective spinup or spindown torque at the place where they dissipate. But even a symmetrically excited wave field will exert a 'viscous' stress by exchanging momentum between parts of the shearing flow (the analogy of wave packets with particles holds here). A related problem in astrophysics is that of the tidal interaction in a binary (Zahn, 1977, Goldreich and Nicholson, 1989 and references therein). Press (1981) has stressed that internal gravity waves generated at the base of the convection zone may be quite important. If ω_0 is the frequency of the dominant convective eddy near the base, N a typical buoyancy frequency in the core just below the convection zone, one estimates a wave energy flux into the core which is a fraction ω_0/N of the solar energy flux itself. This is of the order 10^{-3} ($\omega_0 \sim 10^{-6}s^{-1}$, $N \sim 10^{-3}s^{-1}$). Most of this flux is in the form of waves with a very small vertical length scale however, which dissipate by radiative damping within a very short distance below the convection zone. The interesting waves are those generated by the high frequency, large length corner in the spectrum of motions of the convection zone. The energy in this corner is poorly estimated by current theories of the convection zone. Numerical simulations of wave excitation by the convection zone seem in order (*cf.* Hurlburt *et al.* 1986).

3. Magnetic fields

The role of magnetic fields is much harder to quantify than any of the above processes, because a number of fairly basic things are unknown. On the other hand, magnetic fields may well be of central importance, at least for the angular momentum transport in the core.

From observations at the surface, estimates of the field strength in the convection zone can be made. If most of the field resides near the base of the convection zone during the activity cycle (as seems likely for reasons of stability, Moreno Insertis 1986, Schmitt *et al.* 1984, van Ballegooijen, 1982, Spruit and Roberts, 1983), it would plausibly be of the order 10 000 G, covering a depth of the order 1000-10 000 km. The field strength in the core is essentially unknown, with upper limits of the order 10^7 placed by helioseismology measurements (Dziembowski, this volume). Increasing with its strength, the field has effects of decreasing subtlety. Order of magnitudes can be made based on time scales.

3.1 TIME SCALES

The *dynamic* time scale of the sun (the sound crossing time or the the time for a sattelite to orbit once around its surface) is of the order $t_d = (R/g)^{1/2} \sim 2 \ 10^3$ s, where $R = 7 \ 10^{10}$ cm is the solar radius and $g = 2.7 \ 10^4$ cm s^{-2} its surface gravity. Its *rotational* period $T = 2\pi/\Omega$ is 2 10^6 s (25 d). The *spindown* time scale due to the solar wind torque is of the same order as the age of the sun, at any point in time; hence presently of the order $t_s = 10^{17}$ s.

These time scales can be compared with the magnetic time scale, which is the time it takes an Alfvén wave to cross the core:

$$t_B = R/v_A = \sqrt{4\pi\rho}R/B \qquad \sim 1.5 \ 10^{10}/B \quad \text{s} \tag{1}$$

(with a mean density $\rho \sim 1$ for the core). This is the time scale on which the field can communicate forces across the core. It is also the time scale on which the field will evolve, if it is dynamically unstable. If

$$t_B \sim t_d \quad \text{or} \quad B = B_1 \sim (8\pi P)^{1/2} \quad (\sim 10^8 \text{G}) \tag{2}$$

the field has a significant influence on the shape of the star. From upper limits on distortions measured helioseismologically, such high fields can be excluded. When

$$t_B \sim T \quad \text{or} \quad B = B_2 \sim \Omega R\sqrt{4\pi\rho} \quad (\sim 3 \ 10^5 \text{G}) \tag{3}$$

the influence of the field on the p-mode frequencies is of the same order as the rotational splitting of these modes. This kind of field should be measurable helioseismologically. A field of the order

$$t_B \sim (t_s T)^{1/2} \quad \text{or} \quad B = B_3 \sim (\frac{\Omega}{t_s})^{1/2} R\sqrt{4\pi\rho} \quad (\sim 1 \text{G}) \tag{4}$$

is of special importance; it is the field that is just large enough to keep the core corotating with the surface, under the influence of the solar wind torque. When

$$t_B \sim t_s \quad \text{or} \quad B = B_4 \sim \frac{R}{t_s}\sqrt{4\pi\rho} \quad (\sim 10^{-6} \text{G}) \tag{5}$$

the field redistributes angular momentum on a time scale equal to the age of the sun; fields lower than this are therefore not important in the present context. The values in brackets in these expressions are for the present sun.

3.2 STRENGTH OF THE INITIAL FIELD

In view of the limited information observations provide on the present field strength in the core, it is useful to consider the the field as the sun was born, and speculate a bit on how strong it could have been. Star formation, at least in its later stages, proceeds through accreting gas from a disk surrounding the protostar. These disks collect gas from distances very large compared with the size of the star; the field embedded in it at that distance gets compressed as it drifts in with the accreting gas. Near the star, this field is of a single polarity with respect to the disk plane at any point in time (history of the accretion process or statistics of the field in the original source of the gas determining which polarity). Support for the assumption of such a stable (over many orbital periods) field of one polarity comes from the observed jets of protostars, which are most naturally explained by magnetic acceleration (Blandford and Payne, 1982; Lovelace *et al.* 1987; Pudritz and Norman, 1983; Königl, 1989) in such a configuration. If the strength of the field caught in the gas at large distance exceeds a certain plausible minimum, the strength of the field near the protostellar surface is determined only by the local balance between the magnetic stress (pulling fields away from the star) and the accretion. By equating the magnetic curvature (the dominant stress component in this case) with the radial gas pressure gradient yields a field strength of the order

$$B = \alpha^{-1/2}(GM)^{1/4}r^{-5/4}\dot{M}^{1/2} \quad \text{G} \tag{6}$$

where G is the gravitation constant, M the mass of the protostar, α the disk viscosity parameter, r the radial distance, and $\dot{M}$ the mass acretion rate (all cgs). This holds for a radially selfsimilar disk (*cf.* Blandford and Payne, 1982). For a protostar of 1 $M_{\odot}$, accreting $10^{-5}M_{\odot}\text{yr}^{-1}$, $\alpha = 0.1$ this yields several hundred G at the stellar surface $r = 10^{11}$. So an ordered poloidal field of significant strength is accreted onto the star.

3.3 SIMPLE MODELS FOR DIFFERENTIAL ROTATION WITH B-FIELDS

3.3.1 *Axisymmetric field* As a model problem, consider an axisymmetric (about the rotation axis) initially poloidal field $\mathbf{B}_p$ (field lines in meridional planes) (Mestel and Weiss, 1987 and references therein, Rädler, 1986). Assume the initial field to be weak (but larger than B_4 above). The problem is to find how the differential rotation $\Omega(r,\theta)$ evolves in time. An azimuthal field component develops due to stretching of the field lines by the differential rotation. If $\Delta\Omega$ is a measure of the difference in rotation rate occurring along a particular field line, this component initially grows as

$$B_\phi \sim B_p t \Delta\Omega. \tag{7}$$

The model can be calculated only if it is further assumed that the poloidal field component remains unchanged during the evolution. This way one avoids, in particular, the difficult issue of the stability of the field. The magnetic stress due to the distortion of the field then exerts a torque $B_\phi B_p/4\pi$ which increases linearly with the azimuthal displacement $t\Delta\Omega$.

As a result the motion on each of the axisymmetric magnetic surfaces corresponding to a particular field line of $\mathbf{B}_p$ is that of a harmonic oscillator. The period of oscillation is the Alfvén travel time along the field line.

3.3.2 *Phase mixing*.   In the absence of dissipative processes the oscillation would go on indefinitely. Since each of the magnetic surfaces oscillates completely independently of the others however, and the Alfvén travel times vary between the surfaces, the oscillations on the surfaces will get increasingly out of phase with each other. The length scale across the surfaces on which the phase varies decreases as $1/t$, and after a finite time magnetic diffusion (the fastest relevant dissipative mechanism) becomes effective and damps the motion. This process is called damping by phase mixing, and has been studied in the coronal context (Ionson, 1978; Heyvaerts and Priest, 1983). Calculations in the present context have been made by Roxburgh (private communication). The net effect is that an initial differential rotation, not driven by an external torque, will dissipate on a time scale that is some multiple of the initial Alfvén travel time, depending only weakly on the value of the diffusivity.

When an external torque is applied, transients are set up that damp by the same mechanism. After this, a configuration results in which the azimuthal field B_ϕ has grown to just that amplitude at which the magnetic stress transmits the external torque on the same time scale as the star spins down. For the present sun, this means that the average of $(B_r B_\phi)^{1/2}$ would be of the order 1 G. The total field strength $(B_p^2 + B_\phi^2)^{1/2}$ is minimized, for a given $B_p B_\phi$, when B_ϕ and B_p are of the same order of magnitude. Thus the value Eq. (4) is the minimum field strength required to transmit the external torque, and at this minimum the azimuthal and poloidal components are of similar magnitude.

It must be stressed that this is a simplistic picture, since it does not take into account that dynamical, nonaxisymmetric instabilities may drastically alter the field configuration at some stage of the process.

3.3.3 *Closed field lines*   The above implicitly assumes that all of the core is connected with the surface by a field line. This is not necessarily the case, there may be regions (tori) with closed field lines. If axisymmetric, these could rotate at their own rate, hardly affected by the external torques (*e.g.* Mestel and Weiss, 1987).

3.3.4 *Nonaxisymmetric fields*   Nonaxisymmetric cases have effectively been studied only in the kinematic approximation; that is the Lorentz force is ignored altogether. Hence such models address the evolution of the field but not that of the differential rotation. Rädler (1986) (see also Zeldovich *et al.* 1983) shows that the nonaxisymmetric component of the field is 'wrapped up' on the differential rotation time scale, so again the length scale on which the field changes direction decreases as $1/t$, as in the phase mixing process addressed above. Magnetic diffusion then destroys the nonaxisymmetric part of the field on a time scale of a few differential rotations, depending only weakly on the value of the diffusivity, and an axisymmetric field survives.

3.4 Stability of fields

Dynamical instability of the field (that is, ignoring dissipative processes), is a major concern, since it changes the configuration on an Alfvén time scale. For the field strength that just suffices to transmit the present solar wind torque to the core (B_3 in section 3.1), this time scale is or the order 10^4 yr, much shorter than the time scale on which we want the field to to its work (but see the modification to the instability time scale by rotation, below). Are fields likely to be dynamically unstable? It is known (Tayler, 1980, van Assche *et al.* 1982) that all purely poloidal fields in stars, as well as all purely toroidal fields are dynamically unstable. Stable fields at least require a combination of toroidal and poloidal components of similar strength (Mestel, 1984). No example of a dynamically stable equilibrium field in a star is known (but they are not so easy to construct). One might be tempted to guess that all fields are dynamically unstable in stars. This is probably incorrect though, since (Moffat, private communication) in the absence of diffusion there is a conservation property that limits the amount of energy an equilibrium can release by an instability. If **A** is the vector potential of **B**, the integral of the magnetic helicity **B** · **A** over the field configuration is conserved for frozen in fields (Woltjer, 1958; see Berger, 1986). In mixed poloidal- toroidal fields the net helicity is generically nonzero. Since zero magnetic energy means zero net helicity, there is for each mixed poloidal- toroidal field a nonzero minimum energy configuration. Whether such configurations can develop in a star, and whether they will remain stable when magnetic diffusion is included is another question (Mestel, 1984 and references therein).

3.5 A LOWER LIMIT TO THE PRESENT FIELD STRENGTH IN THE CORE

The field strengths to be found at the bottom line of this text are such that the Alfvén travel time is much larger than the rotation period. In this circumstance, the coriolis force has a strong influence on the growth rate of an instability. Motions parallel to the rotation axis are not affected by the coriolis force, but unstable displacements can generally not be constructed out of such motions alone. As a result, the inclusion of rotation usually reduces the growth rates by a factor $t_B \Omega$ compared with the nonrotating case (see Pitts and Tayler, 1985 for examples). Dynamically unstable fields therefore reconfigure on a timescale

$$t_{inst} = t_B^2 \Omega. \qquad (\Omega t_B \gg 1). \qquad (8)$$

With this, we can do a consistency check on the initial field strength assumed in section 3.2. Inserting B from Eq. (6) we find for the initial instability time scale

$$t_{inst\ 0} \sim \alpha \frac{\Omega}{\Omega_m} \frac{M}{\dot{M}} \qquad (9)$$

where Ω_m is the maximum rotation rate of the protostar, the Kepler rotation rate at its surface. Since the field is built up by accretion on the time scale $t_{acc} = M/\dot{M}$, the value given by Eq. (6) can be sustained against possible dynamic instability if the rotation rate is near its maximum. The stellar wind torque on the young star rapidly reduces its rotation rate, and the growth rate of dynamic instabilities increases. We can now estimate a lower limit on the field strength in the star by assuming that the field *always* is dynamically unstable. Of course, the field may well get stuck at some point in time into a stable configuration, but then the field, limited now by slower diffusive processes, will only be higher than we are estimating here. If the instability time scale is shorter than the spindown time scale, the field will get smaller; if it is longer, the rotation rate decreases until by Eq. (8) the instability time scale has decreased to match the spindown time scale again (Spruit, 1987b). Thus for a wide range of conditions, the field attains a typical value that is largely independent of the initial conditions, and decreases in time such that its dynamical instability time is the same as the spindown time of the star. The field value that follows from the condition $t_{inst} = t_s$ is just B_3 given in section 3.1. As discussed, it is the field strength that is just sufficient to transmit the stellar wind torque to the core, and the azimuthal field component is of the same order as the poloidal component.

At this strength, the magnetic energy density in the field is of the same order, for the present sun, as the kinetic energy in the turbulent motions envisaged in some of the hydrodynamic transport schemes (such as Zahn's scenario). Since our estimate is a lower limit, this shows it may not be realistic to neglect the magnetic field, even in purely hydrodynamic transport schemes.

For the field Eq. (4), the time scale for transmitting torques through the core is small compared with the spindown time (but long compared with the rotation period), by a factor $\sim 10^6$. Hence the level of differential rotation in the core, if due only to the processes described, should be quite small. (But there are obvious other processes, such as magnetic coupling to the differentially rotating convection zone, that will enforce some level of differential rotation).

Concluding, by assuming that the primordial field is dynamically unstable at all times during the spindown of the star, we have estimated a mimimum field strength to be expected in the core of the present sun, of the order of 1G.

References

Baglin, A., Morel, P.J., Schatzman, E. 1985, *Astron. Astrophys.*, **149**, 309

Berger, M.A. 1986, *Geophys. Astrophys. Fluid Dyn.*, **34**, 256

Blandford, R.D., Payne, D.G. 1982, *Mon. Not. R. astron. Soc.*, **199**, 883

Benton, E.R., Loper, D.E. 1970, *J. Fluid Mech.*, **43**, 75

Benton, E.R., Clark, D.E. 1974, *Ann. Rev. Fluid Mech.* **6**, 257

Brown, T.M. Christensen-Dalsgaard, J., Dziembowski, W.A., Goode, P., Gough, D.O., Morrow, C.A. 1989. *Astrophys., J.,* **343** 526

Duvall, T.L. Jr., Harvey, J.W. 1984, *Nature,* **310**, 19

Goldreich, P., Nicholson, P. 1989, preprint

Heyvaerts, J., Priest, E.R. 1983, *Astron. Astrophys.*, **117**, 220

Hurlburt, N.E., Toomre J., Massaguer, J.M. 1986, *Astrophys., J.,* **311**, 563

Ionson, J.A. 1978, *Astrophys., J.,* **226**, 650

Königl, A. 1989, *Astrophys., J.,* **342**, 208

Lovelace, R.V.E., Wang, J.C.L., Sulkanen, M.E. 1987, *Astrophys., J.,* **315**, 504

Mestel, L. 1984, *Astron. Nachr.* **305**, 301

Mestel, L. Weiss, N.O. 1987, *Mon. Not. R. astron. Soc.,* **226**, 123

Moreno Insertis, F. 1986, *Astron. Astrophys.*, **166**, 291

Osaki, Y. 1982, *Publ. Astr. Soc. Japan,* **34**, 257

Pitts, E., Tayler, R.J. 1985, *Mon. Not. R. astron. Soc.,* **216**, 139

Pizzo, V., Schwenn, R., Marsch, E., Rosenbauer, H., Mühlhäuser, K.-H., Neubauer, F.M. 1983, *Astrophys., J.,* **271**, 335

Press, W.H. 1981. *Astrophys., J.,* **245**, 111

Pudritz, R.E., Norman, C.A. 1983, *Astrophys., J.,* **274**, 677

Rädler, K.-H., 1986, preprint

Sakurai, T. 1975, *Mon. Not. R. astron. Soc.,* **171**, 35

Schatzman, E., 1962, *Ann. Astrophys.,* **222**, 317

Schmitt, J.H.M.M., Rosner, R., Bohn, H.U. 1984, *Astrophys., J.,* **282** 316

Soderblom, D. 1988 in *The Impact of very high S/N etc.,* eds. G. Cayrel de Strobel and M. Spite (IAU Symp. 132), Kluwer, Dordrecht, p381

Spruit, H.C. 1987a, in it Physical Processes on Comets, Stars and Active Galaxies, eds. W. Hillebrandt, E. Meyer-Hofmeister and H.-C. Thomas, Springer, Heidelberg, p78

Spruit, H.C., 1987b, in *The Internal Solar Angular Velocity,* eds. B.R. Durney and S. Sofia, Reidel, Dordrecht, p185

Spruit, H.C., Roberts, B. 1983, *Nature,* **304**, 401

Spruit, H.C., Knobloch, E., Roxburgh, I.W. 1983, *Nature,* **304**, 320

Tassoul, J.-L. 1989, *Astron. Astrophys.,*

Tayler, R.J., 1980, *Mon. Not. R. astron. Soc.,* **191**, 151

Vauclair, S. 1988, *Astrophys., J.,* **335**, 971

van Assche, W., Tayler, R.J. Goosens, M. 1982, *Astron. Astrophys.,* **109**, 166

van Ballegooijen, A.A. 1982. *Astron. Astrophys.,* **113**, 99

Woltjer, L. 1958, *Proc. Natl. Acad. Sci. USA* 44, 489

Zahn, J.-P. 1974, in *Stellar Instability and Evolution,* eds. P. Ledoux, A. Noels and R.W. Rogers, Reidel, Dordrecht, p185

Zahn, J.-P. 1983, in Astrophysical Processes in upper Main Sequence Stars, Geneva Observatory, Switzerland, p225

Zahn, J.-P. 1977, *Astron. Astrophys.,* **57**, 383

Zeldovich, Ya. B., Ruzmaikin, A.A., Sokoloff, D.D. 1983, *Magnetic Fields in Astrophysics,* Gordon And Breach, N.Y.

THEORY OF TRANSPORT PROCESSES

J.-P. ZAHN
Observatoire Midi-Pyrénées
14, avenue E. Belin
31400 Toulouse
France
and

Columbia University
New York, N.Y. 10027
U.S.A.

ABSTRACT. This review focuses on the transport of matter and angular momentum in the radiative zones of stellar interiors. The two main causes of such transport are the convective overshooting in the vicinity of convection zones, and the slow motions (meridional circulation and turbulence) due to the rotation of the star. In addition, momentum can be transfered through waves (generated by the motions above) and through magnetic stresses. The characteristics of those processes are examined, with special emphasis on turbulent diffusion.

1.   Introduction

Transport processes play a key role in the structure and the evolution of a star: above all, they deliver the energy which is produced in the deep interior to the surface, from where it is radiated in into space. Quite naturally, the two modes of transport which are responsible for this, namely radiative transfer and thermal convection, are considered by the astrophysicists as two major subjects, and they occupy a prominent place both in research and in teaching.

The purpose of this review to examine some other processes which contribute, in stellar radiation zones, to the *transport of matter and angular momentum*. It is not necessary to explicit here in detail the motivations for such particular interest: they have been outlined in the introductory talk by E. Schatzman. Let us just recall that some chemical abundances at the surface of the Sun can only be interpreted by invoking some transport of matter in the radiative interior (Schatzman 1969, 1977; Schatzman and Maeder 1981). And the internal rotation of the Sun, which is being unveiled by the thrilling results of helioseismology, is also the result of the transport of angular momentum within the radiation zone.

Let us begin by a brief inventory of the transport processes that are likely to occur. One can distinguish two classes among them: the microscopic processes, which operate at the particle level, and the macroscopic ones, which involve motions on larger scales.

G. Berthomieu and M. Cribier (eds.), Inside the Sun, 425–436.
© 1990 *Kluwer Academic Publishers. Printed in the Netherlands.*

1.1. MICROSCOPIC PROCESSES

Microscopic diffusion is caused by the collisions between particles, and by their interactions with the radiation field. In a stellar interior, the ambiant medium is non uniform, due to the stratification of density and temperature; therefore, the interactions between particles of different species tend to separate them. Similarly, the slight anisotropy of the radiation field also to produces a chemical composition gradient.

The theory of such processes has been established by Chapman and Cowling (1970), and their application to stellar envelopes has been developed mainly by Michaud (1970) and his co-workers; an excellent review on the subject has been written by Vauclair and Vauclair (1982).

Thermal conduction in a gas is also due to particle collisions, through which thermal kinetic energy is exchanged; it is very similar to microscopic diffusion.

Radiative transfer is by far the most powerful transport process in a stellar radiation zone, at least as long as matter is non degenerate. Except in the vicinity of the surface, stellar matter is optically thick; radiation and matter interact so strongly that they achieve the so-called local thermodynamic equilibrium (LTE). Then the radiation field becomes very nearly isotropic, and radiative transfer can be treated likewise as a diffusive process.

The governing equation for all those diffusion processes is

$$\frac{\partial c}{\partial t} = \nabla \cdot D \nabla c + \{S\} \tag{1}$$

where c is the quantity being diffused (energy, concentration of chemical species, etc.) and $\{S\}$ the sources and sinks of that quantity. Various notations are used for the diffusion coefficient, in order to distinguish between the different processes when they compete; here it is simply labeled D. Let us recall that in LTE, the radiative diffusivity is related to the opacity κ through

$$K = \frac{4acT^3}{3C_P \rho^2 \kappa},$$

with the usual notations for the physical constants and for the temperature, the density and the specific heat.

Viscous friction also can be considered as a diffusive process, in that case of momentum, the diffusivity then being the viscosity ν. One of the main characteristics of stellar interiors is the great disparity between that diffusion of momentum and the radiative transport of energy: the ratio $Pr = \nu/K$, which is called the *Prandtl number*, is extremely small (typically 10^{-6} or less). In contrast, the fluids which are the most familiar to us, such as air or water, have a Prandtl number of order unity.

1.2. MACROSCOPIC PROCESSES

Advection through large scale circulations is the simplest of macroscopic processes, and we have many examples of it around us, in Nature and in our household (central heating!). If $\mathbf{V}$ represents the velocity field, the governing equation for the advection of a scalar quantity c just expresses the conservation of that quantity:

$$\frac{\partial c}{\partial t} + \mathbf{V} \cdot \nabla c = \{S\}, \tag{2}$$

with the same notations as above.

In radiative interiors, two causes of large scale circulations have been identified. One has been extensively studied: it is the thermal imbalance of a rotating star. When submitted to the centrifugal force, a star can no longer achieve radiative equilibrium, and this induces a meridional advection of heat which is known as the *Eddington-Sweet* circulation (Eddington 1925, Vogt 1925, Sweet 1950, Mestel 1953, Kippenhahn 1958, McDonald 1972, Tassoul and Tassoul 1982, etc.)

Much less attention has been paid so far by the astrophysicists to the so-called *Ekman circulation*, which is generated in the boundary layer connecting two regions rotating at different speeds (Ekman 1905; see also Pedlosky 1979). Such a circulation is likely to occur at the bottom of the solar convection zone, in which a strong differential rotation is maintained by the convective motions, whereas the radiation zone below appears to rotate much more uniformly (Dziembowski *et al.* 1989, Brown *et al.* 1989). The penetration of such Ekman circulation into the convection zone has already been considered by Bretherton and Spiegel (1968).

One of the consequences of meridional circulation is the advection of angular momentum, which modifies the internal rotation rate of the star, which in turn alters the large scale motion. A complete, physically consistent description of this complex feed back is still lacking; it would require to also include the transport processes which will be discussed next.

Turbulent diffusion. When the velocity field has small scale, time-dependent component **u**, which may be considered as *turbulent* (in an intuitive sense – a proper definition of this term is beyond our scope here), the advection equation (2) can be expanded in the following way, by taking suitable averages:

$$\frac{\partial c}{\partial t} = \nabla \cdot D_t \nabla c + \{\text{higher order terms}\} + \{S\} . \tag{3}$$

For the justification of this procedure we refer to Knobloch (1978); in many instances, it suffices to retain the first term of this formal expansion, which is the second order operator explicited here. The turbulent diffusivity D_t can be deduced from the properties of **u**, as we shall do later. We shall devote the next section to this transport process, since it is likely to play a key role in stellar radiation zones.

In stars, thermal convection is the most efficient form of turbulent diffusion. It is not confined to the convection zone (defined as the region of nearly adiabatic stratification), since the turbulent motions penetrate somewhat into the adjacent layers. This convective overshooting is discussed by J. Massaguer (this volume).

Transport through waves. Due to their oscillatory nature, waves do not contribute much to the transport of matter, unless they reach a rather large, finite amplitude (Weiss and Knobloch 1989). But they are very efficient in transporting energy and momentum, and therefore they too deserve a more detailed examination (in section 3).

428

2. Turbulent diffusion

2.1. ESTIMATE OF THE TURBULENT DIFFUSIVITY

We have already stated that, to first approximation, the transport in a turbulent medium of a scalar quantity (such as temperature, chemical species, etc.) can be described as a diffusion process. For simplicity, we assume here the velocity field to be isotropic enough so that the turbulent diffusivity reduces to a scalar, D_t. This coefficient is determined by the characteristics of the turbulence:

$$D_t = \frac{1}{3} u\ell, \tag{4}$$

where, to first approximation, u is the velocity and ℓ is the size of the largest eddies (or, equivalently, the mean free path of those eddies, or mixing length).

It may seem a simple matter to estimate those quantities u and ℓ, but the prescriptions which are used for that purpose vary from case to case.

For the mixing length ℓ, one can often take the dimension of the turbulent region, when the largest eddies are of that size. A more refined recipe, due to Prandtl, is to choose the distance to the nearest boundary. But in some instances the kinetic energy is injected at a scale which is smaller than that of the whole unstable region; the most vigorous eddies, which contribute most to the turbulent transport, are then of intermediate size.

In the strong stratification of a stellar convection zone, it is not clear whether the eddies extend (or travel) over large vertical distances. It is customary then to follow E. Vitense (1953), and to relate the mixing length to the local pressure scale height, $\ell = \alpha H_P$, the coefficient α being calibrated by comparing the numerical models with the observations.

To estimate the velocity u, various prescriptions are available. Let us take for example the turbulent shear flow: u is then of the order of the variation ΔU of the mean flow speed over the considered domain. The simplest case is when the differential velocity is maintained by some external force at a constant level ΔU, as it occurs in Couette flows. But often that ΔU is governed by the strength of the turbulence, which in turn is determined by another condition, such as the momentum flux or the heat flux which has to be carried. Sometimes it is possible to estimate the rate ε_t at which kinetic energy is injected into the turbulent motions at the scale ℓ, from where it cascades down to smaller scales, to be dissipated there through viscous friction. In this case, if one further assumes that the turbulent eddies obey the Kolmogorov law (see Landau and Lifshitz 1987), u can be derived from

$$\varepsilon_t \approx u^3/\ell. \tag{5}$$

An alternate approach is to estimate u through the growth rate $1/\tau$ of the considered instability :

$$u \approx \ell/\tau ; \tag{6}$$

this conditions implies that the growth of the instability saturates when the non-linear term of the momentum equation, $u\nabla u$, reaches the same level as the linear terms.

The growth rate $1/\tau$ is often approximated by the growth rate derived from the linear perturbation theory, for lack of something better. But it is more correct to calculate it, whenever it is feasible, in the non-linear regime which is attained by the instability. This is done for instance in the classical mixing-length theory for convection, where the velocity is estimated by imagining an eddy accelerated in the actual superadiabatic density gradient:

$$u^2 = C_1 \frac{g}{H_P} (\nabla_{rad} - \nabla_{ad}) \ell^2 .$$

The expression for the (prescribed) convective flux is then

$$F_c = C_2 \, \rho u^3 \frac{H_P}{\ell} ; \qquad (7)$$

the constants C_1 and C_2 are numerical coefficients of order 1, depending on phenomenological details of that approach. Notice that one retrieves here the same ratio u^3/ℓ as in eq. 5 above: it is equivalent to specify the convective flux or the kinetic energy injection rate.

2.2. TURBULENT DIFFUSION IN STELLAR RADIATION ZONES

A radiation zone is a region of stable thermal stratification; therefore, the convective instability that causes the turbulent motions which are suspected there cannot be ascribed to the convective instability (except in the vicinity of a convection zone, from which there is some penetration into the stable layers).

What may then be the cause of such turbulence? It turns out that many instabilities are likely to occur in a radiation zone; they have been described in several reviews (see for instance Knobloch and Spruit 1982, 1983, or Zahn 1983). Such instabilities convert into turbulent kinetic energy other forms of energy which is stored in the star (thermal energy, gravitational, magnetic, kinetic energy of large scale flows, etc.)

These instabilities compete with each other, and it is the strongest of them that will control the turbulent transport in the radiation zone. In all likelihood, the most powerful are the *shear instabilities due to differential rotation*, since they are of dynamical nature and therefore have the fastest growth-rate.

Some amount of differential rotation is always present in a stellar radiation zone. It is due to several causes: contraction or expansion of the star while it evolves, angular momentum loss through a wind, coupling with a differentially rotating convection zone, meridional circulation, tidal braking in a binary star.

Let us examine those shear instabilities in more detail. For the discussion, it is convenient to distinguish between vertical differential rotation, in which the angular velocity varies with depth, and horizontal differential rotation, where it varies with latitude.

2.2.1. *Vertical differential rotation.* A typical velocity profile likely to be encountered in the vertical direction, for instance below the convection zone of a solar-type star, is the so-called *mixing layer*: the angular velocity adjusts from a constant value Ω above to a higher value $\Omega + \Delta\Omega$ below, within a layer of thickness L. Since the profile presents an inflexion point, the flow is unstable to infinitesimal perturbations (Rayleigh 1880); the instability is of dynamical type, and its growth rate is of order

$$\tau^{-1} \approx \Delta\Omega s/L ,$$

s being the distance to the rotation axis (the viscosity has been neglected, for it plays here a negligible role).

But in a stratified medium such as stellar interior, the buoyancy force acts to hinder the instability. If there were no radiative damping, the instability would be suppressed for

$$\Delta\Omega \, s/L \; < \; N \; , \tag{8}$$

N being the buoyancy frequency

$$N^2 \; = \; \frac{g}{H_P}(\nabla_{ad} - \nabla_{rad}) \; .$$

However, radiative damping smoothes out the temperature differences, and therefore it lowers the threshold of the instability, which occurs then as soon as

$$(\Delta\Omega \, R/L)^2 \; > \; N^2 \, Pr \, R_c \; . \tag{9}$$

In this criterion R_c is the critical Reynolds number associated with that profile (it is of order 100), and Pr is the Prandtl number (ratio between the viscosity ν and the thermal diffusivity K), which in a radiation zone is of order 10^{-6} or less.

This condition is valid as long as the star has a uniform chemical composition. If there is a (stable) gradient of molecular weight, the buoyancy force is only partly weakened through radiative damping, and one recovers the original criterion (eq. 8), the buoyancy frequency being then reduced to

$$(N_\mu)^2 \; = \; \frac{g}{H_P}\frac{d\ln\mu}{d\ln P} \; . \tag{10}$$

In all cases where the instability occurs, the largest turbulent eddies have a size of order L and a velocity of order $\Delta\Omega \, s$, and therefore the turbulent diffusivity is of order

$$D_t \; \approx \; \Delta\Omega \, sL \; . \tag{11}$$

Those two criteria (eq. 9 and 10) can be extended to smoother velocity profiles, which do not exhibit an inflexion point, although the instability is then of different nature as we shall see next; all it needs is to replace the finite difference $(\Delta\Omega/L)$ by the derivative $\partial\Omega/\partial r$.

2.2.2. *Horizontal differential rotation.* When the rotation rate varies with latitude, the situation presents two main differences with the vertical differential rotation that we have just considered.

First, the instability of a horizontal shear flow cannot be hindered by a stratification, since the buoyancy force acts only in the vertical direction. And it can been shown that the Coriolis force has no influence either on the instability criterion (see Tritton and Davies 1981).

Second, the horizontal velocity profiles that are likely to occur have in general no inflexion point, and therefore they are stable against infinitesimal perturbations. But they are liable to finite amplitude instabilities, as observed in Nature and in the laboratory, provided the Reynolds number reaches some critical value. Work is in progress to determine theoretically the threshold of that instability (see for instance Lerner and Knobloch 1988).

When a horizontal shear flow becomes unstable, it generates eddies which have the same vorticity as the mean flow, and which are therefore horizontal, and two-dimensional. Those billows, in turn, undergo a three-dimensional instability, provided the vertical motions are not hindered by some restoring force.

One such force is the *Coriolis force* in the rotating star, which will dominate the dynamics of all eddies for which (see Hopfinger *et al.* 1982)

$$u/\ell < \Omega \; ; \tag{12}$$

those will remain horizontal, and they will not contribute to turbulent diffusion in the vertical direction. But the smallest eddies are not sensitive to the Coriolis force; they are three-dimensional, and they obey Kolmogorov's law (eq. 5). Their distribution begins at the scale which verifies both

$$u'/\ell' \approx \Omega \quad \text{and} \quad (u')^3/\ell' \approx \varepsilon_t \, ,$$

and therefore the vertical turbulent diffusivity is given by

$$D_t \approx u' \ell \approx \varepsilon_t/\Omega^2 \, . \tag{13}$$

The other restoring force is the *buoyancy*, but it only operates when there is a vertical gradient of molecular weight, since here again the temperature differences are smoothed out by radiative damping. Such a μ-gradient will inhibit three-dimensional turbulence for the eddies whose turn-over rate is less than the residual buoyancy frequency (eq. 10)

$$u/\ell < N_\mu \, , \tag{14}$$

and it will suppress it entirely when (see Zahn 1983)

$$(N_\mu)^2 > \varepsilon_t/\nu \, . \tag{15}$$

A rather small gradient of molecular weight thus suffices to prevent turbulent diffusion in the vertical direction (for instance, that due to the varying composition of ^{3}He in the Sun). However, such a "μ-barrier" (as Mestel called it in an other context) will still allow momentum transport through gravity waves, as we shall see later on.

2.3. TURBULENT DIFFUSION ASSOCIATED WITH MERIDIONAL CIRCULATION

We have seen that the strength of the turbulent motions is determined by the energy injection rate ε_t. As an illustrative example, we shall estimate this rate in the case where the turbulent motions are caused by a meridional circulation.

We have already mentioned that radiative equilibrium can no longer be achieved in a rotating star. From Von Zeipel's famous paradox (1924), Eddington (1925) and Vogt (1925) drew the conclusion that the star must be the seat of a large scale meridional circulation, such that

$$\rho T \mathbf{U} \cdot \nabla S = -\nabla \cdot F \, , \tag{16}$$

with S being the specific entropy and $\mathbf{F}$ the radiation flux. The procedure to determine the velocity $\mathbf{U}$ was established by Sweet (1950); the circulation time may be expressed in terms of the Kelvin Helmholtz time $t_{KH} = GM^2/LR$ and of the oblateness due to the centrifugal force

$$t_{circ} \approx t_{KH} \left(\frac{\Omega^2 R^3}{GM} \right)^{-1} \, . \tag{17}$$

That meridional circulation advects angular momentum, whose conservation requires

$$\frac{\partial}{\partial t} s^2 \Omega + \mathbf{U} \cdot \nabla \left(s^2 \Omega \right) = \Gamma, \tag{18}$$

where Γ is the torque exerted per unit mass by the turbulent motions (as above, s is the distance to the rotation axis). Likewise, we may express the rate of variation of the rotational energy:

$$\frac{\partial}{\partial t} \frac{1}{2} \left(s \Omega \right)^2 + \Omega \, \mathbf{U} \cdot \nabla \left(s^2 \Omega \right) = \Omega \, \Gamma. \tag{19}$$

In a stationary state, the advection term balances the right hand side, which is the work done by the turbulent torque, and therefore the energy injection rate into the turbulence. Splitting the angular velocity into its mean and fluctuating parts (over a level surface) $\Omega(r) + \delta\Omega(r,\theta)$, and subtracting the kinetic energy of the mean flow from that which is advected into the layer, we obtain the following expression for the turbulent energy input, averaged over a horizontal layer (Zahn 1987):

$$\epsilon_t(r) = - \int_0^1 \delta\Omega(r,\theta) \, \mathbf{U} \cdot \nabla \left(\Omega_0(r) \sin^2\theta \right) d(\cos\theta). \tag{20}$$

Let us stress that this expression is valid for any type of meridional circulation, either the Eddington-Sweet or the Ekman circulation. But from now we must specificy which of those we are considering, since we have to provide the value of the large scale velocity.

When dealing with the Eddington-Sweet circulation, a crude approximation of that energy generation rate ε_t is

$$\varepsilon_t \approx C \frac{L}{M} \left(\frac{\Omega^2 R^3}{GM} \right)^2 ,$$

but for most applications it is necessary to use the full equation 20, with suitable expressions of $\mathbf{U}$ and $\delta\Omega$.

One knows how to derive the meridional velocity from the rotation law $\Omega(r,\theta)$; for instance, when the star is homogeneous and the departures from solid rotation are not too large, the vertical component of $\mathbf{U}$ takes the simple form $U_r(r,\theta) = -U_2(r)P_2(\cos\theta)$, $U_2(r)$ being a positive function of r (McDonald 1972).

The difficulty comes from the poor knowledge of the differential rotation $\delta\Omega(r,\theta)$, which is governed by the horizontal transport of angular momentum, through the two-dimensional eddies mentioned above, and also through the internal waves. The modelization of this transport is still an unsolved problem. In the meanwhile, all we can do is to assert $\delta\Omega = -C \, \Omega_0 P_2(\cos\theta)$, introducing a coefficient C to be calibrated with the observations, much as the mixing-length parameter α used to model stellar convection zones.

This type of turbulent diffusion appears to play a major role in the depletion of Li in solar-like stars (Baglin et al. 1985, Vauclair 1988). It may also affect the abundance of Li observed in the old halo stars (Vauclair 1988).

3. Transport by waves

3.1. GENERAL PROPERTIES

Waves occur in any continuous medium which is in stable equilibrium, due to the very existence of a restoring force. The waves encountered in a stellar radiation zone are of several types, depending on the forces that come into play:

Restoring force	Type of waves
pressure	acoustic, p-modes
buoyancy	internal, g-modes
Coriolis	inertial
id. with curvature	Rossby, toroidal
Laplace-Lorentz	Alfvén

This classification assumes that only one restoring force is operating, but often there are two, or more, giving rise to mixed modes (magneto-acoustic waves, for example).

A familiar property of waves is that they transport energy, but they also transport momentum, and thus angular momentum. In fact, the quantity which is conserved in a travelling wave is the action E/σ

$$\frac{\partial}{\partial t}\left(\frac{E}{\sigma}\right) + \nabla \cdot \left\{(\mathbf{U} + \mathbf{c}_g)\left(\frac{E}{\sigma}\right)\right\} = \{S\} \tag{21}$$

(E: energy density; σ: local frequency; $\mathbf{U}$: velocity of the ambient medium; $\mathbf{c}_g$: group velocity).

To study the transport through wave motion, one has to follow the whole history of the wave: How is it generated? How does it propagate? Where is it damped? Is it destroyed? (In other words: where does the wave release what it transports?)

Here we shall focus only on the two first families of waves; for lack of results bearing on stellar interiors, we skip the waves due to the Coriolis force. The magnetic coupling through Alfvén waves will be treated by H. Spruit (this volume).

3.2. WAVE TRANSPORT IN A STELLAR RADIATION ZONE

3.2.1. Acoustic waves. The most powerful source for wave production is thermal convection: as observed on the Sun, a substantial fraction of the turbulent convective energy is converted into acoustic waves. And some of those waves penetrate into the deep interior, as demonstrated by the frequencies and the wave-lengths that are detected by the helioseismological techniques. However they are only slightly damped in the radiative core, due to their short period, and therefore they deposit there very little of the energy (and momentum) which is carried by them. One still lacks of a quantitative treatment of this transport, to check whether it is indeed negligible.

3.2.2. *Gravity waves.* It is expected that internal gravity waves are excited in the overshoot region just below the convection zone, although there is no observational proof of this yet. The periods and wave-lengths of such waves may be infered from the kinematic properties of the lower convection zone (which are model dependent).

It turns out that those internal waves, due to their long period, are severely damped through radiative dissipation (Press 1981); hence they do not penetrate far into the radiation zone, although they may be very effective in coupling the rotation of that region with that of the convection zone.

However, those long period waves will survive in spite of the strong radiative damping if there is a vertical gradient of molecular weight: although the temperature fluctuations are smoothed out, buoyancy will then still operate on the density fluctuations due to the chemical inhomogeneities. In other words, such a μ-gradient can be the site for *isothermal gravity waves.*

Internal waves are likely to be generated also by the *shear instabilities* which have been described above, especially when there is a gradient of molecular weight. This is suggested in particular by an experience performed by Stillinger *et al.* (1983), who studied the turbulence induced by a grid in a stably stratified fluid. They found that the dynamics of the large scales is dominated by the buoyancy force and that they take the form of internal waves; only the smaller scales, for which $u/\ell > N_\mu$ (see eq. 14), participate in what they call "active turbulence", which is responsible for the vertical diffusion of matter. Those internal waves may thus carry energy and momentum through a region which is impermeable to turbulent diffusion; in other words, *angular momentum is transported, but chemicals are not.*

Another interesting property of gravity waves has been discussed by Goldreich and Nicholson (1989) in the context of tidal braking in binary stars: in a differentially rotating star, those waves may dump their energy and momentum at the corotation point. To be specific, the frequency σ of the wave, which is ω in the rest frame, is a function of depth in the local frame, which rotates at the depth-dependent velocity $\Omega(r)$:

$$\sigma = m\left(\Omega(r) - \omega\right), \tag{22}$$

where m is the azimuthal wave-number. At corotation, both the phase velocity σ and the group velocity vanish, and the vertical wavenumber k_r, which obeys

$$\sigma k_r = Nm/r,$$

becomes infinitely large. Thus the wave either breaks there, or it is strongly damped.

4. Conclusion

To summarize, some turbulent diffusion is likely to occur in the radiation zone of any star; it is caused by the instabilities due to the ever present differential rotation. But this type of transport operates in the vertical direction only when the composition gradient is not too large. Presently, the weak point of the theory is the poor knowledge of the horizontal transport of angular momentum, which determines the rate at which kinetic energy is transferred from the differential rotation of the star into the turbulent motions.

On the other hand, transport by waves is a promising mechanism to explain the redistribution of angular momentum within a star, without affecting its chemical composition.

Unfortunately, very little is known so far about the efficiency of this process, although it is clear that it too will be linked with the energy generation rate characterizing the turbulent motions which are responsible for the production of such waves.

Finally, it appears that two regions of a star like the Sun play a key role in those transport processes: the layers located just below the convection zone, where nearly all such processes are likely to operate and to compete (convective overshoot, meridional circulation, turbulent diffusion, waves), and the upper slope of the ^{3}He abundance, which inhibits the vertical diffusion of matter.

References

Bretherton, F.P. and Spiegel, E.A. 1968, *Astrophys. J.* **153**, 277

Brown, T.M., Christensen-Dalsgaard, J., Dziembowski, W.A., Goode, P., Gough, D.O. and Morrow, C.A. 1989 *Astrophys. J.* **343**, 526

Chapman, S. and Cowling, T.G. 1970, The Mathematical Theory of Non-uniform Gases (Cambridge Univ. Press)

Dziembowksi, W.A., Goode, P.R. and Libbrecht, K.G. 1989, *Astrophys. J.* **337**, L53

Ekman, V.W. 1905, *Arkiv Matem. Astron. Fysik* **2**, 11

Eddington, A.S. 1925, *Observatory* **48**, 78

Goldreich, P.A. and Nicholson, P. D. 1989 *Astrophys. J.*, **342**, 1075

Hopfinger, E.J., F.K. Browand and Y. Gagne 1982, *J. Fluid Mech.* **125**, 505

Kippenhahn, R. 1958, *Z. Astrophys.* **46**, 26

Knobloch, E. 1978, *Astrophys. J.* **225**, 1050

Knobloch, E. and H.C. Spruit 1982, *Astron. Astrophys.* **113**, 261

Knobloch, E. and H.C. Spruit 1983, *Astron. Astrophys.* **125**, 59

Landau, L. and E. Lifschitz 1987, Fluid Mechanics (English translation, 2nd edition; Pergamon edit., London)

Lerner, J. and Knobloch, E. 1988, *J. Fluid Mech.* **189**, 117

McDonald, B.E. 1972, *Astrophys. Space Sci.* **19**, 309

Mestel, L. 1953, *Montly. Not. Roy. Astron. Soc.* **113**, 716

Michaud, G. 1970, *Astrophys. J.*, **160**, 641

Pedlosky, J. 1979, Geophysical Fluid Dynamics (Springer edit.; Berlin, Heidelberg, New York), 482

Press, W. H. 1981, *Astrophys. J.*, **245**, 303

Rayleigh, Lord 1880, *Scientific Papers*, **1**, 474 (Cambridge Univ. Press)

Schatzman, E. 1969, *Astrophys. Lett.* **3**, 139

Schatzman, E. 1977, *Astron. Astrophys.* **56**, 211

Schatzman, E. and A. Maeder 1981, *Astron. Astrophys.* **96**, 1

Stillinger, D.C., K.N. Helland and C.W. Van Atta 1983, *J. Fluid Mech.* **131**, 91

Sweet, P.A. 1950, *Montly. Not. Roy. Astron. Soc.* **110**, 548

Tassoul, J.-L. and M. Tassoul 1982, *Astrophys. J. Suppl.* **49**, 317

Tritton, D.J and P.A. Davies 1981, Hydrodynamical Instabilities and the Transition to Turbulence, edit. H.L. Swinney and J.P. Gollub (Topics in Applied Physics; Springer; Berlin, Heidelberg, New York)

Vauclair, S. and Vauclair, G. 1982, *Ann. Rev. Astron. Astrophys.* **20**, 37

Vauclair, S. 1988, *Astron. Astrophys.* (in press)

436

Vitense, E. 1953, *Z. Astrophys.* **32**, 135

Vogt, H. 1925, *Astron. Nachr.* **223**, 229

Von Zeipel, H. 1924, *Monthly. Not. Roy. Astron. Soc.* **84**, 665

Weiss, J.B. and Knobloch, E. 1989, *Phys. Rev. A* **40**, 2579

Zahn, J.-P. 1983, Astrophys. Processes in Upper Main Sequence Stars (Publ. Observatoire Genève), 253

Zahn, J.-P. 1987, The Internal Solar Angular Velocity (eds. B. R. Durney and S. Sofia; Reidel), 201

SURFACE ABUNDANCES OF LIGHT ELEMENTS AS DIAGNOSTICS OF TRANSPORT PROCESSES IN THE SUN AND SOLAR-TYPE STARS.

A. Baglin and Y. Lebreton
DASGAL, URA 335
Observatoire de Meudon
92 195 Meudon Principal Cedex
France

ABSTRACT. Observations of the surface abundances of lithium, beryllium and helium-3 in the Sun and in solar-type stars of different ages should be interpreted in a coherent way. The abundance of lithium at the surface of a star decreases slowly with age; for stars of the same age it decreases with mass and a dependence on the rotation velocity is suggested. The solar surface lithium is depleted by a factor of 100 relative to the cosmic abundance while an He-3 enrichment of 15% at the solar surface during evolution is suggested.

Observations favour the hypothesis of a slow transport process at work between the outer convective zone and the radiative interior of these stars. Orders of magnitude of the transport coefficient as well as its dependence upon the physical parameters can be inferred from surface abundances of light elements, but at the moment we are far from producing a completely consistent modelization.

1. Introduction

Up to now, there are very few observable quantities which can provide us information on transport processes at work inside the stars. Data on the internal rotation rate of the Sun are now available owing to helioseismological measurements, but for other stars the observations only provide the surface rotation velocity and the surface abundances.

The abundances of the light elements observed at the surface of the Sun and solar-like stars are the best tool to understand the processes of mixing acting in the interior of those stars. Although we have more information for the Sun, any theory of the solar mixing should consist in the proper application to the Sun of physical models applicable to other stars.

The observed abundances of the light elements ^7Li, ^9Be, ^{10}B and ^{11}B are always very small; moreover these elements do not contribute to the energy generation rate in stars so that they can be considered as passive contaminants. The case of ^3He is a little different since it enters the network of hydrogen burning reactions. The distribution of ^7Li, ^9Be and ^3He obtained, at solar age, in the standard solar model of Lebreton and Maeder (1986) are given in fig. 1. ^7Li, ^9Be, ^{10}B and ^{11}B only survive in those outer regions of the star where the temperature is low enough to prevent them from being destroyed by nuclear reactions. According to the updated cross-sections of nuclear reactions of Caughlan and Fowler (1988), ^7Li burns at temperatures of about $2.7 \ 10^6 \ ^\circ K$, ^9Be is destroyed at about $3.5 \ 10^6 \ ^\circ K$ and ^{10}B and ^{11}B around $5.0 \ 10^6 \ ^\circ K$; thus in solar-like stars all these elements are confined to a narrow region extending a few pressure scale heights below the convective zone. The sharp decrease with depth of the Li and Be abundances which can be seen in fig. 1 is due to the strong dependence of their destruction rates with temperature (the rate of the $^7Li(p,\alpha)^4He$ reaction is proportional to about T^{30} around $2.5 \ 10^6 \ ^\circ K$). The peaked distribution of ^3He results from the balance between destruction and creation in intermediate regions of the Sun while ^3He is not affected by nuclear reactions in outer regions.

437

G. Berthomieu and M. Cribier (eds.), Inside the Sun, 437–448.
© 1990 *Kluwer Academic Publishers. Printed in the Netherlands.*

The interpretation of the surface abundances of the light elements relies on the physical processes at work in the outer layers of stars and, of course, on the structure of those layers which varies for stars of different masses. Thus we shall only consider here the case of solar-analogues with masses between $0.8M_\odot$ and $1.2M_\odot$ (i.e., 5000 °K$\leq T_{eff} \leq$6200 °K) and are still in the main-sequence stage. In these young population I stars the lithium abundance has been extensively studied but very little information can be obtained from the other species. After a short review of the observational results we shall discuss the existing modelizations, their difficulties and their successes.

2. Observations of surface abundances.

2.1. LITHIUM, BERYLLIUM, BORON

Let us first review the tools available to determine the surface abundances. Since for the light elements, the abundances to be measured are very small (Li/H $\approx 10^{-9}$ in number), only the resonance lines can be used (see for instance TableII in Boesgaard (1976)), which lists the resonance lines of the light elements either neutral or singly ionized. The only lines lying in the visible spectrum are the Li I and marginally the Be II lines. The boron lines are all in the UV spectrum. The ionization potential of Li I is very low and thus lithium is in the form of Li II at about 5800 °K in the photosphere of solar-like stars. The lithium line at 6707 Å is blended by an iron line. Most Be is in the form of Be II in solar type stars, but the UV spectrum is crowded in the ionization region and the stronger line at 3130.4 Å is badly blended. Thus most of the observational data concern lithium abundances while very little is known about beryllium and almost nothing about boron. Moreover the abundance determinations are possible in objects of moderate rotation only.

A lot of work has been done during the last three decades to measure the abundances of the light elements at the surface of stars and to correlate them with stellar parameters. Herbig (1965) was a pioneer in the subject and showed for the first time a clear correlation between stellar age and abundance of lithium in main-sequence stars of spectral type G. Many complementing results have been obtained afterwards by Wallerstein et al. (1965), Kraft and Wilson (1965), Danziger and Conti (1966), Danziger (1967). The observational results were obtained from photographic plates, most of them at the 200 inch Palomar telescope. They all give the ratio [Li/Ca], which means log(Li/Ca)-log(Li/Ca)$_\odot$ (in number), comparing the Li line at 6707 Å to a neighbouring calcium line. All these results clearly establish the general trend of decreasing [Li/Ca] versus age with emphasis on the depletion with respect to the cosmic abundance. Moreover, Bodenheimer (1965) noticed a general decrease of lithium with advancing spectral type, or decreasing mass, which was confirmed by the observations of Zappala (1972), in the Hyades and other open clusters.

Recently, spectroscopy with high-signal-to-noise ratio has become possible due to the great quality of the new instrumentation (see IAU Symposium n°132 on "The impact of very high S/N spectroscopy on stellar physics" edited by G. Cayrel and M. Spite, 1987). This progress has considerably changed the observational landscape since it is now possible to detect very low lithium abundances in faint objects and to separate the iron line from the lithium one.

Duncan and Jones (1983) and Cayrel et al. (1984) observed the lithium line in stars of the Hyades cluster and showed the now very popular decrease of the lithium abundance with the effective temperature (for $T_{eff} \leq$ 6000 °K , see fig. 2b). The lithium abundance appears to depend on the age of the observed object so that its determination in stars of the same (and known) age like the Hyades stars is of great interest. During the last five years the closest clusters and groups have been observed by different authors with the same techniques (see fig. 2a-c and references therein). The following important results can be drawn from fig. 2a-c: 1) the abundance of lithium in the observed stars is either cosmic or depleted. The cosmic abundance inferred from measurements in young hot stars is Li/H $\approx 10^{-9}$ (in number) with little evidence of real variation, as reviewed by Boesgaard et al. (1988). The value found in meteorites (Nichiporuk and Moore, 1974) is a factor 2 larger which is generally attributed to chemical separation; 2) in the older clusters, which have an age greater than a few 10^8 years, the depletion of lithium increases with stellar age for a given mass and at fixed age increases with decreasing mass. Moreover the Li-T_{eff} (i.e., mass) relation for a given cluster is quite-well defined and the intrinsic scatter is rather small,

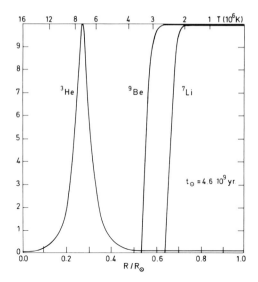

Figure 1 : Abundances of ^{7}Li, ^{9}Be and ^{3}He normalized to their maximum values as a function of the radius and temperature in the standard model of Lebreton and Maeder (1986)

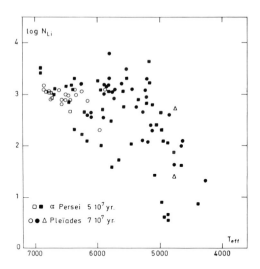

Figure 2a : Observation of the lithium abundance in number of atoms relative to hydrogen, log N_{Li}, in the logarithmic scale where log N_H = 12, as a function of T_{eff} in the two young clusters α Persei and the Pleiades. Observations are from Duncan and Jones (1983,•), Butler et al. (1987, Δ), Boesgard et al. (1988b,o) and Balachandran et al. (1988)

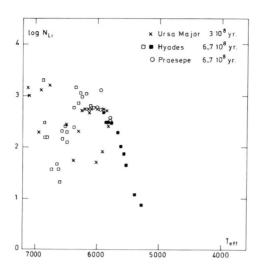

Figure 2b : Same as fig.2a but in the three clusters of intermediate age (Ursa Major, Boesgard et al. (1988a,x) ; the Hyades, Cayrel et al.(1984,), Boesgard and Tripicco (1986a,) and Praesepe, Soderblom and Stauffer (1984,o)).

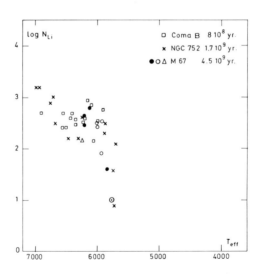

Figure 2c : Same as in fig.2a but in older clusters (Coma Berenices, Boesgard (1987,) ; NGC 752, Hobbs and Pilachowski (1986a,x) and M67, Hobbs and Pilachowski (1986a,•), Spite et al. (1987,o), Garcia Lopez et al. (1988,Δ).

especially in the Hyades. In younger clusters, the dispersion of the Li-abundance at a given T_{eff} may be large but the average depletion is smaller than in the older clusters (see fig. 2a).

Conti (1968) has suggested that rotational braking could be responsible for Li-destruction in stars. Therefore it is worth investigating the possible Li-abundance-rotation and age-rotation relationships in the observed solar-type stars.

In a recent discussion of the available data, Stauffer (1987) points out that while the dispersion among the rotational velocities for a given spectral type is large in the very young cluster α Per, it tends to become less important, as age increases, for clusters of intermediate ages like the Pleiades. At the age of the Hyades, the scatter almost vanishes and the observed relation between rotation and T_{eff} is very well-defined. Moreover, as the age increases, the rapid rotators observed have smaller masses which is probably due to the fact that the time scales of spin down are smaller in higher-mass stars.

Furthermore, recent observations, in α Per and the Pleiades, respectively by Butler et al. (1987) and Balachandran et al. (1988) give some new information on the lithium-rotation relationship. The main result of Balachandran et al. (1988) is that, in the sample observed (46 stars), for a given spectral type all the Li-poor stars are slow rotators while the undepleted stars rotate rapidly. A similar behavior is found in 8 of the 11 Pleiades stars studied by Butler et al. (1987). As discussed by Balachandran et al. (1988), this result could be due either to a process which links the Li-depletion to the rotation rate or to an extended phase of star formation in clusters. Rebolo and Beckman (1988) have found a similar, although more uncertain, trend of "slower rotation less Li" in G stars of similar mass in the Hyades. Whether these results are due to an age spread or to the fact that stars have undergone a different history leading to different initial rotation is still not clear and requires further investigations. It is also worth pointing out that the opposite situation, i.e., fast rotators with no lithium depletion and slow rotators with depletion, at the same spectral type has also been observed in αPer by Butler et al. (1987) and in F and G field stars by Boesgaard and Tripicco (1986,b).

Therefore one has now to be careful when talking about lithium depletion as an age indicator. It is necessary to refer to objects of the same mass or at least same effective temperature and to objects which are rather old (maybe older than the Pleiades).

The Li^6 abundance is very difficult to measure. The Li^6/Li^7 ratio can however be a good indicator as Li^6 is destroyed at much lower temperature than Li^7. Observing several F and G stars Cayrel et al. (1989) have shown that this ratio is always much smaller than the meteoritic value (fig. 3).

2.2. HELIUM-3

The observations of ^{3}He in the solar system are quite difficult to interpret, as reviewed by Lebreton and Maeder (1987). The ^{3}He abundance of the solar atmosphere is extremely difficult to obtain by direct measurements, however the $(^3He/^4He)$ ratio has been measured in many different sites (i.e., planets, solar wind, meteorites) and this provides indirect determinations of the ^{3}He abundance at the solar surface. The comparison of the measurements on the Moon during the Apollo 1969-1972 missions and in the solar wind has lead Geiss (1971) to suggest a possible enrichment of the present wind with respect to meteorites. The recent reanalysis of the observational data by Bochsler et al. (1989) confirms that the zero-age main sequence abundance of ^{3}He might have been slightly lower (by about 15%) than the present day value. It is worth noticing that the solar abundance of ^{3}He itself does not give much indication on the internal structure of the Sun because it is subject to large observational uncertainties. However the ^{3}He abundance puts strong constraints on the mechanisms that can be invoked to explain the lithium depletion at the surface of the Sun. Whatever the chosen mechanism is (for instance any kind of transport processes), it should not lead to an enrichment of ^{3}He at the solar surface greater than about 15%. This puts severe constraints on the so-called non standard solar models.

3. The different theoretical interpretations

Let us review the various scenarios proposed to account for the observed Li-depletion in stars of masses $0.8M_\odot$ to $1.2M_\odot$, close to the main sequence.

442

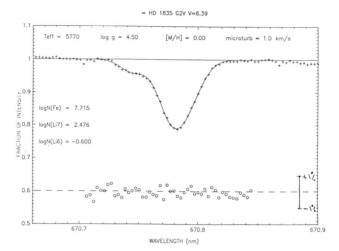

Figure 3 : ^{6}Li/^{7}Li ratio in a G2V star from Cayrel et al. (1989). The best fit of the ^{6}Li, ^{7}LI, Fe blended line at 6707 Å is obtained from a very low value of the ^{6}Li abundance.

3.1. PRE-MAIN SEQUENCE DEPLETION

Bodenheimer (1965) has suggested that the lithium depletion occurs during the pre-main sequence phase and that an abrupt drop of the lithium abundance is observed in low-mass stars (i.e., with B-V> 0.6) because, during the pre-main sequence, their convective zones reach regions sufficiently hot for Li to burn. Bodenheimer (1965) used the available modelization and obtained an acceptable fit to the observations. More recently, D'Antona and Mazzitelli (1984) showed that, with updated models, the convective zones are too shallow and that the fully mixed region has to be extended by an amount of overshooting of $0.7H_p$ (H_p being the pressure scale height) to get reasonable agreement with the new observations. However, the decrease of lithium abundance with stellar age as seen in a large fraction of the stars strongly suggests that only a small fraction of the depletion occurred during the pre-main-sequence stage. In any case, and even if some depletion takes place during the pre-main sequence phase, one will not be able to explain the whole set of observations without invoking a process of depletion taking place during the main sequence stage. However, this interpretationn is probably valid for Li^6, which burns at lower temperature.

3.2. MAIN SEQUENCE DEPLETION.

During the main sequence phase, lithium will be destroyed in stars which have convective zones extending down to the Li-burning region. From the very high temperature sensitivity of the lithium destruction rate, one would expect a very sharp drop of the lithium abundance at the spectral type where the convection zone is just deep enough to reach the Li burning region. There is no evidence for such a sharp decrease in the observations which predict, on the contrary, a very smooth dependence of the Li depletion with mass.

Weymann and Sears (1965) have first studied the effects on the solar lithium depletion of mixing by convective overshoot beneath the convective zone. Straus et al. (1976) and Cayrel et al. (1984) have shown that an agreement with the observations in clusters could be obtained only if the overshooting distance adopted depends on the mass of the star, rather than being simply equal to an arbitrary fraction of the pressure scale height. It is very difficult to accept such an ad-hoc dependence of a physical process with stellar parameters, thus it is reasonable to consider that overshooting alone is not responsible for the observed lithium depletion in stars (although it should be taken into account in the modelization).

A slow and smooth link between the nuclear destruction region and the outer convective zone is then required to account for the observations. This immediatly calls for a diffusion process acting in the radiatively stable zone, as originally proposed by Schatzman (1969). In the process of turbulent diffusion mixing the concentrations of the passive elements at the surface of the stars will then be determined by 1) the structure of the radiative zone, 2) the depth of the convective zone taking into account a reasonable extension due to overshooting processes, 3) the efficiency of the diffusion process characterized by a diffusion coefficient D.

4. Turbulent diffusion mixing during the main sequence

With the assumption that the diffusion timescale is much shorter than the evolution one, the description relies on an underlying model of the star, and on a diffusion equation for the passive contaminants, which does not influence the model itself.

4.1. THE MATHEMATICAL PROBLEM

The concentration of lithium, c, satisfies an equation of conservation of matter

$$\frac{1}{r^2}\frac{\partial}{\partial r}\rho\, r^2 D\frac{\partial c}{\partial r} = \rho\frac{\partial c}{\partial t} + \rho\, K(\rho, t)\, c$$

The diffusion coefficient, D, may be a function of radius, r, temperature, T, density, ρ and age t.
The boundary conditions are important.
At $r=r_m$ one has to insure continuity of concentration and flux

$$4\,\Pi\, r_m^2 D\,\rho \left.\left(\frac{\partial c}{\partial r}\right)\right|_m = \left.\left(\frac{\partial c}{\partial t}\right)\right|_m \int_{r_m}^R 4\,\Pi\, r^2 \rho\; dr + c_m \int_{r_m}^R 4\,\Pi\, r^2 K\,\rho\; dr$$

The lower boundary condition is somewhat arbitrary. The roughest one would be c=o, but one can use an asymptotic value of the derivative obtained through a BKW approximation:

$$(\frac{\partial c}{\partial r}) = (\frac{\partial c}{\partial r})_{asymptotic}$$

4.2. ESTIMATES OF THE DIFFUSION COEFFICIENT

If we assume the diffusion coefficient to be constant the diffusion equation can be written in its simplest form:

$$D\frac{\partial^2 c}{\partial r^2} = \frac{\partial c}{\partial t}$$

with a time variation of the concentration of the form $c = c_o \exp(-\lambda t)$ one gets:

$$\lambda \approx \frac{D}{h^2} \quad\text{and}\quad D \approx \frac{h^2}{T}\ln(\frac{c}{c_0})$$

where T is the duration of the diffusion process and h is the size of the region where the diffusion acts.

In the case of the Sun this simple approximation gives:

$\langle D \rangle = 1000$ cm^{-2} s^{-1} for lithium, with h= 6 10^4 km and c/c$_0$ = 10$^{-2.2}$

$\langle D \rangle = 100$ cm^{-2} s^{-1} for beryllium, with h= 10^5 km and c/c$_0$ =0.3

$\langle D \rangle < 100$ cm^{-2} s^{-1} for helium-3, with h= 3. 10^5 km.

This seems to indicate that the diffusion coefficient in the present Sun decreases with depth, as already stressed by Schatzman (1981).

The values so obtained for the turbulent diffusion coefficient are intermediate between the high diffusion coefficient of the convective zones (D = 10^{13} cm^{-2} s^{-1}) and the small unavoidable microscopic diffusion (D = 1 to 10 cm^{-2} s^{-1}). These values can also be compared to the estimated values required to carry angular momentum and to maintain the almost solid rotation of the Sun between 0.3 and 0.7 R$_\odot$: D = 8000 cm^{-2} s^{-1}.

4.3. MODELIZATION

4.3.1. *parameters.*

Unfortunately many parameters influence the predictions of the time evolution of the surface abundances in a stellar model:
-1) *the input physics* (essentially opacities and nuclear reaction rates) determines the global structure of the model and its evolution, which in turn states the mixing-length value (through the calibration of a solar model);
-2) *the hydrodynamical description of the transport processes*: overshooting, meridional circulation and the various hydrodynamical instabilities (see this Colloqium);
-3) *the macroscopic parameters* of the observed stars. Whereas the effective temperature is now determined with a precision of about 50°K in solar-like stars, luminosities and masses are still badly known. Higher quality data will have to wait for the results of the Hipparcos mission. A 10% variation in the luminosity would lead to a factor 10 fdifference in the predicted Lithium abundance in a solar model.

4.3.2. *the Hyades.*

The well-settled one parameter sequence of the lithium abundance in the Hyades has focused the attention of most of the theoreticians as a challenge. Any theoretical modelization should be able to reproduce the observed slope of the Li-depletion with temperature in the Hyades.

The large sensitivity of the results to the many unknown parameters which enter the modelization explains the long history of apparent successes and failures, parallel to the improvement of our knowledge concerning the physics entering the structure and the hydrodynamical processes.

Baglin et al. (1983) and Cayrel (1983) obtained the correct trend of the Hyades sequence with models including turbulent diffusion but using different numerical procedures. Baglin et al. (1985) have tested the effects on the lithium depletion of different expressions of the turbulent diffusion coefficient. They showed that 1) a diffusion coefficient of the form D= Re*ν where ν is the microscopic viscosity and Re* a pseudo-Reynolds number cannot explain the observed slope because the microscopic viscosity varies too abruptly with effective temperature; 2) the observations were explainable with a diffusion coefficient of the type D= $\alpha(\nabla_{ad}-\nabla_{rad})^{-1}$ which has a sharp decrease just below the convective zone but almost remains constant in a large part of the diffusion region.

However a revision of the input physics (inclusion of low-temperature opacities, updating of the nuclear reaction rates) leads to standard solar models with higher mixing-length values (Lebreton and Maeder, 1986) and with a somewhat different structure of the outer layers. Baglin et al. (1987) have reconsidered the problem using this new input and taking into account the uncertainties in the observed parameters. They conclude that the high and low-mass stars models are difficult to reconcile: convective zones look too small, an important amount of overshooting is needed to extend them and an almost constant diffusion coefficient is required. In particular, the turbulent diffusion coefficient derived by Zahn (1983),

$$D = \frac{4}{5} \frac{r^6}{G^2} \Omega^2 \frac{L^2}{M^3} \frac{1}{\left(\nabla_{rad} - \nabla_{ad}\right)} \left(1 - \frac{\Omega^2}{2\Pi G \rho}\right) \qquad (1)$$

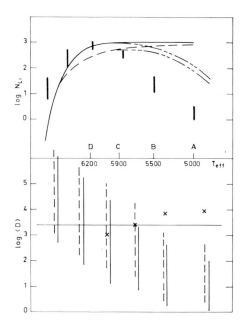

Figure 4a : Predicted surface lithium abundance in the Hyades using a diffusion coefficient given by (1) for the two extreme values of the rotational velocities (continuous line and dashed line correspond respectively to the minimum and maximum value of the observed rotational velocity). The observed values (thick lines) are given for comparison. The effect of an overshooting region expanding over 0.7 H_p is shown (double dashed lines).

Figure 4b : The diffusion coefficient as given by (1) in the diffusion region of the corresponding models in the two hypothesis for rotation, compare to the "effective diffusion coefficient" (crosses) needed to obtain the observed values of the lithium abundance.

which relates the degree of mild turbulence to the rate of differential rotation, decrases very rapidly with the mass in low-mass stars (see fig. 4 where D has been calculated with the most recent determinations of the rotational velocities in the Hyades). According to Baglin and Morel (1989) this coefficient is unable to reproduce the observed slope of the lithium depletion in the cooler Hyades stars although for higher temperatures Vauclair (1988) uses this expression to interpret the lithium dip around $T_{eff}= 6700°K$ discovered by Boesgaard and Tripicco (1986a). It appears then that if (1) were a good candidate for the diffusion coefficient, it should be complemented by an additional mechanism sensitive to the mass, which would be able to decrease the efficiency of the depletion during the main sequence phase.

4.3.3. *lithium abundance in the sun and older clusters.*

To be significant, the comparison of the surface lithium abundance in the Sun and in the Hyades has to be made with a one solar mass star and not at the same T_{eff}, as the evolution from the Hyades age to the solar one corresponds to an increase of T_{eff}. A one solar mass star in the Hyades has a depletion factor of at least 1.5; this would lead to a deficiency of the order of 10^{-3} at the age of the Sun, if the depletion was continuously acting at the same rate. This means that the predicted depletion is too large between 8. 10^8 and 4.5 10^9 years., i.e., that the diffusion coefficient has decreased. As during this phase the star spins down, the velocity field and the turbulence generated by the motions induced by rotation have also probably decreased. These facts are then consistent with a scenario of the kind proposed by Zahn (1987), and suggests to withdraw the assumption of a time independent diffusion coefficient.

The time dependence of the angular velocity is linked to the complex process of spin-down and angular momentum loss, as discussed in this colloquium. In the absence of a precise modelization, only crude estimates are possible. In a first approach Schatzman (1989) suggests a time dependence of the form, $\Omega = \Omega_0(1 + t/t_0)^{-3/4}$, t_0 being the time elapsed between the end of the spin-up and the arrival on the main sequence, and Ω_0 the initial angular velocity. If the diffusion coefficient is assumed to be given by (1) its time dependence can then be written in the following form: $D(t) = D_0 (1 + t/t_0)^{-3/2}$.

With the assumption that the variations of the underlying model are negligible only the timescale is changed. As t increases $D(t)$ becomes very small and the abundance tends to a finite limit. If around one solar mass t_0 is of the order of 3 10^8 years (Schatzman, 1989), then the process is already important for a cluster like the Hyades, and dominates at solar age. The depletion is then $\log (c/c_0)_{\odot} = 1.5\log(c/c_0)_{Hyades}$. This factor is of the right order of magnitude comparing the sequences of the Hyades and of M67. In addition this process might also help solving the difficulty stressed in § 4.3.2. as it introduces an additional mass dependence of Ω through the characteristic time t_0.

4.4 CONSISTENT MODELIZATIONS

As already said, a complete modelization has to take into account the effects of the evolution of the model. The numerical problem becomes then more complicated and the results are less easy to understand and to handle as no algebric dependence exists.

4.4.1. *Helium-3 abundance in the Sun*

For elements like lithium, which do not influence the nuclear evolution, the hypothesis of constancy of the model is reasonable. When studying helium, the model has to be relaxed and one has to solve at the same time the diffusion equation and the equation of evolution.

Lebreton and Maeder (1987) have calculated non-standard models involving mild turbulent diffusion mixing. The turbulent diffusion coefficient chosen is $D = (8r/H_p)\nu$ where ν is the microscopic viscosity. This expression was grossly derived from Zahn's (1983) estimate and is in good agreement with the observational constraints on D. The results show that with this coefficient, diffusion has to be reinforced by a moderate amount of overshooting ($0.6H_p$) in order to be able to give at solar age the observed depletion of lithium at the surface of the Sun. Furthermore the efficiency of mixing in the region associated to the ^{3}He peak is quite moderate and these models lead to a secular enrichment of ^{3}He in the solar atmosphere compatible with the observational limit given by Bochsler et al. (1989).

4.4.2. *The consistent modelization of Pinsonneault et al.*

The most consistent approach at the moment is presented by Pinsonneault et al. (1989). An evolutionary sequence, of a rotating star is followed from the wholly convective phase, including the transport of both angular momentum and chemicals. Diffusion coefficients are estimated from several instabilities and an ad-hoc scaling factor reduces the diffusion coefficient of chemicals with respect to the angular momentum one. Many parameters enter this model. Some of them can be fixed by the available observational constraints : the angular velocity of the Sun at the solar age and the lithium abundance as a function of age. The observed range of Li abundance in cluster stars of similar masses is interpreted as a consequence of a spread in the initial angular momentum.

5. Conclusions

The available observational data have not yet been explained in a completely consistent way. Although it seems reasonable to associate transport processes with the instabilities related to rotation, the physical mechanism is not yet understood. Many parameters enter the stage, and are difficult to include in the simulations: evolution of the star, spin-down, different hydrodynamical instabilities... But, at the moment we know the angular velocity distribution only in the case of the Sun at the solar age and the surface abundances for some other stars, whereas generally mass is unknown and luminosity is very unprecise.

With future space projects like the astrometric satellite HIPPARCOS and the stellar seismology experiment EVRIS we will have considerably more data relevant to this problem. Progress is also needed on the theoretical side to firmly establish the hydrodynamical state of a rotating star.

Acknowledgments

The authors would like to thank E. Schatzman, G. Cayrel de Strobel, R. Cayrel and P.J. Morel for very helpful discussions and communication of unpublished works.

References

Baglin, A., Morel, P. J. 1983, in *Observational Tests of the Stellar Evolution Theory*, IAU Symp. No. **105**, eds A. Maeder, A. Renzini, Reidel Dordrecht 1984, p. 529
Baglin, A., Morel, P.J., Schatzman, E. 1985, *Astron. Astrophys.* **149**, 309
Baglin, A., Morel P.J. 1987, in *the Impact of Very High S/N Spectroscopy on Stellar Physics*, IAU Symp. No. **132**, eds G. Cayrel and M. Spite, Kluwer Academic Publishers 1988, p. 279.
Baglin, A., Morel P.J. 1989, *in preparation*
Balachandran, S., Lambert, D.L., Stauffer, J.R. 1988, *Astrophys. J.* **333**, 267
Bodenheimer, P. 1965, *Astrophys. J.* **142**, 451
Boesgaard, A.M. 1976, *Publ. Astron. Soc. Pacific* **88**, 353
Boesgaard, A.M. 1987, *Astrophys. J.*, **321**, 967
Boesgaard, A.M., Budge, K.B., Burck, E.E. 1988a, *Astrophys. J.* **325**, 749
Boesgaard, A.M., Budge, K.B., Ramsay, M.A. 1988b, *Astrophys. J.* **327**, 389
Boesgaard, A.M., Tripicco M.J. 1986a, *Astrophys. J.* **302**, L49
Boesgaard, A.M., Tripicco M.J. 1986b, *Astrophys. J.* **303**, 724
Bochsler, P., Geiss, J., Maeder, A. 1989, in *Inside the Sun*, Proc. IAU Colloquium 121 , Versailles, France 22-26 May 1989, to be published in Solar Physics
Butler, R.P., Cohen, R.D., Duncan, D.K., Marcy, G.W. 1987, *Astrophys. J.* **319**, L19
Caughlan, G.R., Fowler, W.A. 1988, *Atomic Data and Nucl. Data Tables* **40**, 283
Cayrel, R. 1983 , in *Observational Tests of the Stellar Evolution Theory*, IAU Symp. No. **105**, eds A. Maeder, A. Renzini, Reidel Dordrecht 1984, p.533.
Cayrel, R., Cayrel de Strobel, G., Campbell, B., Däppen W. 1984, *Astrophys. J.* **283**, 205
Cayrel de Strobel, G., Spite, M. 1987, eds. of IAU Symp. No. **132** on *the Impact of Very High S/N Spectroscopy on Stellar Physics*, Kluwer Academic Publishers 1988.
Cayrel R., Cayrel de Strobel G., Bentolila C. 1989 in preparation.
Conti, P.S. 1968, *Astrophys. J.* **152**, 657
D'Antona, F., Mazzitelli I. 1984, *Astron. Astrophys.* 138, **431**
Danziger, I.J. 1967, *Astrophys. J.* **150**, 733
Danziger, I.J., Conti P.S. 1966, *Astrophys. J.* **146**, 392
Duncan, D.K., Jones, B.F. 1983, *Astrophys. J.* **271**, 663
Garcia Lopez, R.J., Rebolo, R., Beckman, J.E. 1988, *Publ. Astron. Soc. Pacific* **100**, 1489
Geiss J. 1971, in *Solar Wind*, NASA SP-308, p. 559
Herbig, G.H. 1965, *Astrophys. J.* **141**, 588
Hobbs, L.M., Pilachowski, C. 1986a, *Astrophys. J.* **309**, L17-L21
Hobbs, L.M., Pilachowski, C. 1986b, *Astrophys. J.* **311**, L37-L40
Kraft, R.P., Wilson, O.C. 1965, *Astrophys. J.* **141**, 828
Lebreton, Y., Maeder, A. 1986, *Astron. Astrophys.* **161**, 119
Lebreton, Y., Maeder A. 1987, *Astron. Astrophys.* **175**, 99
Nichiporuk W., Moore C.B. 1974, *Geoch. Cosmoch. Acta* **38**, 1691
Pinsonneault, M.H., Kawaler, S.D., Sofia, S., Demarque, P. 1989, *Astrophys. J.* **338**, 424
Rebolo, R., Beckman, J.E. 1988, *Astron. Astrophys.* **201**, 267
Schatzman E., 1969, *Astron. Astrophys.* **3**, 339

Schatzman E., 1981, in *Turbulent diffusion and the solar neutrino problem*, CERN 81-11
Schatzman E. 1989, *in preparation*
Soderblom, D.R., Stauffer, J.R. 1984, *Astronom. J.* **89**. 1543
Spite, F., Spite, M., Peterson, R.C., Chaffee, F.H. Jr.: 1987, *Astron. Astrophys.* **171**, L8
Stauffer, J. 1987, in Cool Stars, stellar systems ans the sun. Proceedings of the fifth Cambridge Workshop,Boulder, p; 182.
Straus, J.M., Blake, J.B., Schramm, D.N. 1976, *Astrophys. J.* **204**, 481
Vauclair, S. 1988, *Astrophys. J.* **335**, 971
Wallerstein, G., Herbig, G.H., Conti, P.S. 1965, *Astrophys. J.* **141**, 610
Weymann, R., Sears, R.L. 1965, Astrophys. J. **142**, 174
Zahn, J.P. 1983, in *Astrophysical Processes in Upper Main Sequence Stars*, 13th Advanced Course Saas-Fee, Publ. Geneva Observatory
Zahn, J.P. 1987, in The internal angular velocity of the Sun eds; B.R. Durney and S. Sofia, Reidel, Dordrecht, p. 201.
Zappala, R.R. 1972, Astrophys. J. **172**, 57

CLOSING REMARKS

OPEN QUESTIONS

D. O. GOUGH
Institute of Astronomy and Department of Applied Mathematics
and Theoretical Physics, University of Cambridge, UK; and
Joint Institute for Laboratory Astrophysics, University of
Colorado and National Institute of Standards and Technology,
Boulder, Colorado, USA

The principal aim of this conference has been to address the issues
that pertain to our quest for understanding what is happening in the
very heart of the sun, where nuclear reactions produce the energy
required to replace the emission from the photosphere. As a result of
this energy balance the structure of the sun has hardly varied over
the last 4.6 Gy or so. The sun has belonged, and does still belong,
to what astronomers call the hydrogen-burning main sequence.
    There has been some slight change, however: the mean molecular
weight of the material in the reacting core has been increasing with
time t, as a result of the conversion of hydrogen to helium, and this
has caused a gradual contraction and heating of the central regions of
the star, and a consequent rise of the total luminosity L according to
the formula:

$$L(t) \simeq [1 + \beta(1-t/t_\odot)]^{-1} L_\odot \quad , \tag{1}$$

where $L_\odot$ and $t_\odot$ are the present luminosity and age of the sun. The
coefficient β is a numerical constant proportional to $L_\odot t_\odot/QM \simeq 0.045$,
M being the mass of the sun and $Q \simeq 6.3 \times 10^{18}$ erg g^{-1} being the
energy released per unit mass in the conversion of H to ^{4}He (thus
$QM/L_\odot$ is the time that would be required to convert the entire sun to
helium at the present luminosity assuming that it had started as pure
hydrogen, which it did not); it is also proportional to a dimensionless parameter that depends on the functional form of the opacity $\kappa(\rho,$
T, X) and the gross nuclear energy generation rate $\varepsilon(\rho,$ T, X), where ρ
and T are density and temperature, and X is the hydrogen abundance
(e.g. Gough, 1990). In addition to that, it depends on what has
happened to the material in the core during the main-sequence evolution, but that dependence is quite weak. Thus, if for the moment we
adopt the principal explicit assumptions of the so-called standard
theory of solar evolution, namely that at t = 0 the sun was chemically
homogeneous and that throughout the evolution the core of the sun has
been quiescent, implying that the products of the nuclear reactions
have always remained in situ, then $\beta \simeq 0.40$. (If we were to have
assumed an opposite, quite unrealistic extreme that the entire sun

G. Berthomieu and M. Cribier (eds.), Inside the Sun, 451–475.

were to have been maintained in a chemically homogeneous state by some
mixing process that was too slow to contribute directly to the energy
transport, then the value of β would have been 0.3.) This value is
essentially independent of the presumed initial chemical composition,
provided there are heavy elements enough to dominate the opacity in
the radiative interior (and provided, of course, for given total
heavy-element abundance Z, or given Z/X_0, the initial value X_0 of X is
chosen to ensure $L = L_\odot$ at $t = t_\odot$ as equation (1) implies).

Subject to the assumptions I have mentioned, equation (1) appears
to be the most robust outcome of the theory of solar evolution that is
pertinent to the history of the sun. It is because of that that I
have mentioned it first, to establish some relatively secure starting
position from which to admit our ignorance. It is robust because it
is insensitive to the uncertain details of the internal structure of
the theoretical solar model that produces it, and therefore, of
course, we deduce immediately that were we able to confirm it observationally (which we shall never do directly), it would not be a useful
confirmation of any of the subtle features of that model, particularly
the structure of the core. However, because of that insensitivity it
could in principle be used to test other aspects of the theory upon
which the gross behaviour (namely the gradual rise of L with t from
about 70 per cent of its value today) does depend. Thus, despite my
apparent initial confidence, I must include amongst my open questions:

Is equation (1) correct? (Q1)

And then one is induced immediately to ask:

If it is not, what does that tell us? (Q2)

Of course the trite answer to question (Q2) is that one of the assumptions of the theory is incorrect. 'But which?' is then the natural
response. (I do not include such natural responses to trite replies
amongst my open questions.) I hope that at least the flavour of my
brief introductory description of main-sequence evolution has already
indicated that the answer would not lie in the details of the physics
that is required for establishing such matters as the equation of
state, the opacity or the thermonuclear reaction rates. The qualitative behaviour exhibited by the expression (1) for L(t) is stable to
quite profound, though plausible, modifications to microscopic physics.

As I have already mentioned, the rise in luminosity comes about
because the fusion of hydrogen into helium reduces the number of
particles per unit mass (which increases the mean molecular weight) in
the core, thereby decreasing the pressure at given density and temperature and causing the core to be compressed. In order to sustain the
weight of the star the pressure must be restored, which is accomplished only by establishing a new hydrostatic equilibrium at higher
density and, according to the virial theorem, higher temperature.
This augments the nuclear reaction rates. The only plausible way out
of this situation is to deny that the weight of the star succeeded in
compressing the core. That could come about only if the weight declined with time in step with the diminution of the ability of the

core to sustain pressure. (I am not seriously entertaining so im-
plausible a postulate that many-body physics is so wrong that pressure
does not diminish with decreasing particle number density, nor that
under solar conditions it is not an increasing function of temperature. Neither am I doubting that the nuclear energy generation rate
increases with density and temperature rapidly enough to overcome the
opposing tendency for it to decline as a result of the decreasing
abundance of hydrogen fuel; that does not occur until the end of the
main-sequence phase of evolution when hydrogen is essentially
exhausted from the centre.)

A decline in the weight of the sun could have occurred in either
of two ways: either the total mass M of the sun or the gravitational
constant G (measured in units in which Planck's constant and the speed
of light are invariant) has been decreasing. Both possibilities have
been entertained in the literature, and, if the manners in which they
are presumed to have decreased were such as to produce the same temporal variation of L (which I assume for the moment occurs on a timescale much longer than the characteristic Kelvin-Helmhotz thermal
readjustment time), would have yielded almost indistinguishable structures of the sun today. Thus we are led by question (Q2) to ask:

> Has the solar mass remained constant during   (Q3)
> main-sequence evolution?

and the bigger question:

> Is gravitational physics correct?   (Q4)

The latter question is intended to encompass not only the local law of
gravity expressed by the governing differential equations (if, indeed,
gravity can be described by differential equations), but also the cosmology that may determine the value of G that appears in the Newtonian
approximation.

Of course there is always some level of precision implied by the
questions. Few of them will ever by answered fully, and therefore in
some evolving sense they will always remain open. As partial answers
are provided, the physical implications of the questions will change
as we probe into more and more subtle aspects of the structure of the
inside of the sun.

By way of illustration, let us consider the specific example of
the asymptotic expression for the cyclic frequencies of p modes of low
degree ℓ and high order n in the form:

$$\nu \sim (n + \tfrac{1}{2}\ell + \varepsilon)\nu_o - [A\ell(\ell + 1) - B]\nu_o^2\nu^{-1} + \ldots \quad , \quad (2)$$

where ε, ν_o, A and B are functionals of the equilibrium state which do
not depend on the mode of oscillation. By analogy with question (Q1)
one might then ask:

> Is the asymptotic relation (2) correct?   (Q5)

At some level one can immediately answer in the affirmative, particularly if one retains only the first of the two terms, and, of course,
answers the question only to the precision dictated by the magnitude
of the relatively small second term. One may view the question in
either of two ways, depending on whether one is asking about the
validity of the expression as an approximation to the eigenvalues of a
certain differential boundary-value problem or whether one is asking
if the sun's oscillation frequencies actually satisfy the relationship. Both aspects of the question and its answers need to be understood before measured frequencies can be used to make sound deductions
about the physical state of the solar interior.

Let us therefore begin with just the leading term of the expression (2). It is immediately recognizable (at least to some) as the
leading term in the asymptotic approximation as $n/\ell \to \infty$ to the zeros
of a spherical Bessel function, and therefore represents the high-
order frequencies of relatively low-degree oscillations of a uniform
gas with sound speed c contained in a sphere (e.g. Rayleigh, 1894).
The characteristic cyclic frequency ν_0 is simply the reciprocal of the
time τ_0 taken to traverse a diameter of the sphere at speed c:

$$\nu_0 = \tau_0^{-1} = (2R/c)^{-1} \quad , \tag{3}$$

where R is the radius of the sphere. The constant ε depends on the
conditions imposed by the bounding surface of the sphere, and relates
to the phase shift induced when a wave incident to it is reflected.
If the gas in the sphere is stratified, yet retains spherical symmetry, then provided the wavelength of the oscillations is everywhere
much less than the scale height of variation of the equilibrium state,
JWKB analysis shows that the leading term of equation (2) is unchanged. The expression for ν_0 in terms of τ_0 is unchanged too,
though now that the sound speed is a function of the radial coordinate
r, c must be replaced in equation (3) by its harmonic mean $\bar{c}$:

$$\nu_0^{-1} = \tau_0 = 2R/\bar{c} = 2 \int_0^R c^{-1} dr \quad , \tag{4}$$

as was first shown in the case of adiabatic oscillations of a spherical star by Vandakurov (1967). Of course, a star is not contained
within some well defined boundary, but, as Lamb (1908) first showed
for the case of an isothermal atmosphere, reflection still takes place
in the surface layers provided the wavelength λ of oscillation exceeds
some critical value λ_c, where

$$\lambda_c = 2H(1-2H')^{-1/2} \quad , \tag{5}$$

(e.g. Deubner and Gough, 1984), where H(r) is the density scale height
and a prime denotes differentiation with respect to the argument.

Since λ_c is of the same order of magnitude of H, the condition
$\lambda \ll H$ necessary for the validity of the JWKB approximation is not
satisfied in the reflecting layers. However, those layers are extremely thin compared with a characteristic value of λ in the deep
interior of the star where the JWKB approximation is valid for high-

order modes, and therefore they present themselves to the oscillations of the interior in essentially the same way as a reflecting surface; the phase constant ε can be determined by analysing the solution to the wave equation in the vicinity of reflection in terms of a simple comparison equation using Langer's technique, which has been shown to provide a valid asymptotic approximation to the exact value (Olver, 1974). Vandakurov (1967) determined how the result is related to a polytropic index μ characterizing the stratification of the outer layers of the star: he found

$$\varepsilon = \frac{1}{2}(\mu + \frac{1}{2}) \quad . \tag{6}$$

Thus we see, from a mathematical point of view, how to provide a first-order answer to question (Q5).

The physical answer was first provided by Claverie et al. (1979), from whole-disk Doppler observations of the sun, which are sensitive principally to modes with $\ell \leq 3$. The power spectrum of the oscillations revealed an array of uniformly spaced peaks, whose separation, according to the leading term of expression (2), must be $(1/2)\nu_0$, assuming that modes with both odd and even values of ℓ were present. The absolute values of the frequencies of the peaks determined the value of ε. Thus there were available an estimate of a harmonic mean of the sound speed throughout the sun, and a measure of the density stratification in the outer reflecting layers immediately beneath the photosphere. These were subsequently compared with the properties of theoretical models, but I postpone mentioning the outcome of that until after I have discussed expression (2) more fully.

The second term in expression (2) was obtained first by Tassoul (1980). Its most noteworthy feature, perhaps, is that it has the structure of the corresponding term in the asymptotic expansion of the zeros of the spherical Bessel function that determine the frequencies of high-order acoustic oscillations of a uniform sphere. [In that case $A = (2\pi^2)^{-1}$, and B, like ε, depends on the boundary conditions imposed at $r = R$.] One can show from Tassoul's analysis that

$$A = \frac{1}{4\pi^2 \nu_0} [\frac{c(R)}{R} - \int_0^R \frac{1}{r} \frac{dc}{dr} dr] \quad . \tag{7}$$

The expression for B is complicated, and I shall not reproduce it here; it is sufficient for my purposes to point out that it depends predominantly on conditions in the vicinity of the outer reflecting layers of the star. Once again, assuming no errors have been made, from a mathematical viewpoint one can summon Olver's analysis to affirm that expression (2) formally approximates the eigenvalues of a particular boundary-value problem in the limit $n/\ell \to \infty$. There have been further refinements to the expression, accomplished by replacing the limits of integration in equation (7) by the lower and upper turning points of the governing differential equation (Gough, 1986a), which is equivalent to retaining higher-order terms in the asymptotic sequence and which should improve the accuracy of the expression particularly when n/ℓ is not extremely large. It is important to

point out at this point that for all realistic stellar models the
second term in the square brackets in equation (7) is very much larger
than the (geometrical) first term. Therefore a measurement of A, even
with errors, provides a mean measure of the sound-speed gradient,
weighted by r^{-1} and therefore dominated by conditions in the central
regions of the star.

The second term in expression (2) has been confirmed observationally too, at least for the sun. This was first achieved by Grec
et al. (1980, 1983) who, recognizing the dominance of the leading term
of expression (2) already established observationally by Claverie et
al. (1979), superposed segments of their power spectrum of solar
whole-disk Doppler measurements (with frequency interval close to ν_0)
in order to raise the signature of the second term above the noise.
By measuring the dependence on ℓ it was thus possible both to confirm
the ℓ dependence of the second term and to measure the coefficient A.
Thus, Grec et al. provided the first seismic diagnostic of the state
of the energy-generating core of the sun, which is the principal
subject of this conference.

This brief introductory history of the early days of helioseismic
diagnosis illustrates not only how questions such as (Q5) are answered
progressively, with greater and greater detail and precision, and how
with the answers comes more and more diagnostic information; it also
shows that obtaining diagnostics of the energy-generating core from p-
mode data is much more difficult than obtaining diagnostics of the
rest of the sun. [The most prominent p modes in the solar spectrum
have $n \simeq 25$; furthermore $(1/2)\ell \leq 1.5$ in whole-disk data and $\varepsilon \simeq 1.75$.
Thus both ν_0 and $\varepsilon\nu_0$ are several per cent of the absolute frequencies
$\nu_{n,\ell}$ of typical modes of order n and degree ℓ. However, the frequency
separation $\nu_{n,\ell} - \nu_{n-1,\ell+2} \simeq 2(2\ell+3)(n+\ell/2+\varepsilon)^{-1}A\nu_0$ which measures the
diagnostic A of the core is only about 0.3 per cent of a typical fre-
quency; to measure A to a precision of say a few per cent, which is
necessary to detect the small differences between the sun and theoretical solar models, therefore requires frequencies to be determined to
a part in 10^4.] The reason is quite straightforward, though in two
parts. First, stellar p modes are essentially standing acoustic
waves. The contribution to the frequency from any region in that star
is therefore proportional to the time taken for a sound wave to traverse that region. In the sun, the sound speed at the centre is about
2.5 times the harmonic mean $\bar{c}$, and some 60 times the sound speed in
the photosphere. Therefore the wave spends comparatively little time
in the central regions. The second reason is that there is a central
zone of avoidance by nonradial ($\ell > 0$) modes, whose radius r_t (the lower
turning point of the eigenvalue equation) is given approximately by

$$\frac{c(r_t)}{r_t} = \frac{2\pi\nu}{L} \quad , \tag{8}$$

where

$$L^2 = \ell(\ell+1) \quad , \tag{9}$$

(I hope that this L will not be confused with the luminosity, for
which the same symbol has been used) which therefore hardly

contributes at all to the frequency ν. This latter property can be
used to advantage, however, as I shall explain later. Nevertheless,
it would evidently be of very great advantage to have g-mode frequencies available, in addition to the frequencies we have at hand now,
because g modes sense the central regions preferentially. Indeed,
unpublished inverse calculations by A. J. Cooper and myself have
demonstrated that a very substantial increase in the diagnostic power
of low-degree modes is achieved even with the addition of only a very
few g-mode frequencies (cf. Gough, 1984). Therefore I raise the
question:

Will g modes be measured?   (Q6)

We have heard some discussion at this meeting of the observational
difficulties, and of the recent progress with ground-based networks of
observatories and the suite of helioseismic instruments on the spacecraft SOHO that are being developed to overcome them. If all that
were to fail, Roget Bonnet might have the only practical answer: GONG
on the moon. However, despite my congenital optimism, I am more
doubtful than he that that will come to pass in my lifetime.*
    Before I terminate this discussion of p-mode frequencies, from
which already I seem to have digressed somewhat, I must point out that
unpublished comparisons by Jørgen Christensen-Dalsgaard and myself of
numerically computed frequencies of stellar models with the asymptotic
expression (2) have been rather disappointing. [I should point out
that the asymptotic formula (2) was derived ignoring the perturbation
Φ' to the gravitational potential produced by the density perturbation
associated with the oscillations; comparisons were carried out with
numerical eigenfrequencies computed not only from the full (linearized) adiabatic oscillation equations, which, as Maurice Gabriel has
pointed out at this meeting, are quite poorly represented by the formula, but also from a reduced system from which Φ' had been omitted.]
Although for solar models the value of A inferred from the comparisons
is not very different from that given by equation (7), that is not the
case for main-sequence models with significantly higher or lower
masses than the sun. We found also that replacing the limits of integration by the turning points r_t and R_t, with r_t given by equation (8)
and R_t by the condition $\lambda = \lambda_c$ where λ_c is given by equation (5), does
not improve the situation materially. There appear to be substantial
errors in the asymptotic formula (2) when applied to the finite values
of n/ℓ typical of observed solar p modes, which vary along the main
sequence and which perchance almost cancel for the sun. So perhaps we
have been fooled by the accidental consistency of the asymptotic story
that has been apparently established for the sun. Should we therefore
wonder:

*I contemplated challenging Roget to a wager on this issue, but was
dissuaded by the difficulties of arranging payment were I to win.

Is Nature kind? (Q7)

More specifically, one might ask:

Is asymptotic analysis useful? (Q8)

I believe that neither of these are open questions. I shall address the second first, and postpone the first until later.

Notwithstanding the disappointingly poor correspondence between the absolute values of the true eigenfrequencies of low-degree p modes and their asymptotic approximation, it is likely that the functional dependence of the frequencies on the structure of the sun is given at least qualitatively by the asymptotics. In particular, the quantity

$$d_{n,\ell} = 3(2\ell+3)^{-1}(\nu_{n,\ell} - \nu_{n,\ell+2}) \simeq 6(n + \tfrac{1}{2}\ell + \varepsilon)^{-1}\nu_o A$$

(10)

really is sensitive predominantly to the gradient in the sound speed in the core of the sun, even though the value of A which it measures may not be given precisely by equation (7). Therefore, for example, when comparing the frequencies of a theoretical solar model with those of the sun, we are led by the asymptotics to consider combinations such as $d_{n,\ell}$ to detect errors in the structure of the core of the model; a mere comparison simply of the absolute values of the frequencies alone is far less fruitful. Other more elaborate yet simple ways of comparing frequencies, also based on asymptotic ideas, have been discussed by Christensen-Dalsgaard and Gough (1984) and Christensen-Dalsgaard (1988). I should also mention that quite simple asymptotic expressions for the frequencies of p modes valid also at higher degree have been demonstrated to yield by inversion quite accurate estimates of the sound speed throughout most of the solar interior (Christensen-Dalsgaard et al., 1985). This has stimulated refinements of the inversion procedure (e.g. Gough, 1986b; Sekii and Shibahashi, 1988; Kosovichev, 1988; Vorontsov, 1988; Christensen-Dalsgaard et al., 1989) that have increased confidence in the original inference, more about which I shall discuss later. Finally, permit me to mention also that the asymptotic expression for the frequencies of surface gravity waves (f modes) has served as an important calibrator for both theoretical and observational investigations. For large ℓ the f-mode frequencies satisfy

$$2\pi\nu \sim (LGM/R^3)^{1/2} ,$$

(11)

irrespective of the stratification of the sun. Therefore, one can use this formula for assessing errors in the eigenfrequencies of solar models (e.g. Lubow et al., 1980). I have also used it in the past for assessing errors in observational data, but that becomes more difficult when equation (11) is used to calibrate the spatial scale of the image in the telescope, as sometimes it is. I have mentioned the f modes here simply to reinforce the conclusion that the answer to equation (Q8) is undoubtedly: Yes.

I chose p-mode oscillations to open my discussion because at present they are at the heart of heliophysical research. We have in the sun an extremely valuable physics laboratory, in which many fascinating processes are taking place under conditions that cannot be reproduced on Earth, even by our gracious hosts from CEN Saclay. However, that laboratory is of little use until we have undertaken a precise determination of what those conditions are. Seismic diagnosis is the most powerful tool that we now have at our disposal for accomplishing that task, and therefore it is quite natural that it should be honed to the best of our ability. It is a means to an end, not an end itself, but a challenging means whose promised fruit is not only a knowledge of conditions inside the sun, but also, as a very important byproduct, a more profound understanding of the physics of the dynamical processes involved in the generation, propagation and dissolution of stellar waves. It is partly this double prize that makes the subject particularly satisfying to pursue.

When I was asked to deliver this closing discussion on open questions, I naturally thought of studying the programme of the meeting to judge what issues were most likely to be of interest. Should I try to anticipate what questions would be addressed, and whether they would be answered satisfactorily? After all, most spontaneous remarks need some preparation. However, on reading the title of the very first lecture, by Evry Schatzman, I realised that that would result in duplication of effort, even though the outcome might be quite different. I therefore decided that it would really be best if only during the meeting I planned what matters to discuss, so that the outcome would reflect the flavour of at least one person's reaction to the deliberations that had actually taken place. The wisdom of this decision was confirmed within minutes of the start, for I would never have anticipated the viewpoint that Jean Andouze would take in his excellent prefatory address.

I present in Figure 1 Jean's principal illustration, reproduced to the best of my memory. It shows that the thermonuclear reactions converting hydrogen to helium in the core of the sun were to have been at the centre of attention, as indeed they were. From them issue neutrinos, the detection of which on Earth provides a very important diagnostic, though of quite what we are not sure. Next, our attention is directed to the opacity, whose influence on the overall structure of the sun is greatest in the radiative midregions, between the reacting core and the convection zone. Had Jean anticipated the stir that was to be caused by Carlos Inglesias' announcement of the outcome of most recent refinements to opacity calculations at Livermore? And then we are led to the all-important acoustic oscillations, most of the energy of which resides in the convection zone occupying roughly the outer 64 per cent, by volume, of the sun. We know that there are discrepancies between theoretical computations of the neutrino flux from so-called standard solar models, particularly when they are calibrated to reproduce to the best of our ability the observed frequency spectrum of acoustic oscillations. So, in what respects is the common perception of the solar interior most seriously wrong? Could it be that wimps have been accreted from a universal sea, or not? These questions were indeed subsequently addressed, but what was not discussed explicitly is the most obvious of the questions raised by Figure 1:

wimps, or not?

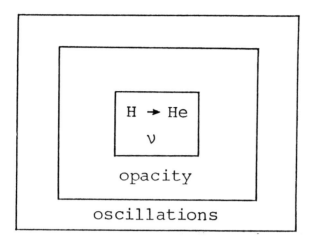

Figure 1. Jean Audouze's principal illustration.

Is the sun square?   (Q9)

Of course, I intend this question to be interpreted in the general
sense: namely, is the structure that of a cuboid (rather than a strict
two-dimensional rectangle with equal sides)? The question must also
be quite profound, for it has been known since time immemorial that
superficially the sun is round: the departure from perfect sphericity,
though not strictly zero as has oftentimes been believed,* has in
modern times been measured to be quite small (e.g. Ambron, 1905; Dicke
and Goldenberg, 1974; Hill and Stebbins, 1975). But what is the shape
of the inside?
     Fortunately, we now know the answer to that question. If the sun
were a cube with edges of semi-length R, for example, the asymptotic
expression for the cyclic frequencies ν of acoustic oscillations would
have essentially the form (cf. Kurtz, 1982):

$$\nu \sim \frac{1}{2}(n^2 + L^2)^{1/2}\ \nu_0\quad,\tag{12}$$

where, once again, ν_0 is given by equation (4) (with r being any of
the Cartesian co-ordinates referred to the principal axes),
$L^2 = \ell^2 + m^2$, and the quantum numbers n, ℓ, and m are integers. For
simplicity I have set to zero a quantity ε arising from the effective

*This belief appears to have arisen in part from Man's interpretation
of God's judgement (Moses, date uncertain) that the sun was good.

phase shift suffered by the waves in the reflecting layers beneath the surface. To include this term is straightforward, as also is the generalization from cubic to cuboidal symmetry; the details are of no matter to my argument. What is important is that the observed frequencies fit the functional form of neither the relation (12) nor its generalizations, whereas they do fit the relation (2), whose functional dependence on n and ℓ (and the degeneracy with respect to m) is the signature of a sphere. Notice how much simpler this argument is than a direct comparison of solar frequencies with the eigenfrequencies of a cubical solar model. More important than simplicity is the appreciation of the signatures of the different symmetries. These can be ascertained quite generally from asymptotic analysis, at least for modes of (sufficiently) high frequency, whereas numerical eigenfrequency computations alone merely provide specific examples.

Many of the discussions throughout this week have quite naturally adopted a standard solar model as a point of reference. The question that must therefore have been raised in the minds of all those who do not already know the answer is:

Is the sun standard? (Q10)

Before even attempting to answer that question one must appreciate what it means, which requires first an answer to the question:

What is the standard solar model? (Q11)

On this issue there is some diversity of opinion, as was evident from the outset from Evry Schatzman's introduction. Evry wishes to include in the standard solar model all the generally accepted physics, including macroscopic motion, reserving for "nonstandard" models only so-called new physics. Thus the standard solar model should include such phenomena as rotation, a magnetic field, large scale circulation, microscopic diffusion, turbulent mixing and material and momentum transport by waves. In particular, it should take into account the nonlinear development of any instabilities that are found. The objective is to obtain a standardized theoretical model, within the framework of standard physics, that provides the most faithful representation of the sun possible.

That is the view of an idealist. The trouble with it in practice is that nobody understands macroscopic physics well enough to carry out the requisite calculations. Therefore each attempt to construct a standard model is likely to be different, and the resulting models would therefore not be standard. Surely it is better to ignore all these complexities, and naively construct a spherically symmetrical well-balanced model that satisfies a simple set of differential equations that can at least be solved. Provided the model did not represent conditions inside the sun too poorly, it would serve as a useful stable basis for comparison of more realistic models. This appears to be the opinion of most of those who actually compute standard solar models (or, perhaps to be more precise, the view expressed in most of the published papers presenting standard solar models), and was well presented in John Bahcall's excellent review of his view of the situ-

ation. John and his collaborators, more than any other group, have
painstakingly assessed the sensitivity of a standard model to every
uncertain parameter they can think of in their description of the
physics, and have addressed how observable quantities (most notably,
the neutrino fluxes) are affected. Their results are summarized in
two mammoth opera (Bahcall et al., 1982; Bahcall and Ulrich, 1988)
which should be compulsory prior reading for anyone contemplating
entering the field. What is abundantly clear from these and the many
other publications on the subject is how the standard model has re-
sponded to new announcements of modifications to nuclear reaction
cross sections, opacities and the equation of state. Thus theoretical
neutrino fluxes, for example, have varied with time, though in recent
years they seem to have hovered within a stable 6-8 snu. So at least
we know the answer to the question:

Is the standard model standard?   (Q12)

One of the sources of variation amongst standard solar models
appears to be numerical imprecision. There is no good excuse for
that. The idealized governing differential equations in the regime
pertinent to the main-sequence history of the sun have no peculiar
properties, and even though they are singular at the centre, that
singularity (which is merely a coordinate singularity) is regular and
is quite straightforward to handle. The same is true of at least the
simple linearized adiabatic pulsation equations that are generally
used for computing oscillation eigenfrequencies. Because of the
assumption of spherical symmetry, the mathematical problem to determine the basic structure of the model is posed in only two dimensions
(spanned by the independent variables r and t), so with modern computers it is in principle quite easy to attain a resolution sufficient to
render truncation error, which I presume, aside from programming
errors, is the major source of imprecision, negligibly small. The
oscillation problem is reduced to only one dimension, and should be
adequately resolved more easily. Modern seismic data, in particular,
have caused us to revise our computational standards, because some
physically important properties of the frequency spectrum depend quite
sensitively on aspects of the model that hitherto have not been considered worthy of accurate modelling. That numerical imprecision is
at least partially responsible for some of the theoretical error
is exhibited, for example, by Ulrich and Rhodes, who in two separate
publications (in 1983 and 1984) presented oscillation frequency spectra (presumably computed separately) of the same solar model, one of
which can be fitted by the asymptotic formula (2) with standard
deviation E less than one tenth of that of the other (Gough, 1986c).
Please note that I use E here merely as a means of indicating the
large variation of a quite subtle feature of the pattern of eigenfrequencies (the relative differences between the corresponding
eigenfrequencies in the two publications is very small). I do not
intend it to be used as a factor for deciding which of the frequency
sets is the more accurate, particularly because at least one of the

authors quite understandably regards the asymptotic formula (2) as
being materially inadequate to describe the numerical results (Bahcall
and Ulrich, 1988). Nor do I intend it to be inferred that the computations by Rhodes and Ulrich are less accurate than others. (Indeed, there is considerable circumstantial evidence to suggest that in
some instances that is very far from the case.) I use this illustration simply because it is the best example I know of an essentially
duplicated published theoretical data set that is relevant to my
discussion. It shows, however, that at least some modern calculations
are too inaccurate. Even though we may not understand the physics of
the solar interior, it is the responsibility of every model-builder in
the subject to find representations of the solutions of the governing
equations that permit the determination of eigenfrequencies to a
precision at least as great as that attained by the observers. Otherwise it will not be possible to know whether discrepancies are indicative of errors in the physics or mere carelessness. It is therefore
of extremely great importance that Jørgen Christensen-Dalsgaard, as
part of the research effort of the Global Oscillations Network Group
(GONG), has undertaken to lead a thorough purge of uncontrolled
numerical error in a group of solar models. The intention is that,
together with his collaborators of like mind in the research group, he
will thus be able to provide a solar model built with clearly defined
physics to a known precision. That standard model will surely become
a standard.

Acquiring a standard standard model does not imply that we would
have a faithful representation of the sun. We already know that
models such as those discussed at this meeting by John Bahcall and by
Sylvaine Turck-Chièze and her collaborators, which I am quite sure
have been computed precisely enough for their purposes, do not agree
with observation (or each other), and are therefore not correct.
Indeed, to my knowledge no standard model that has ever been produced
is correct; in the light of Evry Schatzmann's and André Maeder's
discussions, nobody should expect them to be. That surely answers
question (Q10). But that does not mean that they are not useful. As
André Maeder has reminded us, an essential step towards understanding
Nature is understanding wrong theories.

Without doubt the most extensively discussed discrepancy is the
neutrino luminosity L_ν, which is usually expressed as a flux at Earth,
and often in units of neutrino capture rates in a terrestrial detector
(Bahcall, 1969). One of the most outstanding questions in our subject
is therefore:

What is the value of L_ν? (Q13)

To answer that question we need to know not only the neutrino flux on
Earth, but also what has happened to the neutrinos (ν) during their
passage from the sun. The latter issue has been comprehensively discussed by Haim Harari and Alexei Smirnov in their two excellent reviews. Between them they seem to have raised more open questions
about ν creation, ν types and ν transitions than have been posed on
any other issue discussed at this meeting. What was abundantly clear

from these talks was that understanding of ν physics will come from
combining information from nuclear and particle physics, cosmology and
astrophysics; it cannot be achieved by any one of those branches of
science alone. It is therefore of paramount importance that workers
in these fields be brought together, as has occurred at this conference. Of couse it is necessary for communcation that a common scientific language be spoken, which makes me wonder whether that is in the
minds of those solar physicists who insist on quoting the sun's rotation rate as a cyclic frequency. At present, however, with regard to
the neutrino problem we cannot answer the fundamental questions:

$$\text{Is nuclear physics correct?} \qquad (Q14)$$

$$\text{Is particle physics correct?} \qquad (Q15)$$

The answers to some of the questions raised by the neutrino
physicists will come from the various new ν detectors described early
in the meeting. In some cases the role of the sun will be solely that
of providing a source, whose properties need be known only approximately. The most obvious example is a potential measurement of the
low-energy ν produced by the p + p → D reactions at the beginning of
the proton-proton chain. Although we cannot yet answer the question:

$$\text{Is the sun in thermal balance?} \qquad (Q16)$$

precisely, we are certainly confident that the total rate of generation L_n of thermonuclear energy, almost all of which is a product of
reactions in the proton-proton chain, is presently in approximate
balance with the photospheric luminosity $L_\odot$. Therefore, if a discrepancy between theoretical and measured low-energy ν_e fluxes were
found that is as great as that already encountered for the higher-
energy ν_e, we would surely conclude that ν transitions must have
occurred.

Interpreting ν data to answer more detailed questions will require more detailed and more precise knowledge of conditions inside
the sun, which is partly why the programme to infer the solar structure from seismic observations is so important. The structure of the
reacting core is the most valuable goal, and at this meeting we have
been presented with preliminary and conflicting results of two independent attempts to determine it, one by Wojtek Dziembowski and his
collaborators and the other by Alexander Kosovichev and his collaborator. We do not yet know why the results disagree, but considering
the extreme delicacy of inversions of only high-order p modes to
determine core structure, and bearing in mind that different unproven
procedures on different data sets were employed, that there is disagreement should perhaps be hardly surprising. Nevertheless, it is
important to understand the results, and to add to the excitement of
that challenge I wager Wojtek Dziembowski that conditions in the solar
core will be found to be closer to those estimated by Alexander and
his collaborator than to those by his own group, the measure of close-
ness being the factor by which the neutrino flux (in snu) differs from
that of the appropriate standard model, computed using the same reaction

465

physics as that adopted by Bahcall and Ulrich (1988).* The underlying
implictions of the wager are complicated, because they relate not only
to the influence of possible errors in the data, or of errors in the
inversion procedures, but also to fundamental inconsistencies that may
be present as a result of incorrect assumptions that are embodied in
the frequency constraints that are inverted. I have in mind, for
example, deviations from thermal and nuclear balance that may have
arisen from the nonlinear development of instabilities in the core.
It is interesting to note, for example, that, so far as I am aware,
all dynamical stability studies of the sun published in the last 17
years have found the core to have been unstable to g modes at some
epoch since arrival on the main-sequence, yet modellers, as Ian
Roxburgh complains, almost invariably ignore the consequences. Perhaps their growth has been suppressed at an inconsequential amplitude
by resonant coupling to stable modes (Dziembowski, 1983). But if not,
material and thermal redistribution in the core would have had a profound influence on L_ν. Therefore I include as an important open
question:

Is the core disturbed? (Q17)

I should point out that theoretical solar models are usually
calibrated to reproduce the observed solar radius and luminosity at an
appropriate age $t_\odot$ after arriving on the main sequence. That calibration is an essential feature of any complete description of the sun,
as Jørgen Christensen-Dalsgaard forcefully argued in his model talk.
It is normally accomplished by adjusting the initial helium abundance
Y_0, at given fixed Z_0 or Z_0/X_0, and the mixing-length parameter α
appearing in the formalism determining the entropy gradient in the
convection zone. The calibration is unique, and so yields a one-
parameter sequence of solar models, each of which can be labelled with
the unique value of Y_0. To choose the most appropriate model one
needs answer the question:

What is the value of Y_0? (Q18)

The answer to that question has an obvious important bearing on
theories of helium production during the first fifteen minutes or so
after the Big Bang.
Of course one could ask whether there is a model in the sequence
that reproduces the observed neutrino flux, and if there is, select
that model. There is such a model, but that has a value of Y_0 (about

*At the time of the meeting Kosovichev and his collaborator had made
only a very crude estimate of that factor (0.6) based on a simple
extrapolation from conditions at r = 0 and assuming the entire flux to
scale as the dominant [8]B flux. A more careful estimate (0.7) is
published in these proceedings. The wager, which was accepted by
Dziembowski for the stake of a bottle of cognac, is of course for the
original factor, against Dziembowski's factor of 1.7, which is the
reciprocal of 0.6.

0.15) which is substantially lower than the values observed in the
atmospheres of hot stars, and in any case is in serious conflict with
almost all cosmologies. Moreover, it has also been ruled out by oscillation data. The neutrino problem therefore remains.

Seismic calibrations of solar models to determine Y_O have
revealed since the earliest attempts (Christensen-Dalsgaard and Gough,
1980, 1981) that, even if L_ν is ignored, a model cannot be found that
reproduces the observed data. The situation is basically that if one
approximates the asymptotic expression (2) by its first term only, one
cannot adjust the single parameter Y_O to fit simultaneously both the
global parameter ν_0 and the surface phase parameter ε which characterize the low-degree data. (If one includes the second term as a
guide, and compares more subtle features of the frequencies, additional discrepancies are revealed.) Therefore there must be errors in
the physics of the standard model which, as Christensen-Dalsgaard and
Gough (1984) and Christensen-Dalsgaard (1988) have argued using also
the frequencies of intermediate degree, must be present both in the
radiative interior and in the surface layers. Attempts by others to
fit the seismic data have failed similarly. However, since we do have
a picture from the nature of the discrepancies of where and in what
respect the solar models are in error, and, moreover, since those
errors are of the kind that depend on physics in which I am sure we
should not be confident, I cannot agree with the opinion expressed by
Ulrich and Rhodes (1983), nor with those who have subsequently quoted
them, that the significance of these discrepancies is comparable to
the failure to predict the neutrino flux.

As a topical example I consider the mid-regions of the sun. An
asymptotic inversion of frequencies by Christensen-Dalsgaard et al.
(1985) revealed that the sound speed between $r \simeq 0.3R$ and $r \simeq 0.6R$ is
about 1 per cent greater than in a typical standard model, a result
that has subsequently been confirmed by several other inversions, both
asymptotic and otherwise. Christensen-Dalsgaard et al. pointed out
that that discrepancy could be reduced to an insignificant level in a
standard model if the opacity in the radiative envelope immediately
beneath the convection zone, between temperatures of about 10^6 K and
4×10^6 K, were increased by about 20 per cent. (The temperature at
the base of the convection zone is actually about 2×10^6 K, and since
the opacity in the adiabatically stratified lower regions of the convection zone has essentially no influence on the structure of the
star, one can make no seismic deduction about the value of the opacity
below abut 2×10^6 K.) More recently, Korzennik and Ulrich (1989) and
Cox, Guzik and Kidman (1989; reported by Art Cox at this meeting) have
reached similar conclusions. The report we heard by Carlos Inglesias
that the outcome of the most recent opacity calculations at the
Lawrence Livermore National Laboratory are consistent with these ideas
is therefore of very great interest, for it suggests (and I suspect it
will be shown that it even demonstrates) that the mid-regions of the
sun are in thermal balance. This goes some way towards answering
question (Q16).

The resolution of some of the discrepancies in the outer layers
of the sun appears also to have been found. Beneath the upper super-
adiabatically stratified convective boundary layer, the structure

within the convection zone is quite close to being adiabatic, and we
believe that fluctuations are relatively small. Moreover, it is also
likely that in this region of the sun the magnetic field is dynamically unimportant. Therefore the structure of the convection zone
depends on only a few parameters: principally the value of the entropy
and the helium abundance Y (which is probably the same as Y_0),
provided the equation of state is known. Yet it is required to fit a
whole spectrum of frequencies. Evidently, previous failures to do so
precisely must reflect errors in the equation of state, and it is once
again encouraging that the recent computations discussed by Jørgen
Christensen-Dalsgaard using the new equation of state developed by
Mihalas, Hummer and Dappen have reduced the discrepancies considerably. This is a good illustration of how the sun can reliably be
used as a physics laboratory. In future it is hoped that by studying
the detailed stratification in the ionization zones it will be possible at least to determine Y. My hope is even that it will eventually
be possible also to determine the abundances of carbon and oxygen, and
to test some of the physical assumptions upon which the calculations
of the equation of state depend.

There remain contributions to the frequency discrepancies that
arise from errors in the representation of the superadiabatic convective boundary layer. There, the physics of both the basic state
and the oscillations is very uncertain: convective fluctuations in the
macroscopic state of the gas are substantial, radiative transfer becomes important, at least near the photosphere, the complicated exchange of energy and momentum between the convection and the oscillations is significant, both directly contributing to the dynamics of
the oscillations locally, and indirectly through the nonlocal influence on the eigenfunctions which modifies the dynamics elsewhere, and
magnetic fields, possibly in the form of concentrated fibrils, may
scatter the acoustic waves, to mention but a few of the problems. All
these issues can be raised as open questions.

Finally, let us be reminded that there are errors in the core of
the sun. Continuing to set aside the neutrino problem (for one of our
goals is still to determine the structure of the core independently
of L_ν, in order to ascertain whether the resolution of the neutrino
problem is to be found as a revision of our view of solar structure,
or in nuclear or particle physics), these were first detected as a
discrepancy in the value of A in equation (2) [the mean value of
$d_{n,0}$ is predicted to be about 10 μHz by standard solar models, whereas
the latest observations yield 9.2-9.3 μHz (G. R. Isaak, personal communication; C. Fröhlich, these proceedings)], which implied, according
to equation (7), that a mean value of the sound-speed gradient dc/dr
is greater in the sun than in the theoretical models, the discrepancy
responsible for that being most probably in the core. I have already
discussed the conflicting reports on attempting to elaborate on this
result. I have raised the matter again here partly to complete my
list of errors in current standard solar models (and thus to show that
they exist almost everywhere), though mainly as a means of emphasizing
how remarkable it is that we can even realistically hope to attempt to
estimate the structure of the most inaccessible region of the sun from
only acoustic waves which must propagate from the core through all the

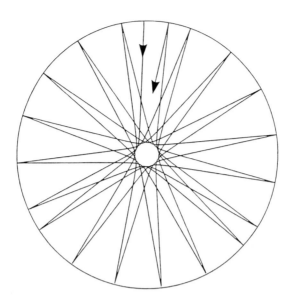

Figure 2. Ray path of an acoustic wave with n/ℓ = 5. The radial co-ordinate has been distorted so as to be uniform in acoustical radius τ. Note that the ray path is not closed, which is generally the case for all nonradial modes.

other erroneously modelled and in some places ill-understood regions of the sun before reaching the surface where they are observed.

Why that is so is best understood in terms of (asymptotic) ray theory. Figure 2 illustrates the path of a typical ray of acoustic waves in a spherically symmetric star whose frequency is presumed to be such that the waves interfere constructively to constitute a reso-nant mode of low degree. The value of the frequency is determined by a quantization condition essentially on the sound travel time along a ray. Therefore the picture has been stretched radially, in such a way that radial distance is proportional to sound travel time $\tau = \int c^{-1} dr$ in the radial direction, in order to provide a more appropriate acous-tical weighting. The most pertinent feature of the pattern is the central zone of avoidance, which is bounded by the envelope of the rays, where they form a caustic surface. This property is important for two reasons. First, the radius of the caustic surface, which depends principally on the ratio ν/L [cf. equation (8)], varies from mode to mode, whereas far from the caustic the rays of all low-degree modes are almost radial and consequently very similar. Therefore, by suitably combining mode frequencies [as a first approximation, merely by subtracting the two almost equal frequencies of modes of like n + (1/2)ℓ (cf. equation (2)) with ℓ differing by 2 (the smallest differ-ence possible), which determines the quantity $d_{n,\ell}$ defined by equation (10)] the large yet nearly identical contributions from the almost radial rays which pervade most of the star can be made to cancel, and the result depends mainly on conditions within the vicinity of the caustic surfaces. Secondly, because the rays are almost horizontal near the caustic surface, they spend much more time in a given radial interval $d\tau$ of τ in the neighbourhood of the caustic than they do elsewhere, and therefore the relative contribution to the frequency from the core, though very small, is greater than one might have expected by naively comparing sound speeds. It is worth noting now that in a body with cubic symmetry, having acoustic frequencies satisfying equation (12) or a generalization of it, there is generally no caustic surface. Therefore, had the sun been square, determination of the structure of the core from acoustic modes would have been much more difficult: probably impossible with observational data at the present level of accuracy. That surely answers question (Q7) affirmatively.

Before I leave the subject of the core, I must at least mention the currently fashionable question:

Does the sun harbour wimps? (Q19)

The question has been asked by several of the contributors to this meeting. It was addressed most extensively in David Spergel's well-balanced presentation, and also by John Faulkner standing uncharacteristically (and metaphorically) at the periphery; John's planned contribution to this meeting was his entertaining discussion of the advantages of infesting other stars with wimps. The attractive feature of imagining the sun to have collected an appropriately adjusted cloud of wimps is that a theoretical model that has been tuned to reproduce the obseved neutrino flux has been found also to more-or-less reproduce the value of the seismic parameter A appearing in equation (2). However, according to Gough and Kosovichev (1988), details of the sound-speed distribution in the solar core inferred from seismic analysis are not in accord with the theoretical model C of Faulkner and Gilliland (1985) which harbours wimps. John Faulkner has chided us at this meeting for misrepresenting the case, because a rise of the relative sound-speed difference (between the sun and a standard solar model) as one approaches $r = 0$, which is exhibited in the solar inversions depicted in Figure 11 of Gough and Kosovichev (1988) and which I reproduce here in Figure 3, is not present in the representation of the wimp model in Figure 9 of Gough and Kosovichev, whereas it should be. I should perhaps first explain the reason. It was difficult to estimate the sound speed in the wimp model because, as Faulker et al. (1986) encountered, insufficient information was provided in the original paper. It was necesary to supplement the information with data from a different model, scaling appropriately, which was not wholly consistent. Moreover, those data were measured from figures, to which it was evident that some draughtsman's licence had been granted, which added to the uncertainty. Finally the hydrostatic equations of stellar structure were integrated using a pocket calculator (which explains why only a simple, though adequate, approximation to the equation of state was employed). In order to produce a solar model which accurately satisfied the hydrostatic equations, Gough and Kosovichev reintegrated those equations using the value of $\Gamma = d\ln p/d\ln\rho$ obtained by drawing a smooth curve through the data of

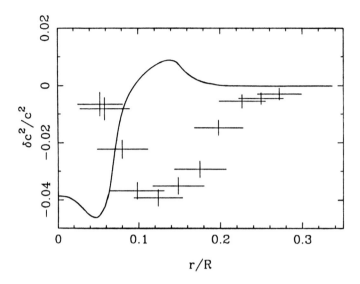

Figure 3. The crosses represent relative differences $\delta c^2/c^2$ in sound speed c between the sun and Christensen-Dalsgaard's (1982) standard solar model 1, inferred by Gough and Kosovichev (1988). The continuous line is the estimate by Faulkner, Gough and Vahia (1986) of the corresponding difference between the wimp infested and the standard model.

Faulkner et al. to define the stratification; they could have chosen c^2 instead, but did not. Consequently the value of c^2 was not identical to that inferred by Faulkner et al. To set the record straight I include in Figure 3 a curve drawn through the values of $\delta c^2/c^2$ taken from the working notes of Faulkner et al. (referred, perhaps not surprisingly, to a different standard model); the curve does show a slight upturn, but it is not as pronounced as that inferred for the sun. Therefore the conclusion that there is a substantial discrepancy is maintained. But that does not imply that the sun does not contain energy-transporting wimps. We have analysed but a single wimpish model. Perhaps some careful tuning of the several free parameters in the wimp theory could result in a theoretical model that is not yet ruled out by observation.

I shall conclude, quite briefly, with the dynamical questions that were discussed in connection with the solar cycle:

> How does the sun rotate? (Q20)

> What controls the solar cycle? (Q21)

We have a partial answer to question (Q20). The angular velocity, averaged in a rather complicated way about the equatorial plane, has

been determined as a function of r by Duvall and his collaborators
(1984). Aside from a shallow maximum at a radius of about 0.9R, this
equatorial average of the angular velocity appeared to be roughly
constant in the convection zone, and then declined slowly with depth
until the edge of the energy-generating core.

> Does the radiative interior of the sun really   (Q22)
> rotate more slowly than the convection zone?

And if so;

> Why does the radiative interior of the sun   (Q23)
> rotate more slowly than the convection zone?

The answer to the first of these two questions can be obtained only
observationally, presumably by more precise measurements.
    More detailed measurements of the nonuniformity of the rotational
splitting of acoustic modes have permitted us to infer the latitudinal
distribution of the sun's angular velocity down to a depth of about
0.5R. The results are summarized in Figure 4. It appears that,
roughly speaking, the latitudinal variation observed in the photosphere is maintained throughout most of the convection zone, and then
there is a transition towards latitudinally independent rotation
beneath. How abrupt is the transition cannot yet be determined, since
it appears to occur on a scale no greater than the resolution length
of the data. Nor do we yet know whether the radiative interior
rotates approximately uniformly on spheres throughout, or whether at
depths greater than that to which the current measurements penetrate
the angular velocity is greater at the poles than it is at the
equator. Roughly speaking, however, the value of the angular velocity
in the radiative interior immediately beneath the convection zone,
though lower than in the convection zone in the equatorial regions, is
greater at high latitudes, and appears to be such as to lead to
approximately no torque being applied between the two regions of the
sun, assuming that the stress between the zones is proportional to the
shear. That condition is just what is required for a steady state,
and therefore goes some way towards both interpreting and answering
question (Q23).
    Understanding what happens at greater depths is perhaps a greater
challenge. There observations are much less secure, and much less
detailed.
    A particularly interesting and challenging question is thus:

> How does the core rotate?   (Q24)

The inversion carried out by Duvall and his collaborators indicates
that the core might be rotating rather more rapidly than the rest of
the sun. But that conclusion depends on the rotational splitting of
the lowest-degree modes, which are the least accurately determined.
There is corroborative evidence for the high rotational splitting of
low-degree p modes from the whole-disk observations of Isaak and his
collaborators (Isaak, 1986), and some evidence from apparent g-mode

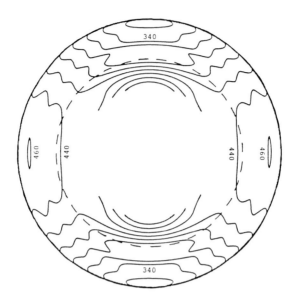

Figure 4. Contour diagram of the sun's angular velocity, estimated from the inversions by Christensen-Dalsgaard and Schou (1988), Dziembowski, Goode and Libbrecht (1988) and Brown et al. (1989).

data, but there is considerable doubt over the interpretation of those observations. The question still is quite open, so much so that Jack Harvey and I have a wager on the issue.* Indeed Claus Frohlich's comment at this meeting that his line-width measurements of dipole and quadrupole p modes are consistent with the core rotating at the same rate as the surface throws some doubt on the previous contradictory claims.

An interesting consequence of a rapidly rotating core is that it is unlikely to be pure steady rotation. Associated with the rotational shear would be a meridional component to the flow, and the whole flow is likely to vary with time. Could it be with a characteristic period of 22 years? The interesting discussions by Jean-Paul Zahn and Henk Spruit at this meeting show that the situation is not yet understood, though the recent observations have certainly triggered a redoubling of theoretical interest.

> Is the motion in the core essentially laminar, and on only a large scale? (Q25)

> On what timescale does it vary? (Q26)

*I have wagered that his observations are basically right, and he that they are wrong.

I was interested that Jean-Paul Zahn appears to be taking the idea of
^{3}He fingers seriously. Surely fingers would act as relatively efficient absorbers of acoustic waves, and so enable p modes to transfer
angular momentum between the core and the convection zone on an interesting time scale. Another possibility, which has attracted the
attention of several people during the last two decades, is that there
are g modes of quite high amplitude trapped in the core. As Ian
Roxburgh has reminded us at this meeting, these will modify the
nuclear reaction rates in the core, both directly through the nonlinearities in the fluctuations, and indirectly due to the modification brought about by nonlinear fluctuation interactions to the
thermal stratification of the core. It now seems not unlikely that
the outcome would be a reduction in the neutrino flux. It would also
lead to a rapidly rotating core, surrounded in the equatorial regions
by a region of relatively slow rotation. That is consistent with the
equatorially averaged angular velocity inferred by Duvall et al.
(1984) from seismic rotational splitting data. Nevertheless, one
cannot allow one's thoughts to wander this far without forgetting
Dziembowski's (1983) conclusion that the modes cannot attain an
amplitude great enough for this phenomenon to be significant.
Associated with all these issues is the question:

What is the geometry and the strength of the   (Q27)
internal magnetic field?

Henk Spruit had concluded, by reasoning I do not fully understand,
that there is a sheet of field at the base of the convection zone of
intensity of about 10^4 G. This is contradicted by the analysis of the
even component of Ken Libbrecht's p-mode splitting data by Wojtek
Dziembowski and Phil Goode, who argue that they have detected a field
a thousand times more intense. This does not prove Henk wrong, however, because it is not yet known whether the p-mode splitting is
produced by a magnetic field. One possibility is that it is a latitudinal variation of the thermal stratification at the base of the
convection zone. Wojtek Dziembowski has argued against this on the
ground that one cannot induce asphericity by purely thermal means
without destroying hydrostatic balance. The ground cannot be denied,
but its relevance can certainly be questioned. Kuhn (1988) implicity
did just this, by disregarding hydrostatic balance when trying to
explain splitting data in terms of sound-speed variations, in the face
of previous calculations that had taken some account of the hydrostatic constraint and had thereby failed to reproduce the earlier
splitting data of Duvall et al. (1986) without producing a photospheric temperature variation that was implausibly large (Gough and
Thompson, 1988). It seems likely, therefore, that the mean stratification in the convection zone is not in strict hydrostatic equilibrium.
A rough estimate of the large-scale flow velocity that would thus
be driven by the hydrostatic imbalance resulting from a thermal distortion of magnitude sufficient to produce the p-mode splittings is
quite moderate, and does not seem to be contradicted by observation.
Such a flow, varying through the solar cycle, is likely to modify the

latitudinal variation of the angular velocity of the sun, on the rectified 11-year period. Whether such a modulation will be detected in rotational splitting data in the foreseeable future is certainly an open question.

References

Ambron, L., 1905, Astr. Mittheilungen (Göttingen Obs.), Part VII.
Bahcall, J. N., 1969, Phys. Rev. Lett., 23, 251-254.
Bahcall, J. N., Huebner, W. F., Lubow, S. H. Parker, P. D. and Ulrich, R. K., 1982, Rev. Mod. Phys., 54, 767-799.
Bahcall, J. N. and Ulrich, R. K., 1988, Rev. Mod. Phys., 60, 297-372.
Brown, T. M., Christensen-Dalsgaard, J., Dziembowski, W. A., Goode, P. R., Gough, D. O. and Morrow, C. A., 1989, Astrophys. J., 343, 526-546.
Christensen-Dalsgaard, J., 1988, Seismology of the sun and sun-like stars (ed. E. J. Rolfe, ESA SP-286, Noordwijk), pp. 431-450.
Christensen-Dalsgaard, J., Duvall Jr., T. L., Gough, D. O., Harvey, J. W. and Rhodes Jr., E. J., 1985, Nature, 315, 378-382.
Christensen-Dalsgaard, J. and Gough, D. O., 1980, Nature, 288, 544-547.
Christensen-Dalsgaard, J. and Gough, D. O., 1981, Astron. Astrophys., 104, 173-176.
Christensen-Dalsgaard, J. and Gough, D. O., 1984, Solar seismology from space (ed. R. K. Ulrich et al.; JPL publication 84-84, Pasadena), pp. 193-201.
Christensen-Dalsgaard, J., Gough, D. O. and Thompson, M. J., 1989, Mon. Not. R. astr. Soc., 238, 481-502.
Christensen-Dalsgaard, J. and Schou, J. 1988, Seismology the sun and sun-like stars (ed. E. J. Rolfe, ESA SP-286, Noordwijk), pp. 149-153.
Claverie, A., Issak, G. R., McLeod, C. P., van der Raay, H. B. and Roca Cortes, T., 1979, Nature, 282, 591-594.
Cox, A. N., Guzik, J. A. and Kidman, R. B., 1989, Astrophys. J., 342, 1187-1206.
Deubner, F-L. and Gough, D. O., 1984, Ann. Rev. Astron. Astrophys., 22, 593-619.
Dicke, R. H. and Goldenberg, H. M., 1974, Astrophys. J. Suppl., 27, 131-182.
Duvall Jr., T. L., Dziembowski, W. A., Goode, P. R., Gough, D. O., Harvey, J. W. and Leibacher, J. W., 1984, Nature, 313, 27-28.
Dziembowski, W. A., 1983, Solar Phys., 82, 259-266.
Dziembowski, W. A., Goode, P. R. and Libbrecht, K. G., 1988, Astrophys. J. Lett., 337, L53-L57.
Faulkner, J. and Gilliland, R. L., 1985, Astrophys. J., 299, 994-1000.
Faulkner, J., Gough, D. O. and Vahia, M. N., 1986, Nature, 321, 226-229.
Gough, D. O., 1984, Solar seismology from space (ed. R. K. Ulrich et al., JPL Publication 84-84, Pasadena), pp. 49-78.
Gough, D. O., 1986a, Hydrodynamic and magnetohydrodynamic problems in the sun and other stars (ed. Y. Osaki, Univ. Tokyo Press, Tokyo), pp. 117-143.

Gough, D. O., 1986b, Seismology of the sun and the distant stars (ed. D. O. Gough, NATO ASI Ser. C, Reidel, Dordrecht) 169, 125-140.

Gough, D. O., 1986c, Highlights of Astronomy (ed. J-P. Swings) 7, 283-293.

Gough, D. O., 1990, Mat. Fys. R. Dan. Acad. Sci., in press.

Gough, D. O. and Thompson, M. J., 1988, Advances in helio- and asteroseismology (ed. J. Christensen-Dalsgaard and S. Frandsen, Reidel, Dordrecht), pp. 175-180.

Grec, G., Fossat, E. and Pomerantz, M., 1980, Nature, 288, 541-544.

Grec, G., Fossat, E. and Pomerantz, M., 1983, Solar Phys., 82, 55-66.

Harvey, J. W. and Leibacher, J. W., 1984, Nature, 310, 22-25.

Hill, H. A. and Stebbins, R. T., 1975, Astrophys. J., 200, 471-483.

Isaak, G. R., 1986, Seismology of the sun and the distant stars (ed. D. O. Gough, Reidel, Dordrecht), pp. 223-228.

Korzennik, S. G. and Ulrich, R. K., 1989, Astrophys. J., 339, 1144-1155.

Kosovichev, A. G., 1988, Seismology of the sun and sun-like stars (ed. E. Rolfe, ESA SP-286, Noordwijk), pp. 533-537.

Kuhn, J. R., 1988, Astrophys. J., 331, L131-L134.

Kurtz, D. W., 1982, Mon. Not. R. astr. Soc., 200, 807-859.

Lamb, H., 1908, Proc. Lond. Math. Soc., 7, 122-141.

Lubow, S. H., Rhodes Jr., E. J. and Ulrich, R. K., 1980, Nonradial and nonlinear stellar pulsation (ed. H. A. Hill and W. A. Dziembowski, Springer, Heidelberg), pp. 300-306.

Moses (date uncertain), The Holy Bible, Genesis, 1, 14-19.

Olver, F. W., 1974, Asymptotics and special functions (Academic Press, New York).

Rayleigh, 1894, The theory of sound (2nd edn, Macmillan, London).

Sekii, T. and Shibahashi, H., 1989, Publ. Astr. Soc. Japan, 41, 311-331.

Tassoul, M., 1980, Astrophys. J. Suppl., 43, 469-490.

Ulrich, R. K. and Rhodes Jr., E. J., 1983, Astrophys. J., 265, 551-563.

Ulrich, R. K. and Rhodes Jr., E. J., 1984, Solar seismology from space (ed. R. K. Ulrich et al., JPL Publ. 84-84, Pasadena), pp. 371-377.

Vandakurov, Yu. V., 1967, Astron. Zh., 44, 786-797.

Vorontsov, S. V., 1988, Seismology of the sun and sun-like stars (ed. E. J. Rolfe, ESA SP-286, Noordwijk), pp. 475-480.

SOUVENIRS

Il était une fois un Roi Soleil ...

Au dix-septième siècle, lorsque le Roi devient Soleil, un nouvel univers s'organise autour de cet astre. On y connaît des planètes proches et influentes, qui sont les familiers, les favorites, tel ou tel grand prêcheur et même quelques nobles, dont il faut apprivoiser la lumière, pour l'atténuer et même l'étouffer au contact d'un éclat supérieur.

Ces astres familiers suivent un cours réglé par une loi précise, et cette loi s'appelle l'étiquette. On ne saurait s'y soustraire, ni la briser, pas plus que les étoiles ne peuvent rêver d'indépendance. Du petit lever au petit coucher , en passant par toutes les phases de la journée royale, les planètes intimes évoluent selon l'ordre fixé. Si par accident, ou par folie, une de ces planètes décidait de rompre cet ordre, elle serait frappée d'extinction, reléguée par décret dans un exil obscur, loin de toute lumière. Cette disparition d'étoile est appelée une disgrâce.

Au-delà des planètes proches, s'organise un univers compliqué, ou tout au moins une galaxie, qu'on appelle la cour. Un siège céleste nouveau a été construit pour cette galaxie. C'est le Château de Versailles, où une galerie des glaces multiplie les distances et chavire les perspectives.

Cette galaxie, qui tourne autour d'un soleil unique - parfaitement nommé monarque, c'est-à-dire le seul - connaît toutes les agitations physiques des corps célestes. On y est aspiré par des tourbillons inexplicables, on y frôle des cataclysmes, des chutes, des vertiges, de soudaines élévations, on y voit des explosions sur la couronne du Soleil, et ces explosions provoquent de profondes brûlures, des rages, des guerres et toutes sortes de soulèvements. En cherchant bien, dans les couloirs labyrinthiques, on peut aussi trouver des naines blanches, des géantes rouges, des quasi-stars et même des poussières interstellaires, qu'on remarque à peine, et qui portent pourtant le monde de demain.

La loi générale est simple, et elle est de type centripète : toute la galaxie tend à se rapprocher du Soleil. L'apercevoir est un privilège, être aperçu par lui est un luxe. Quand il s'arrête un instant dans sa course pour vous dire un mot, c'est le monde entier qui s'arrête, étonné, et vous êtes alors dans un bonheur rare et si enviable qu'il peut devenir dangereux. Si ces quelques mots deviennent une invitation, si vous êtes admis à entrer en vrai contact avec le Soleil, vous subissez une métamorphose, vous devenez une planète familière, et votre vie change.

From the Sun King to the King Sun

In the seventeenth century, when King becomes Sun, a new universe forms around this Star made up of nearby and influential Planets: family members and favourite women, famous preachers, and even a few noblemen, whose light must be tamed, dimmed or even smothered by a brighter light.

These familiar bodies follow a course ruled by an exacting law defining a solar system: a Court protocol by the name of "étiquette". No one would avoid it or break it, any more than stars might dream of independence. From the "little rising" to the "little setting", passing through all phases of the royal day, the intimate Planets must evolve in a set order. If, by accident or through madness, one of these Planets decides to break this order, it is extinguished, sent by decree to some dark and remote exile light-years from the Sun. The disappearance of such a celestial body is then simply called a "disgrace".

Beyond the nearby Planets exists a complicated universe, a Galaxy, called the Court. A new seat has been built to host this Galaxy. This is the Château of Versailles, where a Gallery of Mirrors multiplies distances and distorts perspectives.

This Galaxy, revolving around the one and only Sun (pointedly called Monarch, that is "the Only One"), is subject to the physical movements of all the celestial bodies. One is sucked in by mysterious whirlpools, another grazed by cataclysms, vertiginous falls, sudden depths or risings; explosions can be seen on the Sun's crown-corona, which cause severe burns, rages, wars, and all kinds of uprisings. If one were to peer closely into the Château's labyrinth, one would also discover white dwarfs, red giants, quasi-stars, or even interstellar dust grains which nobody notices, yet which are the seeds of tomorrow's world.

The expression of the general law is simple, and is centripetal : the whole Galaxy inclines ever closer to the Sun. To glance Him is a privilege, to be noticed by Him is a favour. When He stops His course to speak to you, the whole world stops in surprise, and you find yourself overwhelmed by a rare happiness so enviable that it can become dangerous. If those few words become an invitation, if you are allowed into contact with the Sun, you are suddenly turned into a familiar Planet and your life changes forever.

Au-delà même de la cour, on peut dire que s'étendent d'autres espaces, d'autres galaxies plus obscures et de toute manière moins intéressantes, qui gravitent nécessairement autour de l'astre central, mais en reçoivent des rayons affaiblis. Ce sont les villes et les campagnes de France, les peuples, les corporations, les paysans, mais aussi les pays étrangers, l'Espagne, le Saint-Empire, l'Italie, l'Angleterre, sans oublier les terres immenses de l'Afrique ou du Nouveau Monde, où des peuples clairement inférieurs se prosternent devant un soleil frauduleux, en tout cas devant un soleil purement physique, un soleil esclave et non plus roi !

Quelque part à l'Orient, dans l'Empire de la Chine, on sait qu'il existe un autre Soleil, dont le rayonnement ne peut être nié, car tous les voyageurs l'attestent. Il se nomme l'Empereur de Chine, et les terres qu'il gouverne s'appellent assez curieusement l'Empire du Milieu. Le Roi Soleil a daigné reconnaitre l'existence de cet astre oriental. Il lui a même envoyé des ambassadeurs et il a reçu les siens. On dit cet astre pacifique, bien que son empire soit tourmenté, comme tous les empires. On sait aussi qu'en recevant un portrait équestre du Roi d'Occident (car le Soleil fait du cheval !), l'Empereur de Chine, peu habitué aux techniques picturales de l'Ouest et ignorant tout de la relativité des représentations, a demandé: "Mais pourquoi le Roi-Soleil, lorsqu'il se présente à son peuple, se fait-il peindre la moitié du visage en marron ?"

Beyond the Court-Galaxy lie other Spaces, other more obscure Galaxies, clearly less interesting but still gravitating around the central Sun and irradiated by His weakening rays... These are not only the cities and regions of France, its people, its corporations, its peasants, but also foreign countries Spain, the Holy Roman Empire, Italy, England, not to mention the immensities of Africa or of the New World, where clearly inferior people adore a fraudulent, purely physical Sun, a slave of nature not a King...

Somewhere towards the East, in China, resides another Sun, whose brightness cannot be denied, as returning travellers testify. He is the Emperor of China, and his land is called, oddly enough, the Middle Empire. The Sun King has condescended to acknowledge the existence of this oriental Sun. He has even had ambassadors sent to him, and has received his. This other Sun is reputedly peaceful, although his empire is in upheaval as all empires are. Receiving as a present a painting of the Sun King on horseback (since, like Pegasus, the Sun rides), the Emperor of China, unaware of western painting technique and ignorant of relativity, asks: «Why, when showing himself to his people, must the Sun King have half of his face painted brown ?»

Jean-Claude Carrière

Il y a 200 ans à Versailles

Bernard Coppens

Mai 1789

En ce 21 mai, fête de l'Ascension, les députés du Clergé, de la Noblesse & du Tiers-Etat, qui sont réunis à Versailles pour y tenir les Etats Généraux du Royaume, ont suspendu les séances de leurs chambres respectives. Qu'ils se rassemblent par petits groupes, selon leur province d'origine ou selon leurs affinités, ou qu'ils se retirent dans leur logement pour écrire à leurs commettants, tous mettent à profit ce jour de relâche pour faire le point de la situation.

Depuis l'ouverture solennelle faite par le Roi le 5 mai, les Etats Généraux n'ont pas encore pu se constituer, ni entamer leurs travaux, dont toute la France attend pourtant la solution des problèmes accumulés depuis tant de générations, problèmes dont la manifestation la plus douloureuse est l'insoutenable déficit des finances publiques.

Mais pour que les Etats Généraux puissent se constituer, il faut au préalable que les députés se soumettent à la vérification de la validité de leur élection. Or, la Noblesse prétend que cette opération doit se faire à l'intérieur de chaque ordre, alors que le Tiers-Etat réclame la vérification en commun des pouvoirs, s'appuyant sur le fait que les députés sont appelés à travailler en commun.

La Noblesse & le Tiers-Etat ne veulent pas se départir de leur position, parce que la solution de cette question préalable emportera avec elle la solution du problème de fond: les décisions seront-elles prises dans les Etats Généraux en votant par ordre, ou en votant par tête? Dans la première hypothèse, les jeux seraient faits d'avance puisque Clergé & Noblesse -les deux ordres privilégiés- uniront leur voix contre la seule voix du Tiers, ce qui provoquerait une immense déception dans le pays qui attend de ces Etats Généraux la fin de tous ses maux, par la fin des abus qui se sont introduits au cours des siècles dans le corps social. Dans la seconde hypothèse, les décisions iront immanquablement dans le sens du Tiers-Etat puisque cet ordre a obtenu, le 27 décembre 1788, une représentation égale à celle des deux ordres réunis, & qu'il peut compter sur l'appoint des voix de quelques nobles "éclairés", ainsi que sur celles d'une bonne partie du "bas clergé".

May 1789

May 21, Ascension Day. Representatives of the Clergy, of the Nobility, and of the Third Estate, gathering in Versailles to hold the Estates General of the kingdom, have adjourned the sessions of their respective Houses. Whether they meet in small groups, according to their province of origin, or to their affinities, or whether they retire to their rooms to report to their constituents, all take the opportunity of this holiday to assess the situation.

Since the official opening by the King on May 5th, the Estates General have not formed or begun their work on the many problems that have piled up for so many generations. One problem, the deficit of the public treasury, is particularly distressing.

However, in order for the Estates General to form, the validity of the representatives' election first has to be checked. The Nobility claims that this should be done separately within each order, whereas the Third Estate demands a common check of the proxies, arguing that the representatives will have to work in common in the future anyway.

The Nobility and the Third Estate both stand by their positions, because the solution to this prerequisite question will entail the solution to the heart of the problem: will future decisions be made by the Estates General through a vote by order, or by personal votes? In the first case, the game would not be fair since the Clergy and the Nobility, the two privileged orders, would unite their votes against the sole vote of the Third Estate. Clearly, this situation would generate a huge disappointment in the country, which hopes from these Estates General that all problems will be solved simply by bringing to an end all kinds of social abuses having accumulated over the centuries. In the second case, the decisions would conversely be along the lines of the Third Estate, since that order had succeded to obtain, on December 27th, 1788, a number of representatives equal to that of both other orders, and it may count on a few additional votes of a few "enlightened" nobles, and on a significant fraction of those of the lower clergy.

Mais ceci signifierait la fin des privilèges, donc la transformation complète du royaume, & l'on pense bien que les privilégiés ne sont pas décidés à se laisser dépouiller sans opposer la plus ferme résistance.

Le Roi, qui ne se sent pas capable de faire respecter ses décisions, n'a pas osé trancher lui-même la question du mode de scrutin, & en a laissé la responsabilité aux Etats Généraux eux-mêmes. Cette faiblesse de l'autorité royale est la cause de l'inaction de l'assemblée, qui ne peut prendre aucune décision tant qu'elle n'est pas constituée, & qui ne peut pas se constituer tant qu'elle ne peut pas prendre de décision.

Ainsi chaque ordre tient-il une conduite différente, cherchant à dénouer la crise dans le sens de ses revendications.

Le Tiers-Etat s'assemble chaque jour dans la salle des Menus Plaisirs, celle dans laquelle s'est tenue la séance d'ouverture en présence du Roi, prétend ne pouvoir se constituer qu'en présence & avec les deux autres ordres, & ne se considère en attendant que comme une réunion de citoyens qui attend de se réunir à d'autres citoyens. Les députés s'occupent de questions de procédure & essayent de trouver une façon d'agir sans se reconnaître constitués. Leur démarche est toute tracée: inviter les deux autres ordres à se réunir à eux, vérifier les pouvoirs en commun & voter par tête; si la Noblesse et le Clergé refusent, le Tiers est décidé à se proclamer la Nation & à faire des lois qui assujettiront également les classes privilégiées.

La Noblesse se réunit de son côté, vérifie ses pouvoirs (6 mai) & se déclare chambre constituée (11 mai), affirmant de cette façon qu'elle n'entend en rien entrer dans les vues du Tiers.

Le Clergé, divisé entre prélats & curés, veut jouer un rôle conciliateur entre les deux autres ordres. Aussi a-t-il proposé que les pouvoirs de tous les députés soient vérifiés par les commissaires choisis dans les trois ordres. La Noblesse, qui a rejeté cette proposition, a néanmoins accepté de nommer des commissaires afin de conférer avec ceux des autres ordres sur des moyens de conciliation. Ses commissaires & ceux du Tiers ont été nommés le 19. Connus pour la plupart pour leur position intransigeante, leur choix ne permet pas d'augurer qu'une conciliation aura lieu, & on n'attend de ces conférences aucun autre résultat que d'aigrir davantage les esprits.

But that would mean the end of the privileges, hence a complete transformation of the kingdom, and it is a matter of course that those benefiting from them will fight most vigourously to oppose it.

The King, who thinks he is not strong enough to impose his decisions, has not dared decide which ballot system should be used, and has left the question to be settled by the Estates General themselves. This weakness of the royal authority has brought the proceedings to a standstill, since the assembly cannot take decisions if it has not formed, and cannot form without having taken a decision.

Therefore, each order follows a different path, trying to overcome the crisis to his own advantage.

The Third Estate gathers everyday in the "Room of the Little Pleasures", the very room in which the opening session had been held in the King's presence. It claims that it can form only together with the other two orders, and thinks of itself only as a meeting of citizens waiting for other citizens to join them. Its representatives are busy with matters of procedure, and try to find a way to act without being formed. Their plan is straightforward: they want to invite the two other orders to join them, check the proxies in common, and hold personal votes; if the Nobility and the Clergy refuse, then the Third Estate will name itself the Nation, and will make laws to which the privileged classes will also have to be subjected.

The Nobility meets separately, checks its own proxies (May 6th), and declares itself formed, thereby demonstrating that it does not in the least support the views of the Third Estate.

The Clergy, divided between preasts and prelates, wants instead to play a conciliatory role between the two other orders. Accordingly, it has proposed that all proxies be checked by commissioners chosen within the three orders. The Nobility has rejected this proposal, but has agreed to name commissioners to discuss with those of the two other orders about possible means of conciliation. Its commissioners and those of the Third Estate have been appointed on the 19th. Most of them being known for their uncompromising views, their choice in fact leaves no hope as to a future conciliation, and no result is expected from these meetings other than embittering the minds even more.

Adrien Duquesnoy, député du Tiers-Etat de Bar-Le-Duc, écrit ce jour à ses commettants: "Quoiqu'il arrive, nous ne sortirons de la crise actuelle que par une secousse terrible, & après nous être longtemps battus les uns contre les autres avec nos fers, nous nous endormirons de lassitude dans le sein du despotisme le plus absolu.".

Adrien Duquesnoy, representative of the Third Estate of Bar-le-Duc, writes that day to his constituents: ' Whatever happens, we will come out of the present crisis only through a terrible upheaval, and after having fought each other for a long time, we will be so weary that we will sleep into the utmost despotism.'

Vendredi 22 mai 1789

Friday, May 22nd, 1789

Heures	Temper.	Press (*)	Hygrom.	Ciel	Sky
	°R - °C	Pouc.lig. - Torr			
5 AM	7,8 - 9.8	27.11,7- 757.3		Couvert. Gouttes d'eau	Cloudy. Some rain
Noon	18,0 - 22.5	27.11,9- 757.8	77	vers 6h30 de l'après-midi	in the afternoon
Midnight	11,5 - 14.4	27.11,9- 757.8			

(*) A 23 toises au-dessus du niveau moyen de la Seine (alt.70 m)

(*) 138 feet above mean level of river Seine (i.e. 70 m a.s.l.)

Les députés du Tiers-Etat débattent la motion présentée le 20 mai par M. de la Borde qui propose la formation d'un comité de rédaction qui rédigerait ce que la chambre jugera à propos de publier. La majorité se déclare contre cette motion, trop hâtive, & estime qu'il ne faut pas décréter avant d'être constitué ce qu'il ne convient de décréter qu'après s'être constitué.

The representatives of the Third Estate discuss the motion presented by Mr. de la Borde, who suggests the creation of a board of editors to write up what the House will see fit to publish. A majority stands against this hasty motion, and thinks that one cannot decree before the formation takes place what one may decree only afterwards.

La chambre du Clergé se divise en bailliages pour travailler à l'examen de ses cahiers.

The Clergy splits into bailiwicks to work on their records.

Paris. Un des émeutiers arrêtés lors des événements du 28 avril au faubourg Saint-Antoine (affaire Réveillon), Pierre Mary, écrivain, est pendu. Cinq autres émeutiers assistent à l'exécution attachés au carcan. Ils sont ensuite marqués au fer rouge & conduits aux galères (c.à.d. au bagne). La femme Bertin, également condamnée à être pendue, & qui s'est déclarée grosse, a obtenu un sursis.

In Paris, one of the rioters arrested during the events of April 28th in Faubourg Saint-Antoine (the Réveillon case), Pierre Mary, writer, is hanged. Five other rioters witness the execution tied to iron collars. They are then branded with red-hot iron and sent to the galleys. The sentence to also hang a woman Bertin, who declared herself pregnant, is suspended.

Le parlement de Paris enregistre une "déclaration du Roi" qui attribue aux prévôts des maréchaussées la connaissance & le jugement en dernier ressort des particuliers prévenus d'émotions populaires, d'attroupements, d'excès & de violences, qui ont lieu dans différentes provinces. Cette façon de rendre la justice est particulièrement expéditive.

The Paris Parliament records a "King's declaration", who gives to provosts of the constabulary the power to judge in last resort individuals involved in popular uprisings, gatherings, excesses and violent acts having taken place in various provinces. This form of justice is particularly expeditious.

Samedi 23 mai 1789

Saturday, May 23rd, 1789

Heures	Temper.	Press.	Hygrom.	Ciel	Sky
	°R · °C	Pouc.lig. · Torr			
5 AM	11,2 - 14.0	28. 0,1- 758.2		Assez beau toute	Rather fair
Noon	19,9 - 24.9	28. 0,6- 759.4	72	la journée	the whole day
10h30 PM	14,8 - 18.5	27.11,3- 756.4		Aurore Boréale à 11h	Aurora Borealis at 11PM

Versailles. Chambre du Tiers-Etat (qui affecte de ne se désigner que par la dénomination de "communes"). M. Target, s'appuyant sur les alarmes que répandent dans les provinces l'inaction & le silence que gardent les Etats Généraux, reprend la proposition de former un comité de rédaction. Mais cette motion est une nouvelle fois repoussée à la majorité des suffrages.

La chambre de la Noblesse autorise ses commissaires à annoncer au Tiers Etat que la Noblesse renoncera à tous ses privilèges en matière d'impositions dès que chaque ordre, délibérant librement, aura pu établir les principes constitutionnels sur une base solide.

Les 32 commissaires conciliateurs (8 pour le Clergé, 8 pour la Noblesse & 16 pour le Tiers-Etat) s'assemblent vers 6 h du soir. Mais rien n'est conclu au cours de cette première conférence, & les commissaires s'ajournent au 25.

Un arrêt du Conseil d'Etat du Roi renouvelle les défenses de bâtir sur le terrain des Champs Elysées à Paris, en exécution des déclarations de 1724, 1726 et 1728 sur la fixation des limites de Paris.

Paris. Inauguration à la Bourse du buste du Roi exécuté par M. Houdon. Dans son discours, M. Diancourt dit: "Adressons nos vœux à l'Etre suprême pour que le Roi & ce ministre vertueux (M. Necker) soient à jamais inséparables, & qu'ils ne cessent de s'occuper, comme ils le font, de notre repos, de notre bonheur & de la félicité publique."

Versailles. In the House of the Third Estate, which now pretends to be known only under the name of "Commons", Mr. Target, drawing on the concern spreading in the provinces about the idleness and the silence of the Estates General, proposes once more to appoint a board of editors. But this motion is once more rejected by a majority of votes.

The House of the Nobility allows its commissioners to inform the Third Estate that the Nobility will relinquish all its tax privileges as soon as each order, in a free deliberation, establishes the contitutional principles on a sound basis.

The thirty-two conciliatory commissioners (eight for the Clergy, eight for the Nobility, and sixteen for the Third Estate) meet around 6 p.m. But nothing comes out of this first conference, and the meeting is adjourned until the 25th.

A judgment by the Council of State of the King reiterates the interdiction to build on the Champs Elysées grounds, as ruled by the declarations of 1724, 1726, and 1728 on the question of setting the limits of the city of Paris.

Paris. Inauguration at the stock exchange of a bust of the King by Mr. Houdon. In his discourse, Mr. Diancourt says: « Let us send to the Supreme Being our wishes that the King and this virtuous Minister (Mr. Necker) never split, and that they continue to look after our rest, after our happiness, and after the public peace, as they are now doing. »

Dimanche 24 mai 1789.

Sunday, May 24th, 1789

Heures	Temper.	Press.	Hygrom.	Ciel	Sky
	°R · °C	Pouc.lig. · Torr			
5 AM	12,8 - 16.0	27.10,0- 753.5		A demi couvert.	Partially cloudy
1 PM	23,2 - 29.0	27. 8,7- 750.6	75	toute la journée	all the day
9 PM	17,4 - 21.7	27. 7,9- 748.7		N.L. à 10h30 du soir	New Moon 10:30 PM

Versailles. Les députés des trois ordres qui n'étaient pas arrivés lors de la présentation qui a eu lieu le 2 mai, ont

Versailles. The representatives of the three orders who did not arrive in time for the ceremony of May 2nd have the

l'honneur d'être présentés et nommés au Roi par le marquis de Brézé, Grand Maître des cérémonies de France.

Le Bailli de la Brianne, Ambassadeur Extraordinaire de l'Ordre de Malte a, ce jour, en habit de Cérémonie de l'Ordre, une audience particulière du Roi, pendant laquelle il remet sa lettre de créance à Sa Majesté; cet ambassadeur, accompagné de beaucoup de baillis, commandeurs et chevaliers de l'Ordre qui lui font cortège, est conduit à l'audience de Sa Majesté, et à celle de la Famille Royale, par le sieur de Tolozan, Introducteur des ambassadeurs. Le sieur de Séqueville, Secrétaire ordinaire du Roi pour la conduite des Ambassadeur, précède.

Leurs Majestés & la Famille Royale signent le contrat de mariage du Comte de la Brisse d'Amilly, Officier au Régiment du Roi, avec Mademoiselle le Tonnelier de Breteuil, & celui du Comte Maurice de Caraman, Major en second des Carabiniers de Monsieur, avec Mademoiselle de la Garde.

Le soir, Leurs Majestés soupent à leur grand couvert; pendant le repas, la Musique du Roi exécute différents morceaux sous la conduite du sieur Martini, Surintendant de la Musique de Sa Majesté, en survivance.

honour to be presented to the King by the Marquis de Brézé, the Great Master of Ceremonies of France.

The Bailiff de la Brianne, Ambassador-at-large of the Order of Malta, wearing the Order's official dress, is received this day in audience by the King, to remit his credentials. This ambassador, surrounded in retinue by many bailiffs, commanders and chevaliers of the Order, is led to the audience of His Majesty and of the Royal Family by Master de Tolozan, Introductor to the Ambassadors. Master de Séqueville, the King's Ordinary Secretary for Ambassadors, leads the party.

Their Majesties and the Royal Family sign the marriage contract between Count de la Brisse d'Amilly, Officer of the King's Regiment, with Mademoiselle Le Tonnelier de Breteuil, and that of Count Maurice de Caraman, first mate of Monsieur's [the King's brother] Carabineers, with Mademoiselle de la Garde.

In the evening, Their Majesties have supper in grand setting. During the meal, the King's chamber orchestra, conducted by old Master Martini, Superintendant of His Majesty's Music, plays various pieces.

Lundi 25 mai 1789

Monday, May 25ᵗʰ, 1789

Heures	Temper. °R - °C	Press. Pouc.lig. · Torr	Hygrom.	Ciel	Sky
5 AM	14,4 - 18.0	27. 7,6- 748.1		Petite pluie à l'aube	Some rain at dawn
Noon	19,2 - 24.0	27. 7,9- 748.8	73,5	et à 7h30 du soir	and at 7.30 PM
10h15 PM	11,4 - 14.2	27. 8,3- 749.7		Couvert toute la journée	Cloudy all day

Versailles. Constatant que les débats ont été jusqu'à présent bruyants & tumultueux, alors que la liberté suppose la discipline, le comte de Mirabeau propose la formation d'un règlement de police provisoire de l'assemblée des Communes afin "d'établir l'ordre & la liberté des débats, & recueillir les voix dans toutes leur intégrité" & de "prendre un mode de débattre & de voter qui donne incontestablement le résultat de l'opinion de tous". Cette motion passe à la pluralité de 436 voix contre 11.

La chambre de la Noblesse décide que le procès-verbal de ses séances sera imprimé chaque semaine.

Versailles. Observing that the discussions have up to now been tumultuous and noisy, whereas freedom requires discipline, Count de Mirabeau suggests the creation of a provisional internal regulation of the assembly of Commons, in order 'to establish order and freedom in the debates, and collect the votes in all their integrity', and to 'choose a way to debate and to vote which will indisputably reflect everyone's opinion'. The motion passes by a majority of 436 to 11.

The House of Nobility decides that the minutes of its sessions will be printed each week.

Poursuite des conférences entre les commissaires des trois ordres. Une proposition conciliatoire du Clergé est repoussée par les deux autres ordres. Elle portait que les pouvoirs de chaque député seraient vérifiés par la chambre de son ordre, puis confirmés par les autres. Les cas litigieux devaient être examinés par des commissaires pris dans chacun des trois ordres.

The conferences between the commissioners of the three orders continue. A conciliatory proposal of the Clergy is turned down by the two other orders. This proposal suggested that the proxies held by each representative would be checked by his own House, then confirmed by the other Houses, and that dubious cases would be examined by commissioners appointed by the three orders.

Mardi 26 mai 1789

Tuesday, May 26ᵗʰ, 1789

Heures	Temper.	Press.	Hygrom.	Ciel	Sky
	°R·°C	Pouc.lig. · Torr			
5 AM	11,6 - 14.5	27. 8,5 - 750.1		Pluie l'après-midi et	Rain in the afternoon
2.30 PM	14,9 - 18.6	27. 8,9- 751.0	81	dans la soirée	and in the evening
9.15 PM	12,1 - 15.1	27. 9,2- 751.7		Couvert toute la journée	Cloudy all day

Versailles. L'assemblée des communes décide à une très grande majorité que le règlement provisoire sur la police intérieure sera rédigé par le doyen & ses adjoints. Les commissaires conciliateurs nommés par le Tiers font à l'assemblée le rapport des conférences tenues le 23 et le 25 de ce mois. Les députés entendent avec satisfaction que Target, Mounier et Rabaud de Saint-Etienne ont réfuté "avec une excellente logique, & des connaissances très approfondies" les objections de la Noblesse.

La chambre de la Noblesse, après avoir entendu le rapport de ses commissaires, arrête à la pluralité de plus de 200 voix que, pour cette tenue des Etats Généraux, les pouvoirs seront vérifiés séparément, & qu'une nouvelle forme à observer pour l'organisation des prochains Etats Généraux ne pourra être décidée que par l'assemblée constituée suivant les anciennes règles (c'est à dire en votant par ordre, et avec droit de vote pour chaque ordre, donc sans espoir de changement). Par cet arrêté, la Noblesse rompt les négociations avec le Tiers.

Versailles. The assembly of Commons, to a very large majority, decides that the provisional internal regulation will be written up by the dean and his assistants. The conciliatory commissioners appointed by the Third Estate report to the assembly on the conferences held on the 23rd and 25th of this month. The representatives are satisfied to hear that Target, Mounier, and Rabaud de Saint-Etienne have refuted the objections of the Nobility 'with an excellent logic, and a very deep knowledge'.

The House of the Nobility, after having heard the reports by its commissioners, votes by a majority of more than two hundred that, for this session of the Estates General, proxies will be checked separately, and that a new procedure, valid for the next session, can be decided only by an assembly formed according to the older rules. This means voting by order, with a veto for each order, effectively preventing any change to take place. By this decision, the Nobility breaks the negotiations with the Third Estate.

Mercredi 27 mai 1789

Wednesday, May 27ᵗʰ, 1789

Heures	Temper.	Press.	Hygrom.	Ciel	Sky
	°R·°C	Pouc.lig. · Torr			
5 AM	9,8 - 12.2	27. 9,3 - 751.9		Pluie toute la matinée	Rain in the morning
2.30 PM	14,5 - 18.1	27.10,6 - 754.8	80,5	Quelques éclaircies dans	Some clearing in
9 PM	9,5 - 11.9	27.11,6 - 757.1		la soirée	the evening

Versailles. L'arrêté pris la veille par la Noblesse, & l'annonce qu'elle a communiqué au Clergé afin de l'inviter à suivre son exemple, suscitent l'indignation dans la chambre

Versailles. The decision taken on the previous day by the Nobility, and the fact that it has urged the Clergy to follow its example, infuriates the Commons. Count de Mirabeau

des Communes. Le comte de Mirabeau propose que les Communes se tournent vers le Clergé, qui a des prétentions à jouer le rôle de conciliateur, & s'adressent à lui "d'une manière qui ne laisse pas le plus petit prétexte à une évasion". En conséquence, les Communes envoient au Clergé une députation solennelle chargée de transmettre le message suivant: "Les députés des communes adjurent, au nom du Dieu de paix & au nom de l'intérêt national, MM. du Clergé de se réunir à eux dans la salle de l'assemblée générale, pour aviser aux moyens d'opérer l'union & la concorde, si nécessaires en ce moment au salut de la chose publique".

Plusieurs curés proposent de déférer immédiatement à cette invitation. Mais les évêques parviennent à faire retomber l'enthousiasme, & deux prélats vont annoncer au Tiers-Etats que le Clergé s'est occupé avec zèle d'une matière d'un si grand intérêt; mais que la séance s'étant prolongée au-delà de trois heures, ils se sont séparés & ont remis la délibération au lendemain.

suggests that the Commons turn towards the Clergy, which claims to play a conciliatory role, and speak to it «in a manner that leaves absolutely no room for ambiguity». Consequently, the Commons send a solemn delegation in charge of remitting to the Clergy the following message: 'the representatives of the Commons, in the name of the God of peace, and in the name of the nation's interest, beg the members of the Clergy to join them in the room of the general assembly, to study the ways and means to reach union and concord, now so necessary for the salvation of the public interest'.

Several priests propose to immediately accept this invitation. But the bishops succeed in cooling heads down, and two prelates announce to the Third Estate that the Clergy had zealously examined such an interesting matter; but since the session lasted until after three, they had to adjourn and postpone their deliberation until the following day.

Juin - Octobre 1789

June - October 1789

La guerre des ordres qui se déroule à Versailles continuera de fixer l'attention de la France & de l'Europe.

Après plusieurs péripéties, le Tiers-Etat entreprendra seul, le 12 juin, la vérification des pouvoirs de tous les députés. Après avoir été rejoint par plusieurs membres du bas clergé, il se proclamera le 17 juin, "Assemblée Nationale" (c'est à dire les seuls représentants légitimes de la Nation, habilités à faire de nouvelle lois & à donner une constitution au royaume).

Le 20 juin, la salle des Menus Plaisirs ayant été fermée d'autorité, les députés se réunissent dans la salle du Jeu de Paume, & font le serment solennel de ne pas se séparer avant d'avoir donné une constitution à la France.

Le 23 juin, au cours d'une "séance royale", le Roi déclare nuls les actes du Tiers-Etat, & il présente son programme de réformes. Mais le refus opposé par le Tiers aux ordres du Roi lui enjoignant de quitter la salle, montre qu'un nouveau pouvoir est né, face au pouvoir jusque là unique du monarque.

In the following weeks, the war between the orders taking place in Versailles will continue to attract the attention in France and throughout Europe.

After several turns, the Third Estate will undertake to check alone the proxies of all the representatives. After having been joined by several members of the lower clergy, it will proclaim itself "National Assembly" on June 17th, meaning that it comprises the only legitimate representatives of the nation, empowered to decide new laws, and to give a constitution to the kingdom.

On June 20th, the Room of the Little Pleasures having been closed by order of the King, the representatives gather in the Room of the Palm Game, and take the solemn oath not to leave without having given France a constitution.

On June 23rd, during a "royal session", the King declares the acts of the Third Estates void, and presents his own program of reforms. But the refusal of the Third Estate to obey the King's orders to leave the room shows that a new power is born, confronting the until now only power of the monarch.

Les 24 et 25 juin, la majorité du Clergé & une importante minorité de la Noblesse rejoignent l'Assemblée Nationale. Dés lors, le Roi est obligé de céder, & il demande lui-même à la majorité de la Noblesse de rejoindre les autres députés (27 juin).

La révolution est faite.

C'est le refus de la Cour de reconnaître la nouvelle situation, & les tentatives maladroites pour y mettre un terme, qui provoqueront l'insurrection de juillet à Paris dont l'épisode saillant, la prise de la Bastille, n'est que la démonstration symbolique de la fin de l'absolutisme royal.

C'est encore à Versailles que l'Assemblée Nationale proclamera l'abolition du régime féodal & des privilèges (nuit du 4 août), & la Déclaration des droits de l'homme et du citoyen (20-26 août). Pourtant la présence à Versailles, ville née du caprice d'un souverain tout-puissant, des organes de décision n'est plus qu'un anachronisme.

Le 5 octobre, les parisiens, sur l'initiative des femmes du peuple, viennent à Versailles réclamer du pain. Ils ramènent le lendemain dans la capitale la famille royale ("le boulanger, la boulangère, & le petit mitron").

L'Assemblée Nationale, qui n'a vraiment plus rien à faire dans ce décor d'un autre temps, témoignage de ce que pouvait la volonté d'un monarque absolu, suivra de peu le Roi, & ira s'établir à Paris le 19 octobre.

Le rôle de Versailles est terminé.

Une page de l'histoire est définitivement tournée.

On June 24th and 25th, the majority of the Clergy and a significant fraction of the Nobility join the National Assembly. The King is then forced to yield, and, on June 27th, he asks the majority of the Nobility to join the other representatives.

The revolution is over.

It is only because the Court refuses to acknowledge the new situation and tries unsuccessfully to stop it, that riots will take place in Paris in July. The most famous episode, the taking of the Bastille prison, only symbolically demonstrates that royal absolutism has come to an end.

Again in Versailles, the National Assembly will proclaim the abolition of the feudal system and of all privileges (on the eve of August 4th), and the Declaration of the Rights of Man and of the Citizens (August 20th to 26th). The presence of the deciding bodies in Versailles, a city created by the whim of an all-powerful sovereign, is not anachronistic any more.

On October 1st, Parisians, pushed by the women, come to Versailles to ask for bread, and on the following day, they take the Royal Family ("the baker, the baker's wife, and the little baker's boy") to the capital.

The National Assembly, having nothing more to do in this scenery of ancient times testifying of the will of an absolute monarch, soon follows the King and settles in Paris on October 19th.

The role of Versailles is over.

A chapter of history is closed forever.

Bernard Coppens

LIST OF PARTICIPANTS

AGUER Pierre
    CSNSM - Orsay
    Bat. 104
    F - 91405 Orsay
    FRANCE

ALEXEYEV Evgenii
    Institute for Nuclear Research
    Academy of Sciences of the USSR
    60th, October Anniversary prospect 7a,
    Moscow 117312
    URSS

ALVAREZ Manuel
    DASOP
    Observatoire de Paris - Meudon
    5, pl. Jules Janssen
    F - 92195 Meudon Principal Cedex
    FRANCE

ANGUERA Montserrat
    Instituto de Astrofisica de Canarias
    E - 38200 - La Laguna (Tenerife)
    ESPAGNE

APPOURCHAUX Thierry
    ESTEC/ESA
    P.O. Box 299
    NL - 2200 AG Noordwijk
    PAYS BAS

ARPESELLA Cristina
    INFN - Milano
    Via Celoria 16
    I - 2013 MILANO
    ITALIE

ARTZNER Guy
    LPSP
    B.P. n° 10
    F - 91371 Verrières-le-Buisson
    FRANCE

AUDOUZE Jean
    Présidence de la République
    Palais de l'Elysée
    F - 75008 PARIS
    FRANCE

BAGLIN Annie
    DASGAL
    Observatoire de Meudon
    F - 92195 Meudon Principal Cedex
    FRANCE

BAHCALL John
    Institute for Advanced Study
    Bldg. E
    Olden Lane
    Princeton, NJ 08540
    ETATS UNIS D'AMERIQUE

BAKER Norman H.
    Columbia University
    Astronomy Department
    New York, NY 10027
    ETATS UNIS D'AMERIQUE

BALLESTER Jose Luis
    Departament de Fisica
    Universitat de les Illes Balears
    E - 07071 Palma de Mallorca
    ESPAGNE

BALMFORTH Neil
    The Institute of Astronomy
    University of Cambridge
    Madingley Road
    Cambridge, C133 OHA
    ROYAUME UNI

BARLOUTAUD Roland
    8, rue de la Ronce
    F - 92410 Ville d'Avray
    FRANCE

BARRIO David
Department of Physics and Astronomy
University College London
Gower street
London WC1E 6BT
ROYAUME UNI

BAYLEY David
Ecole Normale Supérieure de Lyon
46, rue d'Italie
F - 69007 Lyon
FRANCE

BAYM Gordon Alan
University of Illinois
1110 W. Green st.
Urbana, IL 61801
ETATS UNIS D'AMERIQUE

BELLI Pier-Luigi
Dipartimento di Fisica
II-Universita di Roma "Tor Vergata"
Via Carnevale
I - 00173 Roma
ITALIE

BELVEDERE Gaetano
Istituto di Astronomia
Viale Doria 6 - Citta Universitaria
I - 95125 Catania
ITALIE

BELY-DUBAU Françoise
Université de Nice - Observatoire
B.P. 139
F - 06003 Nice Cedex
FRANCE

BEREZINSKY V.S.
Lebedev Institute
Moscow

URSS

BERTHOMIEU Gabrielle
Université de Nice - Observatoire
B.P. 139
F - 06003 Nice Cedex
FRANCE

BISNOVATYI-KOGAN Gennadii
Space Research Institute
Profsoyuznaya 84/32
Moscow 117810
URSS

BLOM Cathrine
Institute of Physics
University of Bergen
Allegaten 55
N - 5008 Bergen
NORVEGE

BOCHSLER Peter
Physikalisches Institut
University of Bern
Sidlerstr. 5
CH - 3012 Bern
SUISSE

BONETTI Silvia
Dipartimento di Fisica dell'Università
Via Celoria, 16
I - 20133 Milano
ITALIE

BONNET Roger
Agence Spatiale Européenne
8-10, rue Mario Nikis
F - 75738 Paris Cedex 15
FRANCE

BORG Alain
CEN Saclay
DPhPE/SEPh
91191 Gif-sur-Yvette Cedex
FRANCE

BOUQUET Alain
LAPP
Chemin de Bellevue
B.P. 909
F - 74019 Annecy-le-Vieux Cedex
FRANCE

BRADU Pascal
DRET/SDR/G.34
26 Boulevard Victor
F - 75996 Paris Armées
FRANCE

BRANDENBURG Axel
Observatory and Astrophysics Laboratory
University of Helsinki
Tähtitorninmäki
SF-00130 Helsinki
FINLANDE

BROWN Timothy M.
High Altitude Observatory
NCAR
P.O. Box 3000
Boulder, CO 80307
ETATS UNIS D'AMERIQUE

BÜRGI Alfred
Physikalisches Institut
University of Bern
Sidlerstraße 5
CH - 3012 Bern
SUISSE

CACCIANI Alessandro
Dipartimento di Fisica
Università di Roma "La Sapienza"
Piazzale Aldo Moro 2
I - 00185 Roma
ITALIE

CALLEBAUT Dirk K.
Physics Department UIA
University of Antwerp
Universiteitsplein 1
B-2610 Antwerpen (Wilrijk)
BELGIQUE

CASSE Michel
CEN Saclay
DPhG/SAP
F - 91191 Gif-sur-Yvette Cedex
FRANCE

CATALA Claude
Observatoire de Paris - Meudon
5, pl. Jules Janssen
F - 92195 Meudon Principal Cedex
FRANCE

CHEN Biao
Purple Mountain Observatory
Nanjing
REPUBLIQUE POPULAIRE DE CHINE

CHERRY Michael
Department of Physics
Louisiana State University
Baton Rouge, LA 70803
ETATS UNIS D'AMERIQUE

CHRISTENSEN-DALSGAARD Jørgen
Astronomisk Institut
Aarhus Universitet
DK - 8000 Aarhus C
DANEMARK

CLEVELAND Bruce T.
Los Alamos National Laboratory
Mail Stop 0449
Los Alamos, NM 87545
ETATS UNIS D'AMERIQUE

COPLAN Michael
University of Maryland
Institute for Physical Science and Technology
College Park, MD 20742-2431
ETATS UNIS D'AMERIQUE

COURTAUD Didier
DPGAA/PAP
Centre d'Etudes de Limeil-Valenton
B.P. n° 27
F - 94190 Villeneuve St Georges
FRANCE

COX Arthur N.
Los Alamos Laboratory
MSB 288
Los Alamos, NM 87545
ETATS UNIS D'AMERIQUE

CRIBIER Michel
CEN Saclay
DPhPE/SEPh
F - 91191 Gif-sur-Yvette Cedex
FRANCE

CUNY Yvette
DASOP
Observatoire de Paris - Meudon
5, pl. Jules Janssen
F - 92195 Meudon Principal Cedex
FRANCE

DANIEL Hans-Ulrich
Springer-Verlag GmbH & Co. KG
Tiergartenstraße 17
Postfach 10 52 80
D - 6900 Heidelberg 1
REPUBLIQUE FEDERALE D'ALLEMAGNE

DÄPPEN Werner
ESTEC
P.O. Box 299
NL - 2200 AG Noordwijk
PAYS BAS

DAVIS Raymond
University of Pennsylvania
Dep. of Astronomy and Astrophysics
Philadelphia, PA 19104
ETATS UNIS D'AMERIQUE

de BELLEFON Alain
LPC Collège de France
Place Marcelin Berthelot
F - 75231 Paris Cedex 05
FRANCE

de GREVE J.P.
Astrophysisch Institut Vrije
Universiteit Brussel
Pleinlaan 2
B - 1050 Brussel
BELGIQUE

DELACHE Philippe J.
Observatoire de la Côte d'Azur
B.P. 139
F - 06003 Nice Cedex
FRANCE

DEROBERT Dominique

Onera

DEUBNER Franz-L.
Institut für Astronomie und Astrophysik
Universität Würzburg
D - 8700 Würzburg - Am Hubland
REPUBLIQUE FEDERALE D'ALLEMAGNE

DOMINGO Vicente
ESA/ESTEC, Space Science Department
P.O.Box 299
NL - 2200 AG Noordwijk
PAYS BAS

DONATI J.F.
Observatoire de Paris-Meudon
F - 92195 Meudon Principal Cedex
FRANCE

DOSTROVSKY Israel
Weizmann Institute of Science
76100 Rehovot
ISRAEL

DUVALL Thomas L.Jr.
NASA/Goddard Space Flight Center
National Solar Observatory
950 N. Cherry Ave
Tucson, AZ 85726
ETATS UNIS D'AMERIQUE

DZIEMBOWSKI Wojciek
N.Copernicus Astronomical Center
Ul. Bartyca 18
PL - 00716 Warsaw
POLOGNE

EBELING Werner
Humboldt-Universität zu Berlin
Sektion Physik
Invalidenstraße 42
DDR - Berlin 1040
REPUBLIQUE DEMOCRATIQUE ALLEMANDE

ESCOUBES Bruno
Centre de Recherches Nucléaires
B.P. 20
F - 67037 Strasbourg Cedex
FRANCE

ESPIGAT Pierre
LPC Collège de France
Place Marcelin Berthelot
F - 75231 Paris Cedex 05
FRANCE

FAULKNER John
Lick Observatory
University of California
Santa Cruz, CA 95064
ETATS UNIS D'AMERIQUE

FLECK Bernhard
Institut für Astronomie und Astrophysik
der Universität Wurzburg - Am Hubland
D-8700 Wurzburg
REPUBLIQUE FEDERALE D'ALLEMAGNE

FOING Bernard
I.A.S./L.P.S.P.
B.P. 10
F - 91371 Verrières-le-Buisson Cedex
FRANCE

FORESTINI Manuel
Institut d'Astrophysique et de Géophysique
Université Libre de Bruxelles
C.P. 165, Av. F.D. Roosevelt, 50
1050 Bruxelles
Belgique

FOSSAT Eric
Université de Nice - Observatoire
B.P. 139
F - 06003 Nice Cedex
FRANCE

FRÖHLICH Claus
Welstrahlungszentrum/World Radiation Center
Physikalisch Meteorologisches Observatorium
Postfach 173
CH - 7260 Davos Dorf
SUISSE

GABRIEL Maurice
Institut d'Astrophysique de l'Université de Liège
5 Av. de Cointe
B - 4200 Liège
BELGIQUE

GAVRIN Vladimir N.
Institute of Nuclear Research
Academy of Sciences of the USSR
60th October Anniversary prospect 7a
Moscou , 117312
URSS

GERBIER Gilles
CEN Saclay
DPhPE/SEPh
F - 91191 Gif-sur-Yvette Cedex
FRANCE

GEROYANNIS Vassilis S.
Astronomy Laboratory
Department of Physics
University of Patras
GR - 26110 Patras
GRECE

GOODE Philip
Department of Physics
New Jersey Institute of Technology
Newark, NJ 07102
ETATS UNIS D'AMERIQUE

GORISSE Michel
CEN Saclay
DPhG/SAP
F - 91191 Gif-sur-Yvette Cedex
FRANCE

GOUGH D. O.
Institute of Astronomy
Madingley Road
Cambridge, CB3 OHA
ROYAUME UNI

GRANDPIERRE Attila
Konkoly Observatory
P.O. Box 67
1525 Budapest
HONGRIE

HAHN Richard L.
Brookhaven National Laboratory
Chemistry Departement 555
Upton, NY 11955
ETATS UNIS D'AMERIQUE

HAMPEL Wolfgang
Max-Planck-Institut für Kernphysik
P.O. Box 103 980
D - 6900 Heidelberg 1
REPUBLIQUE FEDERALE D'ALLEMAGNE

HARARI Haim
  Weizman Institute
  Rehovot
  ISRAEL

HASHIMOTO Masaaki
  CEN Saclay
  DPhG/SAP
  F - 91191 Gif-sur-Yvette Cedex

HERISTCHI D.
  DASOP
  Observatoire de Meudon
  F - 92195 Meudon Principal
  FRANCE

HILL Frank
  National Solar Observatory
  P.O. Box 26732
  Tucson, AZ 85726-6732
  ETATS UNIS D'AMERIQUE

HUBER Martin C.E.
  ESA / ESTEC
  Space Science Department
  P.O. Box 299
  NL - 2200 AG Noordwijk
  PAYS BAS

IBEN Icko
  Astronomy Department
  Penn State University
  525 Davey Lab
  University Park, PA 16802
  ETATS UNIS D'AMERIQUE

IGLESIAS Carlos A.
  Lawrence Livermore National Laboratory
  P.O. Box 808 L-296
  Livermore, CA 94550
  ETATS UNIS D'AMERIQUE

JASZCZEWSKA Marzanna
  Obserwatorium Astronomicczne
  Uniwersytet Jagiellonski
  ul. Orla 171
  PL - 30-244 Krakow
  POLOGNE

KAPLAN Jean
  LPTHE
  Université de Paris 6, Tour 24-14 5ème étage
  2, place Jussieu
  F - 75251 Paris Cedex 05

KIKO Jürgen
  Max-Planck Institut für Kernphysik
  D - 6900 Heidelberg
  REPUBLIQUE FEDERALE D'ALLEMAGNE

KIRSTEN Till
  Max-Planck-Institut für Kernphysik
  P.O. Box 103 980
  D - 6900 Heidelberg 1
  REPUBLIQUE FEDERALE D'ALLEMAGNE

KOCHAROV Grant E.
  A.F. Ioffe Physical - Technical Institut of
  the Academy of Science of USSR
  Polytechnicheskaya 26
  194021 Leningrad
  URSS

KOSHIBA Masa-Toshi
  Department of Physics
  Tokai University
  Hiratsuka, Kanagawa 259-12
  Tomigaya 2-28 Shibuya, Tokyo 151
  JAPON

KOSOVICHEV Alexander
  Crimean Astrophysical Observatory
  334413, p/o Nauchny
  Crimea
  URSS

KOTOV Valery A.
  Crimean Astrophysical Observatory
  Academy of Science USSR
  Nauchny Crimea 334413
  URSS

LABONTE Barry
  Institute for Astronomy
  2680 Woodlawn Drive
  Honolulu, HI 96822
  ETATS UNIS D'AMERIQUE

LAGAGE Pierre-Olivier
  CEN Saclay
  DPhG/SAp
  F - 91191 Gif-sur-Yvette Cedex
  FRANCE

LAVELY Eugene
  M.I.T. 54-610
  Cambrige, MA 02139
  ETATS UNIS D'AMERIQUE

LEBRETON Yveline
  DASGAL
  Observatoire de Paris
  5, pl. Jules Janssen
  F - 92195 Meudon Principal Cedex
  FRANCE

LIPUNOV Vladimir
  Sternberg State Astronomical Institute
  Universitetskij Prospect 13
  119899, Moscow
  URSS

LOMBARD Roland
  Division de Physique Théorique
  Institut de Physique Nucléaire
  F - 91406 Orsay Cedex
  FRANCE

LUSTIG Günter
  Inst. für Astronomie
  Universität Graz
  Universitätsplatz 5
  A - 8010 Graz
  AUTRICHE

MAEDER André
  Observatoire de Genève
  CH - 1290 Sauverny
  SUISSE

MAK Hay Boon
  Department of Physics
  Stirling Hall
  Queen's University
  Kingston K7L 3N6
  CANADA

MANUS DUBOSC Claude
  CEN Saclay
  DPhG
  F - 91191 Gif-sur-Yvette Cedex
  FRANCE

MARTIN François
  LPTHE
  Université de Paris 6, Tour 24 5ème étage
  2, place Jussieu
  F - 75251 Paris Cedex 05

MASSAGUER J.M.
  ETS de Ingenieros de Caminos
  Universidad Politechnica de Catalunya
  Jorge Girona Salgado 31
  E - Barcelona 34
  ESPAGNE

MERONI Emanuela
  INFN
  High Energy Physics
  Via Celoria, 16
  I - 20133 Milano
  ITALIE

MERRYFIELD William
  Institute of Astronomy
  Madingley Road
  Cambridge, CB3 OHA
  ROYAUME UNI

MITALAS Romas
  Department of Astronomy
  University of Western Ontario
  London
  Ontario N6A 3K7
  CANADA

MONTMERLE Thierry
  CEN Saclay
  DPhG/SAp
  F - 91191 Gif-sur-Yvette Cedex
  FRANCE

MORROW Cherrilynn
  Institute of Astronomy
  University of Cambridge
  Madingley Road
  Cambridge, CB30HA

MOSCA Luigi
  CEN Saclay
  DPhPE/SEPh
  F - 91191 Gif-sur-Yvette Cedex
  FRANCE

MOSSBAUER Rudolf L.
  Physics Department E15
  TU München
  James-Franck Str.
  D - 8046 Garching b. München
  REPUBLIQUE FEDERALE D'ALLEMAGNE

NAKAHATA Masayuki
  Institute for Cosmic Ray Research
  University of Tokyo
  Midori-cho 3-2-1, Tanashi,shi
  Tokyo, 188
  JAPON

NEFF James
  Greenbelt

NOELS Arlette
  Institut d'Astrophysique
  5 Av. de Cointe
  B - 4200 Ougrée-Liège
  BELGIQUE

NOVOTNY Eva
  Institute of Astronomy
  Madingley Road
  Cambridge, CB3 OHA
  GRANDE BRETAGNE

OTMIANOWSKA-MAZUR Katarzyna
  Obserwatorium Astronomicczne
  Uniwersytet Jagiellonski
  ul. Orla 171
  PL - 30-244 Krakow
  POLOGNE

PALETOU Fréderic
  I.A.S./L.P.S.P.
  B.P. 10
  F - 91371 Verrières-le-Buisson Cedex
  FRANCE

PALLE Pere L.
  Instituto di Astrophysica di Canarias
  E - 38200 La Laguna (Tenerife)
  ESPAGNE

PARKINSON John H.
  Mullard Space Science Lab
  University College London
  Holmbury St.Mary
  Dorking, Surrey
  ROYAUME UNI

PATERNÕ Lucio
  Istituto di Astronomia
  Citta Universitaria
  I - 95125 Catania
  ITALIE

PEACH Gillian
  Deptartment of Physics and Astronomy
  University College London
  Gower street, London WC1E 6BT
  ROYAUME UNI

PEAK Lawrence S.
  Falkiner High Energy Department
  School of Physics
  University of Sydney
  Sydney, N.S.W. 2006
  AUSTRALIE

PECKER Jean-Claude
  Pusat-Tasek
  Les Corbeaux
  F - 85350 L'Ile d'Yeu
  FRANCE

PLAGA Rainer
  Max-Planck-Institut für Kernphysik
  P.O. Box 103 980
  D - 6900 Heidelberg 1
  REPUBLIQUE FEDERALE D'ALLEMAGNE

PRADERIE Françoise
  Observatoire de Paris-Meudon
  F - 92195 Meudon Principal Cedex
  FRANCE

PROVOST Janine
Observatoire de Nice
B.P. 139
F - 06300 Nice Cedex
FRANCE

RÄDLER Karl-Heinz
Central Institute for Astrophysics
Rosa Luxembourg Str. 17A
1591 Postdam
REPUBLIQUE DEMOCRATIQUE
D'ALLEMAGNE

RAISBECK Grant
C.S.N.S.M.
Bat.108
F - 91405 Orsay
FRANCE

REGULO ROUEZ Clara
Instituto de Astrofisica de Canarias
38200 - La Laguna (Tenerife)
ESPAGNE

REINES Frederick
Department of Physics
University of California
Irvine, CA 92717
ETATS UNIS D'AMERIQUE

RHODES Edward J.
Department of Astronomy
University of Southern California
Los Angeles, CA 90089-1342
ETATS UNIS D'AMERIQUE

RIBES Elisabeth
DASOP
Observatoire de Paris - Meudon
5, pl. Jules Janssen
F - 92195 Meudon Principal Cedex
FRANCE

RICCI Donatella
Dipartimento di Fisica
Università di Roma "La Sapienza"
Piazzale Aldo Moro 2
I - 00185 Roma
ITALIE

RICH James
CEN Saclay
DPhPE/SEPh
F - 91191 Gif-sur-Yvette Cedex
FRANCE

ROLFS Claus
Westf. Wilhems-Universität
Institut für Kernphysik
Wilhem-Klemm-Str. 9
D - 4400 Münster
REPUBLIQUE FEDERALE D'ALLEMAGNE

ROSATI Piero
Dipartimento di Fisica
Università di Roma "La Sapienza"
Piazzale Aldo Moro 2
I - 00185 Roma
ITALIE

ROXBURGH Ian W.
Astronomy unit, School of Math. Sciences
Queen Mary College
Mile End Road
London, E1 4 NS
ROYAUME UNI

RUSIN V.
Astronomical Institute
Slovak Academy of Sciences
05960 Tatranska Lomnica
Tchécoslovaquie

SCHATZMAN Evry
DASGAL - Bat. Copernic
Observatoire de Paris
5, pl. Jules Janssen
F - 91195 Meudon Principal Cedex
FRANCE

SCHMIEDER Brigitte
DASOP
Observatoire de Paris
5, pl. Jules Janssen
F - 91195 Meudon Principal Cedex
FRANCE

SCHRIJVER Carolus J.
ESA/ESTEC (SC)
P.O. Box 299
NL - 2200 AG Noordwijk
PAYS BAS

SMIRNOV Alexei
    Institute for Nuclear Research
    Academy of Sciences of the USSR
    60th October Anniversary Prospect 7a
    117312 Moscow
    URSS

SODERBLOM David
    Space Telescope Science Institute
    3700 San Martin Dr.
    Baltimore, MD 21218
    ETATS UNIS D'AMERIQUE

SOTIROVSKI Pascal
    Observatoire de Paris - Meudon
    5, pl. Jules Janssen
    F - 92195 Meudon Principal Cedex
    FRANCE

SPERGEL D.N.
    Department of Astrophysical Sciences
    Princeton University
    P.O. Box 708
    Princeton, NJ 08544
    ETATS UNIS D'AMERIQUE

SPIRO Michel
    CEN Saclay
    DPhPE/SEPh
    F - 91191 Gif-sur-Yvette Cedex
    FRANCE

SPRUIT Hendrik
    MPI für Physik und Astrophysik
    Institut für Astrophysik
    Karl-Schwarzchild Straße 1
    D - 8046 Garching b. München
    REPUBLIQUE FEDERALE D'ALLEMAGNE

STOLARCZYK Thierry
    CEN Saclay
    DPhPE/SEPh
    F - 91191 Gif-sur-Yvette Cedex
    FRANCE

SYLWESTER Barbara S.J.
    Space Research Center
    Polish Academy of Sciences
    ul. Kopernica 11
    PL - 51-622 WROCLAW
    POLOGNE

SYLWESTER Janusz
    Space Research Center
    Polish Academy of Sciences
    ul. Kopernica 11
    PL - 51-622 WROCLAW
    POLOGNE

TAO Charling
    CEN Saclay
    DPhPE/SEPh
    F - 91191 Gif-sur-Yvette Cedex
    FRANCE

TASSOUL Jean-Louis
    Département de Physique
    Université de Montréal
    CP 6128, succ. A
    Montréal, Province de Québec
    CANADA

TASSOUL Monique
    Département de Physique
    Université de Montréal
    CP 6128, succ. A
    Montréal, Province de Québec
    CANADA

THOMAS Richard N.
    Astrophysical Nonequilibrium Thermodynamics
    11, square de Port-Royal
    F - 75013 Paris
    FRANCE

THOMPSON Michael
    High Altitude Observatory
    NCAR
    P.O. Box 3000
    Boulder, CO 80307
    ETATS UNIS D'AMERIQUE

TOOTH Patrick
    Department of Applied Mathematics
    and Theoretical Physics
    Silver Street
    Cambridge, CB3 9EW
    ROYAUME UNI

TORELLI Maria
    Osservatorio Astronomico di Roma
    Via del Parco Mellini 84
    I-00136 Roma
    ITALIE

TRAN MINH Françoise
DASOP
Observatoire de Paris - Meudon
5, pl. Jules Janssen
F - 92195 Meudon Principal Cedex
FRANCE

TSCHARNUTER Werner M.
Inst. f. Theoretische Astrophysik der Univ.
Im Neuenheimer Feld 561
D - 6900 Heidelberg 1
REPUBLIQUE FEDERALE D'ALLEMAGNE

TURCK-CHIEZE Sylvaine
CEN Saclay
DPhG/SAp
F - 91191 Gif-sur-Yvette Cedex
FRANCE

VAN SANTVOORT Jacques
ESA
Apartado 54065
E - 28080 Madrid
ESPAGNE

VENTURA Rita
Istituto di Astronomia
Citta Universitaria
I-95125 Catania
ITALIE

VIGNAUD Daniel
CEN Saclay
DPhPE/SEPh
F - 91191 Gif-sur-Yvette Cedex
FRANCE

VIGNERON Caroline
DESPA
Observatoire de Paris - Meudon
5, place Jules Janssen
F - 92195 Meudon Principal Cedex

von FEILITZSCH Franz
Physics Department E15
TU München
James-Franck Str.
D - 8046 Garching b. München
REPUBLIQUE FEDERALE D'ALLEMAGNE

VUILLEMIN Michel
DPGAA/PAP
Centre d'Etudes de Limeil-Valenton
B.P. n° 27
F - 94190 Villeneuve St Georges
FRANCE

WENESER J.
Brookhaven National Laboratory
Physics 510A
Upton, NY 11973
ETATS UNIS D'AMERIQUE

YERLE Raymond
Observatoire Midi-Pyrénées
14, avenue Edouard Belin
F - 31400 Toulouse
FRANCE

YIOU Françoise
C.S.N.S.M.
Bat.108
F - 91405 Orsay
FRANCE

ZAHN Jean-Paul
Observatoire Midi-Pyrénées
14, avenue Edouard Belin
F - 31400 Toulouse
FRANCE

ZATSEPIN George
Institute for Nuclear Research
Academy of Sciences of the USSR
60th October Anniversary Prospect 7a
117312 Moscow
URSS

ZHUGZHDA Yuzef
Izmiran
Troitsk
Moscow 142092
URSS

ZOREC Jean
Institut d'Astrophysique de Paris
98 bis, Bd. Arago
F - 75014 Paris
FRANCE

ZYLBERAJCH Sylvain
CEN Saclay
DPhPE/SEPh
F - 91191 Gif-sur-Yvette Cedex
FRANCE

LIST OF POSTERS

Session on Solar Neutrinos and Models

ALEXEYEV	Evgenii	Search for solar flare neutrinos at the Baksan INR underground scintillation telescope (with L.N. Alexeyeva, A.E. Chudakov, I.V. Krivosheina)
BONETTI	Silvia	BOREX : a future solar neutrino experiment at LNGS (with the Boron Solar neutrino Collaboration)
BOUQUET	Alain	WIMPs in Stars : an analytical approach (with J. Kaplan, F. Martin, G. Raffelt, P. Salati, J. Silk)
BOUQUET	Alain	WIMPs and solar evolution codes (with Y. Giraud-Héraud, J. Kaplan, F. Martin, C. Tao, S. Turck-Chièze)
CHERRY	Michael	Solar neutrino backgrounds : atmospheric effects
DZIEMBOWSKI	W.A.	Solar model from Helioseismology and the Neutrino flux problem (with A. Pamyatnykh and R. Sienkiewicz)
de BELLEFON	Alain	Indium
FAULKNER	John	WIMPs in the Sun and other stars
FINZI	Arrigo	Non-Baryonic matter from Halo and the solar neutrino problem (with A. Harpaz)
GERBIER	Gilles	Experimental search for Cosmions
GEROYANNIS	Vassilis S.	Rotating Viscopolytropic Stellar Models
GRANDPIERRE	Attila	Nuclear instability of the Sun and the Neutrino problem
HAHN	Richard L.	^{71}As as a test of the Gallium neutrino detector
HAUBOLD	Hans J.	Fourier spectrum analysis of the solar neutrino capture rate (with E. Gerth)
KOCHAROV	Grant E.	1) ^{3}He isotope and the solar neutrino puzzle 2) Cosmic ray modulation during the Maunder minimum 3) Increased counting rates in Davis' experiment and nonstationary processes in the Solar Substance (with G. Kovaltsov, I. Usoskin)
MAK	Hay Boon	The Sudbury Neutrino Observatory
MOREL	Pierre	Updated stellar evolution codes and the standard solar model (with J. Provost, G. Berthomieu)

NOELS	Arlette	Galactic cosmions and solar models
PLAGA	Rainer	Violations of the Pauli principle and the interior of the Sun.
RAISBECK	Grant	Detection of ^{7}Be neutrinos from a lithium solar neutrino detector
RAYCHAUDHURI	Probhas	Time variation of solar neutrino flux and its implication in stellar structure
SCHROEDER	Norman	The Molybdenum - Technetium Solar Neutrino Experiment (with K. Wolfsberg, D. Rokop).
TURCK-CHIEZE	Sylvaine	On the accuracy of solar modelling (with M.Cassé)
ZATSEPIN	George	Present status of Baksan gallium detector.

Session on Helioseismology and Diffusion

ANGUERA	Montserrat	An attempt to identify low l - low n solar acoustic spectrum (with P.L. Pallé, F. Perez, T. Roca Cortes)
APPOURCHAUX	Thierry	Observation of low-degree solar modes in intensity fluctuation (with B. Andersen)
APPOURCHAUX	Thierry	Conceptual design of an instrument dedicated to low-l and low-frequency solar modes observation
BALMFORTH	Neil	Mixing-length theory and the excitation of solar acoustic oscillations (with D.O. Gough)
BEL	Nicole	On the influence of the magnetic field on the solar oscillations
BROWN	Timothy M.	An inverse method for p-mode scattering measurements.
CACCIANI	Alessandro	Use and Performance of the Magneto-Optical Filter for low-l mode measurements (with D. Ricci and P. Rosati)
COX	Arthur N.	Period and stability of solar g-modes
DELACHE	Philippe J.	Wavelet analysis of long term solar variability
DZIEMBOWSKI	Wojciek	Magnetic field in the Sun's interior from oscillation data (with P.R. Goode)
GABRIEL	Maurice	The l dependent part of D_{nl} and the structure of the solar core
GOUGH	Douglas O.	Sensitivity of solar eigenfrequencies to the age of the Sun (with E. Novotny).

HILL	Frank	Mapping flows in the Solar convection zone using oscillation ring diagrams.
JEFFERIES	Stuart	Rotational splitting of the low degree solar p-modes
KOSOVICHEV	Alexander	Using helioseismological data to probe chemical composition in the solar core.
LABONTE	Barry	Acoustic Imaging Through the Sun (with D. Braun and T. Duvall)
LAVELY	Eugene	Testing mixing length theory with helioseismology.
LAVELY	Eugene	The Influence of Gary Glatzmaier's Convective Flow model on Solar Oscillations (with M. Ritzwoller)
PALLE	Pere L.	Variations of the low l solar acoustic spectrum correlated with solar cycle (with C. Regulo, T. Roca Cortes)
PROVOST	Janine	Nonequidistent spectrum of gravity modes (with G. Berthomieu, E. Gavryuseva, W. Gavryusev).
REGULO ROUEZ	Clara	Splitting of p-modes of low degree
RHODES	Edward J.	Evidence for degree-dependent variations in the frequency splittings of solar sectoral p-modes (with A. Cacciani, S. Korzennik)
ROCA CORTES	Teodoro	The low frequency solar velocity spectrum (with P. Palle).
THOMPSON	Michael	Solar rotational splitting measurements (with S. Tomczyk)
YERLE	Raymond	Limb darkening Oscillations : Solar and Terrestrial
YIOU	Françoise	Cosmogenic ^{10}Be as a probe of time variations in solar activity
ZHUGZHDA	Yuzef	1) Observations of intensity fluctuations of low-1 modes 2) Seismology on space observatory coronas (with N. Lebedev and I. Kopaev) 3) On the possibility of 160-min resonance oscillations in the earth atmosphere

Session on Convection, Dynamo and Transport

ALLKOFER	Otto Claus	First results from the HEGRA project
ARTZNER	Guy	Solar photographic astrometry
BALLESTER	Jose Luis	Periodicities and asymmetries in solar activity (with G. Vizoso)
BISNOVATYI-KOGAN	Gennadii	Angular velocity distribution in convective regions

BOCHSLER	Peter	The abundance of 3He in the solar wind - a constraint for models of the Solar evolution (with J. Geiss, A. Maeder)
BRANDENBURG	Axel	The nonlinear solar dynamo and differential rotation (with D. Moss,G. Rudiger,I. Tuominen).
RUSIN	V.	Large-scale distribution of the global magnetic field, green corona and prominences during an extended activity cycle (with V. Bumba and M. Rybansky)
BRUECKNER	Guenter	Intermediate-term solar periodicities : 100 to 500 days
CALLEBAUT	Dirk K.	1) Generation of magnetic fields in the sun. 2) Sunspot cycle from solar oscillations
CHAN	Kwing Lam	Differential rotation around Solar Convection Zone
CHRISTENSEN-DALSGAARD J.		The depth of the Solar Convection Zone (with D. Gough and M. Thompson)
COPLAN	Michael	The abundance of minor ions in the solar wind and comparison to solar abundances (with K. Ogilvie, P. Bochsler, J. Geiss)
COURTAUD	Didier	The influence of metallicity on opacity coefficients (with G. Damamme,E. Genot,M. Vuillemin, S. Turck-Chièze)
DÄPPEN	Werner	The equation of state of the solar interior : a comparison of results from two competing formalisms (with Y. Lebreton, F. Rogers)
DERMENDJIEV	Vladimir	Solar activity in the past and the problem of solar dynamo (with Y. Shopov, G. Buyukliev)
DEUBNER	Franz-L.	Comment on Solar Convection
DONATI	J.F.	Zeeman-Doppler imaging : a new option for magnetic field measurement in active solar-type stars (with M. Semel, F. Praderie)
FOING	Bernard	Probing stars with magnetic activity signatures (with S. Char and S. Jankov)
FORESTINI	Manuel	New constraints on the solar Li and Be
GRANDPIERRE	Attila	Explosive convection at the solar core
KOTOV	Valery A.	Pulsation of the sun as a probe of 22-year cycle and central solar core (with T. Tsap).
LIPUNOV	Vladimir	Magnetic field inside the sun and magnetic properties of collapsed stars
LUSTIG	Günter	Solar meridional plasmas motions from 1982 until 1986 (with H. Wöhl)

MERRYFIELD	William	Azimuthal convective rolls and the subsurface magnetic field
MONTESINOS	Benjamin	Magnetic field pattern and transtion region activity.
MORROW	Cherrilynn	Determining solar asphericity by asymptotic inversion
RÄDLER	Karl-Heinz	On the non-axisymmetric magnetic field modes of the solar dynamo (with A. Brandenburg, I. Tuominen).
ROXBURGH	Ian W.	Mixing in the solar interior
RUSIN	V.	1) Periodicities in the green corona (530.3 nm) brightness for the sun as a star (with J.Zverko) 2) Large scale development of the green corona and prominence occurences (with M. Rybansky)
SYLWESTER	Janusz	Possible scenarios for build-up of Calcium abundance differences in flares (with B.Sylwester, R. Bentley)
TOOTH	Patrick	A new test on nonlocal mixing-length theory (with D.O. Gough)
TSCHARNUTER	Werner M.	Instabilities in the Early Protosun
VIGNERON	Caroline	Angular momentum transport in pre-main sequence stars of intermediate mass (with C. Catala,E. Schatzman,A. Mangeney)
VUILLEMIN	Michel	Nouveaux calculs d'opacités spectrales

SPONSORS

The Organizing Committee is glad to acknowledge the moral and financial support of :

- Institut de Recherche Fondamentale du Commissariat à l'Energie Atomique (IRF-CEA)

- International Astronomical Union (IAU)

- Agence Spatiale Européenne (ESA)

- Centre National d'Etudes Spatiales (CNES)

- Centre National de la Recherche Scientifique (CNRS)

- Laboratoire d'Astrophysique Théorique du Collège de France

- IBM

- **MATRA ESPACE** ➷

- Mairie de Versailles

- European Physical Society (EPS)

- Rank Xerox

- Société Française de Physique

- International Union of Pure and Applied Physics (IUPAP)

INDEX OF KEYWORDS